Constructive Quantum Field Theory II

NATO ASI Series

Advanced Science Institutes Series

A series presenting the results of activities sponsored by the NATO Science Committee, which aims at the dissemination of advanced scientific and technological knowledge, with a view to strengthening links between scientific communities.

The series is published by an international board of publishers in conjunction with the NATO Scientific Affairs Division

A	**Life Sciences**	Plenum Publishing Corporation
B	**Physics**	New York and London
C	**Mathematical and Physical Sciences**	Kluwer Academic Publishers Dordrecht, Boston, and London
D	**Behavioral and Social Sciences**	
E	**Applied Sciences**	
F	**Computer and Systems Sciences**	Springer-Verlag
G	**Ecological Sciences**	Berlin, Heidelberg, New York, London,
H	**Cell Biology**	Paris, and Tokyo

Recent Volumes in this Series

Volume 227—Dynamics of Polyatomic Van der Waals Complexes
edited by Nadine Halberstadt and Kenneth C. Janda

Volume 228—Hadrons and Hadronic Matter
edited by Dominique Vautherin, F. Lenz, and J. W. Negele

Volume 229—Geometry and Thermodynamics: Common Problems of Quasi-Crystals, Liquid Crystals, and Incommensurate Systems
edited by J.-C. Tolédano

Volume 230—Quantum Mechanics in Curved Space-Time
edited by Jürgen Audretsch and Venzo de Sabbata

Volume 231—Electronic Properties of Multilayers and Low-Dimensional Semiconductor Structures
edited by J. M. Chamberlain, L. Eaves, and J.-C. Portal

Volume 232—New Frontiers in Quantum Electrodynamics and Quantum Optics
edited by A. O. Barut

Volume 233—Radiative Corrections: Results and Perspectives
edited by N. Dombey and F. Boudjema

Volume 234—Constructive Quantum Field Theory II
edited by G. Velo and A. S. Wightman

Series B: Physics

Constructive Quantum Field Theory II

Edited by

G. Velo

University of Bologna
Bologna, Italy

and

A. S. Wightman

Princeton University
Princeton, New Jersey

Plenum Press
New York and London
Published in cooperation with NATO Scientific Affairs Division

Proceedings of a NATO Advanced Study Institute on
Constructive Quantum Field Theory II,
held July 1-15, 1988,
in Erice, Sicily, Italy

Library of Congress Cataloging-in-Publication Data

NATO Advanced Study Institute on Constructive Quantum Field Theory (2nd : 1988 : Erice, Italy)
Constructive quantum field theory II / edited by G. Velo and A.S. Wightman.
p. cm. -- (NATO ASI series. Series B. Physics ; v. 234)
"Proceedings of a NATO Advanced Study Institute on Constructive Quantum Field Theory II, held July 1-15, 1988, Erice, Sicily, Italy"--T.p. verso.
Includes bibliographical references and index.
ISBN-13: 978-1-4684-5840-4 e-ISBN-13: 978-1-4684-5838-1
DOI: 10.1007/978-1-4684-5838-1
1. Quantum field theory--Congresses. I. Velo, G. II. Wightman, A. S. III. Title. IV. Series.
QC174.45.A1N365 1988
530.1'43--dc20 90-7902
CIP

Softcover reprint of the hardcover 1st edition 1990
A Division of Plenum Publishing Corporation
233 Spring Street, New York, N.Y. 10013

PREFACE

The seventh Ettore Majorana International School of Mathematical Physics was held at the Centro della Cultura Scientifica Erice. Sicily, 1-15 July 1988. The present volume collects lecture notes on the session which was entitled *Constructive Quantum Field Theory II.* The II refers to the fact that the first such school in 1973 was devoted to the same subject. The school was a NATO Advanced Study Institute sponsored by the Italian Ministry of Scientific and Technological Research and the Regional Sicilian Government.

At the time of the 1973 Erice School on Constructive Field Theory, the speakers could summarize a decade of effort on the solution of superrenormalizable models in two dimensional space-time leading to the verification of the axioms of relativistic quantum field theory for these examples. The resulting lecture notes have proved to be exceptionally useful and are still in print. In the decade and a half that have elapsed since that time, there has been much hard work with the ultimate objective of providing a rigorous mathematical foundation for the quantum field theories in four dimensional space-time that summarize a large fraction of our current understanding of elementary particle physics: QCD and the electroweak theory. The lecture notes of the 1988 school record the fact that, although this objective has not been reached, important progress has been made. The ultraviolet stability of Yang-Mills theory in four dimensions has been treated and renormalizable (not superrenormalizable) models in two dimensional space-time, Gross-Neveu models, have been solved. The Gross-Neveu model in three-dimensional space time has been solved for all sufficiently large N; this is the first non-trivial non-renormalizable field theory ever treated exactly and nonperturbatively. A balanced presentation of the present state of the subject requires a general account of rigorous renormalization group methods as well as a survey of their most important applications. Multiscale expansions and perturbative renormalization theory using those expansions are illustrations of the former. The solution of the Gross-Neveu and Wess-Zumino models and the classification of conformally invariant field theories illustrate the latter. In addition, the reader will find a survey of string theory and of its open problems, a discussion of random surface theory and an account of an attempt to study the quantization of gravity within the framework of local quantum theory.

We hope that the lecture notes will provide both a summary of current research in the field and an introduction for those who are just entering it.

G. Velo

A. Wightman

CONTENTS

CONSTRUCTIVE QUANTUM FIELD THEORY FROM I TO II

SEMINARS

CONSTRUCTIVE QUANTUM FIELD THEORY FROM I TO II
REMARKS ON WHAT WE HAVE LEARNED

A.S. Wightman

Departments of Mathematics
Princeton University.
Princeton, NJ 08544

In the decade and a half which separates the first Erice mathematical physics school on *Constructive Quantum Field Theory* (1973) [1] from our present enterprise, there has been striking progress in quantum field theory. The following remarks are intended as a contribution to a general perspective on what we have learned. They may thereby serve as a prelude to the following lectures on new and current developments. For this purpose, I think it is instructive to trace the development of a particular theme from its origin in the study of the perturbation series for Green's functions in the 1950's to the recent non-perturbative treatment of the Gross-Neveu model. The theme may be vaguely expressed as the idea that fermions are better behaved than bosons.

Of course, in a certain sense, this is already evident from the definition of annihilation and creation operators a and a^*, because the number operator, a^*a, has eigenvalues 0 and 1 for fermions, while for bosons its eigenvalues are $0, 1, 2, \ldots$. Thus, for bosons the annihilation operator a is unbounded, while for fermions it is bounded. However, what I have in mind is a more delicate affair having to do with the behavior of the higher order radiative corrections in the perturbation theory of interacting fields. To describe this more precisely, I have to return to the early days of renormalization theory.

The Algebraic Structure of the Unrenormalized Perturbation Series

1951 was a very good year for formal quantum field theory. Two years before Dyson had published his perturbative analysis of the S matrix of quantum electrodynamics, in which the nth order term in the perturbation series in the fine structure constant is expressed as a sum of contributions labelled by Feynman diagrams [2]. The analysis was based on the use of an interaction picture in which the basic fields have a time dependence determined by the free field Hamiltonian.

Constructive Quantum Field Theory II, Edited by G. Velo and
A. S. Wightman, Plenum Press, New York, 1990

More explicitly, if the Hamiltonian of the theory is split into a free and an interaction part:

$$H = H_0 + H_1$$

and the unitary propagators of H and H_0 are introduced as solutions of

$$i\hbar \frac{\partial U(t,t_0)}{\partial t} = HU(t,t_0), \quad U(t_0,t_0) = 1$$
$$i\hbar \frac{\partial U_0(t,t_0)}{\partial t} = H_0 U(t,t_0), \quad U_0(t_0,t_0) = 1$$

then the *interaction picture propagator* is defined

$$V(t,t_0) = U_0(t,t_0)^{-1} U(t,t_0).$$

It satisfies

$$i\hbar \frac{\partial}{\partial t} V(t,t_0) = H_1^I(t) V(t,t_0)$$

where

$$H_1^I(t) = U_0(t,t_0)^{-1} H_1 U_0(t,t_0)$$

If we define

$$V(t,t') = V(t,t_0) V(t',t_0)^{-1}$$

Then we have

$$V(t,t') = V(t',t'') = V(t,t'')$$

and

$$V(t,t) = 1$$

for any times t, t', t''. The S-operator in the interaction picture is the limit

$$S = \lim_{\substack{t \to +\infty \\ t' \to -\infty}} V(t,t')$$

If $O^{Heis}(t)$ is a field operator in the Heisenberg picture and $O^I(t)$ is the corresponding operator in the interaction picture adjusted to agree at time t_0:

$$O^{Heis}(t_0) = O^I(t_0)$$

then

$$O^{Heis}(t) = V(t,t_0)^{-1} O^I(t) V(t,t_0)$$

Dyson's original expansion says

$$V(t,t_0) = 1 + \sum_{n=1}^{\infty} \left(\frac{-i}{\hbar}\right)^n \frac{1}{n!} \int_{t_0}^{t} \cdots \int_{t_0}^{t} dt_1 \ldots dt_n (H_1^I(t_1), \ldots H_1^I(t_n))_+ \qquad (1)$$

where $(\ )_+$ indicates time ordering

$$(H_1^I(t_1), \ldots H_1^I(t_n))_+ = H_1^I(t_{i_1}) H_1^I(t_{i_2}) \ldots H_1^I(t_{i_n})$$
$$\text{if } t_{i_1} \geq t_{i_n} \geq \ldots \geq t_{i_n}$$

and $i_1 i_2 \dots i_n$ is a permutation of $1 \dots n$. Dyson's 1951 formula is [3]

$$O^{Heis}(t) = O^I(t) + \sum_{n=1}^{\infty} \left(\frac{-i}{\hbar}\right)^n \frac{1}{n!} \int_{t_0}^{t} \cdots \int_{t_0}^{t} dt_1 \dots dt_n R(O^I(t); H_1^I(t_1), \dots H_1^I(t_n)) \tag{2}$$

where R is the celebrated retarded function

$$R(O^I(t); H_1^I(t_1), \dots H_1^I(t_n)) = \sum_{i_1 \dots i_n} \theta(t - t_{i_1})\theta(t_{i_1} - t_{i_2}) \dots \theta(t_{i_{n-1}} - t_{i_n})$$
$$[\dots [O^I(t), H_1^I(t_{i_1})], H_1^I(t_{i_2}), \dots H_1^I(t_{i_n})]$$

On the other hand, Gell Mann and Low introduced [4] a relation between the physical vacuum, Ψ_0, and the bare vacuum, Φ_0, in the Heisenberg picture

$$\begin{aligned} \Psi_0 &= c_1 V(t_0, -\infty)\Phi_0 \\ &= c_2 V(\infty, t_0)^{-1}\Phi_0. \end{aligned}$$

Thus,

$$1 = \bar{c}_2 c_1 (V(\infty, t_0)^{-1}\Phi_0, V(t_0, -\infty)\Phi_0)$$

i.e.

$$\bar{c}_2 c_1 = [(\Phi_0, S\Phi_0)]^{-1}$$

Insertion of these expressions in a vacuum expectation value of a time ordered product of Heisenberg picture operators yields

$$(\Phi_0, (O_1^{Heis}(t_1), \dots O_n^{Heis}(t_n))_+ \Phi_0) = \frac{(\Phi_0, (S, O_1^I(t_1), \dots O_n^I(t_n))_+ \Phi_0)}{(\Phi_0, S\Phi_0)} \tag{3}$$

where the numerator is, by definition

$$(\Phi_0, V(\infty, t_1) O_1^I(t_1) V(t_1, t_2) O_2^I(t_2) V(t_2, t_3) \dots O_n^I(t_n) V(t_n, -\infty)\Phi_0)$$

if $t_1 \geq t_2 \geq \dots \geq t_n$. The expansion (1) then yields for the numerator

$$\sum_{l=0}^{\infty} \left(\frac{-i}{\hbar}\right)^l \frac{1}{l!} \int_{-\infty}^{\infty} \cdots \int_{-\infty}^{\infty} d\tau_1 \dots d\tau_l (\Phi_0, (H_1^I(\tau_1), \dots H_1^I(\tau_l), O_1^I(t_1), \dots O_n^I(t_n))_+ \Phi_0). \tag{4}$$

The expression (3) and its expansion (4) constitute the celebrated Gell-Mann Low formula.

The task of evaluating the terms of the series (4) is quite similar to Dyson's work in evaluating S-matrix elements. Consider, for example, the case of quantum electrodynamics, where

$$H_1^I(t) = -\int d^3x j_\mu^I(x) A^{I\mu}(x)$$

and

$$O_1^I(t_1) \dots O_1^I(t_n) = \psi^I(x_1) \dots \psi^I(x_r) \psi^{I\dagger}(y_1) \dots \psi^{I\dagger}(y_r) A_{\nu_1}^I(z_1) \dots A_{\nu_s}^I(z_s)$$

Then the lth order contribution factors because the vacuum, Φ_0, is a product state of the fermion vacuum and the boson vacuum.

$$\Phi_0 = \Phi_{0F} \otimes \Phi_{0B}$$

so

$$\begin{aligned}\int_{-\infty}^{\infty}\int_{-\infty}^{\infty} & d\tau_1 \ldots d\tau_l(\Phi_0, (H_1^I(\tau_1), \ldots H_1^I(\tau_l),\\ & O_1^I(t_1)\ldots O_1^I(t_n))_+\Phi_0)\\ & = (-1)^l \int \ldots \int d^4\xi_1 \ldots d^4\xi_l(\Phi_{0F}, (j^I_{\mu_1}(\xi_1), \ldots j^I_{\mu_l}(\xi_l),\\ & \prod_{j=1}^{r} \psi^I(x_j) \prod_{k=1}^{r} \psi^{I\dagger}(y_k))_-\Phi_{0F})\\ & (\Phi_{0B}, (A^{I\mu_1}(\xi_1), \ldots A^{I\mu_l}(\xi_l), A^I_{\nu_1}(z_1)\ldots A^I_{\nu_s}(z_s))_+\Phi_{0B})\end{aligned} \qquad (5)$$

A systematic evaluation of the free field vacuum expectation values appearing here in terms of the algebraic notions of pfaffian, hafnian, determinant and permanent was the contribution of E. Caianiello in 1953 and 1954 [5]. The vacuum expectation value of the boson field A^I is

$$(\Phi_{0B}, (A^{I\mu_1}(1), \ldots A^{I\mu_l}(l), A^I_{\nu_1}(z_1)\ldots A^I_{\nu_s}(z_s))_+\Phi_{0B}) = [1\ldots l z_1 \ldots z_s]$$

where the square bracket is the hafnian defined recursively in terms of the square bracket with two arguments

$$[1\ldots N] = \sum_{j=2}^{N}[1j][\widehat{1}2\ldots\widehat{j}\ldots N].$$

Here

$$[jk] = (\Phi_{0B}, (A^{I\mu_j}(j), A^{I\mu_k}(k))_+\Phi_{0B})$$
$$[1\ldots N] = 0 \text{ for } N \text{ odd}$$

and the abbreviation j for ξ_j has been used. This definition differs from that of the pfaffian

$$(1\ldots N) = \sum_{j=2}^{N}(-1)^{j+1}(1,j)(\widehat{1}\ldots\widehat{j}\ldots N)$$

only in the extra factor $(-1)^{j+1}$. Vacuum expectation values of time ordered products of fermion field operators can be expressed in a similar way in terms of pfaffians, but the natural definition of the time ordered product turns out to involve an extra sign factor arising from the anticommutativity of the fermion fields. It has been known since the 19th century that pfaffians can be expanded in determinants (E. Arnauldi 1890) and Caianiello showed that hafnians can be expanded in terms of permanents which are the analogues of determinants with the alternating sign removed from their definition. In Caianiello's notation, which coincides with that of Cayley, a determinant of a matrix is written in terms of the labels of its rows and columns

$$\begin{pmatrix} a_1 a_2 \ldots a_n \\ b_1 b_2 \ldots b_n \end{pmatrix} = \sum_{i_1 \ldots i_n} \sigma \begin{pmatrix} 1 \ldots n \\ i_1 \ldots i_n \end{pmatrix} \prod_{j=1}^{n} \begin{pmatrix} a_j \\ b_{i_j} \end{pmatrix}$$

where $\sigma\binom{1\ldots n}{i_1\ldots i_n}$ is the signature of the permutation $\binom{1\ldots n}{i_1\ldots i_n} \downarrow$ and $\binom{a_j}{b_k}$ is the entry of the matrix with a_j as column index and b_k as row index. Similarly the permanent is

$$\begin{bmatrix} a_1 a_2 \ldots a_n \\ b_1 b_2 \ldots b_n \end{bmatrix} = \sum_{i_1 \ldots i_n} \prod_{j=1}^{n} \begin{pmatrix} a_j \\ b_{i_j} \end{pmatrix}$$

In particular, Caianiello showed that

$$(\Phi_{0F}, (: \psi^{\dagger}_{\rho_1}(1)\psi_{\sigma_1}(1) :, \ldots : \psi^{\dagger}_{\rho_l}(l)\psi_{\sigma_l}(l) :, \psi_{\beta_1}(x_1), \ldots \psi_{\beta_r}(x_r)\psi^{\dagger}_{\alpha_1}(y_1), \ldots \psi^{\dagger}_{\alpha_r}(y_r))_+ \Phi_{0F}$$
$$= \begin{pmatrix} 1 \ldots l & x_1 \ldots x_r \\ 1 \ldots l & y_1 \ldots y_r \end{pmatrix} (-1)^{\frac{r(r+1)}{2}}$$

with the conventions

$$\begin{pmatrix} j \\ j \end{pmatrix} = 0, j = 1, \ldots l,$$

and

$$\begin{pmatrix} j \\ k \end{pmatrix} = (\Phi_0, (\psi^{\dagger}_{\rho_\gamma}(\xi_j), \psi_k(\xi_k))_+ \Phi_0)$$

Thus, finally

$$\begin{aligned}
&(\Psi), (\psi_{\beta_1}(x_1), \ldots \psi_{\beta_r}, \psi^{\dagger}_{\alpha_1}(y_1), \ldots \psi^{\dagger}_{\alpha_r}(y_r), A_{\mu_1}(z_1), \ldots A_{\mu_s}(z_s))_+ \Phi_0) \\
&= \sum_{l=0}^{\infty} \left(\frac{ie}{\hbar}\right)^l \frac{1}{l!} \int \cdots \int d1 \ldots dl \gamma^{(1)} \ldots \gamma^{(l)} \\
&\begin{pmatrix} 1 \ldots l & x_1 \ldots x_r \\ 1 \ldots l & y_1 \ldots y_r \end{pmatrix} (-1)^{\frac{r(r+1)}{2}} [1 \ldots l z_1 \ldots z_s] \qquad (6) \\
&\times \left[\sum_{l=0}^{\infty} \left(\frac{ie}{2}\right)^l \frac{1}{l!} \int \cdots \int d1 \ldots dl \gamma^{(1)} \ldots \gamma^{(l)} \begin{pmatrix} 1 \ldots l \\ 1 \ldots l \end{pmatrix} [1 \ldots l] \right]^{-1}
\end{aligned}$$

The matrix indices on the γ's, $\gamma^{\mu_j}_{\rho_j \sigma_y}$, are understood to be contracted with the corresponding indices on the propagators. Caianiello gave the tools required to derive analogous formulae for all the standard field theories of that day.

It is not quite trivial to derive the Feynman rules from this formula. It would be if it were true that in l^{th} order there are exactly $l!$ terms arising from the product of the determinant and the hafnian which give rise to the same Feynman diagram, but that is true only if the Feynman diagram contains no disconnected diagrams, vacuum-vacuum parts. Otherwise the infamous symmetry numbers make their appearance. See A. Buccafurri and G. Fano 1959 [6].

As far as the mathematical rigor of these formulae is concerned there wasn't much. They are, in general afflicted by both infra-red and ultraviolet divergences, the first arising from the integrations over all space-time, the second from the products of singularities in the integrand. Nevertheless they have been a bounteous source of inspiration and the individual terms can be made to make sense by introducing an infrared and an ultraviolet cutoff.

Convergence of the Cutoff Series

The expression (6) is a ratio of two series the lth term of which contains l^{th} order determinants. That is reminiscent of the original form of the Fredholm theory for the resolvent of a Fredholm operator. In fact, it is exactly the Fredholm theory, if one replaces the interaction with a quantized electromagnetic field by an interaction with an external electromagnetic field. This simplified case was worked out in detail by Matthews and Salam and independently by Schwinger in 1954 [7].

Now in Fredholm's original argument a key idea is to use Hadamard's inequality to bound the determinants. It says: If A is an $l \times l$ matrix

$$|\det A| \leq (\sup_{j,k-1,\dots l} |A_{jk}|)^l l^{l/2} \tag{7}$$

Could it be that an analgous argument here would give the convergence of the cutoff perturbation series? That was the idea of Caianiello in 1956 [8]. He took for simplicity the case $s = 0$, and noting that only even orders in l contributed replaced the summation index l by $2l$. Now a hafnian $[1 \dots 2l]$ has $(2l-1)!! = (2l-1)(2l-3)(2l-5)\dots 1$ terms and

$$(2l-1)!! = \frac{(2l)!}{2^l l!}$$

Each term of the hafnian contributes equally so the series in the numerator of (6) is

$$\sum_{l=0}^{\infty} \left(\frac{-e^2}{2\hbar^2}\right)^l \frac{1}{l!} \int \dots \int d1 \dots dl \gamma^{(1)} \dots \gamma^{(2l)} \begin{pmatrix} 1 \dots 2l & x_1 \dots x_r \\ 1 \dots 2l & y_1 \dots y_r \end{pmatrix} \tag{8}$$

Caianiello introduced the space time box of volume $|V|$ to give an infra-red cutoff and bounded the two point functions by constants. The result was a majorant series converging geometrically and therefore for $|e^2| <$ some constant. Since the similar argument works for the denominator, the perturbation series for these cutoff Green's functions converge in a disc. This behavior is in contrast with the divergence of the analogous Green's function in theories of a cutoff self coupled Bose field, $\lambda\phi_4^4$ or $\lambda\phi_3^4$, where the analogous series is

$$\sum_{l=0}^{\infty} \frac{(-\lambda)^l}{l!} \int \dots \int [11112222 \dots llll x_1 \dots x_{2r}] d1 \dots dl$$

with the majorant

$$\sum_{l=0}^{\infty} \frac{|\lambda|^l}{l!} |V|^l C^{l+r} (4l + 2r - 1)!!$$

if $\|[jk]\| \leq C$. Since $\frac{(4l+2r-1)!!}{l!} \sim \frac{(4l+2r)^{2l+2r}}{l^l}$, the majorant diverges. According to Hurst, Thirring and Petermann [9], this is characteristic of the series itself, so we have reached the first precise formulation of the above mentioned theme: *fermions are nicer than bosons because the perturbation series for the Green's functions in these cutoff field theories containing fermions converges in a disc rather than diverging as in pure boson theories.*

Caianiello's 1956 result was improved by Yennie and Gartenhaus in 1958 [10]. They showed that with similar cutoffs the numerator and denominator of the expressions (6) for the Green's function are entire functions of e^2 but that the denominator has an infinite number of zeros in the complex e^2 plane. An interesting feature is that it did not appeal to Hadamard's inequality for determinants. Instead it used the fact that the annihilation and creation operators for fermions are bounded operators. This result of Yennie and Gartenhaus was recovered by Buccafurri and Caianiello in 1958 [11] using methods closer to Caianiello's original arguments with the additional bonus that the argument worked also for a pure fermion theory with Fermi interaction. Here the matter rested for several years while the effects of renormalization were studied both in perturbation theory and non-perturbatively.

Convergence of the Cutoff Rearranged Renormalized Series

The difficulty with the renormalized perturbation series as seen in the 1960's was that the operation of renormalization was defined in so complicated a way that it was hard to estimate the size of the renormalized n^{th} order contributions. However, this complaint was less applicable to the superrenormalizable models in two dimensional space time on which attention began to be focussed as a result of new developments in constructive quantum field theory.

Analogues of QED in this case are Y_2 models, the theory of a Dirac two component spinor field coupled to a boson field with a scalar or pseudoscalar or vector interaction. These models have only two divergent Feynman diagrams

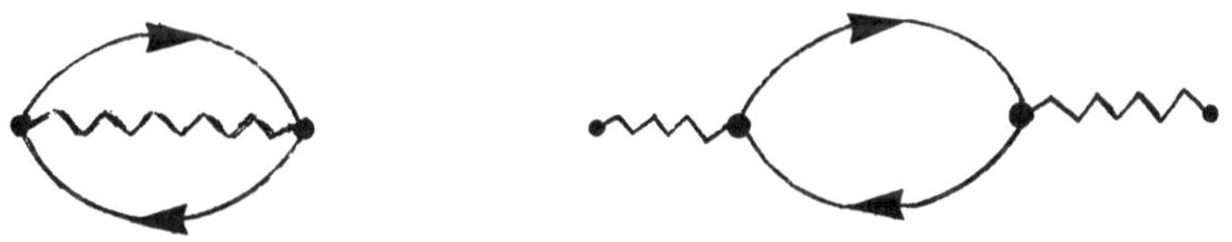

They are therefore superrenormalizable. The task of obtaining an expression for the renormalized perturbation series analogous to (6) for the unrenormalized perturbation series was completed by Barry Simon in 1969 [12]. He noted that the terms requiring renormalization are those arising from two cycles in the expansion of the determinant $\begin{pmatrix} 1\ldots l & x_1\ldots x_r \\ 1\ldots l & y_1\ldots y_r \end{pmatrix}$. The process of renormalization corresponds to replacing the contribution from the bubble, 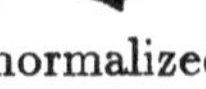 by a renormalized contribution, 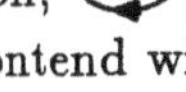 REN. A proof of convergence of the renormalized series would then have to contend with the difficulty that the alteration of the bubble contribution could destroy the cancellations that made Caianiello's proof of the convergence of the cutoff unrenormalized series work. Simon chose to prove a different result. By an infinite rearrangement he brought the series (8) into the form

$$\sum_{l=0}^{\infty}\left(\frac{-e^2}{2\hbar}\right)^l \frac{1}{l!}\int d1\ldots d(2l)\begin{pmatrix} 1\ldots 2l & x_1\ldots x_r \\ 1\ldots 2l & y_1\ldots y_r \end{pmatrix}' [12]'[34]'\ldots[2l-1,2l]' \qquad (9)$$

where $\begin{pmatrix} 1\dots 2l & x_1\dots x_r \\ 1\dots 2l & y_1\dots y_r \end{pmatrix}'$ is a "bubblessian" i.e. a modified determinant from whose expansion all terms containing two cycles have been dropped. $[jk]'$ is the "aerated propagator" obtained by summing all chains of renormalized bubbles

$$[jk]' =$$

Simon showed that the "bubblessian" satisfies an inequality

$$\left| \begin{pmatrix} 1\dots n \\ 1\dots n \end{pmatrix}' \right| \leq \left(\sup_{j,j=1,\dots n} |A_{jk}| \right)^n \left(\frac{9n}{4} \right)^n \tag{10}$$

analogous to Hadamard's inequality for the determinant (7). Thus, making cutoffs analogous to those in Caianiello's discussion: $|(jk)| \leq F, |[jk]'| \leq B$, and space time box of volume $|V|$, Simon proved the convergence of cutoff, rearranged, renormalized perturbation series.

In passing it is worth noting that Simon's bubblessian, $\det'(A)$, is not the same thing as $\det_2 A$, introduced into Fredholm theory early in this century in order to deal with operators $A = 1 + K$ for which trK and trK^2 are infinite. For an $l \times l$ matrix $\det'(A)$ is a polynomial in the matrix elements of K while $\det_2(A)$ is an entire function of infinite degree.

Of course, it would have been much nicer if the rearranged renormalized perturbation series without cutoffs could have been proved to converge. However, Simon's "bubblessian" inequalities do not seem to be an adequate tool for this purpose because they contain the sup norm of $|(a_{jk}|$, which is infinite. As of 1969, there was no technique in sight to overcome this difficulty.

Incidentally, at that time, Simon also looked at Fermi interactions of a massive fermion field (= massive Thirring model) but, remarking correctly that such theories are renormalizable not superrenormalizable, put them aside. Then they sat until a new set of ideas associated with the renormalization group was brought to bear by Feldman, Magnen, Rivasseau and Sénéor [13], and independently by Gawedzki and Kupiainen [14], exploiting brilliant insights of Gross and Neveu [15].

In passing, it is worth noting that there existed, early on, an example of a renormalized perturbation series which is divergent but Borel summable. The example was provided by Schwinger's exact calculation of the vacuum polarization of electrons in an external magnetic field. The expression comes from an effective Lagrangian [16]

$$\mathcal{L} = -\frac{(e|\vec{H}|)^2}{8\pi^2} \int_0^\infty \frac{ds}{s^3} e^{-s/(e|\vec{H}|/m^2)} \left\{ s\, coths - 1 - \frac{1}{3} s^2 \right\}. \tag{11}$$

As was shown by Ioffe, the perturbation series in powers of e is

$$\mathcal{L} = -\frac{1}{8\pi^2} \sum_{n=2}^{\infty} \frac{(2n-3)!}{(2n)!} 2^{2n} B_{2n} |\vec{H}|^{2n} m^{-4n+4} e^{2n}$$

where the B_{2n} are the Bernoulli numbers and

$$B_{2n} \sim (-1)^{n-1} \frac{(2n)!}{2^{2n-1}\pi^{2n}}$$

and it diverges [17]. On the other hand, as was remarked by Ogievetsky, the formula (11) displays $\mathcal{L}$ as the Borel transform (up to a factor) of the function $s^{-3}\left\{s\, coth s - 1 - \frac{1}{3}s^2\right\}$ which is analytic and exponentially bounded in a half strip, so the perturbation series is Borel summable [18]. This is a warning that one cannot expect better behavior, in general.

Modern Times; Cluster Expansions, Mayer Expansions, Phase Cell Expansions

I will continue the story in the context of a pure fermion theory, the Gross-Neveu model. It generalizes the Thirring model to N anticommuting fermion fields, $\psi_{i,a}; i = 1, 2, a = 1, \ldots N$, coupled with an O_N symmetric Fermi interaciton

$$\lambda\left(\sum_{a=1}^{N}\psi_a^\dagger\psi_a\right)^2$$

Here

$$\psi_a^\dagger\psi_a = \sum_{j,k=1}^{2}\psi_{j,a}^*(\gamma^0)_{jk}\psi_{k,a}$$

where

$$\gamma^0 = \begin{Bmatrix} i & 0 \\ 0 & -i \end{Bmatrix}, \gamma^1 = \begin{pmatrix} 0 & 1 \\ -1 & 0 \end{pmatrix}$$

In the original Gross-Neveu paper the fermion fields were taken massless and the main focus was on the phenomena of chiral symmetry breaking and mass generation displayed by the model. On the other hand, the massive model seems more natural if the object is to show that rigorous renormalization group methods suffice to construct the solutions of a renormalizable (*not* superrenormalizable) field theory. To make such a construction was an important unsolved problem of constructive quantum field theory as late as the early 1980's. The work assumes N fixed, taking any value > 1. The case $N = 1$, the Thirring model, is exceptional in this family because the slope of its β function at zero coupling is zero in contrast to the situation for all $N > 1$, for which the slope is negative. That means that for $N > 1$, the model is asymptotically free, a fact first pointed out by P.K. Mitter and P. Weisz [18].

The technique of choice for proving the existence of solutions of the massive Gross-Neveu model was the method of the rigorous renormalization group. It enables one to choose systematically the cutoff dependence of approximate solutions so that the limiting theory exists and so that the errors can be rigorously controlled.

To explain how these ideas permit one to say more about the perturbation series, we follow the argument of the Ecole Polytechnique group [13]. They introduced an ultra-violet cutoff via the Pauli-Villars regularization technique and showed how the results of Caiainello et al can be obtained from Gram's inequality in one fell swoop. The argument is so neat that I cannot resist repeating it. They write for the Fourier transform of the two point function

$$\frac{\eta(p^2, M_1, M_2, M_2, M_4)}{(-\not{p} + m)} \tag{12}$$

with

$$\eta(p^2, M_1, M_2, M_3, M_4) = (p^2 + m^2)\left\{[p^2 + m^2]^{-1} + \sum_{j=1}^{4} J_j [p^2 + M_j^2]^{-1}\right\}$$

and choose the real constants J_i so that

$$\frac{\eta(p^2, M_1, M_2, M_3, M_4)}{p^2 + m^2} = J_5 [p^2 + m^2]^{-1} \prod_{j=1}^{4} [p^2 + M_j^2]^{-1} \tag{13}$$

M_1, M_2, M_3, M_4 are distinct auxiliary masses of the Pauli Villars method; they are made to approach infinity when the cutoff is to be removed. The rapid fall-off displayed by (13) is what makes the method work as a regularization. The space-time propagator takes the form:

$$C(x-y) = \int A(x-t)B(t-y)d^2t$$

where

$$\hat{A}(p) = (\not{p} + m)[p^2 + m^2]^{-1}[p^2 + M_1^2]^{-1}[p^2 + M_2^2]^{-1}$$

and

$$\hat{B}(p) = J_5 [p^2 + M_3^2]^{-1}[p^2 + M_4^2]^{-1}$$

Thus, C is a Hilbert space scalar product of L^2 functions. Therefore, in the expansions of the numerator and denominator of the Gell-Mann low formula, (6), one has a determinant of scalar products. For example, for the denominator

$$\sum_{l=0}^{\infty} \frac{\lambda^l}{l!} \int \cdots \int_{\Lambda} \det\{(f_j, f_k)\} d\xi_1 \ldots d\xi_l \tag{14}$$

where

$$f_j(t) = A(\xi_j - t), \quad g_k(t) = B(t - \xi_k)$$

and the integration has been confined to a large box, Λ. Now Gram's inequality says

$$|\det\{(f_j, g_k)\}| \leq \prod_{j=1}^{l} \|f_j\|_2 \|g_j\|_2$$

so the series (14) is dominated by

$$\sum_{l=0}^{\infty} \frac{|\lambda|^l}{l!} |V|^l \prod_{j=1}^{l} \|f_j\|_2^l \|g_j\|_2^l$$

an entire function, and the Schwinger functions cutoff in this way are ratios of entire functions.

Now any such ratio of entire functions is analytic in a disc centered at the origin and of radius equal to the distance from the origin to the nearest zero of the denominator. The next step is to show that, for fixed ultraviolet cutoff, one can write this analytic function in such a form that the thermodynamic limit $V \to \infty$ can be taken and that the limit is analytic in some sufficiently small disc in the λ plane. This is done by developing a cluster expansion followed by a Mayer expansion. The techniques for carrying out this procedure have developed gradually over the last two decades. They have been an apparently indispensible part of the progress in rigorous statistical mechanics and quantum field theory. Crudely, the idea is to arrange the terms of the series so that all integration variables are connected to external variables and therefore the integrand decays exponentially as the variables approach infinity by virtue of the exponential decay of the free propagator. I refer to Brydges's les Houches lectures for a helpful review [20].

The next step in the argument deals with the dependence on the ultraviolet cutoff. Here a slicing of momentum space is introduced with the scale 1, $M, M^2, \dots M^\rho$ where M is some fixed number greater than one. The cutoff in the propagator is then fixed in terms of (12):

$$\eta_\rho(p) = \eta(p, M^{\rho+1}, 2M^{\rho+1}, 2\sqrt{2}M^{\rho+1}, 4M^{\rho+1})$$

and written as a collapsing sum

$$\eta_\rho(p) = \sum_{k=0}^{\rho} \eta^\rho(p)$$

where $\eta^0(p) = \eta_0(p)$ and for $k \geq 1$

$$\eta^k(p) = \eta_k(p) - \eta_{k-1}(p)$$

η^k determines a propagator for the kth slice, $\eta^k(p)[-\zeta_\rho p + m_\rho]^{-1}$. (The field strength renormalization ζ_ρ, ignored above has been introduced here.) The phase cell expansion is obtained by a simultaneous expansion in coordinate space squares and momentum space slices.

On each scale a cluster and Mayer expansion is performed, and equations derived which relate quantities on that scale to those on other scales. In the resulting expressions it is possible, just as in the preceding case, to pass to the limit $\Lambda \to \infty$. However, the decomposition into momentum slices offers the opportunity for a discussion of renormalization in the spirit of the renormalization group. The four leg and two leg diagrams are separated into a renormalized part and a renormalization correction to the coupling constant and mass respectively. Since this is done on each scale it defines running coupling constants λ_k^ρ, m_k^ρ and ζ_k^ρ. Expressing the Schwinger functions on scale k as an expansion in λ_k^ρ with propagator containing m_{ren} and ζ_k^ρ or λ involves an infinite rearrangement of the expansion in terms of the bare constants λ_ρ, but it converges uniformly in ρ and so enables one to pass to the limit $\rho \to \infty$, and thereby to define Schwinger functions free of cutoff. In [13], the resulting convergent expansion is referred to as the PRPSE (the partially renormalized phase space expansion).

Using the PRPSE to define the solutions of GN_2, one can introduce λ_{ren} and to prove that the perturbation series in the renormalized coupling constant is Borel summable to the solution.

CONCLUSION

The outcome of this study is the first non-perturbative treatment of a renormalizable (not superrenormalizable) quantum field theory: an existence theorm of solution of the Gross-Neveu model for arbitrary values of the renormalized mass and sufficiently small values of the renormalized coupling constant. The Schwinger functions of the solutions can be recovered from their renormalized perturbation series by the Borel summation procedure. The partially renormalized phase space expansion, which is obtained by an infinite rearrangement of the renormalized series converges to the solution for sufficiently small values of the renormalized coupling constant. The convergence of the PRPSE is closely tied to the cancellations arising from the fermionic character of the fields.

Does this behavior justify the view that interacting fermion fields are kinder and gentler than boson fields? Yes, but the lesson appears to be that it is only in the context of rigorous renormalization group methods that the better behavior of fermionic perturbation expansions can be exploited to yield existence theorems for solutions. This contrasts with the puzzling situation for the $\lambda\phi_4^4$ theory of a self interacting boson field in four space-time dimensions. For the "wrong" sign of the coupling constant λ, there is an existence theorem and the perturbation series is asymptotic to the solution [22] but for the "right" sign of λ we have famous no-go theorems [23]. We do not know whether the lack of an existence theorem for solutions with the "right" sign reflects the non-existence of solutions or merely the lack of a technique to construct them.

The situation is complicated further by another notable event recorded in these proceedings: the non-perturbation construction of the first non-trivial non-renormalizable quantum field theory, the Gross-Neveu model in three space-time dimensions, for all sufficiently large N. (See the lectures of R. Sénéor.)

I will not try to summarize all the riches you are about to hear, but it should not go without notice that with the announced results of Balaban on YM_4, Yang-Mills theory in four space-time dimensions, we have got one foot through the narrow gate into the promised land. For more than a decade, a gallant band of workers has been trying to bring quantum gauge field theories under the control of constructive quantum field theory, with the ultimate objective of QCD_4 and electro-weak theory. This was expected to be a very hard problem, and that it has turned out to be. So we have another reason to celebrate at this second school on constructive quantum field theory. Of course, Balaban's theory in its present form cannot quite be regarded as ready for household use, but one can hope that a fearless younger generatior will make it so.

REFERENCES

1. *Constructive Quantum Field Theory*, Ed. G. Velo and A. Wightman, Lecture Notes in Physics 25, Springer Verlag, 1974.
2. F.J. Dyson, The S-Matrix in Quantum Electrodynamics, *Phys. Rev.* *75*: 1736-1755 (1949).

3. F.J. Dyson, Heisenberg Operators in Quantum Electrodynamics I, *Phys. Rev. 82*: 428-439 (1951).
4. M. Gell-Mann and F. Low, Bound States in Quantum Field Theory, *Phys. Rev. 84*: 1350-1354 (1951).
5. E.R. Caianiello, On Quantum Field theory I Explicit Solution of Dyson's Equation in Electrodynamics without the use of Feynman Graphs, *Nuovo Cimento 10*: 1634-1652 (1953); II Nonperturbation Equations and Methods, ibid *11*: 493-529 (1954).
6. A. Buccafurri and G. Fano, Formulae for Feynman Graphs of Arbitrary Topology, *Nuovo Cimento 13*: 628 (1959).
7. A. Salam and P.T. Matthews, The Fredholm Theory of Scattering in a Given Time Dependent field, *Phys. Rev. 90*: 690-695 (1953). J. Schwinger, The Theory of Quantized Fields V, *Phys. Rev. 93*: 615-628, (1954).
8. E. Caianiello, Number of Feynman Graphs and Convergence, *Nuovo Cimento 3*: 223-225 (1956).
9. A. Hurst, An Example of a Divergent Perturbation Expansion in Field Theory, *Proc. Comb. Phil. Soc. 48*: 625-639 (1952). W. Thirring, On the Divergence of Perturbation Theory for Quantized Fields, *Helv. Phys. Acta 26*: 33-52 (1958). A. Petermann, Renormalization dans les series divergentes, *Helv. Phys. Acta 26*: 291-299 (1953).
10. D.R. Yennie and S. Gartenhaus, Convergence of the S. Matrix, *Nuovo Cimento 9*: 59-76 (1958).
11. A. Buccafurri and E.R. Caianiello, On the convergence of perturbative expansion, *Nuovo Cimento 8*: 170-173 (1958).
12. B. Simon, Convergence of the regularized renormalized perturbation series for super renormalizable field theories, *Nuovo Cimento 59A*: 199-214 (1969).
13. J. Feldman, J. Magnen, V. Rivasseau and R. Sénéor, A renormalizable field theory: The massive Gross-Neveu model in two dimensions, *Commun. Math. Phys. 103*: 67-103 (1986).
14. K. Gawedzki and A. Kupiainen, Gross-Neveu model through convergent perturbation expansions, *Commun. Math. Phys. 102*: 1-30 (1985).
15. D. Gross and A. Neveu, Dynamical symmetry breaking in asymptotically free theories, *Phys. Rev. D10*: 3235-3253 (1974).
16. J. Schwinger, On gauge invariance and vacuum polarization, *Phys. Rev. 82*: 664-679 (1951).
17. B.L. Ioffe, On the divergence of perturbation theory in quantum electrodynamics, *Doklady Akad. Nauk. USSR 94*: 437-438 (1954).
18. V.I. Ogievetsky, On a possible interpretation of perturbation theory in quantum field theory, *Doklady Akad. Nank USSR 109*: 912-922 (1956).
19. P.K. Mitter and P. Weisz, Asymptotic scale invariance in a massive Thirring model with $U(n)$ symmetry, *Phys. Rev. D8*: 4410 (1973).
20. D. Brydges, *A Short Course in Cluster Expansions*, pp. 132-183 in K. Osterwalder and R. Stora, eds. *Critical Phenomena, Random systems, Gauge Theories*, Elsevier Science Publishers, 1986.

21. The Federbush model of which the solutions were rigorously constructed by S. Ruijsenaars, *Commun. Math. Phys.* *87*: 181-228 (1982) is renormalizable and has an S matrix $\neq 1$, but is generally regarded as trivial; the proof isn't.
22. K. Gawedzki and A. Kupiainen, Non-trivial continuum limit of a ϕ_4^4 model with negative coupling constant. *Nucl. Phys.* *B257*: 474-504 (1985).
23. M. Aizenman and R. Graham, On the renormalized coupling constant and the susceptibility in ϕ_4^4 field theory and the Ising Model in four dimensions. *Nucl. Phys.* *B225*: 261-288 (1983). J. Fröhlich, On the triviality of $\lambda\phi_4^4$ theories and the approach to the critical point in $d > 4$ dimensions, *Nucl. Phys.* *B200*: 281 (1982).

RENORMALIZATION GROUP *

Roland Seneor

Centre de Physique Théorique

Ecole polytechnique

91128 Palaiseau Cedex

France

INTRODUCTION

Since the first Erice course on Constructive field theory, in 1973, considerable progresses have been done in understanding the exact mathematical meaning of field theory. In particular the renormalization problem is now fully understood as well at the perturbative level than at the constructive one, at least for the so-called asymptotically free theories. It is now a fact that there are no big differences between a super renormalizable theory and an asymptotically free one and I will briefly argue on this point. To construct a field theory one introduces cutoff: space and momentum cutoff; the aim of the games is then to prove the existence of some object of interest, for instance the generating functional, when one removes these cutoff. The removal of the space cutoff is done through an expansion, the Cluster Expansion, which was proposed in 1973 by Glimm, Jaffe and Spencer [1]. No big changes have been introduced since then. The general mechanism which control the removal of the momentum cutoff is the convergence of diagrams in perturbation theory. This idea was already understood by many mathematicians or physicists in the beginning of the seventies.

Let us see roughly how it works for a massive $P(\phi)_2$ theory, to be more specific for the Euclidean ϕ^4 in 2 dimensions. Suppose one has cutoff the momenta at M^ρ, $M > 1$, ρ being some large integer; the removal of the cutoff is then to take the limit ρ going to infinity. Then suppose one knows how to construct Green functions (more precisely, their Euclidean version, the Schwinger functions) in this cutoff theory. One may try to compare the results obtained with two different choices of cutoff indices ρ and ρ'. On the perturbative level the difference will be expressed in term of Euclidean Feynman graphs, convergent ones, since for a scalar field theory in 2 dimensions all diagrams are convergent provided one takes for the interaction Wick

* Supported in part by the Department of Energy under Grant DE-GF02-88ER25065.

Constructive Quantum Field Theory II, Edited by G. Velo and A. S. Wightman, Plenum Press, New York, 1990

ordered monomials with at least one internal line of momentum between M^{ρ} and $M^{\rho'}$. The convergence of the integrals expressing the diagrams means then that their values, because of the above remark on the choice of momentum for some line, will go to zero as $inf(M^{\rho}, M^{\rho'})$ goes to infinity. However, because of the divergence of the perturbative series this fact is not enough to prove the existence of the uncutoff expression. Since one cannot expand completely the exponential of the interaction, one has to do a truncated perturbative expansion, but because of the Wick ordering of the interaction, the exponential of the interaction is bounded by a constant blowing up with the momentum cutoff. The idea is therefore to produce enough convergent diagrams in order to beat this bound by their smallness. There are some care to be taken in generating the perturbative terms; although it is possible to associate one small factor per perturbative vertex, the number of graphs having n vertices grows as n!, therefore there is a balance between this two effects. For a ϕ^4 interaction, this works indeed only in 2 dimensions, in more then 2 dimensions one needs to do a more refined analysis relating space and momentum scales. However the general mechanism is the same, except that

1) for superrenormalizable field theories, the diagrams have to be renormalized in order to be convergent and the counterterms can be put by hand "à la Gellman-Low"

2) for strictly renormalizable theories one needs also to have convergent diagrams, this is done by a partial renormalization i.e. by renormalizing only the diagrams which are divergent. This can be done by introducing counterterms. However the locality of the renormalized interaction which is a general physical requirement imposes the introduction of an infinite number of unnecessary counterterms, for example coupling constant counterterms; since they are of no interest for renormalizing diagrams they are resummed to define effective parameters, for example, an effective coupling constant. Then the theories divided themselves in 2 categories, those for which this parameter is bounded and in general converges (to zero), the so called *asymptotically free* field theories, and those for which it diverges. For the first ones, the analysis of above can be applied, for the last ones nothing is known and it is expected that they did not exist in a meaning acceptable for the physicists (undefined or free field theories).

I. BOUNDS ON CONVERGENT CONNECTED GRAPHS

In this section we will prove bounds on completely convergent diagrams. Let G be a connected graph, $V(G)$ its set of vertices, $L(G)$ the set of internal lines and $E(G)$ the set of external (half-) lines. We define a subgraph G' of G by specifying $V(G'), L(G')$ and $E(G'), V(G')$ and $L(G')$ being respectively subsets of $V(G)$ and $L(G)$ such that $l \subset L(G')$ connects two vertices of $V(G')$; the external lines of $G', E(G')$, are the lines of $L(G) \cup E(G)$ which are attached to the vertices of $V(G')$ and which are not internal to G'.

A vertex of a graph G is *external* if there are some external line(s) hooked to it and internal on the contrary. We denote by $V_E(G)$ (respectively $V_I(G)$) the set of external (respectively internal) vertices of G.

Definition: A connected diagram is convergent if all its connected subdiagrams are convergent.

The notion of convergence is the usual one. Suppose we consider the theory of a scalar field ϕ in d dimensions whose propagator of the line l behave for large p as $p^{-\alpha_l}$ and consider v a vertex to which $2n$ such fields are attached. The power counting $\omega(v)$ of the vertex v is given by

$$\omega(v) = d(n-1) - \sum_l \alpha_l n \tag{1}$$

where the sum is over the lines l attached to the vertex.

Definition: A diagram G is convergent if

$$\omega(G) = \sum_{v \in V(G)} \omega(v) + d = -d(V(G)-1) + \sum_{l \in L(G)} (d - \alpha_l) < 0 \tag{2}$$

In the case of a ϕ^4 field theory with the usual propagator one has

$$\omega(G) = V(G)(d-4) - E(G)(\frac{d-2}{2}) + d \tag{3}$$

when $d = 4$, $\omega(G)$ depends only on the number of external lines of the graph and this is the characteric of a strictly renormalizable theory.

Remark (using a parity argument) that only the graphs with 0, 2 or 4 external lines are divergent.

Similarly one finds that a ψ^4 theory with propagators behaving as $\frac{1}{p}$ (fermion propagators!) for large p gives

$$\omega(G) = V(G)(d-2) - E(G)(\frac{d-1}{2}) + d \tag{4}$$

This time it is for $d = 2$ that $\omega(G)$ depends only on the number of external lines. This last example corresponds for what concerns the ultraviolet power counting to a 2 dimensional Gross-Neveu model. Remark that the divergent diagrams are of the same type as those of ϕ^4 in 4 dimensions.

We will now set a general hypothesis for convergent graphs.

Consider a set Ω of connected graphs such that if $\tilde{G}$ is a subgraph of G, $G \in \Omega$, then $\tilde{G} \in \Omega$.

Hypothesis: $\exists \epsilon, \epsilon > 0$, *such that* $\forall G, G \in \Omega$,

$$\omega(G) \leq -\epsilon E(\tilde{G}) \tag{5}$$

This hypothesis presents only some interest when the set Ω is infinite. In fact, for any connected completely convergent finite graph G, if one takes Ω to be the set of all its connected subgraphs, then there always exists such an ϵ. Moreover, for the completely convergent connected graphs of a ϕ^4 theory, ϵ can be taken equal to $\frac{1}{5}$.

To each connected graph G is associated a function $A(G)$, the amplitude, of the external vertices $x_1, ..., x_n$:

$$A(G)(x_1, ..., x_n) = \int \prod_v dx_v \prod_l C(x_l, x_l') \tag{6}$$

where the first product runs over the internal vertices of G and the second one over the internal lines l of G, x_l and x'_l being the end points of the line l. The functions $C(x,y)$ are the propagators or covariances of the theory and we have assumed that they are independant of the lines as it has to be if we think about these graphs as being the perturbative diagrams of a given field theory.

Remark, and this will be usefull for what follows, $A(G)(x_1,...,x_n)$ is one of the connected term in

$$\int \prod_v v(\phi) d\mu_C(\phi)$$

where $d\mu(\phi)$ is the Gaussian measure of covariance C and mean zero and $v(\phi)$ is some Wick monomial of ϕ. One also introduces the average of $A(G)(x_1,...,x_n)$ over unit cubes:

$$\overline{A}(G) = \int \chi_{\Delta_1} .. \chi_{\Delta_n} A(G)(x_1,...,x_n) dx_1 ... dx_n \tag{7}$$

Let the covariance C be given by

$$C = (-\Delta + m^2)^{-1} \tag{8}$$

then we have

Theorem: Let G be a completely convergent connected graph, then $\exists K, \delta, \delta > 0$, small, such that

$$| \overline{A}(G) | \leq K^{L(G)} \sup_{x_v \in \Delta} \prod_{l \in G} e^{-m(1-\delta)|x_l - y_l|} \tag{9}$$

It is easy from this theorem to get bounds on the Fourier transform of $A(G)$ [2,3].

Proof: First, using a scaling argument, we choose $m = 1$. We then introduce a sequence of momentum scales $\{M^i\}$, $i = 1,2,..$, $M > 2$, and a corresponding decomposition of the covariance

$$C = \sum_0^\infty C^i$$

with

$$\tilde{C^i}(p) = \tilde{C}(p)\eta^i(p) \tag{10}$$

where

$$\begin{aligned} \eta^i(p) &= \eta(p/M^{i+1}) - \eta(p/M^i), \qquad i > 0 \\ \eta^0(p) &= \eta(p/M) \end{aligned} \tag{11}$$

η being a smooth function of the modulus of p such that $\eta(0) = 1$.

Then one can show that in d dimensions, ($d > 2$), ∃K, constant, such that the kernel of C is bounded by

$$C^i(x-y) \leq K M^{i(d-2)} e^{-M^i|x-y|} \tag{12}$$

for $\eta(p) \simeq e^{p^2+m^2}$ and

$$C^i(x-y) \leq K \frac{M^{i(d-2)}}{(1 + M^i|x-y|)^r} \tag{13}$$

for $\eta(p)$ with compact support, r being some positive integer depending on the smoothness of η.

Thus, with $C^i(l)$ a shorthand notation for $C^i(x_l - y_l)$,

$$\overline{A}(G) = \sum_{i_1} .. \sum_{i_L} \int \prod_v dx_v C^{i_1}(l_1)..C^{i_L}(l_L) = \sum .. \sum \overline{A}(G_{i_1..i_L}) \tag{14}$$

Let us now consider the graph $G_{i_1..i_L}$ made of G by restricting the momentum of the lines l_k to be $i(l_k) = i_k$. It can be written as an inductive limit of graphs G_j. Each G_j is an union of connected components G_j^α whose lines are given by the lines of $G_{i_1..i_L}$ whose indices are $\geq j$.

For a vertex v of G (from now on when no confusion is possible, we use G instead of $G_{i_1..i_L}$) let

$$\begin{aligned} h_v &= \{\sup i(l); l \in V(G)\} \\ b_v &= \{\inf i(l); l \in V(G)\} \end{aligned} \tag{15}$$

where by convention $b_v = -1$ if $v \in V_E(G)$.

Because the C's are translation invariant, one vertex location is not fixed. We choose an arbitrary vertex v_0 of G and consider it as fixed. One then fixes inductively, starting from the graphs with low indices, a fixed vertex in each connected component. For $G_0 = G$ one has taken v_0; if $v_0 \in G_1^\alpha$ then the fixed vertex in G_1^α is still v_0; if $v_0 \notin G_1^\alpha$ then one choose an arbitrary vertex of G_1^α as fixed vertex; one does the same analysis for the components G_2^α and so on and so forth. The fixed vertex of G_j^α will be denoted v_j^α.

To estimate the graph the strategy is

1) to replace each propagator by its bound 2) to deduce inductive bounds on the graphs using the subgraph decomposition previously introduced.

One first decomposes the exponential bound of C^i using

$$M^i \geq \frac{1}{2} \sum_{j \geq 0}^{i} M^j \qquad for \quad M \geq 2 \tag{16}$$

thus

$$e^{-M^i|x-y|} \leq e^{-(1-\delta)M^i|x-y|} \prod_{j=0}^{i} e^{-\frac{\delta}{2}M^j|x-y|} \tag{17}$$

Using the fact that $M^i \geq 1$, the first factor on the right hand side gives the factor $e^{-(1-\delta)|x-y|}$ of the bound in the theorem (remember that m has been chosen equal to 1). The second factors of the bound will then be attributed to each subgraph containing the line having C^i as propagator by the formula

$$\prod_{l \in G} \prod_{j=0}^{i(l)} e^{-\frac{\delta}{2}M^j|x_l - y_l|} = \prod_{j,\alpha} \prod_{l \in G_j^\alpha} e^{-\frac{\delta}{2}M^j|x_l - y_l|} \tag{18}$$

We now give the rules to integrate over the vertices. Each vertex has to be integrated. We chose to consider the graph as built starting from the highest momenta and introducing progressively lower ones. Given a subgraph we thus integrate over all non fixed vertices which have not been integrated previously. A vertex v is therefore integrated at level

$$i(v) = sup\{j|\exists l \in v, i(l) = j; v \neq v_j^\alpha, \forall \alpha\} \tag{19}$$

This means that the vertex integration, for a non fixed one, will be controlled by the behaviour of the highest momentum line arriving to it. The contribution of propagators from the other lines arriving at the vertex will be estimated by taking the supremum over the vertex location.

There is still an ambiguity since more than one line can be at the same momentum. We solve this problem by introducing for each connected component G_j^α a tree T_j^α going through all its vertices. It is not uniquely define but this is of no importance. Its length $|T_j^\alpha|$ is given by

$$|T_j^\alpha| = \sum_{l \in T_j^\alpha} |x_l - y_l| \tag{20}$$

Take now the vertices which have to be integrated $V(j,\alpha) = \{v|v \in G_j^\alpha$ such that $i(v) = j\}$ and construct a tree starting from v_j^α and going through all of them. This can be obtained by " turning " around the original tree T_j^α starting from the origin (see the figure). This orders the v's. Let us relabel the vertices $v(j,\alpha,i)$, $i = 1, 2, .., n$, according to this order. Then using the triangular inequality one gets

$$2|T_j^\alpha| > |x_{v_j^\alpha} - x_{v(j,\alpha,1)}| + \sum_{i=1}^{n-1} |x_{v(j,\alpha,i)} - x_{v(j,\alpha,i+1)}| \tag{21}$$

Using these remarks, one performs the integration over the vertices of $V(j,\alpha)$ starting from those vertices which have the highest indices. Using formulas (18), (19) and (20) one gets

$$\int \prod_{l \in L(G_j^\alpha)} e^{-\frac{\delta}{2} M^j |x_l - y_l|} \prod dx_{v(j,\alpha,i)} \leq \int e^{-\frac{\delta}{2} M^j |T_j^\alpha|} \prod dx_{v(j,\alpha,i)} \leq$$

$$\leq \int e^{-\frac{\delta}{4} M^j \sum_0^{n-2} |x_{v(j,\alpha,i)} - x_{v(j,\alpha,i+1)}|} \times$$

$$\times \left(\int e^{-\frac{\delta}{4} M^j |x_{v(j,\alpha,n-1)} - x_{v(j,\alpha,n)}|} dx_{v(j,\alpha,n)} \right) \prod_1^{n-1} dx_{v(j,\alpha,k)} \tag{22}$$

and one bounds the last integral using

$$\int e^{-\frac{\delta}{4} M^j |x_{v(j,\alpha,n-1)} - x_{v(j,\alpha,n)}|} dx_{v(j,\alpha,n)} \leq \int e^{-\frac{\delta}{4} M^j |x|} dx = K(\delta, d)[M^{-j}]^d \tag{23}$$

One then repeats this procedure and gets as a resulting bound

$$\int \prod_{l \in L(G_j^\alpha)} e^{-\frac{\delta}{2} M^j |x_l - y_l|} \prod dx_{v(j,\alpha,i)} \leq [KM^{-jd}]^n \tag{24}$$

where $n = n(j, \alpha)$ is the number of vertices to be integrated at level j.

Remarks:

1) In the bounds of formulas (22) and (24) the domain of integration of a vertex is the whole space if this vertex is internal or the volume of a unit cube if it is external. We bound this last case by integrating over the whole space.

2) The bound (23) is obtained as follows

$$\int e^{-\frac{\delta}{4}M^j|x|}dx = \sum_{\Delta \in \mathcal{D}^j} \int_\Delta e^{-\frac{\delta}{4}M^j|x|}dx \tag{25}$$

where $\mathcal{D}^j$ is a partition of space into an union of disjoint cubes of size M^{-j}. Now

$$\int_\Delta e^{-\frac{\delta}{4}M^j|x|}dx \leq K e^{-\frac{\delta}{4}M^j c_\Delta}[M^{-j}]^d \tag{26}$$

c_Δ being the center of the cube Δ. Finally

$$\sum_{\Delta \in \mathcal{D}^j} e^{-\frac{\delta}{4}M^j c_\Delta} \leq K\delta^{-d} \tag{27}$$

Collecting these bounds for all lines and vertices of G we have

$$\int \prod_{l \in L(G)} e^{-\frac{\delta}{2}M^j|x_l - y_l|} \prod_{v \in V(G)\setminus v_0} dx_v \leq \prod_{j,\alpha} [KM^{-jd}]^{n(j,\alpha)} \tag{28}$$

Introducing the factor in front of the exponential in the bound (12) for the covariance we get that

$$|\overline{A}(G_{i_1..i_L})| \leq \prod_{l \in G_{i_1..i_L}} [KM^{i(l)}]^{d-2} \prod_{v \neq v_0} [KM^{-ji(v)}] \tag{29}$$

Using

$$\begin{aligned} [M^{i(l)}]^{d-2} &= \prod_{j=1}^{i(l)} M^{d-2} = \prod_{1 \leq j \text{ s.t. } l \in G_j} M^{d-2} \\ [M^{-di(v)}] &= \prod_{j=1}^{i(v)} M^{-d} = \prod_{1 \leq j \text{ s.t. } v \in G_j, v \neq v_j^\alpha} M^{-d} \end{aligned} \tag{30}$$

one can rewrite the bound (29) as

$$\begin{aligned} & K^{L(G)} K^{V(G)-1} \prod_{j \geq 1} \prod_\alpha [M]^{-d(V(G_j^\alpha - 1) + \sum_{l \in G_j^\alpha}(d-2)} \\ &= K^{L(G)} K^{V(G)-1} \prod_{j \geq 1} \prod_\alpha [M]^{-d(V(G_j^\alpha - 1) + L(G_j^\alpha)(d-2)} \\ &\leq K^{L(G)} \prod_{j \geq 1} \prod_\alpha [M]^{-\epsilon E(G_j^\alpha)} \end{aligned} \tag{31}$$

where one used formula (2) for $\omega(G)$ and the *hypothesis.*

We now use this bound to perform the sum over the momenta in formula (14). The last inequality in formula (31) shows that one gets a factor $M^{-\epsilon}$ per external line of G_j^α. Given a line l ending at a vertex v, l is an external line of G_j^α, for some α provided $v \in G_j^\alpha$ and $i(l) < j$. This means that given any vertex v and any (half-)line l hooked to it, there is an associated factor $M^{-\epsilon(h_v - i(l))}$. Since each vertex has a lowest momentum line, this means that we have in particular a factor $M^{-\epsilon(h_v - b_v)}$ per vertex. Let us now suppose that there are at most ν lines per vertex (this a convenient but not fondamental hypothesis) and index the lines by $i = 1, ..., \nu$ and their momentum indices by $i(v,1), ..., i(v,\nu)$. Using

$$\sum_{(j,j')} |i(v,j) - i(v,j')| \leq \nu^2 (h_v - b_v) \tag{32}$$

the bound (31) becomes

$$K^{L(G)} \prod_v M^{\epsilon' \sum_{(j,j')} |i(v,j) - i(v,j')|} \tag{33}$$

with $\epsilon' = \frac{\epsilon}{\nu^2}$.

To sum over the momentum indices using formula (32), one will generate a tree structure (in the space of the momentum indices!). The tree is generated as follows: one starts from an external vertex of G; we choose an arbitrary line labelled l_1 hooked to this vertex whose index is $i(l_1)$. The other end is hooked to another vertex to which we associate another line l_2. One then chooses a line l_3 hooked to one of the vertices attached by $l_1 \cup l_2$. Then one chooses l_4 with end point a vertex in $l_1 \cup l_2 \cup l_3$. This operation is done up to exhaustion of all the internal lines of G, i.e. up to $l_{L(G)}$. This gives an ordering of the internal lines of G.

One then starts summing over all possible values of the momentum index $i(l_{L(G)})$, then on the values of $i(l_{L(G)-1})$, ..., using

$$\sum_{i=0}^{\infty} M^{-\epsilon'|i-j|} \leq 2 \sum_{i=0}^{\infty} M^{-\epsilon' i} = 2[1 - M^{-\epsilon'}]^{-1} \tag{34}$$

and the fact that for l_1 the related index is -1. The bound of (34) gives another $K^{L(G)}$ and this ends the proof.

Remarks.

1) By analysing the graphs with the help of momentum slicing and space localization we have seen the appearance of two types of exponential decrease:

a) a decrease between space points due to the short range of the interaction (in the case of momentum cutoff with compact support the decrease is not exponential but power like)

b) a decrease between momentum slices due to the convergence of the subgraphs.

2) There is an exponential decrease (a power like for cutoff with compact support)

$$exp(-(1-\delta)M^j |\text{length of a tree connecting the external vertices}|)$$

j being the lowest internal momentum index of the graph.

3) When bounding the sum over the momentum assignment we could have used $2^{-(i-j)}$, instead of $M^{-(i-j)}$, to sum over the i's, $i \geq j$, it thus remains an overall convergent factor

$$(M/2)^{i_{max}(G)-i_{min}(G)}$$

This means that up to exponential factors the leading terms in graph estimates come from *flat* graphs, i.e. from graphs whose internal momenta have the same index.

4) In bounding connected graphs we introduce factors like

$M^{i(l)(d-2)}$ per internal line

$M^{-i(v)d}$ per integrated vertex

they are equivalent to a factor M^{-j} per external line of momentum index j.

II. RENORMALIZATION OF COMPLETELY CONVERGENT GRAPHS

According to the analysis of chapter I the appearance of a divergent subgraph G at level i is characterized by the fact that, in the remaining factor $[M^{-i}]^{E(G)}$, $E(G)$ is less or equal to four. To recover the required power one will *renormalize* the subgraph.

Let us consider the case of a 2-point function ($E = 2$) for a ϕ^4 in 4 dimensions. It can be written

$$I = \int \phi(x) F_i(x,y) \phi(y) d^4x d^4y. \tag{1}$$

In this formula $F_i(x,y)$ is the amplitude of a convergent graph with 2 external lines whose vertices are localized at x and y, and we suppose that the lowest momentum index of its internal lines is i. In the scheme presented in the previous chapter instead of $\phi(x)$ and $\phi(y)$ one would have some propagators or sums of propagators $C^k(x-.)$ and $C^l(y-.)$ with $k,l < i$; we used the remark folowing formula (I.6) to replace them by two fields with momentum indices less than i. Moreover, by remark 2) at the end of chapter I, $F_i(x,y)$ is bounded by $e^{-(1-\delta)M^i|x-y|}$.

Remark 1: By the way we present the renormalization problem, the renormalization of the 2-point functions appears to be necessary only if the external fields have indices less than i.

To renormalize I we subtract to it

$$\frac{1}{2}\int \phi^2(x)\left(\int F_i(x,y)d^4y\right)d^4x + \frac{1}{2}\int \phi^2(x)\left(\int F_i(y,x)d^4y\right)d^4x \tag{2}$$

This gives

$$-\frac{1}{2}\int (\phi(x)-\phi(y))^2 F_i(x,y) d^4x d^4y. \tag{3}$$

We now use

$$\phi(x)-\phi(y) = \int_0^1 \frac{d}{dt}\phi(y+t(x-y))dt = \int_0^1 \overrightarrow{x-y}.\vec{\nabla}\phi(y+t(x-y))dt \tag{4}$$

to rewrite (3)

$$-\frac{1}{2}\int_0^1\int_0^1\left(\int(\overrightarrow{x-y}.\vec{\nabla}\phi(y+t_1(x-y))\times\right.$$
$$(\overrightarrow{x-y}.\vec{\nabla}\phi(y+t_2(x-y))F_i(x,y)d^4xd^4ydt_1dt_2 \tag{5}$$

Regrouping the two $\overrightarrow{x-y}$ with F_i and using its decrease we have

$$|(x-y)_\alpha(x-y)_\beta F_i(x,y)| \leq |(x-y)^2|e^{-(1-\delta)M^i|x-y|} \leq K(\delta)M^{-2i}e^{-(1-2\delta)M^i|x-y|} \tag{6}$$

What we see is that the amplitude of the 2-point function is now bounded by M^{-2i} times a slightly weaker exponential decrease. The price we pay for that is the appearance of gradients on the external fields. This can be generalized to the following assertion: *for an amplitude of index i, each factor $\overrightarrow{x-y}$ can be converted into M^{-i}.* Suppose that the momentum index of $\phi(x)$ is k; when integrated at level k, $\vec{\nabla}\phi(x)$ will give $\vec{\nabla}C^k(x-.)$ and it is easy to check that

$$|\vec{\nabla}C^k(x-.)| \leq KM^{k(d-2)}M^k e^{-M^k|x-.|} \tag{7}$$

i.e. ∇ has been replaced by M^k. Again this can be replaced by the assertion: *each gradient acting on a field of momentum index k can be replaced in the bounds by M^k.*

The net result of the subtraction is therefore a *transfer* of momentum from the amplitude of the graph to the external legs. We loose a factor M^{2k} but win M^{-2i}, i.e. everything happens as if the number of internal lines has diminished by 2 units but the number of external ones has increased by 2 units $(E(G) \to E(G)+2)$.

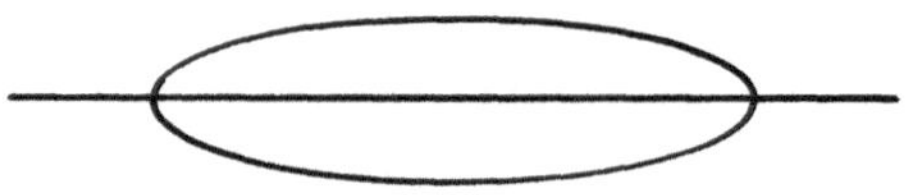

Fig. II.1 The lowest order 2-point graph

Let us look at an example.

Consider (1) with the lowest order 2-point graph (see Fig. II.1) of a ϕ^4 theory in 4 dimensions and suppose that

$$F_i = \sum_{j\geq i} F_{(j)}$$

with the internal lines of the $F_{(j)}$'s all of index j (an exact physical situation would have been to suppose that the infimum over the momentum indices of the internal lines is j). Each propagator is bounded by M^{2j} times the exponential factor and is nearly constant on cubes of $\mathcal{D}_j$, thus the amplitude of the 2-point function $F_{(j)}$ is bounded by

$$KM^{6j}e^{-M^j|c_\Delta - c_{\Delta'}|}. \tag{8}$$

After subtraction it is bounded by

$$KM^{4j}e^{-(1-\delta)M^j|c_\Delta - c_{\Delta'}|} \tag{9}$$

and there are gradients acting on the external legs. This means that the subtracted graph behaves as a four point function. However we are not done since in 4 dimension a ϕ^4 4-point function is still divergent and this will be seen when summing over j.

Consider a term with a given j and suppose for simplicity that the external fields are of index $k, k \leq j$. Because of this hypothesis, the external fields are nearly constant on scale bigger than k. To integrate on space we use

$$\int (\ .\ ) d^4z = \sum_{\Delta' \in \mathcal{D}_k} \ \sum_{\Delta \subset \Delta', \Delta \in \mathcal{D}_j} \int_\Delta (\ .\ ) d^4z \tag{10}$$

The sum over Δ' will treated at levels below k. We are only concerned with the last sum and with the integral. Since the argument of integration is nearly constant on scale j, the integrals over the 2 vertices give $(M^{-4j})^2$. For one of the external vertices, the sum over the cubes of $\mathcal{D}_j$ contained in $\mathcal{D}_k$, and indeed in the whole space, is insured by the exponential decrease of bound (9).

For the sum over the remaining cube, Δ (corresponding to the fixed vertex), there are no convergent factors coming from the amplitude or from the external fields, thus the sum over cubes of $\mathcal{D}_j$ contained in cubes of $\mathcal{D}_k$ leads to a factor $M^{-4(k-j)}$.

Collecting all the power like factors in M, we see that after integrations and summations the factor M^{4j} of the bound (9) is replaced by M^{4k}, i.e. is independent of j. This is the characteristic signature of a logarithmic divergence. In fact for large j, the sum up to j gives j which is, up to a constant, $Log M^j$.

In order to get something convergent we therefore need to subtract from (1) one more term. To do that we will modify formula (4) and rewrite it

$$\phi(x) - \phi(y) = \overrightarrow{y-x}.\vec{\nabla}\phi(x) + \int_0^1 (1-t) \sum_{\alpha,\beta} (x-y)_\alpha (x-y)_\beta \nabla_\alpha \nabla_\beta \phi(y + t(x-y)) dt \tag{11}$$

One then subtracts from (3)

$$-\frac{1}{4} \int \left(\left(\overrightarrow{(y-x)}.\vec{\nabla}\phi(x) \right)^2 + \left(\overrightarrow{(x-y)}.\vec{\nabla}\phi(y) \right)^2 \right) F_i(x,y) d^4x d^4y. \tag{12}$$

The remainder is this times a true renormalized expression. In fact it is made of terms having at least 3 gradients acting on the external fields and the amplitude is multiplied by a corresponding number of factors $x - y$.

Let us see on our example why those terms are good.

Take a term with 3 derivatives. Because we have one more derivative than the case we treat above, the renormalized amplitude is bounded now by M^{3j}. After summation in cubes of size M^{-k}, one gets $M^{-j}M^{-4k}$ which can be rewritten as $M^{-(j-k)}M^{-5k}$. The first factor controls now the sum over $j, j \geq k$, and the last factor will be compensated at level k by the 3 gradients (M^{3k}) and the behaviour of the 2 fields (M^{2k}).

Such an analysis can be repeated for the 4-point functions. To

$$\int \phi(x_1)...\phi(x_4)F_j(x_1...x_4)d^4x_1..d^4x_4 \tag{13}$$

one can subtract (it is possible, as for the 2-point function, to subtract more symmetrical expressions)

$$\int \phi(x_1)^4 F_j(x_1...x_4)d^4x_1..d^4x_4. \tag{14}$$

This gives a renormalized quantity

$$\sum_{j=2}^{4} \int \phi(x_1)..(\phi(x_j) - \phi(x_1))\phi(x_{j+1})..\phi(x_4)F_j(x_1...x_4)d^4x_1..d^4x_4 \tag{15}$$

Again one can write $\phi(x_j) - \phi(x_1)$ as in formula (4). The field is replaced by a derived one and the amplitude is multiplied by $x_j - x_1$ meaning that (15) behaves as if there was 5 external lines.

Remark 2: 1) If in a given graph G we replace, in the analysis of chapter 1, its divergent subgraphs by their renormalized expressions, then the graph can be bounded exactly in the same way as if it was completely convergent.

2) We have subtracted to divergent graphs counterterms which seems to be local since of the form $(\phi(x))^2$, $(\nabla\phi(x))^2$ or $(\phi(x))^4$. However this is not the case because they have momentum indices which are linked to the index of the divergent graph. For example let us look at a 4-point "divergent" amplitude $F_j(x_1, ..., x_4)$ appearing as subgraph in a graph G, the counterterm will be

$$\int \tilde{F}_j(0)(\phi_{j-1}(x))^4 d^4x \tag{16}$$

where we have introduced

$$\phi_j = \sum_0^j \phi^k. \tag{17}$$

Roughly speaking, $\phi_j(x)$ is not localized at x since by definition it represents a field with an ultraviolet cutoff at M^j, therefore its local fourth power is not local in $\phi(x)$.

This remark is very important since it shows that the price to pay for having uniform graph estimates is that the divergent graphs have to be renormalized in a way which does not fulfil the usual requirement of *locality of the counterterms*. We will call this way of renormalizing a *useful renormalization* or *partial renormalization*. The subtracted counterterms are the useful part of the counterterms.

Let us see on an example why it is licit to separate the counterterms into 2 categories: the useful ones and the non-useful ones.

Consider the 4-point function given by the simple bubble (see Fig. II.2) with external momenta k. Its amplitude is formally given by

$$\int \frac{1}{(p-k)^2+1}\frac{1}{p^2+1}d^4p \tag{18}$$

Fig. II.2 The simple bubble

Using the α-parametrization we can rewrite it as

$$\int_0^\infty \left(\int_0^1 e^{-\lambda(\alpha(1-\alpha)k^2+1)}d\alpha\right)\frac{d\lambda}{\lambda} \tag{19}$$

and since the integration over α is on a compact set, we modify it and take a simplified form of (19) having the same ultraviolet behavior:

$$\int_0^\infty \frac{e^{-\lambda(k^2+1)}}{\lambda}d\lambda \tag{20}$$

A cutoff version of (20) is now

$$I_\kappa(k)=\int_{\kappa^{-2}}^\infty \frac{e^{-\lambda(k^2+1)}}{\lambda}d\lambda \tag{21}$$

with $\kappa > k$.

I_κ diverges logarithmically when $\kappa\to\infty$. The usual local counterterm is $I_c = I_\kappa(0)$. Its useful part (since k is the momentum of the external lines) is

$$I_c^u=\int_{\kappa^{-2}}^{k^{-2}} \frac{e^{-\lambda}}{\lambda}d\lambda. \tag{22}$$

Let us now estimate $I_{ren} = I - I_c$:

$$I_{ren}=\int_{\kappa^{-2}}^{k^{-2}} e^{-\lambda}\left(\frac{e^{-\lambda k^2}-1}{\lambda}\right)d\lambda+\int_{k^{-2}}^\infty \frac{e^{-\lambda(k^2+1)}}{\lambda}d\lambda-\int_{k^{-2}}^\infty \frac{e^{-\lambda}}{\lambda}d\lambda=A+B+C \tag{23}$$

where the last term $-C$ is the non-useful part of the counterterm. The term A is bounded

$$A \leq \int_{\kappa^{-2}}^{\infty} e^{-\lambda} k^2 d\lambda \leq 1 \tag{24}$$

uniformly in κ. But B and C are both in $k^2 e^{-\frac{1}{k^2}}$ and do not compensate each other since

$$C - B = \int_{k^{-2}}^{1+k^{-2}} \frac{e^{-\lambda}}{\lambda} d\lambda \tag{25}$$

This has for consequence that when we estimate completely renormalized graphs one part of the estimates is as for completely convergent graphs but there is another part concerning the useless counterterms and their corresponding amplitude. This leads to the following theorem for ϕ^4 in 4 dimensions

Theorem: If G is a connected graph with f subgraphs having 2 or 4 external legs, then

$$|\overline{A}(G)| \leq K^{V(G)} f! K^f \tag{26}$$

For the proof of the theorem we refer to [4]. The factors $f!K^f$ represents the effects of the useless counterterms. Since this is a consequence of the locality of the counterterms we see that the renormalon problem [5] has for origin this requirement.

III. CONSTRUCTION OF MODELS

We now describe what are the general ideas which permit the construction of physically relevant models. However to simplify the discussion our starting point will be an unrealistic field theory, the scalar ϕ^4 in 4 dimensions. This comes from the fact that physically relevant models generally implies complicated structures (for example, fermions and non abelian gauge symmetry) which can be ignored if we stay at the level of principles.

We thus consider a scalar field ϕ and a theory in 4 dimensions whose Lagrangian density $\mathcal{L}$ is given by

$$\mathcal{L}(\phi)(x) = g\,(\phi(x))^4 + \frac{1}{2}\left((\nabla\phi(x))^2 + m^2(\phi(x))^2\right) = \mathcal{L}_I(x) + \mathcal{L}_0(x) \tag{1}$$

What we learn from the construction of superrenormalizable field theories [1] and from the Gell'Man-Low analysis [6] is

1) to have a well defined starting expression we have to introduce cutoff which restrict the theory to finitely many degrees of freedom. We thus restrict the volume of integration to Λ and cutoff the momenta at M^ρ, the aim being to let $\Lambda \to \mathbf{R}^4$ and $\rho \to \infty$.

2) in order to get the correct perturbation series (the renormalized one!) one has to modify the interaction Lagrangian $\mathcal{L}_I$ by adding to it counterterms in the form $\delta\mathcal{L}_I$ and one formally writes the Green functions (Euclidean)

$$< \phi(x_1)...\phi(x_n) >_{\Lambda,\rho} = \frac{\int \phi(x_1)...\phi(x_n) e^{-\int_\Lambda \mathcal{L}_\rho(\phi)(x)} \prod_x D\phi(x)}{\int e^{-\int_\Lambda \mathcal{L}_\rho(\phi)(x)} \prod_x D\phi(x)} \tag{2}$$

In practice, in (2), one separates in the functional integrations the quadratic part in the fields from the rest and takes the limit $\Lambda \to \mathbf{R}^4$. Thus in (2)

$$e^{-\int_\Lambda \mathcal{L}_\rho(\phi)(x)} \prod_x \mathcal{D}\phi(x)$$

is replaced by

$$e^{-\int_\Lambda \mathcal{L}_{I,\rho}(\phi)(x)} d\mu_{C_\rho}(\phi)$$

where $d\mu_{C_\rho}(\phi)$ is the Gaussian measure of mean zero and covariance

$$\tilde{C}_\rho(p) = \frac{\eta_\rho(p)}{p^2 + m^2} \tag{3}$$

and

$$\mathcal{L}_{I,\rho}(x) = (g + \delta g)(\phi(x))^4 + \frac{1}{2}\delta m^2(\phi(x))^2 + \frac{1}{2}\delta\varsigma(\nabla\phi(x))^2 \tag{4}$$

here δg is the coupling constant counterterm, δm^2 is the mass counterterm and $\delta\varsigma$ is the wave function counterterm.

After these replacements, the convergence of (2) is shown by expanding in series the numerator and the denominator. Factorizing the dominant terms (they correspond to terms in which no vertices are created, see below), both of them are shown to be convergent and their ratio can be expressed as a uniformly (with respect to Λ and ρ) convergent expansion. Expressing the ratio as an expansion is what is called performing a *Mayer expansion*[7].

The series produced for the numerator and the denominator are obtained by the way of two expansions one in position space and the other one in momentum space.

The position space expansion is a *cluster expansion.* It has been introduced in [1] by Glimm, Jaffe and Spencer. It allows to take the limit when $\Lambda \to \mathbf{R}^4$; one uses the fact that exponential or large enough power decrease of the covariance permit us to assert that the effects of far distant part of the interaction (from the set of x_i's for the numerator or from an arbitrary fixed point for the denominator) are small (exponentially or power like). A lattice is introduced which defines the unit of length with respect to which distances have to be measured. As we showed in chapter 1, for a field of momentum index i the propagator naturally exhibits a decrease in the scaled distance with respect to this momentum, therefore the natural lattice for it will be $\mathcal{D}_i$.

The momentum space expansion or decoupling expansion (we will describe it in more details in what follows) is done to test the coupling between different momenta. In fact if the interaction is local in x, it is not local in momentum:

$$(\phi(x))^4 = \sum_{i_1,\ldots,i_4} \phi^{i_1}(x)\ldots\phi^{i_4}(x)$$

where the ϕ^j have been defined with respect to

$$C_\rho = \sum_0^\rho C^j$$

Thus at each level, let us say j, one tries to factorize the interaction, i.e. to write

$$\mathcal{L}_{I,\rho}(\phi_\rho) = \mathcal{L}_{I,j-1}(\phi_{j-1}) + \mathcal{L}_{I,\rho}(\phi_\rho - \phi_{j-1}). \tag{5}$$

Again, the natural regions of space in which the testing has to be done are the cubes whose size are related to the momentum index, i.e. here the cubes of $\mathcal{D}_j$.

Both expansions produce polynomial in the fields in front of the exponential either by some derivatives of the field $\frac{\delta}{\delta\phi(x)}$ or by derivatives $\frac{d}{dt}$ with respect to some interpolating parameters t (introduced in the interaction to compare different dependence with respect to momentum levels) acting on the exponential of the interaction. When integrated with respect to the Gaussian measures, these polynomials generate perturbative graphs whose structure is exactly the one described in chapter 1. Moreover if the correct counterterms have been introduced, these graphs are completely convergent ones and are therefore bounded by K^n, n being the number of vertices.

Remark 1: in a scalar field theory, one cannot produce too many fields in front of the exponential of the interaction; this is because when gaussianly integrated the number of produced graphs behaves as some factorial in the number of fields. This is a trace of the unboundedness of the boson fields. Conversely perturbation expansions of fermions do not generate such factorials, a trace of the fact that the fermi fields are bounded operators. In the case of Bose fields, this difficulty is cured first by limiting carefully their number only to what is necessary to explicitly produce the cancellation of divergences of divergent graphs with their counterterms and secondly by using the positivity of the interaction (a physical requirement!) to dominate (like $x^n e^{-x^4} \leq (n!)^{\frac{1}{4}}$) the nonintegrable remaining powers of the fields.

Remark 2: roughly speaking, after these expansions have been performed, a Green function appears as a sum of product of connected graphs, having the structure of completely convergent graphs, characterized by sets of momentum indices and by sets of cubes of lattices (indexed by the momentum scales). The convergence of this sum is ensured if there is a small enough factor per cube. From what we explained above there is at least one interaction vertex produced in each cube, thus the small factor can be obtained by choosing a small enough coupling constant, and this is the case for *all the models built up to now*.

Remark 3: we have seen in chapter I that the sum over momentum assignment of completely convergent graphs produces a bound K^n where n is the number of vertices. Following Remark 2 this factor can be controlled by a small coupling constant.

We now look in more details, in the case of ϕ^4, to the decoupling expansion and its meaning for the coupling constant.

As shown in formula (4) the initial ϕ^4 part is $g_\rho \int \phi^4$ where we have introduced the *effective coupling constant* or *bare coupling constant* g_ρ at level ρ. At this level we write (see formula II.17 for the notation)

$$\phi = \phi_\rho = \phi^\rho + \phi_{\rho-1} \tag{6}$$

and try to decouple ϕ^ρ from $\phi_{\rho-1}$ in cube of size related to the momentum index ρ. So in each cube $\Delta \in \mathcal{D}_\rho$ we introduce an interpolating parameter $t_{\rho,\Delta}$ $(= t)$ as follows:

$$I_\Delta = e^{-g_\rho \int_\Delta (\phi_\rho)^4} \tag{7}$$

becomes

$$I_\Delta(t) = e^{-g_\rho \int_\Delta (\phi^\rho + t\phi_{\rho-1})^4 - (1-t^4) g_{\rho-1} \int_\Delta (\phi_{\rho-1})^4}. \tag{8}$$

It interpolates between $I_\Delta(1) = I_\Delta$ and

$$I_\Delta(0) = e^{-g_\rho \int_\Delta (\phi^\rho)^4} e^{-g_{\rho-1} \int_\Delta (\phi_{\rho-1})^4} \tag{9}$$

which is, up to the mass terms of the form

$$\exp - \left(\int_\Delta \mathcal{L}_{I,\rho}(\phi^\rho) + \int_\Delta \mathcal{L}_{I,\rho-1}(\phi_{\rho-1}) \right).$$

The expansion is then generated by a Taylor formula

$$I_\Delta = I_\Delta(0) + I_\Delta^{(1)}(0) + ... \frac{1}{4!} I_\Delta^{(4)}(0) + \frac{1}{4!} \int_0^1 (1-t)^4 I_\Delta^{(5)}(t) dt \tag{10}$$

Remark 4 At momentum level ρ, formula (6) expresses the decomposition of ϕ into a *high momentum* field ϕ^ρ and a *low momentum* field $\phi_{\rho-1}$. This decomposition generalizes to level i with the high momentum part being $\sum_{j \geq i} \phi^j$.

Remark 5: It is easy to check that by the way the t-dependence has been put each derivative with respect to t produces a ϕ^4 monomial with at least one ϕ^ρ and one $\phi_{\rho-1}$. Consequently the remainder of the Taylor formula, which is of order 5, produces at least 5 low momentum fields.

The Taylor expansion is made in all cubes of $\mathcal{D}_\rho$ (more precisely in $\Lambda \cap \mathcal{D}_\rho$). Similarly the testing of space coupling is made by a cluster expansion in this lattice. The empty cubes, corresponding to $I_\Delta(0)$ and no cluster links, are factorized out and then a Mayer expansion is performed. Both the cluster and the decoupling expansions are parametrized in term of cubes. The numerator and the denominator are expressed as sums of product of connected expressions labelled by cubes. The cubes labelling a connected expression define its support, and in a product the supports are not allowed to overlap (by the way the terms are produced). The ratio of the numerator and the denominator is then similar to the division of 2 formal series labelled by supports; this is the point of view taken in [8]. A more "intuitive" approach is the one mimicking the Mayer expansion of statistical mechanics which introduces a potential V coupling supports, with value $+\infty$ if the support overlap and 0 otherwise. The ratio is than computed by introducing in both the numerator and the denominator and for each pair of supports

$$1 = e^{-V} = (e^{-V} - 1) + 1 \tag{11}$$

A coefficient $(e^{-V} - 1)$ between 2 expressions is called a *Mayer link*. It completes the previous notion of connectivity. The result of the Mayer expansion is a series in term of generalized graphs.

Let us now integrate over $d\mu_{C^\rho}$. The low momentum fields are the external legs of the graphs generated by the integration. Since they have at least 5 external legs, the graphs produced by the remainder are convergent. More generally there are connected graphs (connected by cluster or Mayer links) with more than 4 external legs. They are convergent. But there are also graphs with 4 or less external legs. Let us consider the case of 4 external legs. These graphs are "divergent" (logarithmically).

They need to be renormalized. Take one of these graphs and suppose its external vertices are localized in $x_1 \in \Delta_1, ..., x_4 \in \Delta_4$; according to the way we explain the renormalization (the symmetrical one) in chapter II, we need counterterms localized in $\Delta_1, ..., \Delta_4$. Let us consider Δ_1. The 4-point function counterterm is contained in $I^{(4)}_{\Delta_1}(0)$. It is produced by 4 t-derivatives acting on $I^{(4)}_{\Delta_1}(0)$:

$$-4!(g_\rho - g_{\rho-1}) \int_{\Delta_1} (\phi_{\rho-1})^4 \tag{12}$$

It is natural to identify it with the required counterterm and thus to define $g_{\rho-1}$ by

$$g_\rho - g_{\rho-1} = \delta g_\rho \tag{13}$$

with $\frac{1}{4!}\delta g_\rho$ being the coupling constant counterterm at level ρ. At lowest order in g_ρ, δg_ρ is quadratic and given by the bubble graphs.

More generally one defines an effective coupling constant at level i by

$$g_i = g_\rho - \delta g_\rho - ... - \delta g_{i+1} \tag{14}$$

For the expansion to converge, one requires that all effective coupling constants are small. The renormalized coupling constant is defined to be

$$g_{ren} = \Gamma_4(0,0,0,0) \tag{15}$$

where Γ_4 is the 1-particule irreducible 4-point function in momentum space. If the $g_i^{\prime}$s are small then $g_{ren} = g_0 + O(g_0^2)$.

If we suppose that the $g_i^{\prime}$s are small, then g_{j-1} can be expressed in term of g_j by a formula like (14), i.e. the dominant contribution of δg_j is quadratic in g_j and given by the bubble graphs. But in a given theory, the bubble graphs have a constant sign, thus, with our assumption, the $\delta g_j^{\prime}$s have a constant sign.

Suppose that $\delta g_j > 0, \forall j$, then

$$g_0 < g_1 < ... < g_\rho \tag{16}$$

It easy to see that if $g_\rho \to 0$ when $\rho \to \infty$ then by (16) g_0 and consequently g_{ren} go to zero (i.e. the limit theory is free). If g_ρ is small but independent of ρ, then since usually δg_j behaves as $Cst.g_j^2$ one arrives at the same conclusion.

Definition: An asymptotically safe field theory is a theory for which $\delta g_j < 0, \forall j \in \mathbf{N}$.

It follows from this definition that one has (16) in reverse order. It is then possible to have $g_{ren} \neq 0$ but small and $g_\rho \to 0$ when $\rho \to \infty$. In this case, one speaks about an *asymptotically free field theory.*

Before describing explicit models, let us make 2 remarks.

Remark 6.: Formula (8) generalizes to level i. Let, for $\rho \geq j \geq i$, Δ_j be cubes of D_j such that $\Delta_k \subset \Delta_j$ if $j > k$ and let $t_{\Delta_j} = t_j$ be the corresponding interpolation parameters then

$$I_{\Delta_\rho}(t_\rho, ..., t_i) = e^{-g_\rho \int_{\Delta_\rho} (\phi_\rho(t))^4 - \sum_{j=i-1}^{\rho-1} (1-t_j^4) g_j \int_{\Delta_\rho} (\phi_j(t))^4} \tag{17}$$

where

$$\phi_j(t) = \phi^j + t_j(\phi^{j-1} + t_{j-1}(\phi^{j-2} + ... + (\phi^{i+1} + t_{i+1}(\phi^i + t_i\phi_{i-1}))...)) \tag{18}$$

One immediately sees from this formula that if a vertex has an effective coupling constant g_j, the momentum indices of the fields attached to it are less or equal to j.

Remark 7.: at level i, suppose that we write in formula (18) g_j as $g_i+\delta g_{i+1}+\ldots+\delta g_\rho$, then terms like $\delta g_k \int_\Delta (\phi_i(t))^4$, $k \leq i$, correspond to useless counterterms.

Let us now look at some specific model. None of the possible models are simple. One can think about Φ_4^4 but as we already mention it this model is not asymptotically free for the physical sign of the coupling constant i.e. the positive one (a necessary requirement to the boundedness from below of the energy). Other four dimensional models are more complicated since they are, to be asymptotically free, the so called non abelian gauge theory, and gauge invariance implies new difficulties. The simplest models which remain are 2 dimensional; they are the non linear Sigma models and the Gross-Neveu model. Since A. S. Wightman already introduced the 2 dimensional Gross-Neveu model (G-N_2) in its introduction talk, I will briefly describe it. Clearly, non linear Sigma models can be studied along the lines of the present lecture.

The Lagrangien density defining the model is

$$\mathcal{L}(\bar{\Psi},\Psi) = \bar{\Psi}(i\ \partial\!\!\!/\xi + m)\Psi - \frac{\lambda}{2}(\bar{\Psi}.\Psi)^2 \tag{19}$$

where Ψ_a^α is an N component fermi field, α is the color index running from 1 to N, a is the spinor index with value 1 or 2 and $\bar{\Psi}.\Psi = \sum_{\alpha,a} \bar{\Psi}_a^\alpha \Psi_a^\alpha$. We introduce a momentum cutoff $\eta_\rho(p)$ in a convenient way and start from a theory whose bare parameters are λ_ρ, m_ρ and ξ_ρ.

A basic object of interest is for instance an unnormalized Schwinger function

$$\begin{aligned} &\mathbf{S}_{2p,\Lambda,\rho}(y_1,\ldots,y_p;x_1,\ldots,x_p) = \\ &= \int \Psi_{c_1}(y_1)\ldots\Psi_{c_p}(y_p)\bar{\Psi}_{d_p}(z_p)\ldots\bar{\Psi}_{d_1}(z_1)e^{\lambda_\rho \int_\Lambda (\bar{\Psi}\Psi)^2} d\mu_\rho(\bar{\Psi},\Psi) \end{aligned} \tag{20}$$

where $d\mu_\rho$ is the Berezin measure on Grassman variables which formally corresponds to

$$\frac{1}{Normal.} e^{-\int \bar{\Psi}(\xi_\rho \partial\!\!\!/ + m_\rho)\Psi} \prod d\bar{\Psi} d\Psi$$

It is a Gaussian "measure" of mean zero and covariance

$$S_{\rho;a,b}^{\alpha,\beta}(x-y) = \int \Psi_a^\alpha(x)\bar{\Psi}_b^\beta(y) d\mu_\rho(\bar{\Psi},\Psi) \tag{21}$$

and

$$\tilde{S}_{\rho;a,b}^{\alpha,\beta}(p) = \delta^{\alpha,\beta} \frac{\xi_\rho\ p\!\!\!/ + m_\rho}{\xi_\rho^2 p^2 + m_\rho^2} \eta_\rho(p) \tag{22}$$

with the usual notations:

$$p\!\!\!/ = p_0\gamma_0 + p_1\gamma_1$$

the γ's being 2×2 antihermitian matrices such that

$$\{\gamma_a, \gamma_b\} = -2\delta_{a,b}$$

We will denote by $Z_{\Lambda,\rho}$ the unnormalized Schwinger function with no external fields:

$$Z_{\Lambda,\rho} = \mathbf{S}_{0,\Lambda,\rho} \tag{23}$$

Since the Fermi fields are bounded operators, one may try to expand formally the exponential of the interaction in formula (20), each term being evaluated by integrating them with respect to the gaussian "measure" $d\mu$.

Introducing Cayley notation

$$\begin{pmatrix} u_{1a_1} & \dots & u_{na_n} \\ v_{1b_1} & \dots & v_{nb_n} \end{pmatrix} = \det(S_{\rho;a_i,b_j}(u_i - v_j))$$

one then gets

$$\mathbf{S}_{2p,\Lambda,\rho}(\{y\};\{z\}) =$$
$$= \sum_{n=0}^{+\infty} \int \dots \int_{\Lambda} d^2x_1 \dots d^2x_n \frac{\lambda_\rho^n}{n!} \sum_{a,b} \begin{pmatrix} y_{1c_1} & \dots & y_{pc_p} & x_{1a_1} & x_{1b_1} & \dots & x_{nb_n} \\ z_{1d_1} & \dots & z_{pd_p} & x_{1a_1} & x_{1b_1} & \dots & x_{nb_n} \end{pmatrix} \tag{24}$$

Theorem*(10): in a finite volume Λ, with momentum cutoff M^ρ and given m_ρ and ξ_ρ, the unnormalized Schwinger functions* $\mathbf{S}$ *are entire functions of λ_ρ.*

We give a sketch of the proof for the partition function, i.e. $\mathbf{S}_{0,\Lambda,\rho}$.

One writes

$$\frac{\xi \not p + m}{\xi^2 p^2 + m^2} = \frac{\xi \not p + m}{(\xi^2 p^2 + m^2)^{3/4}} \frac{1}{(\xi^2 p^2 + m^2)^{1/4}} = \tilde{A}(p).\tilde{B}(p) \tag{25}$$

thus, with an obvious definition of A_ρ and B_ρ (the cutoff has been split as the product of 2 square roots), one can rewrite formula (22)

$$S_{\rho;a,b}(x-y) = \int A_{\rho;a,b}(x-z) B_\rho(z-y) dz \tag{26}$$

Then applying Gram's inequality

$$|\det(\int f_i(z) g_j(z) dz)| \leq \prod_{i=1}^{n} ||f_i||_2 ||g_i||_2$$

one has

$$|\mathbf{S}_{0,\Lambda,\rho}| \leq \sum_n \frac{\lambda_\rho^n}{n!} \int \dots \int_\Lambda dx_1 \dots dx_n \left(||A_\rho||_2^2 ||B_\rho||_2^2\right)^n$$
$$\leq \exp 2N\lambda_\rho |\Lambda| ||A_\rho||_2^2 ||B_\rho||_2^2$$

with

$$||A_\rho||_2^2 = \sum_{a,b} \int A_{\rho;a,b}(x) \bar{A}_{\rho;b,a}(x) dx$$

and a similar definition for the norm of B without the sum over the spinor indices since B is diagonal in the spinor indices. To end the proof one remarks that both the norm of A and the norm of B are bounded by $Cst.M^{\rho}$.

The quantities which will have a finite limit when removing the space and momentum cutoff are the normalized Schwinger functions

$$S_{2p,\Lambda,\rho}(\{x\},\{y\}) = \mathbf{S}_{2p,\Lambda,\rho}(\{x\},\{y\})/Z_{\Lambda,\rho} \tag{27}$$

Both the numerator and the denominator are expanded using formula (24).

We now give the main steps in the series of arguments leading to the proof of the convergence of $S_{2p,\Lambda,\rho}(\{x\},\{y\})$.

Two expansions have to be superimpose: one, the *cluster expansion*, testing large space separations and another one, the *phase space expansion*, testing the momentum coupling.

1. The cluster expansion

Consider a lattice $\mathcal{D}$; let Δ be squares of $\mathcal{D}$ and $\mathcal{B}$ the set of pairs of distinct squares of $\mathcal{D}$. For each $b \in \mathcal{B}$, b being the pair (Δ, Δ'), one introduces a parameter s_b, $0 \leq s_b \leq 1$. With

$$\chi_b(x,y) = \chi_{\Delta}(x)\chi_{\Delta'}(y) + \chi_{\Delta}(y)\chi_{\Delta'}(x)$$

one then defines

$$\mathbf{1}_{\mathcal{D}}(\{s\})(x,y) = \sum_{\Delta\in\mathcal{D}} \chi_{\Delta}(x)\chi_{\Delta}(y) + \sum_{b\in\mathcal{B}} s_b\chi_b(x,y) \tag{28}$$

which has the property that

$$\mathbf{1}_{\mathcal{D}}(\{1\})(x,y) = 1.$$

One replaces in formula (26) $A(x-z)$ and $B(z-y)$ by $A(\{s_b\};x,z)$ and $B(\{s_b\};z,y)$ where

$$A(\{s_b\};x,z) = \mathbf{1}_{\mathcal{D}}(\{s\})(x,z)A(x-z)$$

and a similar formula for B. Then in formula (24) everything becomes s- dependant.

The initial expression corresponds to all $s_b \equiv 1$. The expansion is generated inductively. One starts from a square Δ, containing an initial field or, an arbitrary one if there are no initial fields, and tests to which square it is connected; then one tests to which third one these 2 squares are connected, and so on and so forth. To perform the testing one uses an interpolating formula. For s one of the s_b one applies

$$f(1) = f(0) + \int_0^1 \frac{d}{ds} f(s)ds \tag{29}$$

where $f(s)$ stands for $\mathbf{S}_{\mathcal{D}}(\{s\})$, all $s_{b'}$ but s_b being hold fixed. Acting for example on $B(s;x,y)$ this means that one has $B(0;x,y)$ in the first term, i.e. that x and y are in the same square, and one has $\frac{d}{ds}B(s;x,y)$ for the second one, i.e. x and y belongs to distinct squares. In this last case, we say that there are *springs* which connect these squares. These springs are a way to express the fact that $B(x,y)$, as a function of

$|x-y|$, has power like or exponential like (depending on the choice of momentum cutoff) decrease. This ensures (and we have a similar assertion for A) that knowing x we can sum over all possible connections with y's, i.e. all $b \in \mathcal{B}$ such that one of the square of the pair b contains x. The squares, associated to a pair b for which the second term of formula (29) has been chosen, are said to be **connected**. In this way, the inductive rule given above generates a sum of product of terms $K(\Gamma)$ made of connected squares whose union is Γ. In what follows, $K(\Gamma)$ will represent the terms or graphs produced in this way as well as their numerical value (the amplitude), and Γ is the support of the term. For more details the reader should look at [1], [8], [9] or [10].

In superrenormalizable theories, the lattice $\mathcal{D}$ can be the lattice of unit cubes, but for a strictly renormalizable one, one has to introduce a momentum space decomposition $\{M^j\}$ as in chapter I and to do a cluster expansion for each related lattice, i.e. for each $\mathcal{D}_j$. Nevertheless we will go on explaining how to deal with normalized functions as if there was a unique scale $\mathcal{D}$ and therefore a unique cluster expansion.

Omitting all unnecessary subscripts and noting $\Omega = \{x\} \cup \{y\}$, an unnormalized Schwinger function can be written after the cluster expansion

$$\mathbf{S}(\Omega) = \sum_{n\geq 1} \sum_{\substack{\Gamma_1,\ldots,\Gamma_n \subset \mathcal{D}\cap\Lambda \\ \Gamma_i\cap\Gamma_j=\emptyset, i\neq j}} \left(\prod K(\Gamma_i)\right)(\Omega) \tag{30}$$

each Γ having, according to the rules defining the expansion, no overlap with the other ones.

Introducing a potentiel between support of connected sets

$$V(\Gamma_i, \Gamma_j) = \begin{cases} 0 & \text{if } \Gamma_i \cap \Gamma_j = \emptyset \\ +\infty & \text{if } \Gamma_i \cap \Gamma_j \neq \emptyset \end{cases}$$

one can rewrite $\mathbf{S}$ without any constraint of non overlapping

$$\mathbf{S} = \sum_{\{\Gamma_i\}} \prod K(\Gamma_i) e^{-V(\Gamma_i,\Gamma_j)} \tag{31}$$

We now expand (31) replacing each $e^{-V(\Gamma_i,\Gamma_j)}$ by $\left(e^{-V(\Gamma_i,\Gamma_j)} - 1\right) + 1$. The choice of the term in the parenthesis means to introduce a *Mayer link* between Γ_i and Γ_j, it takes the value -1 if Γ_i and Γ_j have a non empty intersection and zero otherwise. The choice of 1 means there are no support restriction between Γ_i and Γ_j.

This allows to define new connected objects $K(M_k)$, the *Mayer graphs*, made of $K(\Gamma_i)$, $i = 1$ to n, which have the following structure. A M_k is an ordered sequence of $\Gamma_1, \ldots, \Gamma_p$, for some p, and its amplitude $K(M_k)$ is $\frac{1}{p!}\prod_{i=1}^{p} K(\Gamma_i)$ times some Mayer links connecting the Γ's. The support of the Mayer graph M_k is $\Gamma_1 \cup \ldots \cup \Gamma_p$. Its connectedness is ensured by the connectedness of the Γ's and by the Mayer links linking different Γ's.

One can then separate among the $K(M)$'s those, $K(M,\omega)$, which contain external fields, i.e. some subset ω of Ω, from those which do not contain them. One also factorizes empty squares, i.e. squares which are not connected to other ones and which contain no initial fields. They correspond to the partition function Z_Δ for a single square of $\mathcal{D}$ (whose choice is irrelevant by translation invariance). This gives

$$\mathbf{S}(\Omega) = (Z_\Delta)^{|\Lambda|/|\Delta|} \sum_{\substack{\omega_1 \ldots \omega_p \\ partition\ of\ \Omega}} \prod K(M_i, \omega_i) \times \sum_n \frac{1}{n!} [\sum_{M \subset \Lambda} K(M)]^n. \tag{32}$$

The partition function is

$$Z = \mathbf{S}(\emptyset) = (Z_\Delta)^{|\Lambda|/|\Delta|} \sum_n \frac{1}{n!} [\sum_{M \subset \Lambda} K(M)]^n. \tag{33}$$

Thus

$$S(\Omega) = \sum_{\substack{\omega_1 \ldots \omega_p \\ partition\ of\ \Omega}} \prod K(M_i, \omega_i). \tag{34}$$

If $\mathcal{D}$ is a lattice of unit cubes, one can use the convergence criterium of D. Brydges [7]:

$$\sum_{\Gamma \in 0} |K(\Gamma)| e^{|\Gamma|} < 1$$

to show (see [10]) that the thermodynamic limit of S_Λ exists, as a convergent series in λ_ρ of radius $r_\rho \leq K(m_\rho) M^{-10\rho}$.

This result will be improved using a phase space analysis.

2. The phase space expansion

As explained in the beginning of the chapter, to do a phase space analysis we introduce a momentum space decomposition through

$$\eta_\rho = \sum_{i=0}^{\rho} \eta^i \tag{35),}$$

a cutoff $\eta^i(p)$ meaning that the momentum p is roughly of order M^i, and

$$\Psi \equiv \Psi_\rho = \oplus \Psi^i \quad \bar{\Psi} \equiv \bar{\Psi}_\rho = \oplus \bar{\Psi}^i \tag{36}$$

corresponding to

$$S_\rho = \sum S^i$$

with

$$\tilde{S}^i(p) = \frac{\xi^i \not{p} + m^i}{(\xi^i p)^2 + (m^i)^2} \eta^i(p) \tag{37}$$

We don't need to introduce t-parameters since the fermion fields are bounded. The decoupling of the interaction is just now the application of the multinomial theorem. In fact, let us introduce the following notation: the interaction

$$V_l(\Gamma) = \int_\Gamma (\bar{\Psi}.\Psi)^2 \, d^2x \tag{38}$$

in a volume $\Gamma \subseteq \Lambda$ and with momentum cutoff M^l, can be written

$$V_l(\Gamma) = \sum_{k=0}^{l} V^k(\Gamma)$$

where

$$V^k(\Gamma) = V_k(\Gamma) - V_{k-1}(\Gamma) = \sum_{k_1,k_2,k_3,k_4}^{k} \sum_{\Gamma} \bar{\Psi}_{k_1}(x)\Psi_{k_2}(x)\bar{\Psi}_{k_3}(x)\Psi_{k_4}(x)d^2x \tag{39}$$

where the sum is such that each $k_i \leq k, i = 1,2,3,4$ and $\sup_i k_i = k$. Then

$$\begin{aligned} e^{\lambda_\rho V_\rho(\Gamma)} &= \sum_n \frac{1}{n!} (\lambda_\rho V_\rho(\Gamma))^n \\ &= \sum_{n(k,\Delta)} \prod_{k,\Delta} \frac{1}{n(k,\Delta)!} (\lambda_\rho V^k(\Delta))^{n(k,\Delta)} \end{aligned} \tag{40}$$

where the product is over all pairs (k, Δ) such that k is an integer, $0 \leq k \leq \rho$, and for k given, $\Delta \in \mathcal{D}_k \cap \Gamma$.

An element of the sum on the r.h.s. of (39) is a vertex of index k. A term in the sum (40) after being "gaussianly" integrated give diagrams as those analyzed in chapter I and II. According to power counting, diagrams with 2 or 4 external "legs" need to be renormalized, more precisely they are renormalized as indicated in chapter II, that is if the momentum indices of the external lines are bigger than the lowest index of the internal ones. This will be done using the usefull part of the counterterms. The counterterms being local, their unusefull part will be summed up in the definition of effective coupling constants as explained in the beginning of this chapter.

3. The complete expansion

The complete expansion is just a combination of the 2 preceding expansions. At each scale i, from $i = \rho$ to $i = 0$, one performs first a cluster expansion in the lattice $\mathcal{D}_i$ then a Mayer expansion. We now describe the content of these steps.

At a given scale, let us say k, one supposes that all "gaussian" integration over fields of indices less than k have been performed. A Schwinger function appears then as a sum of product of connected graphs. Each of these graphs are Mayer graphs from the upper scales indexed by the location of external vertices, i.e. vertices of the Mayer graphs to which are hooked fields of indices bigger or equal to k. Now, one performs the cluster expansion at scale k and the fields of index k are integrated. The result of the cluster expansion can be given, as above, in term of connected graphs; however because of the intertwinning with a multiscale expansion, the notion of connectedness has been enlarged: at level k, there are 2 notions of connectivity for squares of $\mathcal{D}_k$, first the connectivity through upper levels coming from connected Mayer graphs and secondly the connectivity coming from the cluster expansion at level k. The supports Γ of the connected graphs, the $K(\Gamma)$'s, are the squares of $\mathcal{D}_k$ belonging to the graphs. As above, by construction, they do not overlap. One then performs a Mayer expansion which generates through Mayer links new connected graphs, the Mayer graphs, which are labelled by their external vertices, i.e. the vertices of the graph having lines hooked to them with momentum indices less than k. Divergent graphs are renormalized following the rules given at the beginning of the chapter. This leads to introduce mass δm, wave function $\delta\xi$ and coupling constant

$\delta\lambda$ counterterms. If at scale k, the coupling constant is λ_{k+1} and the covariance (or propagator) for momentum k or less is

$$\tilde{S}_k(p) = \frac{\xi_{k+1}\, \not{p} + m_{k+1}}{\xi_{k+1}^2 p^2 + m_{k+1}^2}\eta_k(p) \tag{41}$$

then the new counterterm introduced at this scale, depending on the effective parameters of scales ρ up to $k+1$, add to this parameters to give new effective parameters

$$\lambda_k = \lambda_{k+1} + \delta\lambda_{k+1}; \quad \delta m_k = m_{k+1} + \delta m_{k+1}; \quad \delta\xi_k = \xi_{k+1} + \delta\xi_{k+1} \qquad (42).$$

Remark: formulas (41) and (42) are slightly incorrect; in fact, because the momentum cutoff are not sharp, there is a momentum dependance of the effective parameters (see [10]).

One gets at lowest order in the λ's

$$\delta\lambda_k = \sum_{\substack{i_1,i_2=k,\dots,\rho \\ \inf i_1,i_2=k}} \lambda_{i_{1,2}}^2 \int$$
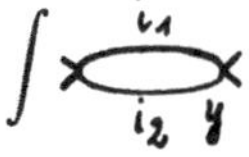

$$d^2y$$

$$+ \sum_{\substack{i_1,i_2,i_3,i_4=k,\dots,\rho \\ \inf i_1,i_2,i_3,i_4=k}} \lambda_{i_{1,2}}\lambda_{i_{1,3,4}}\lambda_{i_{2,3,4}} \int$$
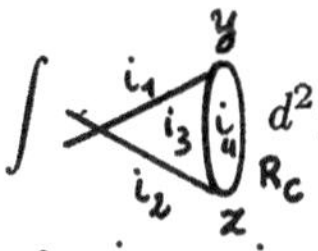

$$d^2y d^2z \tag{43}$$

$$+ \sum_{\substack{i_1,i_2,i_3,i_4=k,\dots,\rho \\ \inf i_1,i_2,i_3,i_4=k}} \lambda_{i_{1,2}}\lambda_{i_{1,2,3,4}}\lambda_{i_{3,4}} \int$$
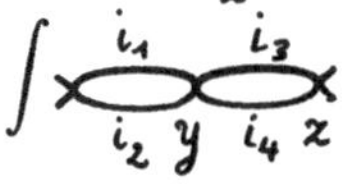

$$d^2y d^2z + O(M^{-k})$$

where R_C means that the bubble is renormalized if $\inf i_3, i_4 > \sup(i_1, i_2)$ and $i_{1,2}$ stands for $\sup(i_1, i_2)$, etc... .

Using recursively formula (42), one can check that

$$\delta\lambda_k = -(\log M)[\beta_2^k \lambda_k^2 + \left(\gamma_3^k - \left(\beta_2^k\right)^2 (\log M)\right)\lambda_k^3 + O(\lambda_k^4)] \tag{44}$$

with

$$(\log M)\beta_2^k = \xi_k^2 \sum_{\substack{i_1,i_2\geq k \\ \inf i_1,i_2=k}} \int$$
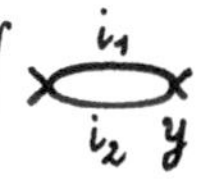

$$d^2y$$

$$(\log M)\gamma_3^k = \xi_k^4 \sum_{\substack{i_1,i_2,i_3 i_4=k,\dots,\rho \\ \inf i_1,i_2=k}} \int$$
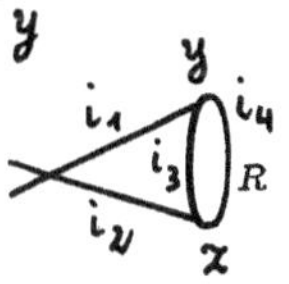

$$d^2y d^2z \tag{45}$$

and

$$\lim_{k\to\infty} \beta_2^k = \beta_2 = -2(N-1)/\pi$$
$$\lim_{k\to\infty} \gamma_3^k = \gamma_3 = (N - \frac{3}{2})/\pi^2 \tag{46}$$

Finally one gets

$$\begin{aligned}
\lambda_{\rho-1} &= \lambda_\rho\{1 - (\log M)[\beta_2(\lambda_\rho/\xi_\rho^2) + (-(\log M)\beta_2^2 + \gamma_3)(\lambda_\rho/\xi_\rho^2)^2] + O\left((\lambda_\rho/\xi_\rho^2)^3\right)\} \\
m_{\rho-1} &= m_\rho\{1 - \gamma(\log M)(\lambda_\rho/\xi_\rho^2) + O\left((\lambda_\rho/\xi_\rho^2)^2\right)\} \\
\xi_{\rho-1} &= \xi_\rho\{1 + \gamma_2(\log M)(\lambda_\rho/\xi_\rho^2)^2 + O\left((\lambda_\rho/\xi_\rho^2)^3\right)\}
\end{aligned} \tag{47}$$

with $\gamma_2 = (2N-1)/(2\pi)^2$ and $\gamma = -(2N-1)/\pi$.

If now we introduce $g_\rho = \lambda_\rho/\xi_\rho^2$, one can rewrite (47)

$$\begin{aligned} g_{\rho-1} &\simeq g_\rho\{1-(\log M)(\beta_2 g_\rho - \beta_2^2(\log M)^2 g_\rho^2 + \gamma_3(\log M)g_\rho^2\}.\{1-2\gamma_2(\log M)g_\rho^2\} \\ &\simeq g_\rho\{1-(\log M)(\beta_2 g_\rho + \beta_3 g_\rho^2 - \beta_2^2(\log M)g_\rho^2)\} \end{aligned} \tag{48}$$

with $\beta_3 = \gamma_3 + 2\gamma_2$.

One can solve equation (48) supposing g_ρ small enough. At lowest order

$$\frac{1}{g_{\rho-1}} = \frac{1}{g_\rho}\frac{1}{(1-\beta_2 g_\rho \log M)} \simeq \frac{1}{g_\rho} + \beta_2 \log M$$

thus

$$\frac{1}{g_0} = \frac{1}{g_\rho} + \beta_2 \log M^\rho$$

from which follows, since $\beta_2 < 0$, that

$$g_\rho = \frac{g_0}{1-\beta_2 g_0 \log M^\rho} \tag{49}$$

If now one adds the next order, one gets

$$\frac{1}{g_0} = \frac{1}{g_\rho} + \beta_2 \log M^\rho + \beta_3 \log M \sum_{j=1}^{\rho} g_j$$

but from (49)

$$\sum_{j=1}^{\rho} g_j \simeq -\frac{1}{\beta_2 \log M}\sum_{j=1}^{\rho}\frac{1}{j} \simeq -\frac{1}{\beta_2 \log M}\log\rho$$

thus

$$\frac{1}{g_0} = \frac{1}{g_\rho} + \beta_2 \log M^\rho - (\beta_3/\beta_2)\log\rho$$

showing that (49) gives the correct asymptotic behaviour. More generally,

$$\begin{aligned} \lambda_\rho &\simeq \varsigma^2[-\beta_2 \log M^\rho + \frac{\beta_3}{\beta_2}\log\log M^\rho + C]^{-1} = f(C,\rho)^{-1} \\ m_\rho &\simeq m\rho^{-(N-1/2)/(N-1)} \\ \varsigma_\rho &\simeq \varsigma \end{aligned} \tag{50}$$

solve asymptotically the equations (47). The constant C is related to g_{ren}^{-1}. Moreover for C large enough the renormalized expansion converges for any $\lambda_\rho \leq f(C,\rho)^{-1}$.

IV. APPLICATION

We will sketch in this last chapter an application of the methods developped in the previous ones.

Our aim is to give an idea of how a *non-renormalizable* model, the *N-component Gross-Neveu model in 3 dimensions, G-N$_3$*, can be shown to exist, for large N, following the methods developped above. For more details see [11] and [12].

1. The formal level

The initial expression is the same as for $G\text{-}N_2$. It is given by the Lagrangian density

$$\mathcal{L}(\bar{\Psi},\Psi) = \bar{\Psi}S^{-1}\Psi - (\lambda/2N)(:\bar{\Psi}.\Psi:)^2 \tag{1}$$

where $\Psi^\alpha(x)$, $x \in \mathbf{R}^3$ and $\alpha = 1,\ldots,N$, is an N-component Fermi field, $:\bar{\Psi}.\Psi: := \bar{\Psi}.\Psi - S$ is the Wick ordered product and $S^{-1} = (i\xi \not\partial + m)$ is diagonal in the color indices. Since we are in a space with an odd dimension we choose the Fermi fields to be 4-component spinors, thus the γ matrices are 4×4 matrices (this allows to have a γ_5, thus to deal, if we want, with chirality). Using that at each point x

$$e^{\frac{\lambda}{2N}(\bar{\Psi}.\Psi)^2} = \int e^{-\frac{\sigma^2}{2}+\sqrt{\frac{\lambda}{N}}\sigma\bar{\Psi}.\Psi}d\sigma \tag{2}$$

and integrating over the Fermi fields we can rewrite the partition function

$$\int e^{-\mathcal{L}(\bar{\Psi},\Psi)}\mathcal{D}(\bar{\Psi},\Psi) = \int e^{-\frac{\sigma^2}{2}}\det(S^{-1}+\sqrt{\frac{\lambda}{N}}\sigma)^N e^{-\sqrt{\lambda N}TrS\sigma}\mathcal{D}\sigma \tag{3}$$

Now from

$$\det(S^{-1}+\sqrt{\frac{\lambda}{N}}\sigma)^N = (\det S)^{-N}\det(1+\sqrt{\frac{\lambda}{N}}S\sigma)^N \tag{4}$$

and

$$\begin{aligned}\det(1+\sqrt{\frac{\lambda}{N}}S\sigma)^N &= e^{NTr\log(1+\sqrt{\frac{\lambda}{N}}S\sigma)}\\ &\simeq e^{\sqrt{N}\sqrt{\lambda}TrS\sigma-(\lambda^2/2)TrS\sigma S\sigma+O(1/\sqrt{N})}\end{aligned} \tag{5}$$

where Tr is the trace with respect to the spinor indices and space points, one sees that, up to the factor $(\det S)^{-N}$, (3) has a $N=\infty$ limit given by

$$\int e^{-\int\frac{\sigma^2}{2}-\frac{\lambda}{2}TrS\sigma S\sigma}\mathcal{D}\sigma \tag{6}$$

We want now discuss this last formula.

First we remember that, in all the formulas of this subsection, S stands for the operator with kernel $S(x-y) = (i\xi\ \not\partial + m)^{-1}(x-y)$ and σ is the operator of multiplication by $\sigma(x)$. Secondly,

$$TrS\sigma S\sigma = \int \sigma(x)\sigma(y)trS(x-y)S(y-x)\ d^3xd^3y \tag{7}$$

here tr means only the trace over the spinor indices.

One sees from (6), that the limit leads to a natural Gaussian measure in σ whose inverse covariance kernel is

$$\delta(x-y) + trS(x-y)S(y-x) \tag{8}$$

Remarks

It is the choice of the $1/N$ dependence of the coupling constant and the fact that the interaction was partly Wick ordered which enable us to get a finite limit for $N \to \infty$.

1. More precisely, in the study of a normalized expression the N-dependent scalar normalization factor factorizes out between the numerator and the denominator, and the limiting measure is given by (7), provided the kernel (8) is of positive type. This measure has a particulary simple expression, it is a Gaussian one, when expressed in term of the auxilliary field σ.

2. The effect of Wick ordering $\bar{\Psi}.\Psi$ is to substract $\sqrt{N} Tr S\sigma$ from the logarithm of the determinant, i.e. a term linear in $\sqrt{N}$ which would have made the measure in σ to blow up for large N.

3. The auxilliary field σ has initially a covariance whose Fourier transform is 1, i.e. it couples ultralocally, in a Fermi type interaction, 2 pairs of Ψ's. In the large N limit one sees that the coupling is no more ultralocal since the initial δ-function has been replaced by (8); the nonlocal correction is given by $trS(x-y)S(y-x)$ which is nothing else than the simple 2 fermion bubble.

4. The rigorous construction of the G-N_3 model for large N will be done by an expansion built around the $N = +\infty$ limit as given by the new measure on σ.

2. The level of intuition

As seen for an usual theory in the previous chapters, in order to start from a well defined expression, we replace the formal level of section 1 by introducing cutoff: momentum cutoff, that we choose to be M^ρ, and space cutoff Λ. This means that we have a set of initial (bare) parameters that we label according to the index of the momentum cutoff (indeed they are also N- and Λ-dependent, this last dependence being small):λ_ρ, ξ_ρ and m_ρ. Accordingly the fermion covariance will be given by

$$S_\rho(x-y) = \frac{1}{(2\pi)^3} \int e^{ip.(x-y)} \frac{\eta_\rho(p)}{\xi_\rho \not{p} + m_\rho} d^3p \tag{9}$$

and for this section, we choose, as cutoff $\eta_\rho(p)$, the step function $\theta(1-(p/M^\rho)^2)$.

We then introduce the ultralocal σ field and integrate out the fermions as in section 1. Using the following standard notation

$$\det(1+K) e^{-TrK + (1/2)TrK^2 + \ldots + (-1)^n (1/n) TrK^n} = \det{}_{(n+1)}(1+K) \tag{10}$$

and the Gaussian measure $d\mu_{C_\rho}(\sigma)$ of mean zero and covariance

$$\tilde{C}_\rho(p) = (1 + \lambda_\rho \pi_\rho(p))^{-1} \tag{11}$$

where

$$\pi_\rho(p) = \frac{1}{(2\pi)^3} \int Tr \tilde{S}_\rho(k) \tilde{S}_\rho(k-p) d^3k \tag{12}$$

an unormalized Schwinger can be given the form (with test functions f_i's and g_i's)

$$\begin{aligned} &\mathbf{S}_{\Lambda,\rho}(f_1, \ldots, f_p; g_1, \ldots, g_p) \\ &= \int \Psi(f_1) \ldots \Psi(f_p) \bar{\Psi}(g_1) \ldots \Psi(g_p) e^{\lambda_\rho \int_\Lambda (:\bar{\Psi}.\Psi:)^2} d\mu_\rho(\bar{\Psi}, \Psi) \\ &= Cst \int \det{}_{i,j}(f_i \frac{1}{1+K_\rho} g_j)\, (\det{}_3(1+K_\rho))^N \, d\mu_{C_\rho}(\sigma) \end{aligned} \tag{13}$$

here K_ρ is the operator with kernel

$$K_\rho(x,y) = \sqrt{\frac{\lambda_\rho}{N}} S_\rho(x-y)\Lambda(y)\sigma(y) \tag{14}$$

Formula (13) expresses the fact that the initial theory appears now as a theory of a scalar field σ with covariance C_ρ and a non local interaction given by $\det_3(1+K)$. The interaction can be expanded (at least formally) using (5) as a sum of monomials in σ:

$$N \log \det_3(1+K) = N \sum_{n=0}^{+\infty} \frac{(-1)^n}{n+3} Tr K^{n+3} \tag{15}$$

or may be considered as given $\sqrt{\frac{\lambda_\rho}{N}} : \bar{\Psi}.\Psi : \sigma$ with the condition that simple bubbles cannot be formed with it.

In the form given by (15), one sees that because K is in $1/\sqrt{N}$, each interaction monomial has a small parameter at least in $1/\sqrt{N}$. We thus expect that the interaction is only a small perturbation of the $N = +\infty$ limit.

However such an assertion depends on the power counting of the theory. Thus our first task is to estimate the large momentum behavior of the σ covariance. In fact we will see that $\xi_\rho^2 \pi_\rho(p)$ behaves as $|p|$ for large p (up to M^ρ). A vertex $S\sigma$ of the interaction corresponds to two Fermi fields (the S part) and one σ-field. The propagators of these three (free) fields behave in the same way, like $1/|p|$, for large p. The diagrams which are made with this vertex satisfy a fermion number conservation property, i.e. for a connected diagram there are as many Ψ than $\bar{\Psi}$.

A standard power counting analysis taking into account the behavior of the corrected σ-field gives for a vertex $D = 0$ and for a diagram G, $D_G = 3 - E_G$, where E_G is the number of external lines. This means that the theory is now a strictly renormalizable field theory. It is easy to check that apart from vacuum diagrams, diagrams with one σ external line are quadratically divergent, diagrams with 2 Fermi fields or 2 σ-fields are linearly divergent and diagrams with 2 Fermi fields and one σ-field are logarithmically divergent. Diagrams with 3 σ-external fields, even if they are formally logarithmically divergent, are in fact finite because of the vanishing of the trace of an odd number of γ matrices.

Remarks

1. Since the bubble with two lines of fermions, the simple bubble, have been taken off the determinant and add to the σ-field propagator, they are excluded from the 2 external σ-field diagrams which can be generated by the interaction.

2. Adding the simple bubble, i.e. π_ρ, to the initial σ-field covariance is the same, on the perturbative level, as building an effective σ-propagator by resumming all these simple bubble contributions.

3. A strictly renormalizable field theory means, as we have seen above, the need of an infinite number of perturbative counterterms. Locality of these counterterms, which is expressed by the fact that there are not more bare parameters than renormalized ones and that there is a relationship between these bare parameters and the renormalized ones, means that the useless parts of the counterterms will generate a

renormalization group flow. As we saw, for the convergence of the expansion, we need the theory to be asymptotically safe. We will see in the next section that this results from the sign of $\pi_\rho(0)$.

To summarize: *we have transformed a non renormalizable field theory into a strictly renormalizable one by summing the simplest most divergent coupling constant contributions.*

2.1 The $N = +\infty$ limit

The limit $N \to \infty$ will be more properly defined if we remove the λ_ρ dependance of the interaction. This is obtained by a scaling of σ: $\sqrt{\lambda_\rho}\sigma \to \sigma$. The σ-covariance is now $(\lambda_\rho)^{-1} + \pi_\rho(p)$ and in order to have a well defined limit when $N \to \infty$, it has to be positive and finite when $\rho \to +\infty$. Let us introduce the decomposition

$$\eta_\rho(p) = \sum_{i=0}^{\rho} \eta^i(p) \tag{16}$$

in (9) and associated to it a splitting of S_ρ as a sum of S^j's. Then we can write

$$\pi_\rho(p) = \sum \pi^{i,j}(p) = \sum \pi^{(k)}(p)$$

where

$$\pi^{(k)}(p) = \sum_{\sup(i,j)=k} \pi^{i,j}(p) \tag{17}$$

The behavior of $\pi^{(i)}$ and of course of π_ρ is controlled by it 0-momentum value. One finds that $\pi^{(i)}(0)$ is negative and proportional to M^i, that $\pi^{(i)}_{ren}(p) \stackrel{\text{def}}{=} \pi^{(i)}(p) - \pi^{(i)}(0)$ is positive and that

$$\pi^i(p) \leq |\pi^i(0)| \tag{18}$$

Introducing $\tilde{m}_\rho = m_\rho/\xi_\rho$ and assuming that $\tilde{m}_\rho M^{-\rho}$ is negligeable, one gets (with the step function cutoff):

$$\xi_\rho^2 \pi_\rho(0) = -M^\rho \frac{2}{\pi^2} \quad \text{and} \quad \xi_\rho^2 \pi^{(\rho)}(0) = -M^\rho \frac{2}{\pi^2}(1 - M^{-1}) \tag{19}$$

To simplify this part of the discussion we assume that $\pi^{(\rho)}(0)$ is of the form $-M^\rho e$ with e constant (in reality it is slightly ρ-dependent through $\tilde{m}_\rho$).

Now, we can rewrite the covariance of σ

$$\frac{\lambda_\rho}{1 + \lambda_\rho \pi_\rho(p)} = \frac{\lambda_\rho^{eff}}{1 + \lambda_\rho^{eff} \pi_{\rho;ren}(p)} \quad \text{with} \quad \lambda_\rho^{eff} = \frac{\lambda_\rho}{1 + \lambda_\rho \pi_\rho(0)}$$

and we ask λ_ρ^{eff} to have a finite positive limit c when $\rho \to \infty$.

This means that

$$\lambda_\rho = (c^{-1} + \sum_0^\rho M^i e)^{-1} \simeq (c^{-1} + M^\rho \frac{1}{1 - M^{-1}} e)^{-1} \tag{20}$$

showing that λ_ρ has to behave asymptotically as

$$M^{-\rho}\frac{1}{e}(1-M^{-1}) \tag{21}$$

One can rediscuss this result in an equivalent form but more suited to the nature of the constructive analysis. Remember that π_ρ is a sum of $\pi^{(i)}$ and think that $\pi^{(i)}$ is the typical contribution from scale i (remark that with our choice of momentum cutoff $\pi^{(i)} \simeq \pi^{i,i}$). Then define a running coupling by

$$\frac{1}{\lambda_i} + \pi^{(i)}(0) = \frac{1}{\lambda_{i-1}} \quad i = 1, \ldots, \rho$$

This is a kind of discrete renormalization group flow. We solve it by

$$\frac{1}{\lambda_\rho} = \sum_0^\rho M^i e + \frac{1}{\lambda_{-1}}$$

from which we recover the above result for λ_ρ and can interpret c as λ_{-1}, i.e. nearly the renormalized coupling constant.

2.2 The case $N \neq +\infty$

Because there is now an interaction given by $\sqrt{\lambda_\rho/N} : \bar{\Psi}\Psi : \sigma$, all parameters may be now N-dependent. We therefore introduce an N-dependent coupling constant in the form $\lambda_\rho g_\rho^2$, where λ_ρ behaves as in (21) up to $1/N$ corrections and g_ρ converges to 1 for $N \to \infty$. We then scale σ by $\sqrt{\lambda_\rho}$, thus the σ-covariance is now $\tilde{C}_\rho(p) = \eta_\rho(p)(\lambda_\rho^{-1} + g_\rho^2 \pi_\rho(p))^{-1}$ and the interaction $(g_\rho/\sqrt{N}) : \bar{\Psi}\Psi\sigma$.

We will do a momentum scale analysis and expect corrections to the initial parameters. In particular, since the theory is just renormalizable, the coupling constant, the fermion mass and wave function and the sigma propagator will become momentum index and N-dependent. These corrections will vanish with N going to infinity and therefore we will give only the leading terms.

To write the equations governing the evolution of the bare parameters we will suppose that the fermion and the boson propagators are decomposed according to the various momentum scales. That is, using the partition of the cutoff (16), $S_\rho = \sum S^j$, where S^j is with cutoff η^i and parameters ξ_i and m_i, and the inverse σ-covariance is, up to the momentum cutoff, given by

$$\lambda_\rho^{-1} + \sum_1^\rho g_j^2 \pi^{(j)}(p)\eta_j(p) = D_\rho(p)^{-1} \tag{22}$$

where each $\pi^{(j)}$ is in term of the bare parameters ξ_j and m_j. For the fermion propagator this is a natural hypothesis since according to the analysis we did in the previous chapter, at each momentum scale j, we will perform the necessary wave function and mass renormalization and therefore define new effective parameters. The modification of the σ-covariance comes from the fact that at scale j the effective parameters, and in particular the coupling constant, are of index j, thus the contribution, from this scale, of the simple bubbles to the covariance will be given in term of g_j, ξ_j and m_j.

We now look at the corrections, at lowest order, induced to level $\rho - 1$ by the renormalization of level ρ.

For this purpose, one introduce the decompositon of $\tilde{C}_\rho(p) = \eta_\rho D_\rho(p)$ as $\tilde{C}^\rho(p) + \tilde{C}_{\rho-1}(p)$ where $\tilde{C}^\rho(p) = \eta^\rho(p) D_\rho(p)$ using the cutoff partition (16).

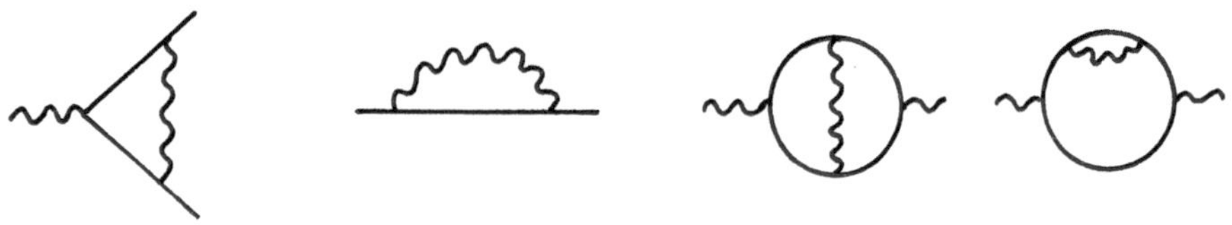

1.1 1.2 1.3

Fig. IV.1 The lowest order divergent diagrams

The contribution to the coupling constant comes (to dominant order) from diagram 1.1 . We get for the new coupling constant at level $\rho - 1$

$$g_{\rho-1} = g_\rho + \delta g_\rho$$

where δg_ρ is the correction produced by the coupling at level ρ. The correction to the fermion mass and to the wave function comes (to dominant order) from diagram 1.2 which we write, k being the external momentum, $-\delta m_\rho - \delta\xi_\rho \not{k} + O(k^2)$. We then define $m_{\rho-1} = m_\rho + \delta m_\rho$ and $\xi_{\rho-1} = \xi_\rho + \delta\xi_\rho$.

One gets

$$\begin{aligned}
\delta g_\rho &= \frac{1}{N^{3/2}} \frac{g_\rho^3}{\xi_\rho^2} \frac{1}{(2\pi)^3} \int \frac{-p^2 + \tilde{m}_\rho^2}{(p^2 + \tilde{m}_\rho^2)^2} \tilde{C}^\rho(p) (\eta^\rho(p))^2 d^3p \\
\delta\xi_\rho &= \frac{1}{N} \frac{g_\rho^2}{\xi_\rho} \frac{1}{(2\pi)^3} \lim_{k\to 0} \left(\frac{1}{k^2} \int \frac{p.k}{(p^2 + \tilde{m}_\rho^2)} \tilde{C}^\rho(p-k) \eta^\rho(p) d^3p \right) \qquad (23) \\
\delta m_\rho &= -\frac{1}{N} \frac{g_\rho^2}{\xi_\rho} \frac{1}{(2\pi)^3} \tilde{m}_\rho \int \frac{1}{(p^2 + \tilde{m}_\rho^2)} \tilde{C}^\rho(p) \eta^\rho(p) d^3p
\end{aligned}$$

Similarly there will be σ-mass corrections, $\delta\kappa_{\rho-1}$, for the boson propagator at level $\rho - 1$. They are given (to dominant order) by the diagrams 1.3 and their contribution is in $1/N$. The dominant contribution is given by

$$\delta\kappa_{\rho-1} = -(\frac{g_\rho}{\xi_\rho})^4 \frac{1}{N} \frac{4}{(2\pi)^6} \int \frac{(-p^2+\tilde{m}_\rho^2)(-(k-p)^2+\tilde{m}_\rho^2)+4\tilde{m}_\rho^2 p.(k-p))}{(p^2+\tilde{m}_\rho^2)^2((p-k)^2+\tilde{m}_\rho^2)^2} \times$$
$$\tilde{C}^\rho(k)(\eta^\rho(p)\eta^\rho(p-k))^2 d^3p d^3k$$
$$-(\frac{g_\rho}{\xi_\rho})^4 \frac{1}{N} \frac{8}{(2\pi)^6} \int \frac{(-p.(p-k)+\tilde{m}_\rho^2)(-p^2+\tilde{m}_\rho^2)+2\tilde{m}_\rho^2(k.p.-2p^2))}{(p^2+\tilde{m}_\rho^2)^3((p-k)^2+\tilde{m}_\rho^2)} \times$$
$$\tilde{C}^\rho(k)(\eta^\rho(p))^3\eta^\rho(p-k)d^3p d^3k \tag{23'}$$

Equations (23,23') were written for the scale ρ. One can write similar equations for any scale j, $1 < j \leq \rho$; to the dominant order they are given by the same diagrams. The boson propogator is now given by $\tilde{C}^j(p) = \eta^j(p)D^j(p)$ where

$$D^j(p)^{-1} = \lambda_\rho^{-1} + \sum_j^\rho g_k^2 \pi^{(k)}(p)\eta_k(p) + \sum_j^{\rho-1} \delta\kappa_k \eta_k(p) \tag{24}$$

2.3 The fixed point analysis

To complete the set of equation (23) one adds the recurrence relation between boson covariances of consecutive index at 0- momentum:

$$D^{j-1}(0)^{-1} = D^j(0)^{-1} + g_{j-1}^2 \pi^{(j-1)}(0) + \delta\kappa_{j-1} \tag{25}$$

Also to be able to handle the system we replace in the analytical expression of the corrections, i.e. (23) and (23') for the index j, the covariance $D^j(p)$ by its 0-momentum value $D^j(0)$.

The difference $D^j(p)-D^j(0)$ is given by $-D^j(0)D^j(p)\sum_{k=j}^\rho \pi_{ren}^{(k)}(p)$ which shows, since $\pi_{ren}^{(j)}$ is positive, that $D^j(0)$ is an upper bound. Conversely, one can assume that there is a constant K, $K \leq 1$, such that $D^j(p) \geq KD^j(0)$, assumption which can be easily checked with the values obtained at the end of this subsection.

Our system of equations is then

$$\begin{aligned}
g_{j-1} &= g_j(1 + \frac{1}{N}(\frac{g_j}{\xi_j})^2 M^j D^j(0)\alpha_j) \\
\xi_{j-1} &= \xi_j(1 + \frac{1}{N}(\frac{g_j}{\xi_j})^2 M^j D^j(0)\beta_j) \\
m_{j-1} &= m_j(1 + \frac{1}{N}(\frac{g_j}{\xi_j})^2 M^j D^j(0)\gamma_j) \\
D^{j-1}(0)^{-1} &= D^j(0)^{-1} - (\frac{g_{j-1}}{\xi_{j-1}})^2 M^{j-1} e_{j-1} + \frac{1}{N} M^{2j}(\frac{g_j}{\xi_j})^2 D^j(0)\delta_j
\end{aligned} \tag{26}$$

where for example

$$\alpha_j = \frac{1}{(2\pi)^3} M^{-j} \int \frac{-p^2+\tilde{m}_j^2}{p^2+\tilde{m}_j^2} (\eta^j(p))^3 d^3p \tag{27}$$

To solve this set of equations we will suppose that the coefficients $\alpha_j, \beta_j, \gamma_j, \delta_j$ and e_j are in fact independent of j. This assumption means that we can neglect $\tilde{m}_j M^{-j}$.

In fact, one finds, with our choice of cutoff, that

$$\alpha_j = -\frac{1}{2\pi^2}(1 - M^{-1}) + O(\tilde{m}_j^2 M^{-2j}) \simeq \alpha$$

with

$$\alpha = -\frac{1}{2\pi^2}(1 - M^{-1}) \tag{28}$$

Similarly, one has that $e_j \simeq e$ and, for example $\beta = \alpha/3$.

We will verify this assumption a posteriori.

Let us now introduce $a_j = (g_j/\xi_j)^2$ and the dimensionless parameter

$$b_j = a_j M^j D^j(0)$$

Then the system (26) is

$$\begin{aligned}
g_{j-1} &= g_j(1 + \frac{1}{N} b_j \alpha) \\
\xi_{j-1} &= \xi_j(1 + \frac{1}{N} b_j \beta) \\
m_{j-1} &= m_j(1 + \frac{1}{N} b_j \gamma) \\
D^{j-1}(0)^{-1} &= D^j(0)^{-1} - D^{j-1}(0)^{-1} b_{j-1} e + \frac{1}{N} D^j(0)^{-1}(b_j)^2 \delta
\end{aligned} \tag{29}$$

from which one gets

$$\frac{b_{j-1}}{1 + b_{j-1}e} = b_j M^{-1} \frac{1}{1 + \frac{1}{N} b_j^2 \delta} \left(\frac{1 + \frac{1}{N} b_j \alpha}{1 + \frac{1}{N} b_j \beta} \right)^2$$

We now look at the leading $1/N$ contribution since, at this level of approximation, we have only taken the leading diagrams and we rewrite the equation in a more convenient form

$$\frac{b_{j-1}}{1 + b_{j-1}e} = M^{-1} \frac{b_j}{1 - \frac{1}{N} b_j 2(\alpha - \beta) + \frac{1}{N} b_j^2 \delta} \tag{30}$$

This equation has a fixed point which satisfies

$$\frac{\delta}{N} b^2 - (M^{-1} e - \frac{2(\beta - \alpha)}{N}) b + 1 - M^{-1} = 0$$

From the two roots we choose the one which converges to the solution of the limiting equation, i.e.

$$\bar{b} = \frac{M-1}{e} \left(1 + \frac{1}{N} (\frac{2(\beta - \alpha)M}{e} + \frac{(M-1)M\delta}{e^2}) \right) \tag{31}$$

Looking now for a solution of the form $\bar{b}(1 + u_j(N))$, where $u_j(N)$ goes to 0 for $N \to \infty$ or $j \to \infty$, one finds that

$$b_j = \bar{b}(1 + \bar{u}M^{-j}e^{\frac{\bar{b}^2\delta}{N}j} + O(M^{-2j+O(1/N)})) \simeq \bar{b}(1 + \bar{u}M^{-j(1+\epsilon)}) \tag{32}$$

with $\bar{u}^{-1} = u_0^{-1} + 1 + \bar{b}^2\delta/N(1 - M^{-1})$, u_0 being some constant that we will choose, to simplify the formulae, very small (it is a free parameter fixing the value of $b_0 = \bar{b}(1 + u_0)$) and $\epsilon = -\bar{b}^2\delta/N(\log M)$.

Then

$$\frac{D^0(0)^{-1}}{D^j(0)^{-1}} = M^j \frac{b_0}{b_j}\left(\frac{1+\frac{\bar{b}}{N}\beta}{1+\frac{\bar{b}}{N}\alpha}\right)^{2j} e^{2\frac{\bar{b}}{N}(\beta-\alpha)\sum_1^j u_k} \tag{33}$$

To our order of approximation the sum in the exponential behaves as a constant $c(u_0)$ which vanishes as u_0 goes to 0, we therefore set

$$\begin{aligned} D^j(0) &= b_j M^{-j}\left(\frac{1+\frac{\bar{b}}{N}\beta}{1+\frac{\bar{b}}{N}\alpha}\right)^{-2j} \\ D^0(0) &= b_0 e^{2\frac{\bar{b}}{N}(\beta-\alpha)c(u_0)} \end{aligned} \tag{34}$$

We then get that

$$\lambda_\rho^{-1} = M^\rho\left(\frac{1+\frac{\bar{b}}{N}\beta}{1+\frac{\bar{b}}{N}\alpha}\right)^{2\rho}\left(\frac{1}{b_\rho} + e\right) \tag{35}$$

which converges when $N \to +\infty$ to the asymptotic value given by (21).

We can now express all the other parameters and get in particular

$$a_\rho = \left(\frac{1+\frac{\bar{b}}{N}\beta}{1+\frac{\bar{b}}{N}\alpha}\right)^{2\rho}$$

which converges to 1 when $N \to +\infty$ and

$$\tilde{m}_\rho = \tilde{m}_0\left(1 + \frac{\bar{b}}{N}(\gamma - \beta)\right)^{-\rho} e^{\frac{\bar{b}}{N}(\gamma-\beta)c(u_0)} \tag{36}$$

It follows from this last equation that for N large enough, whatever is the sign of $\gamma - \beta$, $\tilde{m}_j M^{-j} \ll 1$ (at least for large j). It proves the assumption that we made at the beginning of this section. It extends to all indices provided $\tilde{m}_0$ is small enough.

This study we made for the flow of the paramaters can be extended to the complete set of equation and not only to the dominant contribution. The results will be essentially the same, all the quantities we obtained being defined up to $O(1/N^2)$.

To summarize: *the behaviour of the theory is governed by a dimensionless parameter b_ρ which converges as $\rho \to \infty$ toward a value $\bar{b}$. This ultraviolet fixed point $\bar{b}$ is unique provided one requires some continuity with respect to N.*

3. The existence of the theory

The existence of the theory is proven by the mean of expansions as the ones described in the previous chapters. One first defines an initial expression in term of the bare parameters. One then performs for each momentum level a cluster expansion to control the thermodynamic limit and a decoupling expansion to test the coupling between low and high momenta. This last expansion is done in each phase space cell and requires that the divergent diagrams which may be produced during these steps are renormalized with the help of the counterterms. Since we have eliminated the simple bubbles by inserting them in the σ-field propagator, a vertex has at least a convergent factor $N^{-1/6}$, this is *the small factor per vertex* necessary to the convergence of our expansions.

The local accumulation of fermions is cured by the Pauli principle, i.e. in operator language by the fact that the fermions are bounded operators or in our framework by the fact that derivatives of determinant can be expresssed in term of finite linear combination of determinant [13]. The only problem comes from the local accumulation of σ-fields. Since they are bosons they cannot be integrated gaussianlyas explained below. Because of the resummation of bubbles, a generic expansion step produces diagrams with at least 3 σ-fields. If the expansion is at level i and if the low momentum fields are σ_j, $i \gg j$, then we may face, for example, a term of the expansion in which in each cube $\Delta \in \mathcal{D}_i$ contained in a given cube of $\Delta\prime \in \mathcal{D}_j$ one has produced a triangular diagram which can be shown to be bounded by $M^{-4i}\tilde{m}_i\sigma_j^3/\sqrt{N}$. Thus in $\Delta\prime$ there is a product of $n = |\Delta\prime|/|\Delta|$ such terms. If we then integrate gaussianly the σ_j's it diverges as $n!^{3/2}$, a number which cannot be controlled by the other factors.

This difficulty is cured by looking at the effective potentiel, i.e. the average behavior of the interaction in term of σ.

To conclude we precise the form of the initial expression and give some explanation on the effective potentiel.

3.1 The initial expression

We choose the cutoff $\eta(p)$ such that in the decomposition (16) the support of η^i does not overlap over the support of η^{i-1} and η^{i+1} and is 1 in a neighborhood of $p = M^i$.

Our initial expression will be slightly different from formula (13). We first rescale into σ the λ_ρ dependence of K_ρ and introduce the new parameter g_ρ. We then incorporate in advance in the boson and in the fermion propagators the modifications to the bare parameters which will be introduced during the expansion. For example $\xi \not p + m$ will be replaced by

$$\xi \not p + m + \delta S^{-1} = S^{-1} = (\xi + \delta\xi) \not p + m + \delta m \tag{37}$$

A consequence of this formula is that instead of $(\det_3(1 + K_\rho))^N$ one has

$$(\det_3(1 + K_\rho - \delta K_\rho))^N e^{-NTr\delta K_\rho - N/2\left(Tr(\delta K_\rho^2 - 2K_\rho\delta K_\rho)\right.} \tag{38}$$

with $\delta K_\rho = \frac{g_\rho}{\sqrt{N}} S\delta S^{-1}$ and $\delta S^{-1} = \delta\xi \not p + \delta m$.

Similarly one includes the σ-mass corrections, $\delta\kappa_{\rho-1}(p)$, in the σ-propagator. This gives

$$\tilde{C}_\rho(p) = \frac{\eta_\rho(p)}{\lambda_\rho^{-1} + g_\rho^2 \pi_\rho(p) + \delta\kappa_{\rho-1}(p)} \tag{39}$$

Associate to the momentum slicing defined by (16) there correspond decompositions of the fermion and boson propagators

$$\begin{aligned} \tilde{S}^i &= \frac{\eta^i(p)}{(\xi_i + \delta\xi_i(p))\, \not{p} + m_i + \delta m_i(p)} \\ \tilde{C}^i(p) &= \frac{\eta^i(p)}{\lambda_\rho^{-1} + \sum g_j^2 \pi^{(j)}(p) + \delta\kappa_i(p)} \end{aligned} \tag{40}$$

where

1) $\delta\xi_i(p) = \sum \delta\xi_j \eta_j(p)$

2) $\delta m_i(p) = \sum \delta m_j \eta_j(p)$

3) $\delta\kappa_i(p) = \sum \delta\kappa_j \eta_j(p)$

$\delta\xi_j$ and δm_j being respectively the wave function and the mass counterterms of the fermion propagator at level j, $\delta\kappa_j$ the mass counterterm of the boson propagator at level j (with the convention that $\delta\kappa_\rho = 0$) and the sums are over the j's such that $\rho \geq j \geq i$.

In practice, the definitions given in 1), 2) and 3) are a little more complicated because of the overlap of the momentum cutoff. Also (38) is not the final form, one has to substract from the interaction the vacuum energy terms, the tadpoles in σ and the correction to the resummation of the bubble done by replacing g_ρ by the effective coupling constant g_j (see formulae (22) and (40)).

3.2 The effective potential

We now give some detail on the effective potentiel. At level j, to control the possible accumulation of σ-low momentum field coming from diagrams produced at this level or above, we will use the part of the interaction, i.e. of K, containing a low momentum σ and high momentum S (in term of the notation of chapter III, that will be the K's of the form $\sum_{l,m\geq i;k<i} A^l \sigma^k B^m$). Then one does the following analysis:

1) in small cubes (corresponding to large indices) the low momentum σ-fields are nearly constant

2) by choosing the size of the cubes in the lattices $\mathcal{D}_i$ to be slightly bigger than the natural associated scale, for example for a cube Δ of $\mathcal{D}_i$, $|\Delta|^{1/3} = aM^{-i}$, $a \gg 1$, the high momentum fields,i.e. the S's, do not feel the boundary of the cubes.

We can therefore, with a good level of confidence, replace the low momentum fields by their average $\bar{\sigma} = \frac{1}{\Delta}\int_\Delta \sigma(x) d^3x$ in the cubes in which they are produced and the fermion propagator S by the fermion propagator S_P with periodic boundary conditions on the cube of production. We are then in a situation where it is possible to estimate exactly $\log(\det_3(1+K))^N$. We get

$$\log(\det_3(1+K))^N \leq \begin{cases} -Cst.M^{-3i}\bar{\sigma}^2 M^i & \text{for } \bar{\sigma} \geq \sqrt{N}M^i \\ -Cst M^{-3i}\bar{\sigma}^4 M^{-i}/N & \text{for } \bar{\sigma} \leq \sqrt{N}M^i \end{cases} \tag{41}$$

The quadratic bound we get for large σ comes from the fact that in $\log(\det_3(1+K))^N$ it is the quadratic term $-(N/2)TrK^2$ which dominates. It has to be compared with the behaviour of the Gaussian measure, which for fields of frequency i is $-M^i \int_\Delta(\sigma^i)^2$, i.e. roughly the same behaviour but only for σ^i and not σ_i.

Each low momentum field σ is then written as $\sigma = \bar{\sigma} + \delta\sigma$. The $\delta\sigma$ part behaves as a well localized field [9] and therefore can be integrated gaussianly. Then for the $\bar{\sigma}$ part we have two cases, either $\bar{\sigma}$ is large and we have at our disposal the quadratic effective potential bound, or $\bar{\sigma}$ is not large and we have the quartic bound.

In the first case, one dominates gaussianly each low momentum field with the effective potential of the cube of production using $x^n e^{-x^2} \leq (n!)^{1/2}$. It can be easily seen that each $\bar{\sigma}$ is bounded by M^{-i} if i is the index of the cube of production and there are no accumulation effects since there are only a finite number of cluster or decoupling expansion derivatives acting in a given cube.

In the second case, one uses the bound $x^n e^{-x^4} \leq (n!)^{1/4}$ for a part of the low momentum fields and performs a Gaussian integration for the other ones: this is due to the fact that there is a balance to be done between Gaussian integration which produces square roots of factorial and domination with the quartic potential which produces only factorial to the power 1/4 but also a $N^{1/4}$ per dominated σ. We let the reader to check on the case of accumulation of triangular graphs (see the beginning of section 3) that the integration of half of the σ' and the domination of the other half (the effective potential is $exp(-Cst.M^{-3j}M^{-i}\bar{\sigma}^4/N)$, i.e. the contribution of the effective potential in each cubes of $\mathcal{D}_i$ contained in $\Delta\prime \in \mathcal{D}_j$) gives at the end a small factor $N^{-1/24}$ per σ.

The other diagrams produced by the expansion can be estimated in the same way (there is still a little subtlety for the square diagram with 4 low momentum σ).

Finally we mention another fine technical point. It has to do with the Wick bound, i.e. the final bound on $\log(\det_3(1+K))^N$ and the control of the error done by replacing the initial fermion propagator by an operator with periodic boundary conditions and the σ-field by a constant field. The difficulty comes from the contribution of vertices (or K) where the σ-field are low momentum ones but neverless close the momenta of the fermions.

A standard bound for $\log(\det_2(1+K))^N$ (see [13]) is $(1/2)TrKK^*$ thus

$$\log(\det_3(1+K))^N \leq \frac{1}{2}Tr(K^2 + KK^*) \tag{42}$$

It is easy to see that $Tr(K^2+KK^*)$ is in fact TrK^2_{ren}, but K is now full of the interpolating parameters of the expansions and TrK^2_{ren} is not obviously bounded by its expression without the interpolating parameters, one has therefore to estimate it carefully enough to prove that, especially for the configuration of momenta mentionned above, $exp(TrK^2_{ren})$ is measurable with respect to the σ-covariance.

REFERENCES

1. G. Velo and A. S. Wightman, "Constructive Quantum Field Theory",Springer-Verlag,Berlin/New-York, Erice (1973)

2. J. Feldman, J. Magnen, V. Rivasseau and R. Sénéor, Bounds on completely convergent Euclidean Feynman graphs, Comm. Math. Phys., 98:273 (1985)

3. J. Feldman, J. Magnen, V. Rivasseau and R. Sénéor, Bounds on renormalized graphs, Comm. Math. Phys., 100:23 (1985)

4. C. de Calan and V. Rivasseau, Local existence of the Borel Transform in Euclidean Φ_4^4, Comm. Math. Phys., 82:69 (1981)

5. F. David, J. Feldman and V. Rivasseau, On the Large Order Behavior of Φ_4^4, Comm. Math. Phys., 116:215 (1988)

6. M. Gell-Mann and F. E. Low, Phys. Rew., 95:1300 (1954)

7. D. C. Brydges, A short course on cluster expansion, in "Les Houches 1984: Critical phenomena, random systems, gauge theories", K. Osterwalder and R. Stora, ed., North-Holland, Amsterdam/New-York (1986)

8. J.-P. Eckmann, J. Magnen and R. Sénéor, Decay properties and Borel summability for the Schwinger functions in $P(\Phi)_2$ theories, Comm. Math. Phys., 39:251 (1975)

9. R. Sénéor, Théorie Constructive des Champs, 3me cycle de la physique en Suisse Romande, Geneve (1987)

10. J. Feldman, J. Magnen, V. Rivasseau and R. Sénéor, A Renormalizable Field Theory: The Massive Gross-Neveu Model in Two Dimensions, Comm. Math. Phys., 103:67 (1986)

11. C. de Calan, P. Faria de Veiga, J. Magnen and R. Sénéor, to be published

12. P. Faria de Veiga, thesis

13. J. Magnen and R. Sénéor, Yukawa quatum field theory in three dimensions (Y_3), in "Third International Conference on Collective Phenomena", J. Lebowitz, J. Langer and W. Glaberson ed., The New-York Academy of Sciences, New-York (1980)

CONSTRUCTIVE GAUGE THEORY II

T. Balaban

Rutgers University
Department of Mathematics
New Brunswick, New Jersey 08903 U.S.A.

These lecture notes are continuations of the lectures [1] from the previous Erice summer school. We assume that the reader is familiar with those, and we refer to them. In [1] we have described the lattice regularizations of gauge field theories, and their basic general properties. The renormalization group approach has been applied to the lattice theories, and explained in detail for the small field approximation. Within this approximation the renormalization group flow is determined by a mapping in the space of effective actions. The basic part of this mapping is the mapping of the coupling constant, obtained by solving the renormalization group equation. In these notes we discuss the complete model, including large field regions, and we use the above results in the small field regions. In fact even in the complete model the above described features of the renormalization transformations are preserved, with some modifications.

Let us recall some basic definitions from [1]. There the renormalization transformation was defined as the integral transformation

$$(1) \qquad (T\rho)(V) = \int dU\, \delta(\overline{U}V^{-1})\rho(U),$$

where $\rho(U)$ is a function of the gauge field variables U on a lattice T, $\overline{U}$ is the averaged field on the lattice $T^{(1)}$, and V is a new gauge field on the lattice $T^{(1)}$. The averaging operation was discussed in [1]. Here we have to apply an additional operation after the operation T. This operation is denoted by R, and it changes effective densities on large field regions. Thus, we construct a sequence of effective densities $\{\rho_k\}$ by applying successively the operations RT to the initial density $\rho_0 = \exp[-\frac{1}{g_0^2} A-E]$, where A is the Wilson action, and E is a normalization constant. We get

$$(2) \qquad \rho_k = RT\rho_{k-1} = (RT)^k \rho_0,$$

and we finish the inductive procedure when we reach the unit lattice. Our basic goal is to give a precise inductive description of the densities ρ_k, and

Constructive Quantum Field Theory II, Edited by G. Velo and
A. S. Wightman, Plenum Press, New York, 1990

to prove that it is preserved by the renormalization transformation RT. How precise the description should be depends, to some extent, on what properties of the densities ρ_k we want to prove. Here we consider the simplest property only, the uniform boundedness in the lattice spacing. The lattice spacing plays the role of the ultraviolet cutoff, and the uniform boundedness is called also the ultraviolet stability. Even this basic property demands a very long and complicated description. We will not give all the details of it; we will sketch only main features. This description, formulated in terms of an expansion of the density ρ_k is the main result presented in these lectures. Unfortunately, it is impossible to formulate a precise theorem describing it, so we formulate a statement which gives only an idea of the result.

<u>Theorem</u>. The densities ρ_k are given by inductively defined and uniformly convergent expansions defined below (uniformly in scales of proper lattices).

In fact, a construction of this expansion is the fundamental result, which implies immediately the ultraviolet stability and several other results. In the rest of these lecture notes we will describe general features of this expansion.

To introduce basic ideas we sketch the steps in the first renormalization transformation. We consider the integral

$$\int dU\,\delta(\bar{U}V^{-1})\exp[-\frac{1}{g_0^2}A(U)-E]. \tag{3}$$

The action $A(U)$ is a sum of non-negative terms, namely the plaquette variables $\frac{1}{2}|U(\partial p)-1|^2$ (here $|A|^2 = \mathrm{tr}A^*A$ with the trace normalized to unity, i.e. $\mathrm{tr}I=1$). We use this positivity property to divide the space of all field configurations U into "small field regions" and "large field regions." This is done by introducing the partition of unity

$$1 = \sum_{P_0}\prod_{p\in P_0}\chi(\{|U(\partial p)-1|<\epsilon_0\})\prod_{p\in P_0^c}\chi(\{|U(\partial p)-1|\geq\epsilon_0\}) \tag{4}$$

$$= \sum\chi_0(P)\chi_0^c(P^c),$$

where the sum is over all subsets P_0 of the set of plaquettes. We choose the constant ϵ_0 as $\epsilon_0 = g_0 p(g_0)$, $p(g_0) = A_0(\log g_0^{-1})^p$, where A_0 and p are sufficiently large constants. We take the partition Π_1 of the lattice T_1 into large cubes of the size M. Denote by Q_0 the union of the cubes from Π_1 contained in P_0, and take the second partition of unity

$$1 = \sum_{P_1\subset Q_0'}\prod_{p'\in P_1}\chi(\{|V(\partial p')-1|<\epsilon_1\})\prod_{p'\in P_1^c\cap Q_0'}\chi(\{|V(\partial p')-1|<\epsilon_1\}) \tag{5}$$

$$= \sum_{P_1\subset Q_0'}\chi(P_1)\chi_1^c(P_1^c\cap Q_0'),$$

where the sum is over all subsets P_1 of the set of plaquettes contained in the subset Q_0' of the lattice $T_L^{(1)}$. We take $\epsilon_1 = g_1 p(g_1)$. Actually it is necessary to introduce this decomposition in a more complicated way, for technical reasons which we will not go into, but the above one describes well the essential idea.

Let us make a more general comment here. There are many divisions of different spaces of fields into subsets corresponding to some smallness conditions in this construction. Usually the corresponding definitions of partitions of unity are quite involved technically, because many properties of the effective densities and actions have to be preserved. We will not go into these details; we will describe only basic aspects of most important partitions.

Let us define the set Q_1 as a union of cubes from Π_1 contained in P_1. In this set we introduce the axial gauge conditions in the blocks using the identity

$$(6) \qquad 1 = \prod_{y \in Q_1'} \prod_{\substack{x \in B(y) \\ x \neq y}} \int du(x) \delta(U^u(\Gamma_{y,x})).$$

Let us recall that the contour $\Gamma_{y,x}$ connects the point y with the point x, and $U^u(<x,x'>) = u(x)U(<x,x'>)u^{-1}(x')$, so $U^u(\Gamma_{y,x}) = u(y)U(\Gamma_{y,x})u^{-1}(x)$. We insert it under the integral, and we apply the Faddeev–Popov procedure. This yields the equality

$$(7) \qquad \rho_1(V) = \sum_{P_1 P_0} \chi_1(P_1)\chi_1^c(P_1^c) \int dU \delta(\overline{U}V^{-1}) \delta_{Ax(Q_1')}(U) \cdot$$

$$\cdot \; \chi_0(P_0)\chi_0^c(P_0^c) \exp[-\frac{1}{g_0^2} A(U) - E],$$

where $\delta_{Ax(Q_1')}(U) = \prod_{y \in Q_1'} \prod_{\substack{x \in B(y) \\ x \neq y}} \delta(U(\Gamma_{y,x}))$.

It is easy to see that the characteristic functions together with the δ–functions in (7) yield a lot of restrictions on the field variables considered on the set Q_1. They yield also restrictions on "fluctuation fields." There are many ways of introducing these fields. We define them in the following way. At first we introduce configurations $U_1(V)$ as solutions of the variational problem:

(8) find a minimal point of the action $A(U)$ considered on configurations U having fixed values on some domain Ω_1^c, and satisfying the conditions $\overline{U} = V$ on Ω_1, and also regularity conditions introduced by the functions $\chi_0(\Omega_1)$.

We assume here that Ω_1 is a union of blocks from Π_1. The above problem does not have a unique solution, because the functional $A(U)$, the conditions $U = V$ and the characteristic functions χ_0 are invariant with respect to the gauge transformations u, satisfying the conditions $u(y) = 1$ for $y \in \Omega_1'$. Therefore we have to consider the above problem on orbits of this group of gauge transformations. One of the fundamental technical results is that the variational problem has a unique solution on the space of orbits. Usually we denote by $U_1(V)$ the element of the minimal orbit satisfying the axial gauge conditions in blocks of Ω_1'. We define the fluctuation field in the first step is defined by the equality

$$(9) \qquad U = U'U_1(V), \quad \text{or} \quad U' = UU_1^{-1}(V).$$

This fluctuation field U' is small, by the restrictions on the fields U,V introduced above. This means that $|U'-1| < C'\epsilon_0$ for an absolute constant C'. Introducing a Lie algebra valued field A' by the equality $A' = \frac{1}{i} \log U'$, or $U' = \exp iA'$, we get that A' is small also, i.e. it satisfies the same bound.

Now we expand the expressions in the integrals in (7) with respect to A'. We expand the action $A(U)$ up to third order terms

$$(10) \qquad A(U) = A(\exp iA'U_1) = A(U_1) + \langle A',J_1\rangle + \tfrac{1}{2}\langle A',\Delta A'\rangle + + V_0(A'),$$

and the expressions $(\overline{\exp iA'U_1})\,\overline{U}_1^{-1}$ up to second-order terms: we get $QA' + C(A')$. Next we linearize these expressions performing a properly chosen change of variables of the form $A' = A-hD(A)$, where $D(A)$ is an analytic function with an expansion beginning with second-order terms. Finally, we fix domains Ω_1 and resum over P_0,P_1 determining Ω_1 as a maximal domain contained in Q_1 together with a neighbourhood of some fixed thickness. All these operations yield the following equality:

$$(11) \qquad \rho_1(V) = \sum_{\Omega_1} \chi_1(\Omega_1) \int dU_{|\Omega_1^c}\, \delta(\overline{U}V^{-1})\zeta(\Omega_1^c) \cdot$$
$$\cdot \exp\Big[-\frac{1}{g_0^2} A(U_1) - E + |\Omega_1|\log \sigma_0\Big] \cdot$$
$$\cdot \int dA_{|\Omega_1}\, \delta(QA)\delta_{Ax(\Omega_1')}(A)\chi_0'(\Omega_1)\exp[\log \sigma(A-hD(A)) +$$
$$+ \operatorname{Tr}\log(I-h(\tfrac{\delta}{\delta A}D)(A)) - \frac{1}{g_0^2}\{\langle A-hD(A),J_1\rangle +$$
$$+ \tfrac{1}{2}\langle A-hD(A),\Delta(A-hD(A))\rangle + V_0(A-hD(A))\}\Big],$$

where $\sigma(A)$ is a density of the Haar measure dU on the group G with respect to the Lebesgue measure dA on the Lie algebra $\mathfrak{G}$ of the Lie group G. The integrals above are slightly changed again. At first we introduce an additional partition of unity restricting even more the fluctuation field A. The final small field restriction has the form $|A| < \delta_0$, where δ_0 is a constant of the form $\delta_0 = g_0 p_1(g_0)$, with $p_1(g_0)$ smaller than $p(g_0)$. We do not discuss here why we need this restriction. The second operation is the scaling transformation replacing A by $g_0 A$. We gather together terms of the same order, and we obtain the equality

$$(12) \qquad \rho_1(V) = \sum_{\Omega_1, \Lambda_1} \chi_1(\Omega_1) \int dV_0|_{\Omega_1^c} \delta(\bar{V}_0 V^{-1}) \zeta(\Omega_1^c) \cdot$$

$$\cdot \exp\Big[-\frac{1}{g_0^2} A(U_1) - E + |\Omega_1| \log \sigma_0 + |\Omega_1| d(\mathfrak{G}) \log g_0 \Big] \cdot$$

$$\cdot \int dA|_{\Omega_1} \delta(QA) \delta_{Ax(\Omega_1')}(A) \chi_0'(\Lambda_1^c \cap \Omega_1) \chi^c(\Omega_1 \cap \Lambda_1^c) \chi(\Lambda_1) \exp[-\tfrac{1}{2} <A, \Delta_1 A>$$

$$+ v(g_0 A) - \frac{1}{g_0^2} V(g_0 A) \Big],$$

where the field U restricted to Ω_1^c was denoted by V_0. The domain $\Lambda_1 \subset \Omega_1$ is the final small field domain. We perform the integral above over this domain, leaving the other integrations unchanged. The effect of this integration can be represented in an exponentiated form, using a simple form of a cluster expansion (polymer expansion). This yields the expression

$$(13) \qquad \exp\Big[\log Z^{(0)}(\Lambda_1) + \tfrac{1}{2} <\Lambda_1^c A, \Delta_1 C^{(0)}(\Lambda_1) \Delta_1 \Lambda_1^c A> +$$

$$+ \mathcal{E}^{(1)}(\Lambda_1, g_0, \Lambda_1^c A, U_1) \Big],$$

where $C^{(0)}(\Lambda_1)$ is the covariance of the Gaussian part of the integral (12), and where the function $\mathcal{E}^{(1)}$ can be represented as a sum of localized terms satisfying some exponential decay properties. This will be explained later in a general case.

The final step is renormalization of the effective action. We renormalize it in the region Λ_1 subtracting the value of the action at the configuration $U_1 = 1$, and the counterterm $\beta_1(g_0) A(\phi_1, U_1)$, where the fundamental β-functions were defined in [1],[2], $\phi_1 \in C_0^\infty(\Lambda_1)$, $\phi_1 = 1$ on the domain obtained from Λ_1 by removing the boundary layer of large blocks. The symbol $A(\phi_1, U_1)$ means that we multiply the term in the Wilson action corresponding to a plaquette p by $\phi_1(x(p))$, $x(p)$ is the initial point of ∂p (i.e. $p = p_{\mu\nu}(x) = <x, x+e_\mu, x+e_\mu+e_\nu, x+e_\nu>$ for some $\mu<\nu$). Let us write the renormalization group equation

$$(14) \qquad \frac{1}{g_0^2} = -\frac{1}{g_1^2(x)} + \beta_1(g_0)\phi_1(x).$$

To write the final form of the new effective density we introduce some new notations. We write

$$(15)\qquad \rho_1(V) = \sum_{\{\Omega_1,\Lambda_1\}} \chi_1(\Omega_1)\mathcal{I}_1(\{\Omega_1,\Lambda_1\})\exp A_1(\tfrac{1}{g_1^2},U_1),$$

where $\mathcal{I}_1$ is the integral operation defined by the integration

$$(16)\qquad \mathcal{I}_1(\{\Omega_1,\Lambda_1\}) = \int dV_0\big|_{\Omega_1} \delta(\overline{V}_0V^{-1})\zeta(\Omega_1^c)\,\cdot$$

$$\cdot\, dA\big|_{\Omega_1\cap\Lambda_1^c}\delta(QA)\delta_{Ax}(A)\chi_0'(\Omega_1\cap\Lambda_1^c)\chi^c(\Omega_1\cap\Lambda_1^c)\,\cdot$$

$$\cdot\,\exp\Big[-\tfrac{1}{2}\langle A,\Delta_1A\rangle + \tfrac{1}{2}\langle A,\Delta_1C^{(0)}(\Lambda_1)\Delta_1A\rangle\Big].$$

The effective action A_1 has the form

$$(17)\qquad A_1(\tfrac{1}{g_1^2},U_1) = -A(\tfrac{1}{g_1^2},U_1) + \{\mathcal{E}^{(1)}(\Lambda_1,g_0,U_1) -$$

$$- \mathcal{E}^{(1)}(\Lambda_1,g_0,1) - \beta_1(g_0)A(\phi_1,U_1)\} + \mathcal{B}^{(1)}(U_1,A) - E_1.$$

The boundary term $\mathcal{B}^{(1)}(U_1,A)$ contains all the terms depending on the field A left unintegrated in the domain $\Omega_1 \cap \Lambda_1^c$. The constant E_1 is obtained from E by subtracting all the constants generated in the procedure. Thus E_1 depends on Ω_1,Λ_1.

Now the representation (15) – (17) will be generalized to an inductive description of the k–th effective density. This density $\rho_k(V_k)$ is a function of the new field V_k defined on the lattice $T^{(k)}$. It is represented by an expansion generalizing the expansion (15). Terms of this expansion are parametrized by sequences of domains in T. Let us start with the description of this geometric setting. We consider sequences of localization domains $\{\Omega_j\}$, $j = 1,2,\cdots,k$, such that Ω_j is a union of large cubes in the L^{-j}–scale, or simply $ML^j\eta$–cubes, $\eta = L^{-k}$, and $\Omega_1 \supset \Omega_2 \supset\cdots\supset \Omega_k$. For such a sequence we consider sequences of domains $\{\Lambda_j\}$ such that

$$(18)\qquad \Omega_1 \supset \Lambda_1 \supset \Omega_2 \supset \Lambda_2 \supset\cdots\supset \Omega_k \supset \Lambda_k.$$

We denote

$$(19)\qquad \Gamma_0 = \Omega_1^c,\ \Gamma_j = \Omega_j^{(j)}\backslash\Omega_{j+1}^{(j)},\ j = 1,\cdots,k-1,\ \Gamma_k = \Omega_k^{(k)},$$

$$\mathcal{B} = \bigcup_{j=0}^{k} \Gamma_j\ ,\quad Z_j = \Lambda_j^c.$$

The set $\mathcal{B}$ determines the sequence $\{\Omega_j\}$ and is called the determining set. It plays the basic role in a variational problem generalizing the problem (8). To define this problem we have to introduce corresponding spaces of gauge fields. The field V_j introduced by the j–th renormalization transformation is defined on the lattice $T^{(j)}$. After the integrations it remains on the subset $\Omega_{j+1}^{(j)c} = \Omega_{j+1}^c \cap T^{(j)}$, and it is regular on Γ_j in the sense that $|\partial V_j - 1| < 0(L^2)\epsilon_j$, $\epsilon_j = g_j p(g_j)$. The fields V_j determine a field V defined on $\mathcal{B}$:

(20) $\qquad V = V_j$ on Γ_j, $j = 0,1.\cdots,k.$

It is convenient also to introduce the averaging operation $M_{\mathcal{B}}$ associated with the determining set $\mathcal{B}$. It transforms a gauge field U on the lattice T_η into the gauge field $M_{\mathcal{B}}(U)$ defined on bonds of the set $\mathcal{B}$ by the equalities

(21) $\qquad M_{\mathcal{B}}(U) = M^j(U) = \bar{U}^j$ on Γ_j, $j = 0,1,\cdots,k.$

Now we can formulate the variational problem:

(22) given a regular configuration V on the determining set $\mathcal{B}$, find minima of the action A(U) considered on configurations U satisfying the conditions $M_{\mathcal{B}}(U) = V$ and some regularity properties.

The above problem has similar properties to the problem (8); it is invariant with respect to the group of all gauge transformations u defined on T and satisfying the conditions $u(y) = 1$ for $y \in \mathcal{B}$. Thus we consider the above problem in the space of orbits of this group. The basic result is given by

Proposition 1. In the space of orbits there exists exactly one minimal orbit.

We denote by $U_{\mathcal{B}}(V)$, or $U_k(V)$, or simply U_k, the element of the minimal orbit satisfying axial gauge conditions in blocks.

Let us write now a general form of the expansion of the density $\rho_k(V_k)$. It can be represented as

(23) $$\rho_k(V_k) = \sum_{\{\Omega_j\},\{\Lambda_j\}} \chi_k(\Omega_k)\mathcal{I}_k(\{\Omega_j\},\{\Lambda_j\})\exp A_k(\tfrac{1}{g_k^2},U_k)$$

where the summation is over the admissible sequences of domains. The characteristic function $\chi_k(\Omega_k)$ introduces small field restrictions on the variables V_k on the domain Ω_k. The restrictions are essentially given by the inequalities $|V_k(\partial p)-1| < \epsilon_k$, $\epsilon_k = g_k p(g_k)$, $p \subset \Omega_k^{(k)}$. The next part of the formula (23) is the operation $\mathcal{I}_k$. It is an integral operation determined by all the integrations left at all renormalization steps. It was described in detail in the first step, and a complete definition follows from the inductive procedure. We will not discuss it in detail; we will only mention most important general properties. The last large field region is Z_k, and $\mathcal{I}_k$ is supported in it in the sense that it involves integrations and variables restricted to this region. If Z_k is represented as a union of disjoint regions, e.g. as a union of connected components, $Z_k = X_1 \cup \cdots \cup X_n$. $X_i \cap X_j = \phi$ for $i \neq j$, then

$$(24) \qquad \mathcal{I}_k(Z_k) = \prod_{i=1}^{n} \mathcal{I}_k(X_i).$$

The operations corresponding to disjoint regions commute, i.e. $\mathcal{I}_k(X_i)\mathcal{I}_k(X_j) = \mathcal{I}_k(X_j)\mathcal{I}_k(X_i)$. For a given large field region X the operation $\mathcal{I}_k(X)$ can be factored into a product of one-step operations, and has the form

$$(25) \qquad \mathcal{I}_k(X) = \prod_{j=k-1}^{0} \mathcal{I}^{(j)}(Z_{j+1} \cap X).$$

This is an ordered product, the order indicated in the product symbol. The operation $\mathcal{I}^{(j)}$ involves integration with respect to the gauge field variables V_j on $\Omega_{j+1}^c \cap X$, and with respect to the fluctuation field variables A_j on $Z_{j+1} \cap \Omega_{j+1} \cap X$, if the last set is nonempty. This operation comes from the renormalization transformation T in the j+1-st step, and it has a form corresponding to (16).

The most important part of the inductive assumption is a description of the effective actions A_k. They have the following general form

$$(26) \qquad A_k(\tfrac{1}{g_k^2}, U_k) = -A(\tfrac{1}{g_k^2}, U_k) + \mathcal{E}_k(U_k) + \mathcal{R}_k(U_k) + \mathcal{B}_k(U_k, A) - E_k.$$

All the expressions above depend on the sequences $\{\Omega_j\},\{\Lambda_j\}$. Before we describe them let us explain briefly their meaning. The term $\mathcal{E}_k$ is the regular part of the action. It has the same properties as the effective action constructed in [1] for the small field approximation, but localized properly in different scales to domains determined by $\{\Lambda_j\}$. This expression is fully renormalized, i.e. vacuum energy and coupling constant renormalization counterterms are included into it. The term $\mathcal{R}_k$ arises as the effect of R-operations, and has localization properties similar to those of the term $\mathcal{E}_k$. For this term we perform the vacuum energy renormalization only. The term $\mathcal{B}_k$ includes various expressions localized closely to large field regions. It does not require any renormalization.

We start a more detailed description of the terms in (26) with the definition of the function $g_k^2(x)$. It is defined by the recursive renormalization group equations generalizing the equation (14):

$$(27) \qquad \frac{1}{g_{j-1}^2(x)} = \frac{1}{g_j^2(x)} + \beta_j(g_{j-1})\phi_j(x), \; j = 1,2,\cdots,k.$$

Here the coupling constants g_{j-1} are defined as in [1], and $\phi_j \in C_0^\infty(\Lambda_j)$, $\phi_j = 1$ on Λ_j without a boundary layer.

The term $\mathscr{E}_k$ has the following representation:

$$(28) \qquad \mathscr{E}_k(U_k) = \sum_{j=1}^{k} \Big[\mathscr{E}^{(j)}(\Lambda_j, g_{j-1}, U_k) - \mathscr{E}^{(j)}(\Lambda_j, g_{j-1}, 1) - \beta_j(g_{j-1}) A(\phi_j, U_k) \Big].$$

Localization properties of $\mathscr{E}^{(j)}$ are crucial for our analysis. The following representations hold

$$(29) \qquad \mathscr{E}^{(j)}(\Lambda_j, U_k) = \sum_{z \in \Lambda_j^{(j)}} \mathscr{E}^{(j)}(\Lambda_j, U_k, z)$$

$$(30) \qquad \mathscr{E}^{(j)}(\Lambda_j, U_k, z) = \sum_{z \in X \subset \Lambda_j} \mathscr{E}^{(j)}(X, U_k, z),$$

where the last sum is over domains X which are unions of large cubes in j-th scale (or $ML^j\eta$-cubes in η-scale). The term $\mathscr{E}^{(j)}(X, U_k, z)$ depends on U_k restricted to the domain X, and is a gauge invariant and analytic function of U_k, satisfying the bound

$$(31) \qquad |\mathscr{E}^{(j)}(X, U_k, z)| \le E_0 \exp(-\kappa d_j(X)).$$

Here E_0, κ are sufficiently large constants, and $d_j(X)$ was introduced in [1]. (Let us recall that it is the shortest length of tree graphs contained in X and intersecting all the large cubes of X, measured in L^{-j}-scale.) The last important property is the Euclidean covariance. It holds for $\mathscr{E}^{(j)}(T, U_k, z)$, and it can be formulated that this function is invariant with respect to Euclidean transformations of U_k and z, which leave the lattice $T^{(j)}$ invariant. This implies that the function $\mathscr{E}^{(j)}(T, U_j, z)$ is invariant with respect to the Euclidean transformations of U_j leaving the point z invariant.

Obviously we are interested not only in getting representations of the effective actions, but also in proving uniform bounds for them. We state now a first result of this type. Let Ω be a domain contained in Λ_j, and $\phi \in C_0^\infty(\tilde{\Omega})$, $\phi = 1$ on Ω.

Proposition 2. There exists a constant E_1 independent of j,k,Ω $\{\Omega_j\}$, $\{\Lambda_j\}$, T such that

$$(32)\qquad \Big| \sum_{z\in\Lambda_j^{(j)}\cap\Omega} \mathscr{E}^{(j)}(\Lambda_j,U_k,z) - \mathscr{E}^{(j)}(\Lambda_j,1,z)) -$$

$$- \beta_j(g_{j-1})A(\phi_j,U_k)\Big| \le E_1 \sum_{n=j}^{k} (L^{j-n})^{\beta}|\Gamma_n\cap\Omega|$$

for $\beta < 1$. The constant E_1 depends on β and grows to ∞ if $\beta \to 1$. The volumes are taken in the corresponding scales, i.e. $|\Gamma_n\cap\Omega|$ means the number of points in the set $\Gamma_n\cap\Omega \subset T_1^{(n)}$.

The inequality (32) is the most important property of the action $\mathscr{E}_k$. It holds because the terms of $\mathscr{E}_k$ have been renormalized properly. Taking $\Omega = \Lambda_j$, $\phi = \phi_j$ in (32), and summing the obtained inequalities over j, we get

$$(33)\qquad |\mathscr{E}_k(U_k)| \le E_1 \sum_{n=1}^{k} |\Gamma_n| \sum_{j=1}^{n} (L^{j-n})^{\beta} < E_1(1-L^{-\beta})^{-1} \sum_{n=1}^{k} |\Gamma_n|.$$

This is the required bound for the effective action.

Consider now the term $\mathscr{R}_k(U_k)$ in (26). It has the representation

$$(34)\qquad \mathscr{R}_k(U_k) = \sum_{j=1}^{k} [\mathscr{R}^{(j)}(\Lambda_j,U_k) - \mathscr{R}^{(j)}(\Lambda_j,1)],$$

where each function $\mathscr{R}^{(j)}(\Lambda_j,U_k)$ has the localized representation of the form (30). Terms of this representation have the same properties as the terms in (30), but they satisfy the following stronger bound

$$(35)\qquad |\mathscr{R}^{(j)}(X,U_k)| \le g_j^{\kappa_0} \exp(-\kappa d_j(X)).$$

Here κ_0 can be chosen arbitrarily large, similarly as κ, if the other parameters are fixed properly. The above inequality is connected with the fact that the terms are defined by integrals including large field domains of integrations. After the vacuum energy renormalization we obtain a sum of marginal terms, i.e terms with bounds $0(1)(L^j\eta)^4 g_j^{\kappa_0}\exp(-\kappa d_j(X))$. The sum over X is controlled by the exponential factor, and by the factor $(L^j\eta)^4$. The sum over j is controlled by $g_j^{\kappa_0}$. This yields the following:

Proposition 3. There exists an absolute constant R_1 such that

$$(36)\qquad \Big|\sum_{X\subset\Lambda_j,\,X\cap\Omega\neq\phi} [\mathscr{R}^{(j)}(X,U_k)-\mathscr{R}^{(j)}(X,1)]\Big| \le$$

$$\le R_1 g_j^{\kappa_0} \sum_{n=j}^{k} |\Gamma_n\cap\Omega| \ .$$

(In fact the constant R_1 can be taken as equal to 1 for g_j sufficiently small.)

Similarly as for $\mathscr{E}_k$ we take $\Omega = \Lambda_j$ in (36); we sum over j and we get

$$(37)\qquad |\mathscr{R}_k(U_k)| \le R_1 \sum_{n=1}^{k}|\Gamma_n| \sum_{j=1}^{n} g_j^{\kappa_0} < R_1 \sum_{n=1}^{k}|\Gamma_n| g_j^{\kappa_0-6} <$$

$$< \sum_{h=1}^{k}|\Gamma_n|,$$

for $\kappa_0 \ge 7$ and g sufficiently small.

The term $\mathscr{B}_k$ is the simplest term in (26) from the point of view of the renormalization; it does not need any renormalization. It is a sum of localized terms, with localizations close to large field regions. More precisely we have

$$(38)\qquad \mathscr{B}_k(U_k) = \sum_{j=1}^{k}\mathscr{B}^{(j)}(U_k,A) = \sum_{j=1}^{k}\sum_{X}\mathscr{B}^{(j)}(X,U_k,A),$$

where the last sum is over domains X such that $X \cap \Omega_j \neq \phi$, and $X \cap Z_j \neq \phi$. The term in this sum depends on U_k,A restricted to X. It is an analytic and gauge invariant function of U_k,A, and it satisfies the bound (31) with a different constant, i.e. with B_0 instead of E_0. A bound of $\mathscr{B}_k(U_k)$ is completely elementary. We use the exponential factor $\exp(-(\kappa-1)d_j(X))$ to control the sum over X in (38). The remaining factor $\exp(-d_j(X))$ is used to get an additional small factor, e.g. for $j > n$ it can be bounded by $2^{-(j-1-n)}$. Thus we obtain

$$(39)\qquad |\mathscr{B}^{(j)}(X,U_k,A)| \le B_1 2^{-(j-n)}|\Gamma_n|$$

for $j > n$, and this inequality with $|\Gamma_n\backslash\Lambda_k|$ for $j = n$. Summing over j from n to k, and then over n from 1 to k, we get

(40) $$|B_k(U_k)| \leq 2B_1 \sum_{n=1}^{k} |\Gamma_n|.$$

The constants E_k (depending on $\{\Omega_j\},\{\Lambda_j\}$ also) are obtained by subtracting one-step vacuum energy expressions, generated in small field regions, from the initial constant E. This initial constant is defined in fact as a sum of all such expressions for all the lattices $T^{(k)}$ as the small field regions. The constants E_k satisfy the bound of the form (40).

The above bounds imply easily the following

Proposition 4 (Ultraviolet stability). There exist constants E_-,E_+ independent of η and T, but depending on g_k, such that

(41) $$\chi_k(T)\exp\left[\frac{1}{g_k^2} A(U_k) - E_-|T_\eta|\right] \leq \rho_k(V_k) \leq e^{E_+|T_\eta|}.$$

The basic part of the whole construction is a proof that the inductive assumptions formulated and discussed above are preserved under the renormalization transformation RT. We have discussed the transformation T in the first step. In the general case we repeat the same operations, only in a slightly more complicated setting. The most important complication is connected with the fact that we have to deal with the solutions U_k of the general variational problem. We will not discuss these purely technical problems here. We have not described yet the operation R, so the remaining paragraphs will be devoted to a brief and general discussion of some basic ideas connected with this operation. Let us explain at first what is the nature of the large field problem. Consider a large plaquette variable int he first step. We have $|U(\partial p)-1| \geq g_0 p(g_0)$ for a plaquette $p \in T_1$. The term in the Wilson action, corresponding to the plaquette p, gives the estimate

(42) $$\exp\left[-\frac{1}{g_0^2}[1-\text{Retr } U(\partial p)]\right] \leq \exp(-\tfrac{1}{2}p^2(g_0^2)) < g_0^{p(g_0)}.$$

For $d < 4$ we have $g_0 = g\,\epsilon^{1/2(4-d)}$, and the bound above can be estimated by an arbitrarily large power of ϵ. This is enough to control expressions arising in the large field regions surrounding the plaquette p for all steps of the procedure, i.e., until we reach the unit lattice. For $d = 4$ the bare coupling constant behaves asymptotically as $(a+b \log \epsilon^{-1})^{-1/2}$, for $\epsilon \longrightarrow 0$, with some positive constants a,b hence the bound does not give any positive power of ϵ. It is still small for ϵ small, and it controls a large number of steps, but this number is a small fraction of the total number of steps. Thus, for some large field regions there is a difficulty in continuing the procedure; the small factor arising from large fields in this region does not control further steps. In such situations we have to change the procedure in order to improve the small factor, i.e., we have to be able to renormalize the expression corresponding to the large field region. There are several possible ways of doing it; the one chosen in this paper is closest to the method of G. Gallavotti et al. in [3] and can be described in the simplest way as follows: a large field expression is replaced by the corresponding small field expression in such a way that integrals of the densities are unchanged. Let us elaborate this description. A density ρ after some number of steps is represented in the form

$$(43) \qquad \rho(V) = \sum_{Z} \rho(Z,V),$$

where the sum is over large field regions Z, and V is a gauge field variable. A region Z is decomposed into disjoint subregions Z', Z'', $Z = Z' \cup Z''$, in the following way: Z'' is a union of components of the region Z, for which the small factors connected with large fields control some number of next steps, Z' is a union of remaining components, i.e., components for which the corresponding expressions require a renormalization. For such a decomposition we take the density $\rho(Z'',V)$, and we define the operation R as follows:

$$(44) \qquad (R\rho)(V) = \sum_{Z} \rho(Z'',V) \frac{\int dV_{|Z'} \rho(Z,V)}{\int dV_{|Z'} \rho(Z'',V)}.$$

We will prove that the densities are positive, and the integration domains in the integrals above are nonempty, hence the denominators are positive, and the operation R is well defined. It satisfies the basic normalization property

$$(45) \qquad \int dV (R\rho)(V) = \int dV \rho(V).$$

Consider now the expression on the right–hand side of the definition. It can be written as a double sum over domains Z', Z'', $Z' \subset Z''^{c}$, and the summation over Z' can be applied to the quotients. The quotients are still small, because some small factors in the regions Z' are left for the densities in the numerator. We localize them, trying to decouple components of Z', i.e., we write a polymer expansion, and then we exponentiate it. Thus, we obtain the representation

$$(46) \qquad \sum_{Z' \subset Z''^{c}} \frac{\int dV_{|Z'} \rho(Z' \cup Z'',V)}{\int dV_{|Z'} \rho(Z'',V)} = \exp \sum_{X} R(X,V),$$

where the last sum is over X such that $X \cap Z''^{c} \neq \phi$. Using this representation, we rewrite the definition of the operation R:

$$(47) \qquad (R\rho)(V) = \sum_{Z''} \rho(Z'',V) \exp \sum_{X} R(X,V).$$

Now the advantages of applying such an operation are clear; the densities on the right–hand side still have enough small factors to control the given number of steps, and the expression in the exponential can be treated in the same way, as the small field effective actions are treated in [1]; in particular it can be renormalized in the same way. This renormalization is the necessary renormalization of the expressions connected with the large field regions, and it makes the whole renormalization group procedure convergent, i.e., we can apply all the transformations needed to reach the unit lattice, and we control all the steps of the procedure.

The above description stresses only some general ideas underlying the method used in this paper. The actual procedure is much more complicated, and it also differs from the one presented above in some technical aspects, for example in (44), (46) we take the denominators equal not the integrals of the whole densities $\rho(Z'',V)$, but to the integrals of some parts of these densities. More precisely, we take the parts localized in neighborhoods of the domains Z', so they do not depend on the large field regions Z'', and they are determined by small field effective actions only. These general ideas are very simple and natural. They have many possible variations, and they can be realized in many different ways.

One of the fundamental important problems in constructive quantum field theory is to find better solutions of the large field problem. There are several attempts in this direction, some of which have been mentioned by other lecturers.

REFERENCES

[1] T. Balaban and A. Jaffe, Constructive guage theory, in: "Fundamental Problems of Gauge Field Theory," G. Velo and A. Wightman, eds., Erice lectures, Plenum Press (1986).

[2] T. Balaban, Renormalization group approach to lattice gauge field theories. I., Commun. Math. Phys. 109:249–301 (1987).

[3] G. Benfatto, M. Cassandro, G. Gallavotti, F. Nicolò, E. Olivieri, E. Presutti, and E. Scacciatelli, Ultraviolet stability in Euclidean scalar field theories, Commun. Math. Phys. 71:95–130 (1980).

THE BETA-FUNCTION METHOD FOR RESUMMATIONS IN FIELD THEORY

Giovanni Gallavotti

Centro Interdisciplinare Linceo B. Segre
Accademis dei Lincei, via della Lungara 10, 00165 Roma

1 Perturbation Theory: Feynman graphs, Scales and Trees

We study the functional integral:

$$\int g(d\varphi)\exp\{\int_\Lambda(\lambda:\varphi_x^4:+\mu:\varphi_x^2:+\alpha:(\partial\varphi_x):^2+\nu)d\xi\} \tag{1.1}$$

where Λ is a d–dimensional torus, λ,μ,α,ν are constants and g is a gaussian random field over Λ with a covariance obtained by periodizing the function on R^4 whose Fourier transform is:

$$\sum_{k=0}^{N} C_k(\gamma^{-k}p)\gamma^{-2k} \tag{1.2}$$

where $N<+\infty$ is a cut-off parameter and $C_k(p)$ is holomorphic for $|\Im p_j|<\kappa$, $\kappa>0$, and:

$$\sum_{k=0}^{\infty} C_k(\gamma^{-k}p)\gamma^{-2k}=\frac{1}{1+p^2}$$

$$|C_k(p)|\le\frac{B_\alpha}{1+|p|^\alpha},\qquad \forall\alpha\le A,k,\ |\Im p_j|<\kappa,\ A>8 \tag{1.3}$$

For instance:

Constructive Quantum Field Theory II, Edited by G. Velo and
A. S. Wightman, Plenum Press, New York, 1990

$$\frac{1}{1+p^2} = \frac{e^{-(1+p^2)}}{1+p^2} + \sum_{k=0}^{\infty} \frac{e^{-(1+p^2)\gamma^{-2k-2}} - e^{-(1+p^2)\gamma^{-2k}}}{1+p^2} \qquad \gamma > 1$$

$$C^{(k)}(p) = \frac{e^{-(\varepsilon_k+p^2)\gamma^{-2}} - e^{-(\varepsilon_k+p^2)}}{\varepsilon_k + p^2} \qquad \varepsilon_k = \gamma^{-2k} \tag{1.4}$$

Formula (1.2) allows us to think φ as a sum:

$$\varphi_x \equiv \varphi_x^{(\leq N)} \equiv \sum_{h=0}^{N} \varphi^{(h)} \tag{1.5}$$

where the $\varphi^{(h)}$ are gaussian random fields with covariance given by the periodization of the covariance with Fourier transform $\gamma^{-2k} C_k(\gamma^{-k}p)$. Hence the average value of $\varphi^{(\leq N)2}$ is of the order of $\gamma^{(d-2)N}$: one thus says that the field has *dimension* $-(d-2)/2$ in length units (because γ^{-N} is a typical cut off length scale).
The dots in (1.1) denote the ***Wick ordering*** and are used only for convenience. In fact the coming algebra becomes cosiderably simpler if one uses the martingale property of the Wick ordered polynomials. We remind it: let $x_1, \ldots, x_n$ be n gaussian random variables (not necessarily independent: so that their covariance will be a matrix $C_{i,j}$) and let $y_1, \ldots, y_n$ be n other gaussian variables independent form the former ones (but not necessarily independent: so that their covariance will be a matrix $\bar{C}_{i,j}$). Then, calling P the distribution of $y_1, \ldots y_n$, the martingale property is:

$$\int P(dy_1 \ldots dy_n) \,:(x_1+y_1)^{k_1} \ldots (x_n+y_n)^{k_n}: \equiv :x_1^{k_1} \ldots x_n^{k_n}: \tag{1.6}$$

Denoting $\mathcal{E}_k$ the integration with respect to $\varphi^{(k)}$ the integral (1.1) can be written:

$$\mathcal{E}_0 \mathcal{E}_1 \cdots \mathcal{E}_N \{\exp \int_\Lambda dx \,(\lambda_N \gamma^{(4-d)N} :\varphi_x^4: + \gamma^{2N} \mu_N :\varphi_x^2: + \alpha_N :(\partial\varphi_x)^2: + \nu\gamma^{dN})\} \tag{1.7}$$

having introduced the ***dimensionless couplings*** $\lambda_N, \mu_N, \alpha_N, \nu_N$ (note that the field φ_x has dimension $-(d-2)/2$ so that λ_N is dimensionless because $-(4-d)+d-4(d-2)/2 = 0$, μ_N because $d-2-2(d-2)/2 = 0$, and α_N because $d - 2(1+(d-2)/2) = 0$, and ν_N because $d - d = 0$). To analyze the integral (1.7) we define recursively the *effective potential on scale k* via:

$$V^{(k)}(\varphi^{(\leq k)}) = \log \mathcal{E}_{k+1}(\exp V^{(k+1)}(\varphi^{(\leq k+1)})) \tag{1.8}$$

and we introduce the basic notion of truncated expectations for n general random variables (functions $f_1, \ldots, f_n$ of $\varphi^{(k)}$):

$$\mathcal{E}_k^T(f_1,\ldots,f_n) \equiv \frac{\partial^n}{\partial\varepsilon_1\ldots\partial\varepsilon_n}\log\mathcal{E}_k\big(\exp\sum_1^n \varepsilon_i f_i\big)\Big|_{\varepsilon_i=0} \tag{1.9}$$

This definition allows us to see the following formal expression:

$$V^{(N-1)} = \sum_{n=1}^{\infty}\frac{1}{n!}\mathcal{E}_N^T(V^{(N)},\ldots,V^{(N)}) \tag{1.10}$$

which we represent graphically as:

$V^{(N-1)} =$ 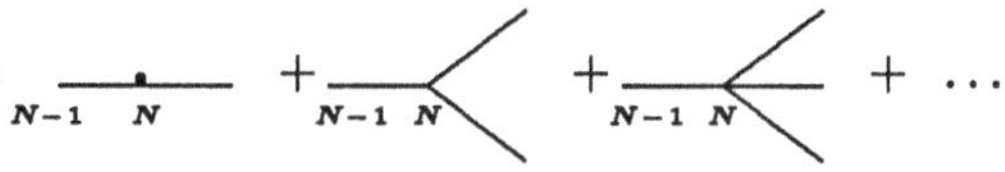 (1.11)

where  represents $V^{(N)}$ and:

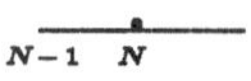 represents $\frac{1}{1!}\mathcal{E}^T(V^{(N)}) \equiv \mathcal{E}_N(V^{(N)})$

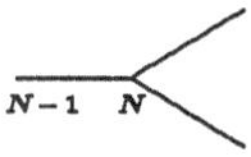 represents $\frac{1}{2!}\mathcal{E}^T(V^{(N)},V^{(N)})$

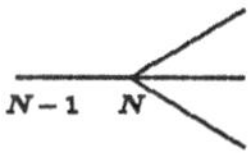 represents $\frac{1}{3!}\mathcal{E}^T(V^{(N)},V^{(N)},V^{(N)})$ (1.12)

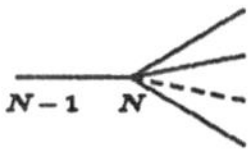 represents $\frac{1}{s!}\mathcal{E}^T(V^{(N)},V^{(N)},\ldots,V^{(N)})$

We see that the above graphs represent functions of $\varphi^{(N-1)}$. It is easy to iterate the above formula. However it is convenient to reorganize its terms before the iteration. For this purpose we introduce two operators acting on the functions of $\varphi^{(\leq N)}$ which we call $\mathcal{L}_N$, $\mathcal{R}_N \equiv (1-\mathcal{L}_N)$. Such operators will be projection operators and the range of $\mathcal{L}_N$, *localization operator*, will be the 4–dimensional manifold of the linear combinations of the functions:

$$\int_\Lambda :\varphi_x^4:,\ \int_\Lambda :\varphi_x^2:,\ \int_\Lambda :(\partial\varphi_x)^2:,\ \int_\Lambda 1 \tag{1.13}$$

which we call the *space of the local interactions.*

The action of the operators $\mathcal{L}_N, \mathcal{R}_N$ on a function 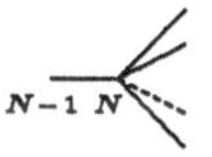 can be simply represented graphically by a superscript L or R as  or  ; therefore we rewrite $V^{(N-1)}$ as:

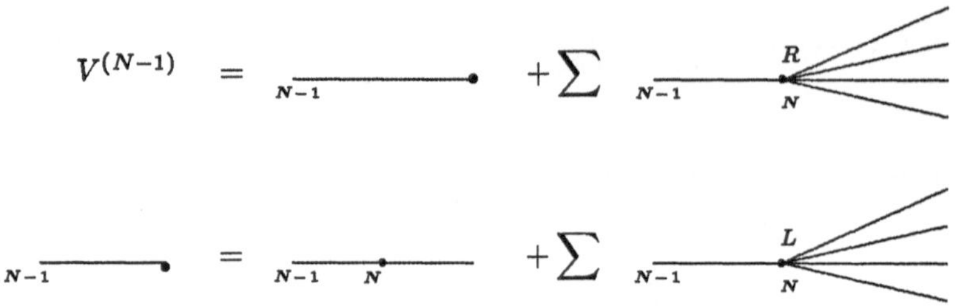

$$V^{(N-1)} = \underset{N-1}{\rule{3cm}{0.4pt}\bullet} \quad + \sum \underset{N-1}{\rule{2cm}{0.4pt}}\underset{N}{\overset{R}{\diagup\!\!\!\!\diagup}}$$

$$(1.15)$$

$$\underset{N-1}{\rule{3cm}{0.4pt}\bullet} = \underset{N-1}{\rule{1.5cm}{0.4pt}}\underset{N}{\bullet}\rule{1.5cm}{0.4pt} \quad + \sum \underset{N-1}{\rule{2cm}{0.4pt}}\underset{N}{\overset{L}{\diagup\!\!\!\!\diagup}}$$

For the time being it is not necessary to specify the form of $\mathcal{L}_N$ and $\mathcal{R}_N$ explicitly: we shall however require that $\mathcal{L}_k$ is defined for all $k = 0,1,\ldots$ and *commutes* with the expectations, *i.e.*:

$$\mathcal{L}_k \mathcal{E}_{k+1} \ldots \mathcal{E}_p = \mathcal{E}_{k+1} \ldots \mathcal{E}_p \mathcal{L}_p \qquad (1.16)$$

which is important to simplify the coming algebra.

Clearly by iteration of (1.14) we find a *simple* expression for $V^{(h)}$:

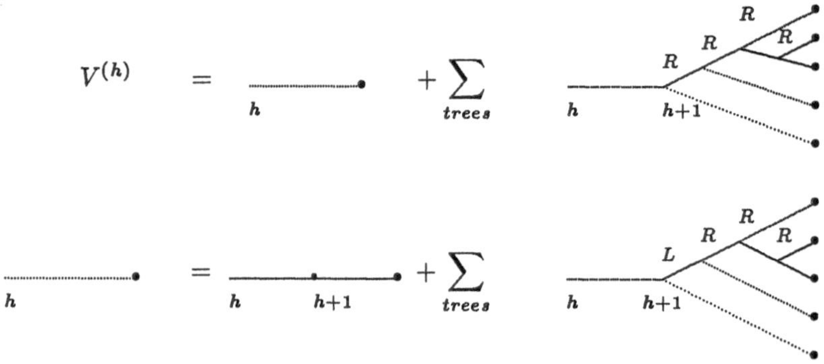

$$V^{(h)} = \underset{h}{\rule{3cm}{0.4pt}\bullet} \quad + \sum_{trees} \underset{h}{\rule{2cm}{0.4pt}}\underset{h+1}{\overset{R}{\cdots}}$$

$$(1.17)$$

$$\underset{h}{\rule{3cm}{0.4pt}\bullet} = \underset{h}{\rule{1.5cm}{0.4pt}}\bullet\underset{h+1}{\rule{1.5cm}{0.4pt}} \quad + \sum_{trees} \underset{h}{\rule{2cm}{0.4pt}}\underset{h+1}{\overset{L}{\cdots}}$$

Here each vertex v of a tree bears a *scale index* h_v and symbolizes the truncated expectation, with respect to $\varphi^{(h_v)}$, $\mathcal{E}_{h_v}^T(F_{v_1}, \ldots, F_{v_{s_v}})$ where $(F_{v_1}, \ldots, F_{v_{s_v}})$ are the functions of $\varphi^{(h_v)}$ described by the trees with root scale h_v following v in the directions of $v_1, \ldots, v_{s_v}$. The R or L superscripts mean that the vertex v really represents:

$$\mathcal{R}_{h_v-1} \mathcal{E}_{h_v}^T(F_{v_1}, \ldots, F_{v_{s_v}}) \qquad \text{or} \qquad \mathcal{L}_{h_v-1} \mathcal{E}_{h_v}^T(F_{v_1}, \ldots, F_{v_{s_v}}).$$

Clearly (1.13) implies that $\underset{h}{\rule{1cm}{0.4pt}}$ can be written:

$$\int_\Lambda \left(\lambda_h \gamma^{(4-d)h} : \varphi_x^{(\leq h)4} : + \mu_h \gamma^{2h} : \varphi^{(\leq h)2} : + \alpha_h : (\partial \varphi_x^{(\leq h)})^2 : + \gamma^{dh} \nu_h \right) dx \qquad (1.18)$$

where $\underline{r}_h \equiv (\lambda_h, \mu_h, \alpha_h, \nu_h)$ are suitable constants.

The main result of renormalization theory for $d = 4$ is that, given $\lambda_o = \lambda$, $\mu_o = \mu$, $\alpha_o = \alpha$, $\nu_o = \nu$ one can define a formal power series:

$$\begin{aligned}
\lambda_h =& \lambda + \sum_{n=2}^{\infty} \sum_{p+q+r=n} b_{pqr}^{(h)} \lambda^p \mu^q \alpha^r \\
\mu_h =& \gamma^{-2h}\mu + \sum_{n=2}^{\infty} \sum_{p+q+r=n} m_{pqr}^{(h)} \lambda^p \mu^q \alpha^r \\
\alpha_h =& \alpha + \sum_{n=2}^{\infty} \sum_{p+q+r=n} a_{pqr}^{(h)} \lambda^p \mu^q \alpha^r \\
\nu_h =& \gamma^{-4h}\nu + \sum_{n=2}^{\infty} \sum_{p+q+r=n} n_{pqr}^{(h)} \lambda^p \mu^q \alpha^r
\end{aligned} \tag{1.19}$$

with coefficients $b^{(h)}, m^{(h)}, n^{(h)}, a^{(h)}$ converging as $\Lambda \to \infty$, $N \to \infty$ and uniformly bounded by:

$$|b_{pqr}^{(h)}|,\ |m_{pqr}^{(h)}|,\ |a_{pqr}^{(h)}|,\ |n_{pqr}^{(h)}| \leq DC^{n-1}(n-1)! \sum_{j=0}^{n-1} \frac{(bh)^j}{j!} \tag{1.20}$$

where D, b, C are suitable computable constants, [1],[2].
Other results will be:

$$\begin{aligned}
\lambda_{h+1} =& \lambda_h + \sum_{n=2}^{\infty} \sum_{p+q+r=n} \beta_{pqr}^{(h)} \lambda_h^p \mu_h^q \alpha_h^r \\
\mu_{h+1} =& \gamma^{-2}\mu_h + \sum_{n=2}^{\infty} \sum_{p+q+r=n} \beta'^{(h)}_{pqr} \lambda_h^p \mu_h^q \alpha_h^r \\
\alpha_{h+1} =& \alpha_h + \sum_{n=2}^{\infty} \sum_{p+q+r=n} \beta''^{(h)}_{pqr} \lambda_h^p \mu_h^q \alpha_h^r \\
\nu_{h+1} =& \gamma^{-d}\nu_h + \sum_{n=2}^{\infty} \sum_{p+q+r=n} \beta'''^{(h)}_{pqr} \lambda_h^p \mu_h^q \alpha_h^r
\end{aligned} \tag{1.21}$$

with $\beta_{pqr}^{(h)\alpha}$ converging as $\Lambda \to \infty$, $N \to \infty, h \to \infty$ and:

$$|\beta_{pqr}^{\cdot}| \leq D\, b^{n-1}(n-1)! \tag{1.22}$$

for all Λ, N *and* h, if $p+q+r=n$, [1],[2].

This theorem says that there is a perturbatively well defined *beta function* allowing us to describe the dependence of $\lambda_h, \mu_h, \alpha_h, \nu_h$ on h via the iteration of a (formal) dynamical system in R^4, or in fact in R^3 as the last equation clearly decouples from the first three. Another result is that if $\lambda_N, \mu_N, \alpha_N$ are chosen as given by (1.18), then all the Schwinger functions admit a formal expansion in λ, α, μ with coefficients uniformly finite in Λ, N. In fact a stronger result holds for the effective potentials $V^{(h)}$. Unfortunately its formulation is slightly involved, although eventually it turns out natural. We have therefore to introduce a few more symbols to describe properly the results on the effective potentials. However it is worth to make the effort of understanding the results because the definitions below will be encountered again when we give a precise definition of the $\mathcal{L}$, $\mathcal{R}$ operators, and by that time will already be somewhat familiar.
If $\varphi \equiv \varphi^{(\leq h)}$, we shall show that:

$$V^{(h)}(\varphi) = \sum \int V_{\Phi}^{(h)}(\xi_1, \ldots, \xi_n) \; : \Phi_{\xi_1, \ldots, \xi_n} : \; d\xi_1 \ldots d\xi_n \tag{1.23}$$

where $V_{\Phi}^{(h)}$ is a suitable coefficient and $: \Phi_{\xi_1 \ldots \xi_n} :$ is a monomial in the fields:

$$\begin{gathered}
\varphi_x, \quad \varphi_y - \varphi_x = D_{xy}, \quad \varphi_y - \varphi_x - (y-x)\partial\varphi_x = S_{yx} \\
\varphi_y - \varphi_x - (y-x)\partial\varphi_x - \frac{1}{2}(y-x)^2\partial^2\varphi_x = T_{yx} \\
\partial\varphi_x, \quad \partial\varphi_y - \partial\varphi_x = D^1_{yx}, \quad \partial\varphi_y - \partial\varphi_x - (y-x)\partial\partial\varphi_x = S^1_{yx} \\
(y-x)D^1_{yz} = \mathcal{D}^1_{zyx}
\end{gathered} \tag{1.24}$$

where $\xi \equiv x$ or $\xi \equiv (x,y)$ or $\xi \equiv (x,y,z)$ and $d\xi = dx$, $d\xi = dxdy$, or $d\xi = dxdydz$ respectively. We call the fields of the first two lines *fields of type 0* and the ones of the last two *fields of type 1*.
Futhermore if Φ is a monomial in the above fields containing n_0 factors of type 0, and n_1 factors of type 1, we define:

$$\rho(\Phi) = n_0 \frac{d-2}{2} + n_1 \frac{d}{2} - d + \omega \tag{1.25}$$

where ω is the order of zero of Φ when its arguments coincide (*i.e.* the sum of the order of zero of its non local terms).
And Φ can appear in (1.21) only if $\rho(\Phi) > 0$ unless Φ is local and in the latter case it can only be one among $: \varphi_x^4 :, : \varphi_x^2 :, : (\partial\varphi_x)^2 :, 1$.
With the above notations one proves that the coefficient V_Φ of Φ is expressed as a power series in $\lambda, \alpha, \mu, \nu$ in which any m^{th}–order term $V_{\Phi}^{(h),m}$ is bounded in the sense that if one considers the expression:

$$I = \int_{\Delta_1 \times \ldots \times \Delta_{\tilde{n}}} |V_{\Phi}^{(h),m}(\xi_1, \ldots, \xi_n)| \sup |\Phi(\xi_1, \ldots, \xi_n)| \, d\xi_1 \ldots d\xi_n \tag{1.26}$$

where $\tilde{n}$ =(total number of points appearing in the indices $(\xi_1, \ldots, \xi_n)$). Then I verifies, for suitably chosen constants D, C, b, κ and for *typical* fields Φ (see comments preceding (2.7)):

$$|I| \le \mathcal{N}(\tilde{n}) D C^{m-1} (m-1)! \sum_{j=0}^{m-1} \frac{(bh)^j}{j!} \left[\exp -\kappa \gamma^h d(\Delta_1, \ldots, \Delta_n)\right] \tag{1.27}$$

where $d(\Delta_1, \ldots, \Delta_n)$=(legth of the shortest curve connecting $\Delta_1, \ldots, \Delta_n$), and $\mathcal{N}(\tilde{n})$ depends on the degree $\tilde{n}$ of the monomial Φ (and in fact $\mathcal{N}(\tilde{n}) \le DC^{\tilde{n}-1}(\tilde{n}-1)!$). This shows that in some sense the effective potential stays local, independently on the scale h, [1],[2].

2 Bounds

To prove the above statements one has, first, to give a precise definition of the operators $\mathcal{R}$, $\mathcal{L}$.
Suppose that $V^{(h)}$ has the form (1.22) subject to the conditions stated after it.
Consider $\mathcal{E}_h^T(V^{(h)}, \ldots, V^{(h)})$. We describe it in terms of the Wick rule for the evaluation of gaussian integrals, which can be stated as follows.
Given $\Phi(\xi_1, \ldots, \xi_n)$ we represent it as a ball with n points inside it which have the purpose of symbolizing $\xi_1, \ldots, \xi_n$; and, when $\xi_i = (x, y)$, (*i.e.* the corresponding factor in the monomial Φ is one of the non local fields), we think of linking by a wiggly line the point y to x so that .ξ_i becomes $x \approx y$; and we use a similar notation when $\xi_i = (x, y, z)$. Then out of each ξ_i we draw as many lines as there are fields in Φ with the index ξ_i.
Each of such lines is regarded as distinct from the others and carries a label apt to identify the corresponding field: for instance if we imagine to label $\alpha = 1, 2, \ldots, 8$ the eight fields in (1.23) then a line with a label α will then denote the corresponding field
The above lines will be thought as *half lines* sticking out of the ball representing Φ. Then we decompose $V^{(h)}$ as in (1.21) and use the multilinearity of $\mathcal{E}^T(F_1, \ldots, F_s)$ in the variables $F_1, \ldots, F_s$ to reduce the evaluation of $\mathcal{E}^T(V^{(h)}, \ldots, V^{(h)})$ to that of $\mathcal{E}_h^T(: \Phi(\xi_1^1, \ldots, \xi_{n_1}^1) :, \ldots, : \Phi(\xi_1^s, \ldots, \xi_{n_s}^s) :)$.
The latter is simply evaluated by pairing (*Wick theorem*: see the lectures of L. Rosen) a few of the half lines sticking out of the s balls representing the s functions in $\mathcal{E}_h^T$, and attributing a label h (standing for *hard*) or s (standing for *soft*) to each pair of lines contracted.
No contraction can take place between half lines emerging from the same ball.
Then we are left with a set of half lines which we interpret as a monomial $\tilde{\Phi}(\tilde{\xi}_1, \ldots, \tilde{\xi}_p)$: and a set of lines bearing an index h or s and joining pairs ξ_i, ξ_j which are interpreted as describing the coefficient in front of : $\tilde{\Phi}(\tilde{\xi}_1, \ldots, \tilde{\xi}_p)$: . The coefficient is simply obtained by considering the product over the lines of the covariances:

$$\begin{cases} C^{(<h)}_{\xi\xi'} = \mathcal{E}_{(<h)}(\Phi^{(<h)}_{\xi}\Phi^{(<h)}_{\xi'}) & \text{if the line } (\xi,\xi') \text{ is soft} \\ C^{(h)}_{\xi\xi'} = \mathcal{E}_h(\Phi^{(h)}_{\xi}\Phi^{(h)}_{\xi'}) & \text{if the line } (\xi,\xi') \text{ is hard} \end{cases} \tag{2.1}$$

Finally one should integrate over the ξ's and sum over all graphs of the above type, discarding those for which the hard lines do not allow to form a path connecting any pair of balls.

This completes the description of the Wick rule. It also makes it clear that $V^{(h)}$ has the form (1.22) with the exception that the Φ in it no longer necessarily verify (1.26). For instance one can generate, from $\lambda_h : \varphi_x^4 :$ and $\lambda_h : \varphi_y^4 :$ the terms $\lambda_h^2 : \varphi_x\varphi_y :$ for which $\rho(\Phi) = -2$, *i.e.* it is not positive, and therefore is not compatible with the prescription following (1.24).

From (1.24) it appears that the above problem can arise only if the degree in φ of $\tilde{\Phi}$ is ≤ 4. To remedy this situation is exactly the purpose of the introduction of the operators $\mathcal{L}, \mathcal{R}$. The $\mathcal{R}$ will be so defined that its action on such $\tilde{\Phi}$ will turn them into a linear combimation of a few monomials $\tilde{\Phi}^R$ with $\rho(\tilde{\Phi}^R) > 0$; this can be simply achieved by subtracting from $\tilde{\Phi}$ a local term which, after integration over the space–time indices ξ, takes the form of a linear combination of $: \varphi_x^4 :, : \varphi_x^2, (\partial\varphi_x)^2, 1$.

This is a rather obvious remark and it can be easily checked by making a complete list of all the possible cases.

The list is given in the tables below (eq. (2.4),(2.5)), describing the action of $\mathcal{R}$ and that of $\mathcal{L}$; the latter is more conveniently described by the operation $\bar{\mathcal{L}}$ related to $\mathcal{L}$ by the relation:

$$\mathcal{L}\int \tilde{\Phi} V \equiv \int \bar{\mathcal{L}}\tilde{\Phi} V \tag{2.2}$$

The list is given in detail because it allows to check that each $\tilde{\Phi}$ is changed by $\mathcal{R}$ into a finite sum of Φ–monomials in each of which one can think that some field of type (1.23) in $\tilde{\Phi}$ is changed into a new field which is on a *higher* level in the partial order:

$$\begin{aligned} &\varphi \le D \le \begin{cases} S \le T \\ D^1 \le S^1 \end{cases} \\ &\partial\varphi \le D^1 \le S^1 \le \mathcal{D}^1 \end{aligned} \tag{2.3}$$

The list describing $\mathcal{R}$ is the following:

$$
\begin{aligned}
\mathcal{R} : \varphi_1\varphi_2 : &=: \varphi_1 T_{21} : \\
\mathcal{R} : \varphi_2 D_{12} : &=: \varphi_2 T_{12} : \\
\mathcal{R} : \varphi_2 S_{12} : &=: \varphi_2 T_{12} : \\
\mathcal{R} : \varphi_1 S_{23} : &=: D_{12} S_{23} : + \mathcal{R} : \varphi_2 S_{23} : \\
\mathcal{R} : \varphi_1 S_{12} : &=: D_{12} S_{12} : + : D_{21} T_{12} + : \varphi_1 T_{12} : \\
\mathcal{R} : \varphi_1 D_{32} : &= \mathcal{R} : D_{12} D_{32} : + \mathcal{R} : \varphi_2 D_{32} : \\
&= - : S_{12} S_{32} : + : S_{12} D_{32} : + : D_{12} S_{32} : + \\
&\quad + : D_{21} T_{32} : + : \varphi_1 T_{32} : \\
\mathcal{R} : D_{12} D_{32} : &= - : S_{12} S_{32} : + : S_{12} D_{32} : + : D_{12} S_{32} : \\
\mathcal{R} : D_{12} D_{34} : &= - : S_{12} \mathcal{D}^1_{434} : + : D_{12} \mathcal{D}^1_{234} : - : S_{12} S_{34} + \\
&\quad + : S_{12} D_{34} : + : D_{12} S_{34} : \\
\mathcal{R} : \varphi_1 \partial\varphi_2 : &=: \varphi_1 S^1_{21} : \\
\mathcal{R} : \partial\varphi_1 \partial\varphi_2 : &=: \varphi_1 D^1_{21} : \\
\mathcal{R} : \partial\varphi_1 D_{21} : &=: \partial\varphi_1 S^1_{21} : \\
\mathcal{R} : \varphi_1 D^1_{21} : &=: \partial\varphi_1 S^1_{21} : \\
\mathcal{R} : \varphi_3 D^1_{21} : &=: D_{31} D^1_{21} : + : D_{13} S^1_{21} : + : \varphi_3 S^1_{21} : \\
\mathcal{R} : \partial\varphi_1 D_{23} : &=: D^1_{13} D_{23} : + : \partial\varphi_3 S_{23} : \\
\mathcal{R} : \varphi_1^2 \varphi_2^2 : &=: \varphi_1^3 D_{21} : + : \varphi_1^2 \varphi_2 D_{21} : \\
\mathcal{R} : \varphi_1 \varphi_2^3 : &=: D_{12} \varphi_2^3 : \\
\mathcal{R} : \varphi_1 \varphi_2^2 \varphi_3 : &=: D_{12} \varphi_2^2 \varphi_3 : + : D_{21} \varphi_2^2 D_{32} : + : \varphi_1 \varphi_2^2 D_{32} : \\
\mathcal{R} : \varphi_1\varphi_2\varphi_3\varphi_4 : &=: \varphi_1 D_{21} \varphi_3 \varphi_4 : + : \varphi_1 D_{12} D_{31} \varphi_4 : + \\
&\quad + : \varphi_1 \varphi_2 D_{31} \varphi_4 : + : \varphi_1 D_{12} D_{13} D_{41} : + \\
&\quad + \varphi_1 \varphi_2 D_{13} D_{41} : + : \varphi_1 D_{12} \varphi_3 D_{41} : + \\
&\quad + : \varphi_1 \varphi_2 \varphi_3 D_{41} :
\end{aligned}
\tag{2.4}
$$

and the action of $\mathcal{R}$ on monomials which do not differ by a sign from the ones in the above list leaves them unchanged, while $\mathcal{R}\Phi = -\mathcal{R}(-\Phi)$ if Φ differs by a sign from one of the above monomials. The operation $\mathcal{R}$ is defined to commute with the scalars. We see that $\mathcal{R}$ acts on $: \varphi_1\varphi_2$ (first line of (2.4) by *changing the meaning* of φ_2 into that of the higher level field T_{21}; and from the second line we see that $\mathcal{R}$ acts on $: \varphi_2 D_{12} :$ by changing D_{12} into the higher level field T_{12}, etc..
The list describing the corresponding action of $\mathcal{L} \equiv (1 - \mathcal{R})$ is:

$$
\begin{aligned}
\bar{\mathcal{L}}1 &= 1\\
\bar{\mathcal{L}} : \varphi_1\varphi_2 :&= : \varphi_1(\varphi_1 + \delta_{21}\partial\varphi_1 + \frac{1}{2}(\delta_{21}^2 \times \partial^2\varphi_1) :\\
\bar{\mathcal{L}} : \partial\varphi_1\partial\varphi_2 :&= : \partial\varphi_1\partial\varphi_1 :\\
\bar{\mathcal{L}} : \varphi_1\partial\varphi_2 :&= : \varphi_1(\partial\varphi_1 + \delta_{21}\cdot\partial\partial\varphi_1) :\\
\bar{\mathcal{L}} : \varphi_1 D_{21} :&= : \varphi_1(\delta_{21}\cdot\partial\varphi_1 + \frac{1}{2}(\delta_{21}^2 \times \partial^2\varphi_2)) :\\
\bar{\mathcal{L}} : D_{13}D_{23} :&= : \delta_{13}\cdot\partial\varphi_3\,\delta_{23}\partial\varphi_3 :\\
\bar{\mathcal{L}} : \varphi_1 D_{23} :&= : \delta_{13}\cdot\partial\varphi_3\,\delta_{23}\partial\varphi_3 : + : \varphi_3(\delta_{23}^2 \times \partial^2\varphi_3) :\\
\bar{\mathcal{L}} : D_{12}D_{34} :&= : \delta_{12}\cdot\partial\varphi_2\,\delta_{34}\cdot\partial\varphi_2 :\\
\bar{\mathcal{L}} : \varphi_1 S_{21} :&= \frac{1}{2} : \varphi_1\,(\delta_{21}^2 \times \partial^2\varphi_2) :\\
\bar{\mathcal{L}} : \varphi_1 S_{23} :&= \frac{1}{2} : \varphi_3\,(\delta_{23}^2 \times \partial^2\varphi_3) :\\
\bar{\mathcal{L}} : \varphi_1 D_{21}^1 :&= : \varphi_1\delta_{21}\cdot\partial\,\partial\varphi_1 :\\
\bar{\mathcal{L}} : \varphi_1 D_{23}^1 :&= : \varphi_3\delta_{23}\cdot\partial\,\partial\varphi_3 :\\
\bar{\mathcal{L}} : \partial\varphi_1 D_{21} :&= : \partial\varphi_1\delta_{21}\cdot\partial\varphi_1 :\\
\bar{\mathcal{L}} : \partial\varphi_1 D_{23} :&= : \partial\varphi_3\delta_{23}\cdot\partial\,\varphi_3 :\\
\bar{\mathcal{L}} : \varphi_1\varphi_2^3 :&= : \varphi_2^4 :\\
\bar{\mathcal{L}} : \varphi_1^2\varphi_2^2 :&= : \varphi_1^4 :\\
\bar{\mathcal{L}} : \varphi_1\varphi_2^2\varphi_3 :&= : \varphi_2^4 :\\
\bar{\mathcal{L}} : \varphi_1\varphi_2\varphi_3\varphi_4 :&= : \varphi_1^4 :
\end{aligned}
\tag{2.5}
$$

and the action of $\bar{\mathcal{L}}$ on monomials which do not differ by a sign from the ones in the above list changes them into 0, while $\bar{\mathcal{L}}\Phi = -\bar{\mathcal{L}}(-\Phi)$ if Φ differs by a sign from one of the above monomials.

We now proceed to estimate recursively the coefficients in (1.22) using their definition via (1.15) and the above definitions of $\mathcal{R}$, $\mathcal{L}$.

We fix a tree γ with n endpoints and we draw a Feynman graph $\bar{G}$ containing n vertices with $n = m_0 + m_2 + m_{2'}$ where m_0 =(number of vertices in $\bar{G}$ with four lines), m_2 =(number of vertices with 2 lines not representing gradients), $m_{2'}$ =(number of vertices with 2 lines representing gradients).

Then we isolate in $\bar{G}$ clusters of vertices hierarchically ordered so that the partial ordering in γ is *represented* by the inclusion relation between the clusters. This allows

us to identify the clusters of vertices of $\bar{G}$ with vertices of the tree γ and, if $v \in \gamma$ has scale index h_v, we think, to help the intuition, of the corresponding clusters of vertices as enclosed ideally in a ball which has roughly scale γ^{-h_v}. In this picture we identify the clusters of the $\bar{G}$ graph and the corresponding vertices of the tree γ.
Remark that there are $O(n!)$ topologically distinct graphs G: but the same *topological* graph (*i.e.* a Feynman graph with vertices and lines carrying no labels) can be labeled in many ways (putting space time indices $x_1, \ldots, x_n$ on its vertices, h or s indices on its internal lines and all the extra indices needed to identify the fields associated with each of the half lines of the graph) giving rise to a labeled graph $\bar{G}$ which can arise when evaluating the contributions to the effective potentials or to the running couplings, of the tree γ under consideration.
We remind that, in a labeled graph $\bar{G}$, labels h, s are attached, by construction, to the lines of $\bar{G}$ so that the points inside a cluster, of any scale, are connected by a path of lines with label h.
Each line is given a frequency index corresponding to the frequency label of the tree vertex v such that the corresponding cluster contains the endpoints of the line but no other inner vertex does.
Then we look at how many ways there are to append the labels $x_1, \ldots, x_n$ to the vertices of a graph G with the same topological structure as a fixed $\bar{G}$ so that the resulting graph is still compatible with the tree γ (*i.e.* it can arise in the way just described).
A lemma (Felder lemma) shows that if one denotes n_v^e the number of lines emerging from the subgraph of any possible labeled graph associated with the subtree $\gamma_v \subseteq \gamma$ starting with root in v, then the number of ways in which one can put labels on G compatibly with γ and with the given n_v^e is bounded simply by:

$$C_\varepsilon^n \prod_{v \in \gamma} e^{\varepsilon n_v^e} s_v! \qquad \forall \varepsilon > 0 \tag{2.6}$$

for a suitably chosen C_ε, if s_v=(number of tree branches emerging from v in γ).
We now imagine to fix γ, G and one of the labelings of G. Starting from the innermost clusters we look at the clusters with only 2 or 4 (or 0) external lines and we modify the meaning of some of the external lines according to the prescriptions given by the operation $\mathcal{R}$. And we continue iteratively by examining the clusters which are next to the innermost, *etc.*
Since the operation $\mathcal{R}$ replaces a product of fields by a finite linear combination of them we see that any labeling of G consistent with γ generates at most C^n new graphs in which the meaning of the lines is different (and $C = 7$ as in the list (2.4) no monomial generates more than 7 other monomials).
We fix the attention on one of such terms: finding a bound on the contribution I of the selected term will, mean to replace the external fields by an estimate of their typical

size which will be supposed to be described by the weights in the norms:

$$||\varphi^{(\leq k)}||_\Delta = \sup_{\xi\in\Delta} |\varphi_\xi^{(\leq k)}|\gamma^{-k(d-2)/2}$$

$$||\partial\varphi^{(\leq k)}||_\Delta = \sup_{\xi\in\Delta} |\partial\varphi_\xi^{(\leq k)}||\gamma^{-kd/2}$$

$$||D^{(\leq k)}||_\Delta = \sup_{\substack{x\in\Delta \\ y\in R^4 \\ |x-y|\leq\gamma^{-h}}} \frac{|D_{xy}^{(\leq k)}|}{(\gamma^k|x-y|)}\gamma^{-k(d-2)/2} \tag{2.7}$$

$$||S^{(\leq k)}||_\Delta = \sup_{\substack{x\in\Delta \\ y\in R^4 \\ |x-y|\leq\gamma^{-h}}} \frac{|S_{xy}^{(\leq k)}|}{(\gamma^k|x-y|)^2}\gamma^{-k(d-2)/2} \qquad etc.$$

This means that in estimating I we replace the Wick ordered product of the fields representing the external lines by the product of the norms of the fields times the reciprocal of the corresponding weight factors and, finally, we replace the norms by 1 (which is their typical size in our normalizations). The result is the following bound (where the origin of the various terms is rather easy to recognize):

$$I \leq \gamma^{n_o^e\, h\,(d-2)/2+n_1^e\, h\, d/2} \int_{\Delta_1\times\cdots\times\Delta_n}$$

$$\prod_{v\in\gamma}\Big(\frac{1}{s_v!}\gamma^{(n_{o,v}^{inner}(d-2)/2+n_{1,v}^{inner}d/2)\,h_v} e^{-\kappa\sum_{\lambda\ hard\ \in v}\gamma^{h_v}d(\lambda)}\Big) \tag{2.8}$$

$$\prod(\gamma^{h_{v_j}}d(\tilde v_j))^{\omega_j}\prod_{i\in I_2}(\gamma^{2h_i}|\mu_{h_i}|)\prod_{i\in I_4}(\gamma^{(4-d)h_i}|\lambda_{h_i}|)(\prod_{i\in I_{2'}}|\alpha_{h_i}|)$$

where $n_{\alpha,v}^{inner}$ =(number of lines of type α inner to the cluster v but not to any smaller one), ω_j =(degree of zero carried by the fields which are outer lines to some $\tilde v_j$ and as such were changed by the $\mathcal{R}$ action for the first time and, furthermore, become inner at the vertex v_j (possibly $v_j = r$, when they are in fact external lines)). Finally we remark that the combinatorial factor $\prod_v s_v!$ arises from the factorials in (1.10), (1.12).
The integral is over boxes $\Delta_1,\ldots,\Delta_n$ of size γ^{-h} and the above formula is also valid if the tree starts with a first non trivial vertex v_o bearing an index L rather than R: however in the latter case it must be, by construction, $h_{v_o} \equiv h+1$.
The exponential decay is used to replace $\gamma^{h_{v_j}}d(\tilde v_j)$ by $const\cdot\gamma^{-(h_{\tilde v_j}-h_{v_j})\omega_j}$ and, conse-

quently, to bound the result of the integrals by the expression $\gamma^{-dh}\prod_{v\geq v_o}\gamma^{-d\,h_v(s_v-1)}$, where s_v=(number of branches emerging from the vertex v in the tree γ). Using $\prod_{v\geq v_o}\gamma^{-d\,h_v(s_v-1)} = \gamma^{-d\,h(n-1)}\prod_{v\geq v_0}\gamma^{-d(h_v-h_{v'})(n_v-1)}$, with n_v=(number of end-points of the tree γ that can be reached from v, and adding and subtracting h to each h_v we find, after introducing the numbers $m_{o,v}, m_{2,v}, m_{2',v}$ of vertices with four lines, with two non gradient lines or with two gradient lines in the subgraph G_v of G constructed with the vertices of the cluster v only and after some elementary algebra:

$$I \leq \gamma^{[(4m_o+2m_2)(d-2)/2+2m_{2'}d/2]h}\gamma^{2m_2\,h}\gamma^{-m_o(d-4)h}\int_{\Delta_1\times\ldots\Delta_n}$$

$$\prod_{v\in\gamma}\frac{1}{s_v!}\gamma^{\left[(4m_{o,v}+2m_{1,v}-n^e_{vo})(d-2)/2+(2m_{2',v}-n^e_{v1})d/2-\omega_v\right](h_v-h'_v)} \tag{2.9}$$

$$e^{\left\{-\frac{\kappa}{1-\gamma}\sum_{\lambda\in\ hard\ \in v}(\gamma^{h_v}+\gamma^{h_v-1}+\ldots+\gamma^{h})d(\lambda)\right\}}\prod|\lambda_{h_j}\mu_{h_i}\alpha_{h_k}|$$

$$\leq\prod_{v\in\gamma}\frac{1}{s_v!}\gamma^{-(h_v-h_{v'})(n^e_{vo}(d-2)/2+n^e_{v1}d/2-d+\omega_v)}e^{-\kappa' d(\Delta_1,\ldots,\Delta_n)}\prod|\lambda_{h_j}\mu_{h_i}\alpha_{h_k}|$$

where the identities:

$$n_v = m_{0v}+m_{2v}+m_{2'v},$$

$$\sum_{v'\geq v} n^{inner}_{v'0} = 4m_{v0}+2m_{v2}-n^e_{v0}, \tag{2.10}$$

$$\sum_{v'\geq v} n^{inner}_{2'v'} = 2m_{2'v}-n^e_{1v}$$

have been used, $\kappa' = \kappa/(1-\gamma)$, $d(\Delta_1,\ldots,\Delta_n)$ is the length of the shortest path connecting $\Delta_1,\ldots,\Delta_n$, and ω_v, by our choice, is such that:

$$\rho_v = n^e_{vo}(d-2)/2+n^e_{v1}d/2+\omega_v-d\geq 1 \qquad \text{for } v>v_o \tag{2.11}$$

Hence we can combine the bound (2.9) togheter with (2.6) and remark that $-\rho_v(h_v - h_{v'})+\varepsilon n^e_v \leq -(h_v-h_{v'})/2$ if ε is small enough (here it is essential that $(h_v-h_{v'})\geq 1$ and $n^e_v = n^e_{v0}+n^e_{v1}$).

We also remark that the number of topologically distinct graphs with n vertices is bounded proportionally to $n!$. The consequence of the above remarks is, as is easily seen, that, for some $|\vartheta_a|<1$, it is:

$$\lambda_h = \lambda_{h+1} + \vartheta_l \sum_{\substack{G,\gamma;\, h_{v_o}\equiv h+1 \\ \{n^e_{vo}\}\{n^e_{v1}\}}} C^n_\varepsilon \prod_{v\geq v_o} \gamma^{-(h_v - h_{v'})/2} \left|\prod \lambda_{h_{v_i}} \mu_{h_{v_j}} \alpha_{h_{v_k}}\right|$$

$$\mu_h = \gamma^2 \mu_{h+1} + \vartheta_m \sum_{\substack{G,\gamma;\, h_{v_o}\equiv h+1 \\ \{n^e_{vo}\}\{n^e_{v1}\}}} C^n_\varepsilon \prod_{v\geq v_o} \gamma^{-(h_v - h_{v'})/2} \left|\prod \lambda_{h_{v_i}} \mu_{h_{v_j}} \alpha_{h_{v_k}}\right|$$

$$\alpha_h = \alpha_{h+1} + \vartheta_a \sum_{\substack{G,\gamma;\, h_{v_o}\equiv h+1 \\ \{n^e_{vo}\}\{n^e_{v1}\}}} C^n_\varepsilon \prod_{v\geq v_o} \gamma^{-(h_v - h_{v'})/2} \left|\prod \lambda_{h_{v_i}} \mu_{h_{v_j}} \alpha_{h_{v_k}}\right| \tag{2.12}$$

$$\nu_h = \gamma^4 \nu_{h+1} + \vartheta_n \sum_{\substack{G,\gamma;\, h_{v_o}\equiv h+1 \\ \{n^e_{vo}\}\{n^e_{v1}\}}} C^n_\varepsilon \prod_{v\geq v_o} \gamma^{-(h_v - h_{v'})/2} \left|\prod \lambda_{h_{v_i}} \mu_{h_{v_j}} \alpha_{h_{v_k}}\right|$$

We write $(\lambda_h, \mu_h, \alpha_h, \nu_h) = \underline{r}_h$ and $\underline{r}_h$ as a formal power series in $(\lambda_o, \mu_o, \alpha_o, \nu_o)$; we denote $\underline{r}_h^{pqrs}$ the coefficients of $\lambda^p \mu^q \alpha^r \nu^s$ and suppose that:

$$\sum_{p+q+r+s=n} |\underline{r}^{pqrs}| \leq DC^{n-1}(n-1)! \sum_{j=0}^{n-1} \frac{(bh)^j}{j!} \tag{2.13}$$

this property is clearly true for $p+q+r=1$ and since the r.h.s. involves, in the non trivial part (*i.e.* in the non linear part) terms of degree less than n in the variables $\underline{r}$: hence one can try some kind of induction; see §3 below where an essentially inductive proof is sketched.

Let us suppose (2.13) valid. Then, since (2.8), (2.9) hold also for $v = v_0$ when the tree has index R over v_0 one also finds that the coefficients of the $m-th$ order contribution to the effective potential is bounded by:

$$I \leq e^{-\kappa' d(\Delta_1,\ldots,\Delta_n)} \sum_{\sum n_j = m} \sum_{(\underline{h})} \prod_{v\geq v_o} \gamma^{-(h_v - h_{v'})/2}$$

$$\prod_{j=1}^{n} (n_j - 1)! \sum_{q=0}^{n_j-1} \frac{(bh_j)^q}{q!} \tag{2.14}$$

which gives (1.26), via the inequality:

$$\sum_{h_1,\ldots,h_m \geq h} \prod_{i=1}^{m} (n_j - 1)! \sum_{q=0}^{n_j-1} \gamma^{-(h_i - k)} \frac{(bh_j)^q}{q!} \leq \bar{D}^m (n-m)! \sum_{p=0}^{n-m} \frac{(bk)^p}{p!} \tag{2.15}$$

valid for suitably chosen $b, \bar{D}$. This is, essentially, the $n!$ bound of De Calan-Rivasseau, see [2], ((19.8) and appendix F).

Since (1.21) is a special case of (1.19) when $h = 1$ and since the convergence statements (as $\Lambda, N \to \infty$) turn out to be a trivial consequence of the arguments leading to the bounds (1.21) we see that the basic bound is really (2.11)

In the next section we analyze (1.20), (1.21).

3 The beta function

The relation (2.12) can be graphically rewritten as:

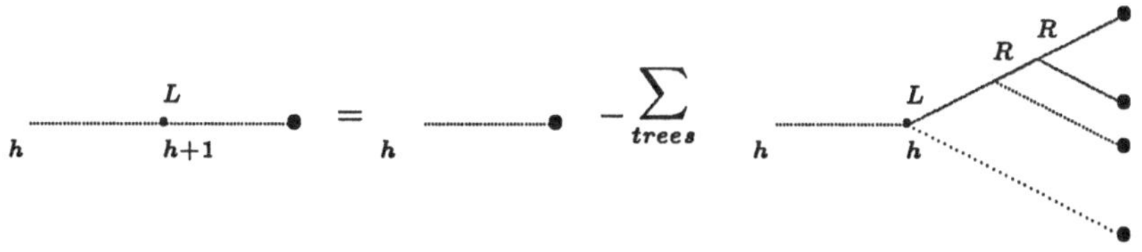

$$\underset{h}{\underset{\;}{\rule{0pt}{0pt}}} \overset{L}{\underline{\qquad}}\!\!\bullet \;=\; \underset{h}{\underline{\qquad}}\!\!\bullet \;-\sum_{trees}\; \underset{h}{\underline{\qquad}}\!\!\overset{L}{\diagup}\!\!\!\!\diagup \qquad (3.1)$$

where the $\sum_{trees}$ means summation over the trees with R labels on all the non trivial vertices except the first and over all the scale indices. We want to deduce (1.20) from the above relation, by suitably iterating it replacing the endlines $\underline{\qquad}\!\bullet$ of the tree in the r.h.s. by the (3.1) itself. For this purpose it will be convenient to introduce the operation $\mathcal{U}_p$ which acting on a function of the fields on scale $q \le k$ simply changes it in the same function of the fields on scale k: $\mathcal{U}_k \Phi^{(\le q)} = \Phi^{(\le k)}$. The new notion permits us to rewrite (3.1) as:

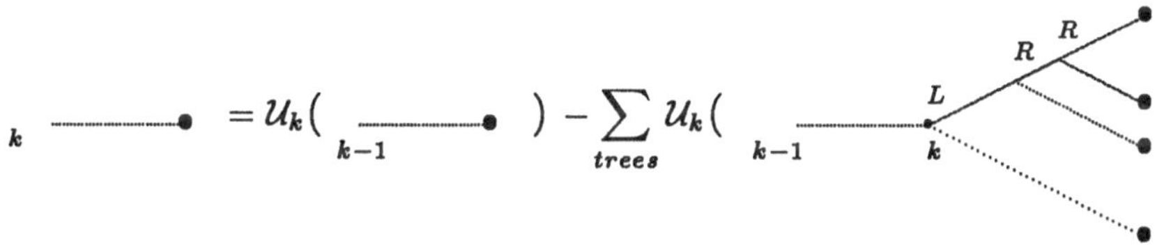

$$\underset{k}{\underline{\qquad}}\!\!\bullet \;=\; \mathcal{U}_k \Big(\underset{k-1}{\underline{\qquad}}\!\!\bullet \Big) - \sum_{trees} \mathcal{U}_k \Big(\underset{k-1}{\underline{\qquad}}\!\!\overset{L}{\underset{k}{\diagup}} \Big) \qquad (3.2)$$

We reexpress $\underset{k-1}{\underline{\qquad}}\!\bullet$ via (3.2) itself $k - 1 - h$ times obtaining:

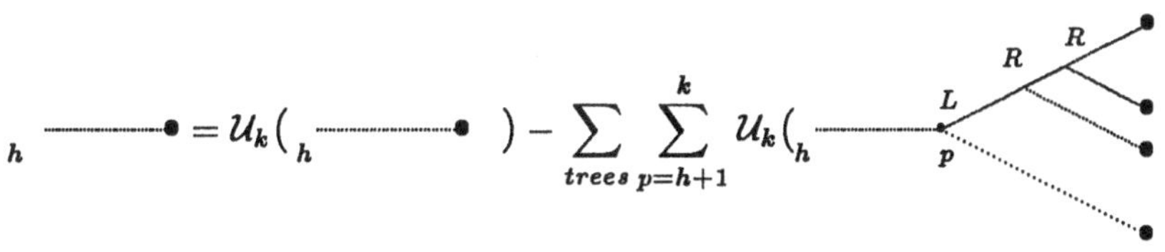

$$\underset{h}{\underline{\qquad}}\!\!\bullet \;=\; \mathcal{U}_k \Big(\underset{h}{\underline{\qquad}}\!\!\bullet \Big) - \sum_{trees} \sum_{p=h+1}^{k} \mathcal{U}_k \Big(\underset{h}{\underline{\qquad}}\!\!\overset{L}{\underset{p}{\diagup}} \Big) \qquad (3.3)$$

If we call $\Lambda =$ (diagonal matrix with diagonal elements $(1, \gamma^2, 1, \gamma^4)$) we can write (3.3) as:

$$\Lambda^k \underline{r}_k = \Lambda^h \underline{r}_h - \sum_{trees} \sum_{p=h+1}^{k} \Lambda^{p-1} (\qquad)^* \tag{3.4}$$

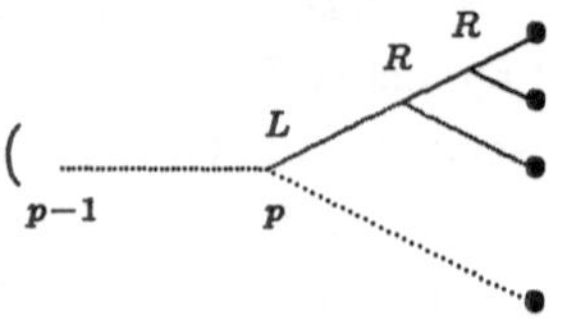

where ( )* denotes the contribution to $\underline{r}_{p-1}$ of the tree in parenthesis.

By iteration (3.4) leads to identify the nth order of the expansion of $\underline{r}_k$ in powers of $\underline{r}_h$ as a sum over all trees with m endlines bearing index L on the first non trivial vertex and indices R or L on the other inner vertices. If v is a vertex bearing a frequency index h_v and an index L then h_v is now subject to the constraint $h_v \leq h_{v'}$ where v' denotes the vertex preceding v in the tree: in this rule one gives frequency $h_r = k$ to the root. Also the last non trivial vertices bear a scale index h. Choosing $k = h+1$ we can therefore express this by a relation:

$$\underline{r}_{h+1} = \Lambda \underline{r}_h + B(\underline{r}_h) \tag{3.5}$$

where B is a formal power series in $\underline{r}_h$. Its coefficients are expressed in terms of trees and graphs as explained in the definition of the symbols; they can be shown (see below) to be bounded uniformly in the cut off N by using (2.11), (2.12) thus defining the *beta function*.

In fact we denote σ one of the possible tree shapes (*i.e.* a tree with no scale labels but only with the L, R labels subject to the above constraints) and we shall call $r_k^{(\alpha)pqrs}(\sigma)$ its contribution to the coefficient of the expansion of $\underline{r}_k$ in powers of $\underline{r}_h$ of order p, q, r, s in $\lambda_h . \mu_h, \alpha_h, \nu_h$ respectively.

Let:

$$r_k^n(\sigma) = \sup_{\alpha} \sup_{p+q+r+s=n} |r^{(\alpha)pqrs}(\sigma)|$$

then we deduce from the above mentioned iteration of (3.4) (and from $r_k^1 = 1$) the:

$$r_k^n(\sigma) \leq \bar{D}\bar{C}^{m-1}(m-1)! \sum_{\underline{h}} \prod_{\substack{v > v_0 \\ v \in \mathcal{R}}} \gamma^{-(h_v - h_{v'})/2} \prod_{i=1}^{m} r_{h_i}^{n_i}(\sigma_i) \tag{3.6}$$

where the sum runs over all the scale indices that can be put on the set $\mathcal{R}$ of vertices of σ which are between the first non trivial vertex and the first next vertices (supposed to be m in number, and denoted $1, 2, \ldots, m$) bearing an index L and σ_i denote the m subtrees that follow the mentioned m vertices. Of course $\sum n_i = n$.

The expression (3.6) can be shown to verify (2.13) by a simple induction argument based on the following key inequalities:

$$\sum_{\underline{h}_v}\prod_{v>r}\gamma^{-(h_v-h_{v'})/2}\prod_{j=1}^{m}[(n_j-1)!\sum_{p=0}^{n_j-1}\frac{(b(h_j-h)^p}{p!}]$$

$$\le \tilde D^m(n-m)!\sum_{p=0}^{n-m}\frac{(b(k-h)^p}{p!}] \tag{3.7}$$

and

$$\prod_{s=1}^{q}[a_s!\sum\frac{(bh)^j}{j!}]\le(\sum a_j)!\sum_{r=0}^{\sum a_j}\frac{(bh)^r}{r!} \tag{3.8}$$

and:

$$\frac{n_1!\dots n_m!}{(n_1+\dots+n_m)!}\sum_{\substack{j_1=0\\ j_1+\dots+j_m=q}}^{n_1}\dots\sum_{j_1=0}^{n_1}\frac{q!}{j_1!\dots j_m!}\le 1\quad\forall n_i,q \tag{3.9}$$

for the details see [2],§19 ((19.10) – (19.17) and appendix E).
The basic problem in the above proof is in the correct guess of the bound (2.13) to be proved: once it is known the check is routine work.
This completes the inductive argument and the estimates leading to (2.13) and determines the expansion of $\underline{r}_{h+1}$ in powers of $\underline{r}_h$ and shows that the computation of the n-th order coefficients of the beta function can be performed in terms of Feynman graphs in a way essentially identical to the one of the effective couplings and potentials.
It is now easy to prove that the *planar* φ_4^4 *theory* can be defined by a convergent expansion if $\lambda_0 > 0$.
In fact the number of topologically distinct planar Feynman graphs is simply bounded by C^n for some C. This eliminates the $n!$ in (2.13),(1.21), which from the previous discussion (c.f.r. the discussion before (2.6)) was coming from the count of the topologically different Feynman graphs with n vertices of type φ^4, φ^2, $(\partial\varphi)^2$. As a consequence, if $|\underline{r}_h| = \max(|\lambda_h|,|\mu_h|,|\alpha_h|,|\nu_h|) < \varepsilon$ and ε is small enough, the power series expressing the effective potentials in powers of the coupling constants $\{\underline{r}_h\}_{h=0}^{\infty}$ found in §2 is convergent and, if the $\{\underline{r}_h\}_{h=0}^{\infty}$ are defined separately and $< \varepsilon$, this can be used as a definition of the theory.
There is a natural way of defining $\underline{r}_h$ if $\underline{r}_0$ is small enough; one simply sets:

$$\underline{r}_h = (\Lambda + B)^h(\underline{r}_0) \tag{3.10}$$

where Λ denotes the linear part in the beta function (3.5).
In order to make this definition correct one has to check that the r.h.s. never looses meaning if $\underline{r}_o$ is small enough: *i.e.* one has to check that the points on the trajectory

$\underline{r}_h$ of the dynamical system defined by the map $(\Lambda + B)$, given in (3.4), stay less than ε for all h if $\underline{r}_0$ is small enough.

One writes, for the analysis of the latter question, the first and second order parts of $\Lambda + B$ considered as a map of R^4 into itself; one finds that the non vanishing terms are *essentially*:

$$\begin{aligned} \lambda' &= \lambda - \beta_0\lambda^2 - \beta_1\lambda\mu \\ \mu' &= \gamma^{-2}\mu + \beta_2\lambda^2 + \beta_3\mu^2 - \beta_4\alpha\mu \\ \alpha' &= \alpha + \beta_5\lambda^2 + \beta_6\mu^2 - \beta_7\mu\alpha \\ \nu' &= \gamma^{-4}\nu - \beta_8\lambda^2 - \beta_9\mu\alpha - \beta_{10}\alpha^2 - \beta_{11}\mu^2 \end{aligned} \tag{3.11}$$

where "essentially" means that in fact the coefficients depend on h but the $\beta_i^{(h)}$ converge to their limits β_i as $h \to \infty$ (exponentially fast): we neglect here this dependence on h as it would only change the discussion in a trivial way.

It is an essential property of the scalar theory that $\beta_0 > 0$: it implies that if $\lambda < 0$ and small then the iterations of (3.5) make it grow. This makes it difficult to study the case $\lambda < 0$, considered the physically interesting one (see below).

The above system, given $\underline{r}_0 = (\lambda_0, \mu_0, \alpha_0, \nu_0)$, admits, as it is easy to verify, a trajectory $\underline{r}_h \xrightarrow[h\to\infty]{} 0$ if $\lambda_0 > 0$ and $|\underline{r}_0| < \varepsilon_1 \leq \varepsilon$ provided μ_0, ν_0 are suitably chosen as functions of λ_0, μ_0. And the same holds, on the basis of general perturbation theory and stability theory in dynamical systems, for the trajectory of the true dynamical system $(\Lambda + B)$ if $|\underline{r}_0| < \varepsilon_2$ (possibly $\varepsilon_2 << \varepsilon_1$).

But the $\underline{r}_h$ depend on $\underline{r}_0$ in a non analytic way. Consider, for simplicity the case $\lambda_o = \lambda > 0$, $\alpha_0 = 0$. Then $\lambda_h(\lambda)$ is not analytic in λ: this can be seen by considering a simplified version of the dynamical system (3.11), namely the map $\lambda' = \lambda - \beta_0\lambda^2$ or, even simpler, the well known evolution equation:

$$\dot{\lambda} = -\beta_0\lambda^2 \tag{3.12}$$

whose solution is the function:

$$\lambda_h = \frac{\lambda}{1 + \beta_0 h\lambda} \qquad h \geq 0 \tag{3.13}$$

and we see that the family of functions $\{\lambda_h(\lambda)\}$ has singularities in λ as close to 0 as wanted, but it is well defined for $\Re\lambda > 0$.

In fact it is possible to study the analitycity properties of the true trajectory (starting from the (3.5) and using somewhat in detail the structure of B and the detailed bounds (2.12) and prove that the $\underline{r}_h$ are analytic functions of λ for $\Re\lambda > 0$ and, furthermore, they are Borel summable at the origin.

It follows that also the effective potentials computed on smooth samples of the external fields, supposed to have the norms (2.7) equal to 1 are analytic in λ for $\Re\lambda > 0$ and Borel summable, (which can be expressed as convergent power series in the $\underline{r}_h$). This shows that the planar theory is fully under control. Unfortunately it has a not very clear physical relevance.

The main problem left concerns the full, non planar, theory: can one give a meaning

to the beta function as a dynamical system in R^4? It seems evident that as soon as one succeeds in this , by the same token, one will be able to give a meaning to the effective potentials and, hence, to the theory itself. The basic difficulty is that the above scheme seems to require $\lambda > 0$ in order to work, and in turn this looks quite unacceptable. It led Landau to conjecture that a scalar theory verifying (in modern terminology) the Osterwalder-Schrader axioms and depending on a parameter λ so that the Schwinger functions have an asymmptotic series decribed by the perturbation theory of φ_4^4, developed above, is necessarily trivial, *i.e.* it has vanishing truncated functions of order ≥ 4 (in other words is gaussian), [3].
However this conjecture is far from being proved: it can, nevertheless, be checked (see, for instance, [4]), under suitably strong extra requirements, which can be considered too restrictive for a truly general statement on the Landau conjecture.
The reason for the hardness of the problem is that one has necessarily to understand what happens when one starts constructing the theory with a bare interaction which is not small: if one believes that the (1.20) has at least an asymptotic validity when the interaction is small, then one should believe that in the same situation the (3.5) should describe quite accurately the flow of (1.20) with which it coincides (essentially) up to second order. But the (3.5) shows that the origin is repulsive along the positive λ axis: hence if $\lambda_0 > 0$ it is inconsistent to suppose that for large h the $\underline{r}_h$ can become smaller and tend to 0 as $h \to \infty$ as it should be necessary in order to prove that the errors introduced by using perturbation theory to some finite order are neglegible (one says that φ_4^4 *is not asymptotically free*). But if $\underline{r}_h$ does not vanish as $h \to \infty$ it follows that one cannot use perturbation theory to study the ultraviolet region where, in fact very strange and new phenomena could happen which could depend quite dramatically also on the type of regularization one has chosen or, more generally, on the chosen bare interaction. This is the point where the triviality proofs make special assumptions which it would be desirable to avoid.
Unfortunately there seems to be no way, yet, to study the beta function and its regularization dependence in a non perturbative fashion and it would be nice to understand the above point at least in the frame of the hierarchical models where, recently, important results have been found and important techniques have been developed, [5].

4 *Remarks on non renormalizable theories*

If $d > 4$ the theory is *non renormalizable*: but most of the discussion in the above sections goes unchanged [6].
In particular one can prove the bounds (2.12) and the (2.13) with the essential difference that λ_{h+1}, ν_{h+1} are now replaced in the r.h.s. of (2.12) by $\gamma^{(4-d)}\lambda_{h+1}, \gamma^d \nu_{h+1}$.
We could then try to deduce formally (3.5) from (2.12) as explained in the beginning of §3. However the matrix Λ^{-1} has now matrix elements $\gamma^{(d-4)}, \gamma^{-2}, 1, \gamma^{-d}\nu_{h+1}$ so that it is no longer ≤ 1 and we could no longer prove uniform bounds on the coefficients of B because Λ^{p-k-1} in (3.4) cannot be bounded by 1 as it is in fact necessary to do in bounding B as described after (3.4) using the bounds (2.11), (2.12), [6].

Hence in the non renormalizable case one has to stop at the (2.11), (2.12).
This however shows that the divergences that arise in the non renormalizable cases are all summarized in the fact that $\underline{r}_h$ do not admit a formal power series expansion in $\underline{r}_0$: the effective potentials are well behaved in the $\underline{r}_h$ (*i.e.* in all dimensions $d \geq 4$ they have the same behaviour and bounds as formal power series in the full family of $\underline{r}_h$ (with $h = 0, 1, \ldots$)).
One could try to use (2.12) as a substitute for the non existent beta function: in fact if one could find a sequence $\{\underline{r}_h\}_{h=0}^{\infty}$ of constants verifying (2.12) at least in the sense of formal power series then one could give a meaning to the perturbation theory by saying that the effective potentials are given by the (perfectly well defined and finite) expressions in terms of trees and graphs as formal power series in the running coupling constants.
A simple way of producing a sequence verifying (2.12) would be to find a fixed popint: $\underline{r}_h \equiv \underline{r}_0, \quad \forall h = 0, \ldots$. However there seems to be no way to attack this problem other than giving to the r.h.s. of (2.12) a non perturbative meaning. This can be done in the case of the *planar* φ^4 theory, [7], where the r.h.s. of (2.12) is easily shown to be analytic in $\{\underline{r}_h\}_{h=0}^{\infty}$ if $|\underline{r}_h| \leq \eta$ for η small enough.
In this way in [7] one succeeds in giving a meaning to φ_d^4, $d = 4 + \varepsilon$, if ε is small enough and if the theory in $4 + \varepsilon$ dimensions is defined as a theory in dimension 4 with a free propagator which in momentum space is $(1 + p^2)^{-2+2\varepsilon}$ (which changes the third diagonal element of the matrix Λ from 1 to γ^{ε}).

References

[1] Gallavotti, G, Nicolò, F.:*Renormalization theory for four dimensional scalar fields, I, II*; Comm. Math. Physics.:100, 545-590, 1985; and 101, 1-36, 1985;

[2] Gallavotti, G.: *Renormalization theory and ultraviolet stability for scalar fields via renormalization group methods*; Rev. Mod. Phys.: 57, 471-562, 1985

[3] Gallavotti, G., Rivasseau, V.: *φ^4-theory in dimension four: a modern introduction to its open problems*; Ann. Inst. H. Poincarè, 40, 185-220, 1984.

[4] Fröhlich, J.:*On the triviality of $\lambda\varphi_d^4$ theories and the approach to the critical point in ($d \geq 4$) dimensions*; Nucl. Phys.: B200, 281- , 1982.

[5] Koch, H., Wittwer, P.: *A non gaussian renormalization group fixed point for hierarchical scalar lattice field theories*; Comm. Math. Phys.: 106, 495-532, 1986.

[6] Felder, G., Gallavotti, G.: *Perturbation theory and non renormalizable scalar fields*, Comm. Math. Phys. 102, 549-571, 1986.

[7] Felder, G.: *Construction of a non trivial planar field theory with ultraviolet stable fixed point*, Comm. Math. Phys., 102, 139-155, 1985

WESS-ZUMINO-WITTEN CONFORMAL FIELD THEORY

Krzysztof Gawędzki

C.N.R.S., I.H.E.S.
91440 Bures-sur-Yvette
France

CONTENTS

These lectures are designed as a possible introduction to conformal field theory (CFT), a subject which has been developing fast in recent years, stimulated by its applications to critical phenomena in two dimensions and to string theory . For the sake of concreteness, we shall concentrate on a specific case of the theory: the Wess-Zumino-Witten (WZW) models of two-dimensional quantum field theory. These are the conformal invariant versions of the sigma models with fields taking values in a compact Lie group $\mathbf{G}$. The plan of the course is as follows:

1. AXIOMS OF CFT

It is convenient to consider CFT from the beginning in the geometric setup, that of the geometry of Riemann surfaces. This should be contrasted with more traditional (and more physical) approach, exposed in J. Fröhlich's lectures [1]. In the latter, one starts with the theory in the Minkowski plane and the Euclidean regime, with possible geometric ramifications, is obtained via analytic continuation. In our approach, patterned on the one advocated by G. Segal [2], see also [3], we shall work directly in the Euclidean space of arbitrary topology. More concretely, we shall consider compact (not necessarily connected) Riemann surfaces Σ with boundary $\partial\Sigma = \bigcup_{i\in I}(\partial\Sigma)_i$, where $(\partial\Sigma)_i$ are the connected components of $\partial\Sigma$, see Fig. 1.

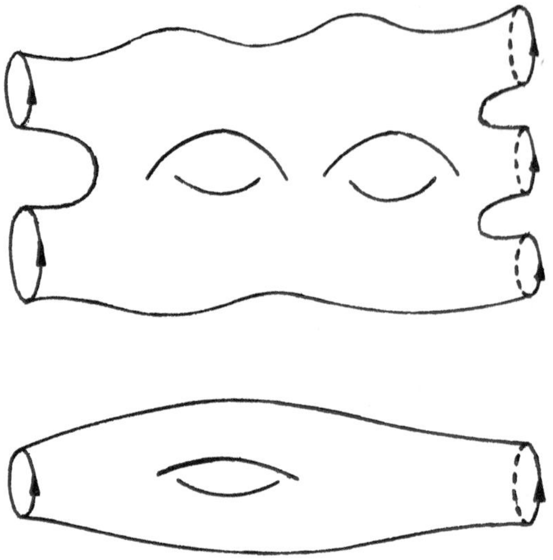

Fig. 1. Riemann surface with boundary

We shall choose (real analytic) parametrizations $p_i : S^1 \longrightarrow (\partial\Sigma)_i$ of the boundary loops and a Riemannian metric g on Σ compatible with the complex structure of Σ (i.e. $g = g_{z\bar{z}}dzd\bar{z}$ locally) and trivial around the boundary (i.e. $p_i^*g = |z|^{-2}dzd\bar{z}$ under the analytic continuation of p_i to a neighborhood of the circle $|z| = 1$ in $\mathbf{C}$). We shall distinguish the positive and negative parametrizations p_i, $i \in I_\pm$, depending on whether p_i respects or not the orientation of $(\partial\Sigma)_i$.

Among many different approaches to quantum field theory, the functional integral formalism is the easiest one to extend , at least formally, to geometrically nontrivial situations. The basic objects of the theory should be the amplitudes

$$\mathbf{A}(\Sigma, p_i, g)\,((\phi_i)_{i\in I_-},\,(\phi_i)_{i\in I_+}) \;=\; \int_{\Phi\circ p_i=\phi_i} exp[-S_\Sigma(\Phi)]\;D\Phi \tag{1}$$

given formally by the functional integration over fields $\Phi : \Sigma \rightarrow M$ fixed on the boundary ($S_\Sigma(\Phi)$ is the Euclidean action of Φ and $D\Phi$ is the formal product $\prod_{\xi\in\Sigma} d\Phi(\xi)$ of volume measures on M). These amplitudes may be viewed as kernels of operators $\mathbf{A}(\Sigma, p_i, g)$ mapping from the space of "in" states $\bigotimes_{i\in I_-} H$ to the space of "out" states $\bigotimes_{i\in I_+} H$ where H is a space of functionals of fields ϕ defined on S^1 . In the conformal invariant case, the amplitudes $\mathbf{A}$ should be (projectively, i.e. up to a factor) invariant under (local) rescalings of the metric g . Their expected properties motivate the following axiomatic definition.

A CFT will be specified by giving a Hilbert space (of states) H with an anti-unitary involution P and trace-class operators

$$\mathbf{A}(\Sigma, p_i, g) \;:\; \bigotimes_{i\in I_-} H \;\longrightarrow\; \bigotimes_{i\in I_+} H$$

with the following properties:

1. Multiplicativity. If (Σ, p_i, g) is a disjoint union of (Σ_1, p_{i_1}, g_1) and (Σ_2, p_{i_2}, g_2) then

$$\mathbf{A}(\Sigma, p_i, g) = \mathbf{A}(\Sigma_1, p_{i_1}, g_1) \otimes \mathbf{A}(\Sigma_2, p_{i_2}, g_2) .$$

2. General covariance. If $D : \Sigma_1 \to \Sigma_2$, $p_{2i} = D \circ p_{1i}$, $D^* g_2 = g_1$ then

$$\mathbf{A}(\Sigma_1, p_{1i}, g_1) = \mathbf{A}(\Sigma_2, p_{2i}, g_2) .$$

3. Conformal invariance. If σ is a real function on Σ vanishing in a neighborhood of $\partial\Sigma$ then

(2)
$$\frac{\delta}{\delta\sigma(\xi)}\bigg|_{\sigma=0} \mathbf{A}(\Sigma, p_i, e^\sigma g) = \frac{ci}{24\pi} R_g \, \mathbf{A}(\Sigma, p_i, g)$$

where c is a real number (called the central charge) and R_g is the curvature form of metric g normalized so that $\int_\Sigma R_g = 2\pi i(2genus + |I| - 2)$.

4. CPT invariance. Denote $p_i^\vee(z) = p_i(\bar{z})$.

(3)
$$(\bigotimes_{i\in I_+} P) \, \mathbf{A}(\bar{\Sigma}, p_i^\vee, g) \, (\bigotimes_{i\in I_-} P) = \mathbf{A}(\Sigma, p_i, g)$$

where $\bar{\Sigma}$ is the Riemann surface complex conjugate to Σ .

5. Unitarity. If Σ' is obtained from Σ by identifying the boundary loops $i_1 \in I_-$ and $i_2 \in I_+$ via $p_{i_2} \circ p_{i_1}^{-1}$, see Fig. 2, then

(4)
$$\mathbf{A}(\Sigma', p_{i'}, g) = Tr_{i_1 i_2} \, \mathbf{A}(\Sigma, p_i, g)$$

where $i' \neq i_1, i_2$ and $Tr_{i_1 i_2}$ is the trace of maps between the factors i_1 and i_2 in the tensor product of spaces H .

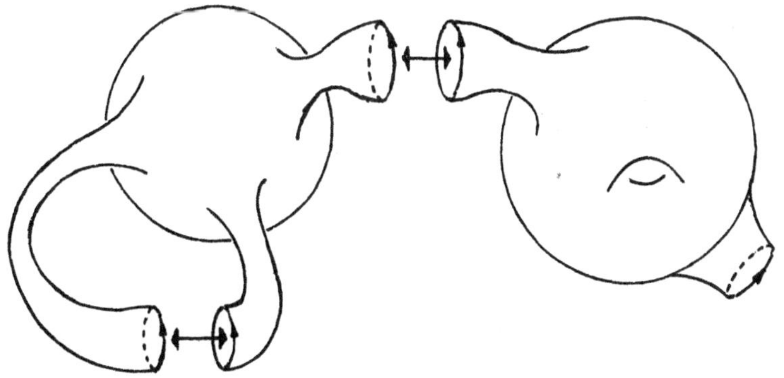

Fig. 2. Gluing Riemann surfaces

REMARK. 1. Formally, the operator P is given by $(PF)(\phi) = \overline{F(\phi^\vee)}$. 2. The least intuitive property above is the conformal anomaly (2) specifying the way in which the normalization of the amplitudes changes with the rescaling of the metric. This type of behavior may be unvealed by a study of specific models.

In some situations, the functional integration can be performed explicitly. The simplest case is that of

Toroidal compactifications.

In this case $\Phi : \Sigma \rightarrow \mathbf{T}^n/\mathbf{Z}^n$ with the action

$$S_\Sigma(\Phi) = \pi i \int_\Sigma g_{ij}\, (\partial\Phi^i)(\bar\partial\Phi^j) + \pi i \int_\Sigma b_{ij}\, (d\Phi^i)(d\Phi^j) \tag{5}$$

where (g_{ij}) is a positive symmetric and (b_{ij}) an antisymmetric matrix. $\mathbf{T}^n$ is taken with its translation-invariant volume form. We shall study below an example of such a system.

A simple modification is the case of

Toroidal orbifolds

where Φ takes values in $\mathbf{T}^n$ divided by a finite group acting on $\mathbf{T}^n$ possibly with fixed points and $S(\Phi)$ is as before.

The simplifying feature of both toroidal compactifications and toroidal orbifolds is that the functional integral may be computed by the semiclassical approximation which gives an exact result. The other systems where the exact result, due to high symmetry, is close to the semiclassical approximation are the WZW theories [4-7] .

2. WZW MODELS

The field of a WZW model takes values in a compact Lie group $\mathbf{G}$. For simplicity, we shall assume that $\mathbf{G}$ is connected, simple and simply-connected. The Euclidean action of a field $\Phi \equiv G : \Sigma \longrightarrow \mathbf{G}$ on a Riemann surface without boundary is defined as

$$S_\Sigma(G) = -\frac{ik}{4\pi}\int_\Sigma \langle G^{-1}\partial G,\, G^{-1}\bar\partial G\rangle - \frac{ik}{24\pi}\int_B \langle \tilde G^{-1}d\tilde G,\, [\tilde G^{-1}d\tilde G, \tilde G^{-1}d\tilde G]\rangle \tag{6}$$

where $\partial = dz\,\partial_z$, $\bar\partial = d\bar z\,\bar\partial_{\bar z}$, $\langle .,. \rangle$ is the (bilinear) Killing form on the complexified Lie algebra $\mathbf{g}^{\mathbf{C}}$ of $\mathbf{G}$ normalized so that the long roots have length squared two (for SU(N) , $\langle X,Y\rangle = tr\, XY$ in the defining representation). $\tilde G$ is an extension of G to a 3-dimensional manifold B with boundary $\partial B = \Sigma$. The ambiguity in (6) is given by

$$\frac{ik}{24\pi}\int_{\tilde B} \langle \tilde G^{-1}d\tilde G,\, [\tilde G^{-1}d\tilde G, \tilde G^{-1}d\tilde G]\rangle$$

for $\tilde B$ without boundary and takes values in $2\pi i k\mathbf{Z}$ (it gives $2\pi i k$ times the generalized degree of the map $\tilde G : \tilde B \longrightarrow \mathbf{G}$) . As a result, the probability amplitudes $exp[-S_\Sigma(G)]$ are well defined for $k \in \mathbf{Z}$. We shall assume below that $k \in \mathbf{N}$ to assure the stability of the theory.

REMARK. The second, topological term in the action is required to obtain a conformal invariant quantum theory [4]. Without this term, the renormalization would break the classical conformal invariance resulting, according to the conventional wisdom, in a massive, asymptotically free quantum field theory [8].

The fundamental property of the action (6) is its behavior under the pointwise multiplication of fields [9]:

(7) $$exp[-S_\Sigma(G_1G_2)] = exp[-S_\Sigma(G_1) - S_\Sigma(G_2) + \Gamma_\Sigma(G_1, G_2)]$$

where

(8) $$\Gamma_\Sigma(G_1, G_2) = \frac{ik}{2\pi}\int_\Sigma \langle G_1^{-1}\bar\partial G_1,\ G_2\partial G_2^{-1}\rangle\ .$$

We shall call identity (7) the Polyakov-Wiegmann (P-W) formula.

EXERCISE. Prove (7) . Indication:

$$\frac{d}{dt}\int_B \langle \tilde G^{-1}d\tilde G,\ [\tilde G^{-1}d\tilde G, \tilde G^{-1}d\tilde G]\rangle = 3\int_\Sigma \langle \tilde G^{-1}\frac{\partial \tilde G}{\partial t},\ [\tilde G^{-1}d\tilde G, \tilde G^{-1}d\tilde G]\rangle\ .$$

In (6) , we may take fields G with values in the complexified group $\mathbf{G}^{\mathbf{C}}$, $S_\Sigma(G)$ remaining well defined modulo $2\pi i$. The stationary points of this action satisfy the (Euclidean) classical equations of motion

(9) $$\partial(G_{cl}^{-1}\bar\partial G_{cl}) = 0\ .$$

Relation (9) is a nonlinear generalization of the Laplace equation. Locally, its solutions have the form

$$G_{cl} = G_1 G_2^*$$

where G_a, $a = 1,2$, are holomorphic fields with values in $\mathbf{G}^{\mathbf{C}}$ and $*$ denotes the anti-involution of $\mathbf{G}^{\mathbf{C}}$ which leaves $\mathbf{G}$ invariant (the hermitian conjugation for SU(N)) .

A. Kac-Moody symmetries

The classical theory possesses rich symmetry structure.

1. Conformal symmetry.

If f is a (local) analytic transformation of Σ then

$$G_{cl} \longmapsto G_{cl}\circ f$$

and

$$G_{cl} \longmapsto (G_{cl}\circ \bar f)^*$$

map classical solutions into classical solutions.

2. Kac-Moody symmetry.

If G_a , $a = 1,2,$ are (local) holomorphic $\mathbf{G}^{\mathbf{C}}$–valued fields then

$$G_{cl} \longmapsto G_1 G_{cl} G_2^*$$

also maps classical solutions into classical solutions.

We shall see below that the Kac-Moody symmetry survives quantization of the model and assures the conformal invariance of the quantum WZW theory.

Up to now, we have defined the action functional $S_\Sigma(G)$ only for fields G defined on Riemann surfaces Σ without boundary. We would like to extend this definition to fields on surfaces with boundary.

Consider first fields

$$G : D \longrightarrow \mathbf{G}^{\mathbf{C}}$$

where $D = \{z \in \mathbf{C} \mid |z| \leq 1\}$.
Let $G' : D' \longrightarrow \mathbf{G^C}$, where $D' = \{z \in \mathbf{C} \mid |z| \geq 1\} \cup \{\infty\} \subset \mathbf{C}P^1$ be a (smooth) extension of G . Let $G \vee G' : \mathbf{C}P^1 \longrightarrow \mathbf{G^C}$ be the field sewn from G and G' . $exp[-S_{\mathbf{C}P^1}(G \vee G')]$ could be taken to represent the action of G except that it depends also on G' . If $G'H'$ is another extension of G , then, by virtue of (7) ,

$$\begin{aligned}(10) \qquad & exp[-S_{\mathbf{C}P^1}(G \vee (G'H'))] \\ &= exp[-S_{\mathbf{C}P^1}(G \vee G')]\, exp[-S_{\mathbf{C}P^1}(1 \vee H') + \Gamma_{D'}(G', H')] \,.\end{aligned}$$

We may use the transition rule (10) to define a complex (holomorphic) line bundle $\mathcal{L}$ over the loop group

$$L\mathbf{G^C} \;=\; Map(S^1 \equiv \partial D,\ \mathbf{G^C})$$

of (smooth) maps from the circle to $\mathbf{G^C}$. $L\mathbf{G^C}$ may be naturally identified with the space $Map(D', \mathbf{G^C}) \,/\, Map_1(D', \mathbf{G^C})$ where $Map_1(D', \mathbf{G^C})$ is the subset of maps which may be smoothly continued by 1 to whole $\mathbf{C}P^1$.

$$\mathcal{L} \;=\; (Map(D', \mathbf{G^C}) \times \mathbf{C}) \,/\, \sim'$$

with

$$(11) \qquad (G', z') \;\sim'\; (G'H',\ z'\, exp[-S_{\mathbf{C}P^1}(1 \vee H') + \Gamma_{D'}(G', H')])$$

for $H' \in Map_1(D', \mathbf{G^C})$. The projection $\pi : \mathcal{L} \longrightarrow L\mathbf{G^C}$ is obviously given by $G' \longmapsto G'|_{\partial D}$. The comparison of (10) and (11) suggests the following definition:

$$exp[-S_D(G)] \;=\; [\,(G',\ exp[-S_{\mathbf{C}P^1}(G \vee G')])\,]_{\sim'} \;\in\; \mathcal{L}_{G|_{\partial D}} \;\equiv\; \pi^{-1}(G|_{\partial D})\,.$$

The line bundle dual to $\mathcal{L}$ may be described as

$$\mathcal{L}^{-1} \;=\; (Map(D, \mathbf{G^C}) \times \mathbf{C}) \,/\, \sim$$

with

$$(12) \qquad (G, z) \;\sim\; (GH,\ z\, exp[-S_{\mathbf{C}P^1}(H \vee 1) + \Gamma_D(G, H)])$$

for $H \in Map_1(D, \mathbf{G^C})$ and the duality is given by

$$(13) \qquad <[(G', z')]_{\sim'},\ [(G, z)]_{\sim}> \;=\; z'z\, exp[S_{\mathbf{C}P^1}(G \vee G')]\,.$$

Notice that we may define

$$exp[-S_{D'}(G')] \;=\; [\,(G,\ exp[-S_{\mathbf{C}P^1}(G \vee G')]\,]\ \in\ \mathcal{L}^{-1}_{G|_{\partial D}}$$

and that

$$(14) \qquad <exp[-S_D(G)],\ exp[-S_{D'}(G')]> \;=\; exp[-S_{\mathbf{C}P^1}(G \vee G')]\,.$$

The anti-involution $*$ may be lifted to $\mathcal{L}$ and $\mathcal{L}^{-1}$ by defining

$$(\,z\, exp[-S_D(G)]\,)^* \;=\; \bar{z}\, exp[-S_D(G^*)]$$

and similarly for $\mathcal{L}^{-1}$ with D replaced by D'.

EXERCISE. Show that $*$ is well defined on $\mathcal{L}$.

Consider now a general compact Riemann surface with boundary parametrizations p_i, $i \in I = I_+ \cup I_-$, see Section 1. One may obtain from Σ a compact surface without boundary Σ^c by sewing a copy D_i of disc D along $(\partial\Sigma)_i$ for $i \in I_-$ and a copy D_i' of D' for $i \in I_+$, see Fig. 3.

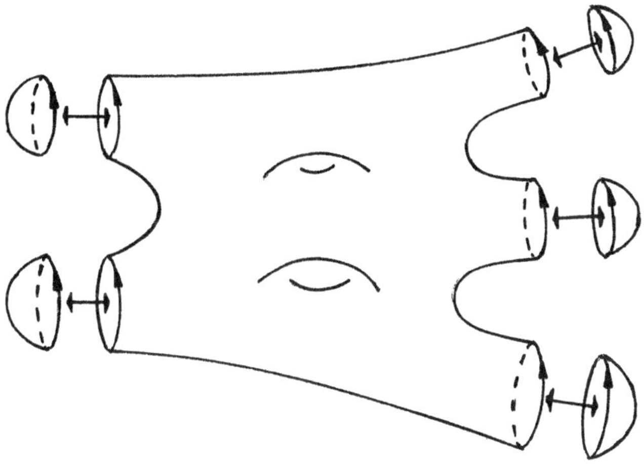

Fig. 3. Completing a Riemann surface

Let $G : \Sigma \longrightarrow \mathbf{G^C}$ and $G^c : \Sigma^c \longrightarrow \mathbf{G^C}$ be its extention. We shall define $exp[-S_\Sigma(G)]$ as the unique element of $(\bigotimes_{i\in I_-} \mathcal{L}^{-1}_{Gop_i}) \bigotimes (\bigotimes_{i\in I_+} \mathcal{L}_{Gop_i})$ such that

$$(15) \qquad < exp[-S_\Sigma(G)],\ (\bigotimes_{i\in I_-} exp[-S_{D_i}(G^c)]) \otimes (\bigotimes_{i\in I_+} exp[-S_{D_i'}(G^c)]) > \\ = exp[-S_{\Sigma^c}(G^c)]$$

(compare (14)).

The amplitudes of the quantum WZW theory are formally given by the functional integrals as in (1)

$$(16) \qquad \mathbf{A}(\Sigma, p_i, g)\ ((G_i)_{i\in I_-},\ (G_i)_{i\in I_+}) = \int\limits_{Gop_i = G_i} exp[-S_\Sigma(G)]\ DG \\ = \int\limits_{Gop_i=1} exp[-S_\Sigma(\tilde{G}G)]\ DG \ \in\ (\bigotimes_{i\in I_-} \mathcal{L}^{-1}_{G_i}) \otimes (\bigotimes_{i\in I_+} \mathcal{L}_{G_i})$$

where $\tilde{G} : \Sigma \longrightarrow \mathbf{G}$, $\tilde{G} \circ p_i = G_i$ is fixed and DG stands for the formal product of Haar measures $\prod_{\xi\in\Sigma} dG(\xi)$ on $\mathbf{G}$. Notice that, by the invariance of the Haar measure,

$$\int\limits_{\mathbf{G}} \mathcal{F}(\tilde{G}G)\ dG$$

does not depend on the choice of $\tilde{G}$ even if $\tilde{G} \in \mathbf{G}^{\mathbf{C}}$ provided that $\mathcal{F}$ is a holomorphic function on $\mathbf{G}^{\mathbf{C}}$. Thus the integral (16) should be defined even for $G_i \in L\mathbf{G}^{\mathbf{C}}$ and should give holomorphic sections of bundles $\mathcal{L}$: formally

$$(17) \qquad \mathbf{A}(\Sigma, p_i, g) \;\in\; \Big(\bigotimes_{i\in I_-} \Gamma_{hol}(\mathcal{L}^{-1})\Big) \otimes \Big(\bigotimes_{i\in I_+} \Gamma_{hol}(\mathcal{L})\Big) .$$

Consequently, the space $\Gamma_{hol}(\mathcal{L})$ of the holomorphic sections of $\mathcal{L}$ (or rather its completion in an appropriate scalar product) appears as the natural space of states of the WZW model. This complication (one could have naively expected a space of standard functions of field values on S^1 = compactified one-dimensional space) is due to the presence of the topological term in the action of the model.

The quantum symmetries of the theory should act (projectively) in its space of states. We shall realize first the Kac-Moody symmetries. They give rise to a projective action of two copies of the loop group $L\mathbf{G}^{\mathbf{C}}$ in $\Gamma_{hol}(\mathcal{L})$ or to an action of two copies of a central extension $\hat{L\mathbf{G}}^{\mathbf{C}}$ of the loop group therein. The extension $\hat{L\mathbf{G}}^{\mathbf{C}}$, which we shall call the Kac-Moody group, is easy to construct. As a set,

$$\hat{L\mathbf{G}}^{\mathbf{C}} \;=\; \mathcal{L} \setminus \text{zero section} .$$

In $\mathcal{L}$, we may define the product by putting

$$(18) \qquad \begin{aligned} &(z_1\, exp[-S_D(G_1)]) \bullet (z_2\, exp[-S_D(G_2)]) \\ &= \; z_1 z_2\, exp[-S_D(G_1G_2) - \Gamma_D(G_1, G_2)] . \end{aligned}$$

EXERCISE. Show that relation (18) defines a map from $\mathcal{L} \times \mathcal{L}$ into $\mathcal{L}$ which provides $\hat{L\mathbf{G}}^{\mathbf{C}}$ with the group structure.

The sequence of group homomorphisms

$$1 \longrightarrow \mathbf{C}^* \longrightarrow \hat{L\mathbf{G}}^{\mathbf{C}} \longrightarrow L\mathbf{G}^{\mathbf{C}} \longrightarrow 1 ,$$

where the second arrow maps z to $z\, exp[-S_D(1)]$ and the third one is induced by the bundle projection π, is exact. $\hat{L\mathbf{G}}^{\mathbf{C}}$ acts on $\mathcal{L}$ by left and right multiplication. This gives rise to a representation $l \times r$ of $\hat{L\mathbf{G}}^{\mathbf{C}} \times \hat{L\mathbf{G}}^{\mathbf{C}}$ in $\Gamma_{hol}(\mathcal{L})$. If $\psi \in \Gamma_{hol}(\mathcal{L})$, $\hat{G}_1$, $\hat{G}_2 \in \hat{L\mathbf{G}}^{\mathbf{C}}$, $\pi(\hat{G}_a) = G_a \in L\mathbf{G}^{\mathbf{C}}$, $G \in L\mathbf{G}^{\mathbf{C}}$, then

$$(19) \qquad (l(\hat{G}_1)\, r(\hat{G}_2)\, \psi)(G) \;=\; \hat{G}_1 \bullet \psi(G_1^{-1} G G_2^{*\,-1}) \bullet \hat{G}_2^* .$$

The above construction may be repeated for the dual bundle $\mathcal{L}^{-1}$ by replacing D by D'.

EXERCISE. Prove the generalized P-W formula

$$(20) \qquad exp[-S_\Sigma(G_1)] \bullet exp[-S_\Sigma(G_2)] \;=\; exp[-\Gamma_\Sigma(G_1, G_2)]\, exp[-S_\Sigma(G_1G_2)] .$$

Now it is easy to see in what sense the representation $l \times r$ realizes the Kac-Moody symmetries on the quantum level. If $G_a : \Sigma \longrightarrow \mathbf{G}^{\mathbf{C}}$ are holomorphic maps then formally

(21)
$$l(exp[-S_\Sigma(G_1)])\ r(exp[-S_\Sigma(G_2))\ \mathbf{A}(\Sigma,p_i,g) = \int exp[-S_\Sigma(G_1 G G_2^*)]\ DG \ = \ \mathbf{A}(\Sigma,p_i,g)\ .$$

It is more customary to describe the Kac-Moody symmetry on the infinitesimal level. Let (t^α) be a basis of generators of Lie algebra $\mathbf{g}$, $\langle t^\alpha, t^\beta\rangle = \frac{1}{2}\delta^{\alpha\beta}$, $[t^\alpha, t^\beta] = i\sum_\gamma f^{\alpha\beta\gamma} t^\gamma$. Define

$$J_n^\alpha \ = \ -i\frac{d}{d\epsilon}\Big|_{\epsilon=0} \ l(exp[-S_D(e^{i\epsilon t^\alpha z^n})])$$

for $n \geq 0$ and similarly for $n \leq 0$ with $\bar{z}^{-n}$ replacing z^n and for $\bar{J}_n^\alpha$ with the right representation r replacing the left one l.

EXERCISE. (Kac-Moody algebra commutation relations). Show that

(22)
$$[J_n^\alpha,\ J_m^\beta] \ = \ i\sum_\gamma f^{\alpha\beta\gamma}\ J_{n+m}^\gamma \ + \ \frac{k}{2}\, n\, \delta^{\alpha\beta}\, \delta_{n+m,0}$$

and the same for $\bar{J}_n^\alpha$'s .

The commutation relations (22) describe a central extension $\hat{L\mathbf{g}}^{\mathbf{C}}$ of the loop group Lie algebra $L\mathbf{g}^{\mathbf{C}}$. Thus $\Gamma_{hol}(\mathcal{L})$ carries a representation of $\hat{L\mathbf{g}}^{\mathbf{C}} \oplus \hat{L\mathbf{g}}^{\mathbf{C}}$.

B. Spectrum

We would like to decompose this representation into irreducible components. Because of physical reasons, we shall be interested only in the so called "positive energy" components, see below. These representations may be generated from the set of primary vectors. $\psi \in \Gamma_{hol}(\mathcal{L})$ is called primary iff

(23)
$$J_n^\alpha\,\psi \ = \ \bar{J}_n^\alpha\,\psi \ = \ 0 \qquad \text{for } n > 0\ .$$

Let

$$N_+ \ = \ \{\ exp[-S_D(G)] \ \mid\ G:\ D \longrightarrow \mathbf{G}^{\mathbf{C}} \text{ is holomorphic and } G(0) = 1\ \}\ .$$

N_+ is a subgroup of $\hat{L\mathbf{G}}^{\mathbf{C}}$ which we can also identify with the subgroup $\{\ G|_{\partial D} \ \mid\ G$ as above $\}$ of $L\mathbf{G}^{\mathbf{C}}$. It is easy to see that $\psi \in \Gamma_{hol}(\mathcal{L})$ is primary iff

(24)
$$l(N_+)\ r(N_+)\ \psi \ = \ \psi\ .$$

For ψ primary, $G_a:\ D \longrightarrow \mathbf{G}^{\mathbf{C}}$ holomorphic, $G_a(0) = 1$, and $G_0 \in \mathbf{G}^{\mathbf{C}}$,

$$\begin{aligned} &\psi(G_1 G_0 G_2^*) \\ &= \ [\ l(exp[-S_D(G_1)])\ r(exp[-S_D(G_2)])\ \psi]\ (G_1 G_0 G_2^*) \\ &= \ exp[-S_D(G_1)] \ \bullet\ \psi(G_0)\ \bullet\ exp[-S_D(G_2)]\ . \end{aligned}$$

Setting for constant G_0

$$\psi(G_0) = \psi_0(G_0)\ exp[-S_D(G_0)]$$

where ψ_0 is an analytic function on $\mathbf{G}^{\mathbf{C}}$, we obtain

$$\psi(G_1 G_0 G_2^*) = \psi_0(G_0)\ exp[-S_D(G_1 G_0 G_2^*)]\ .$$

Since $G_1 G_0 G_2^*$ is a general classical solution on D , the last relation may be rewritten as

$$\psi(G_{cl}|_{\partial D}) = \psi_0(G_{cl}(0))\ exp[-S_D(G_{cl})]\ . \tag{25}$$

By the Birkhoff theorem [10], the boundary values of the classical solutions on D form an open dense set in $L\mathbf{G}^{\mathbf{C}}$. Thus (25) fixes ψ completely.

Notice that the formal expressions

$$\psi = \int \psi_0(G(0))\ exp[-S_D(G)]\ DG \tag{26}$$

satisfy (24) . Thus (25) means that the functional integral (26) can be essentially computed semiclassically (demanding that functions ψ_0 in (25) and (26) be the same would impose normalization conditions on the renormalization procedure which (26) certainly requires).

The above situation is characteristic of how the functional integrals for the WZW model may be computed from general symmetry principles without ever appealing to the usual regularization-renormalization scenario.

Somewhat unexpectedly, we may learn even more from the solution (25) : not all holomorphic functions ψ_0 are allowed. Notice that the space of primary vectors is invariant under the left and right action of $\mathbf{G}^{\mathbf{C}}$ embedded into $\hat{L\mathbf{G}}^{\mathbf{C}}$ by $G_0 \longmapsto exp[-S_D(G_0)]$. This action becomes the left and right regular action of $\mathbf{G}^{\mathbf{C}}$ on analytic functions ψ_0 defined on $\mathbf{G}^{\mathbf{C}}$ whose space decomposes into the direct sum $\bigoplus_{\omega} h_\omega \otimes \bar{h}_\omega$ of irreducible components. h_ω carries the highest weight (HW) representation D^ω of $\mathbf{G}^{\mathbf{C}}$ (we label it by its HW $\omega \in \mathbf{g}$) . More concretely, $h_\omega \otimes \bar{h}_\omega$ is the space of analytic functions

$$\mathbf{G}^{\mathbf{C}} \ni G_0 \longmapsto tr\ (D^\omega(G_0)\ \rho)$$

where ρ is an operator in h_ω ($\dim h_\omega < \infty$) .

We should check whether the sections ψ defined by (25) over an open dense set extend to holomorphic sections over $L\mathbf{G}^{\mathbf{C}}$. Call the (highest) weight ω integrable if $\langle \phi, \omega \rangle \leq k$ where ϕ is the highest root of $\mathbf{g}$. For SU(N) , the HW's are of the form $diag(j_1, ..., j_N)$, $j_l - j_{l+1} \in \mathbf{N} \cup \{0\}$ and $j_1 + ... + j_N = 0$. $\phi = diag(1, 0, ..., 0, -1)$ and the integrable weights satisfy $j_1 - j_2 \leq k$. For SU(2) , the HW's are $diag(j, -j)$ where $j = 0,\ 1/2,\ 1,\ 3/2, ...$ is the spin, and the integrable weights satisfy $j \leq k/2$.

PROPOSITION. Eq. (25) defines an element of $\Gamma_{hol}(\mathcal{L})$ iff $\psi_0 \in \bigoplus_{\omega \text{ integrable}} h_\omega \otimes \bar{h}_\omega$.

S k e t c h o f p r o o f . Let us consider for simplicity the case of SU(2) . The complement of $\{G_{cl}|_{\partial D}\}$ in $L\mathbf{G}^{\mathbf{C}}$ falls, by the Birkhoff theorem, into the union of complex submanifolds

$$\{\ G_1 \begin{pmatrix} z^n & 0 \\ 0 & \bar{z}^n \end{pmatrix} G_2^*\ \mid\ G_a :\ D \longrightarrow \mathrm{SL}(2, \mathbf{C}) \text{ is holomorphic}\ \}\ ,$$

$n \in \mathbf{N}$, of finite codimention. The obstruction to extension of ψ to $L\mathbf{G}^{\mathbf{C}}$ may come only from a singularity when we approach the codimension 1 stratum which corresponds to $n = 1$. Consider the complex one-parameter family of loops

$$\mathbf{C} \ni c \longmapsto G_c = \begin{pmatrix} z & 0 \\ c & \bar{z} \end{pmatrix} \in L\mathrm{SL}(2, \mathbf{C}) . \tag{27}$$

For $c \neq 0$

$$G_c = G_{1c} G_{2c}^* \tag{28}$$

where

$$G_{1c} = \begin{pmatrix} 1 & z/c \\ 0 & 1 \end{pmatrix} , \quad G_{2c}^* = \begin{pmatrix} 0 & -1/c \\ c & \bar{z} \end{pmatrix} \tag{29}$$

so that G_{ac} are holomorphic on D . Let $G_a : D \longrightarrow \mathrm{SL}(2,\mathbf{C})$ be another (c-independent) pair of holomorphic maps. Then

$$\begin{aligned} &\psi(G_1 G_c G_2^*) \\ &= \psi_0(G_1(0) \begin{pmatrix} 0 & -1/c \\ c & 0 \end{pmatrix} G_2(0)^*) \ exp[-S_D(G_1)] \\ &\qquad \bullet \ exp[-S_D(G_{1c}G_{2c}^*)] \ \bullet \ exp[-S_D(G_2^*)] . \end{aligned}$$

It is easy to check that when $c \to 0$

$$exp[-S_D(G_{1c}G_{2c}^*)] \sim \mathcal{O}(c^k) . \tag{30}$$

Since for $\psi_0 \in h_j \otimes \bar{h}_j$, $\psi_0(G_1(0) \begin{pmatrix} 0 & -1/c \\ c & 0 \end{pmatrix} G_2(0)^*)$ behaves at worst as c^{-2j} , we obtain the condition $j \leq k/2$.

Notice that in general there is only a finite number of integrable (highest) weights. Let us define for integrable ω ,

$$\Gamma_\omega(\mathcal{L}) = span \, \{ \, J^{\alpha_1}_{-n_1} ... J^{\alpha_p}_{-n_p} \bar{J}^{\beta_1}_{-m_1} ... \bar{J}^{\beta_q}_{-m_q} I_\omega(h_\omega \otimes \bar{h}_\omega) \mid n_r, \ m_s > 0 \, \} \subset \Gamma_{hol}(\mathcal{L})$$

where I_ω denotes the embedding of $h_\omega \otimes \bar{h}_\omega$ into $\Gamma_{hol}(\mathcal{L})$ defined by (25) . It follows from the theory of the Kac-Moody algebra representations [11,10] that $\Gamma_\omega(\mathcal{L})$ are disjoint irreducible (HW) subrepresentations of $\Gamma_{hol}(\mathcal{L})$ which are moreover unitarizable by imposing $(J^\alpha_n)^* = J^\alpha_{-n}$, $(\bar{J}^\alpha_n)^* = \bar{J}^\alpha_{-n}$. By completion in the scalar product, one can turn $\bigoplus_{\omega \text{ integrable}} \Gamma_\omega(\mathcal{L})$ into a Hilbert space H . The representation of $\hat{L\mathrm{g}}^{\mathbf{C}} \oplus \hat{L\mathrm{g}}^{\mathbf{C}}$ in $\oplus \Gamma_\omega(\mathcal{L})$ integrates to a representation of the unitary part of $\hat{L\mathbf{G}}^{\mathbf{C}} \times \hat{L\mathbf{G}}^{\mathbf{C}}$ in H [12] .

The crucial point about the WZW model is that its Kac-Moody symmetry content, studied above with the help of harmonic analysis on the loop group, determines its dynamics. In this sense, the WZW model may be viewed as a quantum-field-theoretic generalization of the quantum mechanics of the particle on the group manifold. In the latter case the hamiltonian equals the Laplacian = the quadratic Casimir operator on the group and its spectrum may be studied by the harmonic analyses on the group. Let us describe the analogous construction for the WZW theory.

In the space $\Gamma_\omega(\mathcal{L})$ of each HW representation of $\hat{L\mathbf{g}}^{\mathbf{C}} \oplus \hat{L\mathbf{g}}^{\mathbf{C}}$ acts a pair of Virasoro algebras, a central extension of the algebra of conformal vector fields around the circle $|z| = 1$ with generators $L_n = i\, z^{n+1}\partial_z$ and $\bar{L}_n = -i\, \bar{z}^{n+1}\partial_{\bar{z}}$. It is given by the so called Sugawara construction [13] setting

$$L_n \;=\; \frac{1}{k + h^\vee} \sum_{\alpha,n} : J^\alpha_{n-m} J^\alpha_m : \tag{31}$$

and similarly for $\bar{L}_n$ with $\bar{J}^\alpha_n$'s replacing J^α_n's . The normal ordering " : : " inverts the order of operators when $n - m > 0$ and $m < 0$. $h^\vee$ is the dual Coxeter number (the quadratic Casimir of the adjoint representation). $h^\vee = N$ for SU(N) . L_n's (and $\bar{L}_n$'s) satisfy the Virasoro algebra commutation relations:

$$[L_n,\, L_m] \;=\; (n - m)\, L_{n+m} \;+\; \frac{c}{12}\,(n^3 - n)\, \delta_{n+m,0}$$

where the central charge

$$c \;=\; k\; dim\mathbf{G} \;/\; (k + h^\vee)\;.$$

Moreover

$$[L_n,\, J^\alpha_m] \;=\; -m\; J^\alpha_{n+m}\;. \tag{32}$$

The energy and momentum operators of the WZW model are given by

$$\mathcal{H} \;=\; L_0 + \bar{L}_0 - c/12\;, \qquad \mathcal{P} \;=\; L_0 - \bar{L}_0$$

and we may easily read off their spectrum. For ψ a primary vector in $h_\omega \otimes \bar{h}_\omega$

$$L_0\psi \;=\; \bar{L}_0\psi \;=\; \frac{\langle \omega + \rho, \omega + \rho\rangle - \langle \rho, \rho\rangle}{2(k + h^\vee)}\;\psi$$

where ρ is half the sum of the positive roots of $\mathbf{g}$. Due to (32) , the action of J^α_{-n} $(\bar{J}^\alpha_{-n})$ lifts the eigenvalue of L_0 $(\bar{L}_0)$ by n . Thus the spectrum of L_0 and $\bar{L}_0$ is in the non-negative half-line ("positive energy"). The multiplicities of the eigenvalues are somewhat more complicated. They are encoded in the Kac-Moody characters

$$\chi^k_\omega(\tau, G_0) \;=\; (dim\; h_\omega)^{-1} \sum_{\Delta \in spec L_0} Tr\;\; exp[2\pi i \tau \Delta]\; l(G_0)|_{E_\Delta} \tag{33}$$

where E_Δ is the subspace corresponding to eigenvalue Δ of L_0 in $span\,\{\, J^{\alpha_1}_{-n_1} \ldots J^{\alpha_p}_{-n_p} I_\omega(h_\omega \otimes \bar{h}_\omega)\,\} \subset \Gamma_{hol}(\mathcal{L})$, $\tau \in \mathbf{C}$, $Im\,\tau > 0$ and $G_0 \in \mathbf{G}$. The characters χ_ω are known [14].

The knowledge of the energy-momentum operators allows us to compute the amplitudes $\mathbf{A}(\Sigma, p_i, g)$ of the WZW models for the simplest geometries using the Schrödinger picture. Let us consider the cylinder

$$C_\tau \;=\; \{\, z \in \mathbf{C} \mid 0 \le Im\, z \le 2\pi\; Im\,\tau \,\}\; /\; (2\pi\mathbf{Z})$$

with the boundaries parametrized by

$$p_1(z) = i \; log \; z \,, \quad p_2(z) = i \; log \; z + 2\pi\tau \,, \quad |z| = 1$$

and with the metric $g_0 = dz d\bar{z}$ inherited from $\mathbf{C}$. Alternatively, C_τ may be described via the map $z \longmapsto q \; exp[-iz]$, where $q \equiv exp[2\pi i z]$, as the annulus $\{z \in \mathbf{C} \mid |q| \leq |z| \leq 1\}$ with the parametrizations

$$p_1(z) = qz \,, \quad p_2(z) = z$$

and metric $g_0 = \frac{1}{|z|^2} dz d\bar{z}$, see Fig. 4.

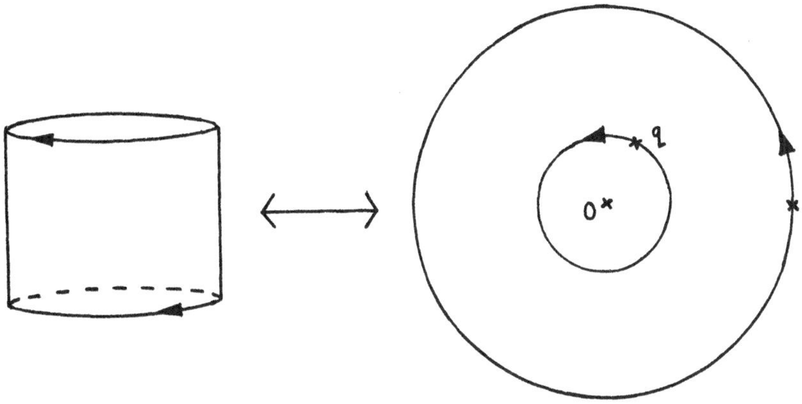

Fig. 4. Cylinder $\simeq$ annulus

The amplitude

$$\mathbf{A}(C_\tau, p_i, g_0) = exp[-2\pi(Im \; \tau)\mathcal{H}] \; exp[2\pi i(Re \; \tau)\mathcal{P}] = q^{L_0 - c/24} \; \bar{q}^{\bar{L}_0 - c/24} \,.$$

The amplitudes for other annular geometries are related to the other operators L_n. For $\Sigma_{z_0,q,n}$ being the annulus (if such exists) contained between $p_i(S^1)$, $p_1(z) = zq(1 + nz_0 z^{-n} q^{-n})^{1/n}$, $p_2(z) = z$,

$$\mathbf{A}(\Sigma_{z_0,q,n}, p_i, g) = const. \; exp[z_0 L_{-n}] \; q^{L_0} \; exp[\bar{z}_0 \bar{L}_{-n}] \; \bar{q}^{\bar{L}_0} \,, \tag{34}$$

the constant depending on the metric.

Using the unitarity axiom 5 of Section 1, we may immediately obtain the amplitudes for complex tori $T_\tau = \mathbf{C}/(2\pi\mathbf{Z} + 2\pi\tau\mathbf{Z})$ (with the metric $g_0 = dz d\bar{z}$). They are known under the name of (toroidal) partition functions:

$$Z(\tau) \quad \left(= \int exp[-S_{T_\tau}(G)] \; DG \right)$$
$$= Tr_H \; \mathbf{A}(C_\tau, p_i, g_0) = (q\bar{q})^{-c/24} \; Tr_H \; q^{L_0} \; \bar{q}^{\bar{L}_0} = (q\bar{q})^{-c/24} \sum_{\omega \text{ integrable}} |\chi^k_\omega(\tau, 1)|^2 \,.$$

They possess the important property of modular invariance

$$Z(\tau) \quad = \quad Z(\frac{a\tau + b}{c\tau + d})$$

for $\begin{pmatrix} a & b \\ c & d \end{pmatrix} \in \mathrm{SL(2,\mathbf{Z})}$. This is due to the conformal isomorphism between T_τ and $T_{\tau'}$, $\tau' = \frac{a\tau+b}{c\tau+d}$, given by the mapping $z \longmapsto z' = \frac{z}{c\tau+d}$ which multiplies the metric g_0 by a constant, preserving the amplitude (see the general covariance and the conformal invariance axioms of Section 1) .

C. Green functions

It is customary to encode the data of quantum field theory in Green functions of a certain set of fields, often called in the Euclidean region the correlation functions [1]. In the WZW theory (as in other CFT's) the set of the primary vectors is in one-to-one correspondence with the set of primary fields with tensorial transformation properties.

For $z_i \in \mathbf{C}$, $i \in I$, $z_i \neq z_{i'}$ for $i \neq i'$ and for $0 \neq q_i \in \mathbf{C}$ sufficiently small, consider the Riemann surface $\Sigma_{z_i,q_i} = \mathbf{C}P^1 \setminus \bigcup_i \{z \mid |z - z_i| < |q_i| \}$ with the boundary parametrized by p_i , $p_i(z) = z_i + zq_i$. Let g be a metric on $\mathbf{C}P^1$ trivial around $\partial\Sigma_{z_i,q_i}$. Let g_i be the induced metric on $\mathbf{C}P^1$ which for $|z| < 1$ $(|z| > 1)$ is the pullback of g under the map $z \longmapsto z_i + zq_i$ $(z \longmapsto z_i + z^{-1}q_i)$. Choose a sequence (ω_i) of integrable weights. The (Euclidean) Green functions are defined as

$$(35) \qquad \mathcal{G}(z_i, \omega_i) \quad = \quad \mathcal{A}(\Sigma_{z_i,q_i}, p_i, g) \, (\bigotimes_i q_i^{-L_0} \bar{q}_i^{-\bar{L}_0} I_{\omega_i})$$

$$(\mathbf{A}(\mathbf{C}P^1, g) \prod_i \mathbf{A}(\mathbf{C}P^1, g_i)^{-1/2})^{-1} .$$

$\mathcal{G}(z_i, \omega_i)$ is a linear form on $\bigotimes_i (h_{\omega_i} \otimes \bar{h}_{\omega_i})$.

EXERCISE. Show that $\mathcal{G}(z_i, \omega_i)$ depends neither on q_i nor on g .

It is easy to see (compare the relation between (25) and (26)) that formally

$$(36) \qquad < \bigotimes_i \psi_{0i} \, , \, \mathcal{G}(z_i, \omega_i) > \; = \; \int \prod_i \psi_{0i}(G(z_i)) \, exp[-S_{\mathbf{C}P^1}(G)] \, DG$$

$$(\int exp[-S_{\mathbf{C}P^1}(G)] \, DG \,)^{-1} .$$

The Green functions satisfy a differential equation obtained by Knizhnik and Zamolodchikov [6]. Let us consider $\mathcal{G}(z_{i_0} + \Delta z, z_{i'}, \omega_i)$ $(i' \neq i_0)$. Decomposing $\Sigma_{z_{i_0}+\Delta z, z_{i'}, q_i}$ into $\Sigma' = \Sigma_{z_{i_0}, z_{i'}, 1, q_{i'}}$ and $\Sigma" = \{z \mid |z - z_{i_0}| \leq 1 \text{ and } |z - z_{i_0} - \Delta z| \geq |q_{i_0}| \}$ (if this is geometrically acceptable, see Fig. 5) with natural parametrizations of the boundary, we may write

$$\mathcal{G}(z_{i_0} + \Delta z, z_{i'}, \omega_i)$$

$$= \quad \mathbf{A}(\Sigma', p_i', g) \, \mathbf{A}(\Sigma", p_i", g) \, (\bigotimes_i q_i^{-L_0} \bar{q}_i^{-\bar{L}_0} I_{\omega_i}) \; / \; \mathcal{N}$$

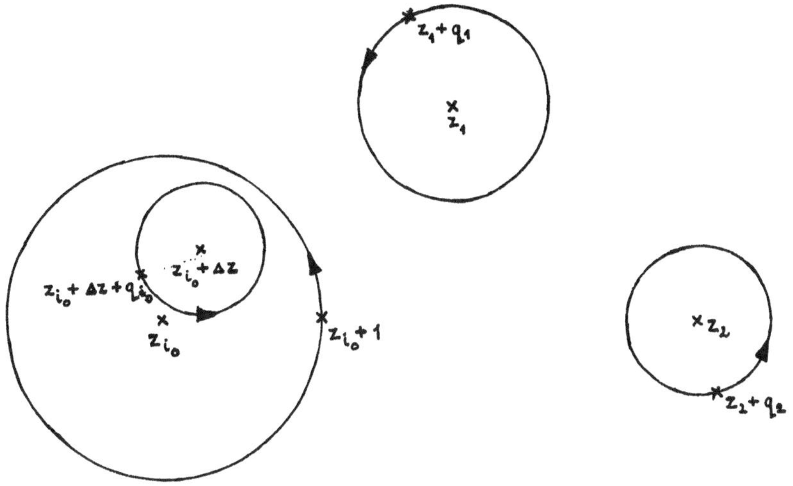

Fig. 5. $\mathbf{C}P^1$ without discs

where $\mathcal{N} = \mathbf{A}(\mathbf{C}P^1, g)\prod_i \mathbf{A}(\mathbf{C}P^1, g_i)^{-1/2}$ and g is trivial around $p_i(S^1)$ and $|z| = 1$. Using (34), we obtain

$$(37)\qquad \begin{aligned} &\mathcal{G}(z_{i_0} + \Delta z, z_{i'}, \omega_i) \\ &= \mathbf{A}(\Sigma', p_i', g)\ (exp[\Delta z L_{-1} + \overline{\Delta z}\bar{L}_{-1}]I_{\omega_{i_0}} \otimes (\bigotimes_{i'} q_{i'}^{-L_0}\bar{q}_{i'}^{-\bar{L}_0} I_{\omega_{i'}}))\ /\ \mathcal{N}' \end{aligned}$$

where $\mathcal{N}' = \mathbf{A}(\mathbf{C}P^1, g)\ \mathbf{A}(\mathbf{C}P^1, g'_{i_0})^{-1/2}\prod_{i'}\mathbf{A}(\mathbf{C}P^1, g_{i'})^{-1/2}$ with g'_{i_0} being for $|z| \leq 1$ ($|z| \geq 1$) the pullback of g by the map $z \longmapsto z_{i_0} + z$ ($z \longmapsto z_{i_0} + z^{-1}$).

EXERCISE. Check that the normalization in (37) is correct by considering the case $\omega_{i_0} = 0$.

Eq. (37) implies that

$$\frac{\partial}{\partial z_{i_0}}\mathcal{G}(z_i, \omega_i) = \mathbf{A}(\Sigma', p_i', g)\ (L_{-1}I_{\omega_{i_0}} \otimes (\bigotimes_{i'} q_{i'}^{-L_0}\bar{q}_{i'}^{-\bar{L}_0} I_{\omega_{i'}}))\ /\ \mathcal{N}'$$

and a similar expression for $\frac{\partial}{\partial \bar{z}_{i_0}}\mathcal{G}$. But by (31),

$$L_{-1}I_{\omega_{i_0}} = \frac{2}{k + h^\vee}\sum_\alpha J^\alpha_{-1}J^\alpha_0\ I_{\omega_{i_0}}\ .$$

On the other hand, for $G : \Sigma' \longrightarrow \mathbf{G}^{\mathbf{C}}$ analytic, by (21),

$$l(exp[-S_{\Sigma'}(G)])\ \mathbf{A}_{\Sigma'} = \mathbf{A}_{\Sigma'}\ .$$

Taking $G(z) = exp[i\epsilon(z - z_{i_0})^{-1}t^\alpha]$ and differentiating over ϵ at $\epsilon = 0$, we obtain

$$\mathbf{A}(\Sigma', p_i', g)\,(J^\alpha_{-1} I_{\omega_{i_0}} \otimes (\bigotimes_{i'} I_{\omega_{i'}}))$$

$$= \sum_{i'} \frac{1}{z_{i_0} - z_{i'}} \mathbf{A}(\Sigma', p_i', g)\,(I_{\omega_{i_0}} \otimes J^\alpha_0 I_{\omega_{i'}} \otimes (\bigotimes_{i'' \neq i_0, i'} I_{\omega_{i''}}))\,.$$

Hence

(38) $$\frac{\partial}{\partial z_{i_0}} \mathcal{G}(z_i, \omega_i) = \frac{2}{k + h^\vee} \sum_{i' \neq i_0} \frac{1}{z_{i_0} - z_{i'}} \mathcal{G}(z_i, \omega_i)\,(J^\alpha_0)_{i_0} (J^\alpha_0)_{i'}$$

where $(J^\alpha_0)_{i_0}$ acts on the i_0'th component of the tensor product $\bigotimes_i (h_\omega \otimes \bar{h}_\omega)$. Similarly for $\frac{\partial}{\partial \bar{z}_{i_0}} \mathcal{G}$ with $\bar{J}^\alpha_0$ replacing J^α_0.

These equations were used to solve for the Green four-point functions in some special cases by Knizhnik and Zamolodchikov [6], Christe and Flume [15] and Zamolodchikov and Fateev [16]. The four point functions determine in particular the three-point functions (by taking one of the weights $\omega_i = 0$). The dependence of the three-point function on (z_i) is fully fixed by the Möbius-group invariance (i.e. invariance under the holomorphic transformations of CP^1). Thus the three point function is given by its value at three fixed points, i.e. by a linear form on $\bigotimes_{i=1}^{3} (h_{\omega_i} \otimes \bar{h}_{\omega_i})$ (the tensor of operator product coefficients [17]).

D. Fusion rules

The Green functions describe the action of the operators $\mathbf{A}(\Sigma_{z_i, q_i}, p_i, g)$ on the primary vectors. They allow, however, to recover the complete operators. To see how this works, let us couple the WZW model to the external gauge field A with values in the Lie algebra $\mathbf{g}^{\mathbf{C}}$. $A = A^{10} + A^{01}$ where $A^{10} = -(A^{01})^*$ (A^{10}, A^{01} are the dz and $d\bar{z}$ parts of A). We put (for a Riemann surface Σ without boundary)

(39) $$S_\Sigma(G, A) = S_\Sigma(G) + \frac{ik}{2\pi} \int_\Sigma [\langle A^{10}, G^{-1} \bar{\partial} G\rangle + \langle G \partial G^{-1}, A^{01}\rangle + \langle G A^{10} G^{-1}, A^{01}\rangle - \langle A^{10}, A^{01}\rangle]\,.$$

The action $S_\Sigma(G, A)$ has a simple transformation properties under the complex gauge transformations

(40) $$A^{10} \longmapsto (H^*)^{-1} A^{10} H^* + (H^*)^{-1} \partial H^* \equiv ({}^H A)^{10}\,,$$
$$A^{01} \longmapsto H A^{01} H^{-1} + H \bar{\partial} H^{-1} \equiv ({}^H A)^{01}\,,$$

where $H : \Sigma \longrightarrow \mathbf{G}^{\mathbf{C}}$ (the proper gauge transformations correspond to H $\mathbf{G}$-valued). Namely

(41) $$S_\Sigma(HGH^*, {}^H A) = S_\Sigma(G, A) - S_\Sigma(H^{-1}(H^*)^{-1}, A)\,.$$

Notice that the term $S(H^{-1}(H^*)^{-1}, A)$ vanishes for the proper gauge transformations (the action (39) is gauge-invariant).

Let us consider the formal functional integral

$$\int \prod_i \psi_{0i}(G(z_i))\, exp[-S_{CP^1}(G,A)]\, DG \;/\; \int exp[-S_{CP^1}(G)]\, DG \;\equiv\; < \bigotimes_i \psi_{0i}\;,\; \mathcal{G}(z_i,\omega_i,A) >$$

for $\psi_{0i} \in h_{\omega_i} \otimes \bar{h}_{\omega_i}$, compare (36) . From (41) we infer the Kac-Moody (global) Ward identity

$$\mathcal{G}(z_i,\omega_i,A) = exp[S_{CP^1}(H^{-1}(H^*)^{-1},A)]\;\; \mathcal{G}(z_i,\omega_i,A) \bigotimes_i (\; l(H(z_i)^{-1})\; r(H(z_i)^{-1})\;)\;. \tag{42}$$

In particular, eq. (42) allows to express $\mathcal{G}(z_i,\omega_i,(H^*)^{-1}\partial H^* + H\bar{\partial}H^{-1})$ by $\mathcal{G}(z_i,\omega_i)$. Since, on $\mathbf{C}P^1$, the set of complex gauge transforms of $A = 0$ is open and dense in the set of all (smooth) gauge fields (again a consequence of the Birkhoff theorem), eq. (42) shows that the Green functions without the gauge field determine the Green functions in the presence of the gauge field, if the latter are regular in A , which is a natural assumption. By subseqent differentiation with respect to A^{01} and A^{10} at $A = 0$ one produces the descendant Green functions with the insertions of the current fields $J = G\partial G^{-1}$ and $\bar{J} = G^{-1}\bar{\partial}G$

$$\mathcal{G}(z_i,\omega_i,u_r,w_s) = \prod_r \frac{\delta}{\delta A^{01}_r(u_r)} \prod_s \frac{\delta}{\delta A^{10}_s(w_s)}\, \mathcal{G}(z_i,\omega_i)\;.$$

They are analytic (anti-analytic) functions of u_r (w_s) as long as the points do not coincide (why?). Their Taylor coefficients at u_r, w_s $= z_i$ allow to recover the action of the amplitudes $\mathbf{A}(\Sigma_{z_i,q_i},p_i,g)$ on the (descendant) states in $\Gamma_{\omega_i}(\mathcal{L})$ (why?) . In particular, from the three-point function given by the tensor of the operator product coefficients, we may determine the amplitudes $\mathbf{A}$ for the "pants": $\mathbf{C}P^1$ without three discs, see Fig. 6.

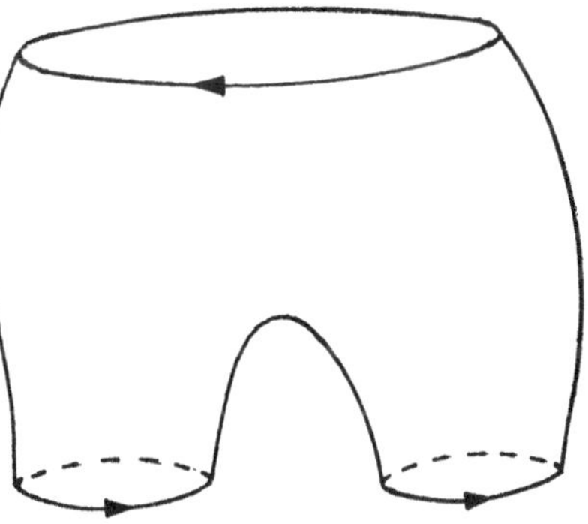

Fig. 6. Pants

Those, together with the amplitudes for the disc and the annuli, should determine, via the unitarity axiom of Section 1 , any amplitudes. That this can be done in a consistent way leading to a CFT satisfying all axioms of Section 1 remains, strictly speaking, still to be proven, although the main ideas of the proof have been already understood [18].

There is a somewhat unexpected consequence of the introduction of an external gauge field into the Green functions. Regularity of $\mathcal{G}(z_i,\omega_i,A)$ in A requires that some of the Green functions vanish. To see how this works, it is convenient (although not necessary) to separate in (42) the dependence on A^{10} and A^{01} .

Consider linear forms σ on $\bigotimes_i h_{\omega_i}$ depending holomorphically on A^{01} and satisfying the Ward identity

$$\sigma({}^H A^{01}) = exp[S_{\mathbf{C}P^1}(H^{-1}, A^{01})] \; \sigma(A^{01}) \bigotimes_i l(H(z_i)^{-1}) . \tag{43}$$

They can be viewed as holomorphic sections of a vector bundle over the space of the gauge fields A^{01} modulo the complex gauge transformations. The space of such sections has been identified in the recent paper by E. Witten [19] with the Schrödinger picture space of states for a three-dimensional topological gauge theory with action given by the Chern-Simons form and in the presence of Wilson lines in representations ω_i crossing the fixed time surface at points z_i . Given σ_1, σ_2 satisfying (43) ,

$$\sigma_1(A^{01}) \otimes \sigma_2(A^{01}) \; exp[-\frac{ik}{2\pi}\int \langle (A^{01})^*, A^{01}\rangle] \tag{44}$$

transforms like $\mathcal{G}(z_i, \omega_i, A)$ under the complex gauge transformations (see (42)) .

EXERCISE. Prove this.

We shall show that for some choices of representations ω_i , there are no non-zero σ's satisfying (43) . Our argument will be a simple modification of a similar one presented in [20]. For simplicity, we shall study the case of the SU(2) WZW model. Consider the loop $\begin{pmatrix} z & 0 \\ c & z^{-1} \end{pmatrix}$ in SL(2,C) , $c \in \mathbf{C}$, and its extension H_c to disc D

$$H_c = \begin{pmatrix} z + cz(1-f(|z|^2))^{1/2} & (1-f(|z|^2))^{1/2} \\ cf(|z|^2) - (1-f(|z|^2))^{1/2} & f(|z|^2)z^{-1} \end{pmatrix}$$

where f is a smooth function, $f(x) = 0$ for $x \leq 1/3$, $f(x) = 1$ for $x \geq 1/3$.
Define

$$A_c^{01} = \begin{cases} H_c^{-1}\bar\partial H_c & \text{on } D , \\ 0 & \text{on } D' . \end{cases}$$

Notice that for $c \neq 0$

$$\begin{pmatrix} z & 0 \\ c & z^{-1} \end{pmatrix} = \begin{pmatrix} 1 & z/c \\ 0 & 1 \end{pmatrix} \begin{pmatrix} 0 & -1/c \\ c & z^{-1} \end{pmatrix} \equiv H_{1c}H_{2c}^{-1}$$

where H_{1c} is analytic on D and H_{2c} is on D' .
Let

$$H'_c = \begin{cases} H_{1c}^{-1}H_c & \text{on } D , \\ H_{2c}^{-1} & \text{on } D' . \end{cases}$$

H'_c is smooth on $\mathbf{C}P^1$ and

$$A_c^{01} = H_c'^{-1}\bar\partial H'_c .$$

Thus for each $c \neq 0$ (but not for $c = 0$) , A_c is a complex gauge transform of $A = 0$.

Let us fix $z_1 = 0$, $z_2 = 1$, $z_3 = \infty$ (this is always possible due to the Möbius invariance) and the HW's corresponding to spins j_i, $j_i \leq k/2$. From (43) , we infer that

$$\sigma(A_c^{01})_{m_1m_2m_3}$$
$$= exp[S_{CP^1}(H_c')] \sum_{m_i'} D^{j_1}_{m_1'm_1}(H_c'(0)) \; D^{j_2}_{m_2'm_2}(H_c'(1)) \; D^{j_3}_{m_3'm_3}(H_c'(\infty)) \;\; \sigma(0)_{m_1'm_2'm_3'}$$

where $D^j_{mm'}(G_0)$ are the matrices of spin j representation in the standard magnetic number basis.

Since $A_c^{01} = (G_0H_c')^{-1}\bar{\partial}(G_0H_c')$ for any $G_0 \in SL(2,\mathbf{C})$ and $S_{CP^1}(G_0H_c') = S_{CP^1}(H_c')$, $\sigma(0)$ has to be an invariant tensor so that it is given by the Clebsch-Gordan coefficients

$$\sigma(0)_{m_1'm_2'm_3'} = C\,(-1)^{j_1-m_1'} < j_1\,(-m_1') \mid j_2\, m_2'\; j_3\, m_3' >$$

where C is a constant. Since

$$D^{j_1}_{m_1'm_1}(H_c'(0)) = (-1)^{j_1+m_1}\,\delta_{m_1+m_1',0}\,,$$

$$D_{m_2'm_2}(H_c'(1)) = c^{m_2-m_2'}\, D^{j_2}_{m_2'm_2}\begin{pmatrix} 0 & -1 \\ 1 & 1 \end{pmatrix},$$

$$D^{j_3}_{m_3'm_3}(H_c'(\infty)) = c^{m_3-m_3'}\,\delta_{m_3+m_3',0}\,,$$

$$\sigma(A_c^{01})_{m_1m_2m_3} = C\,(-1)^{j_3-m_3}\, exp[S_{CP^1}(H_c')] \sum_{m_2'} c^{m_2-m_2'+m_3}$$
$$D^{j_2}_{m_2'm_2}\begin{pmatrix} 0 & -1 \\ 1 & 1 \end{pmatrix} < j_1\, m_1 \mid j_2\, m_2'\; j_3\,(-m_3) >$$

$$= C\,(-1)^{j_3-m_3}\, c^{-m_1+m_2+m_3}\, D^{j_2}_{(m_1+m_3)m_2}\begin{pmatrix} 0 & -1 \\ 1 & 1 \end{pmatrix}$$
$$< j_1\, m_1 \mid j_2\,(m_1+m_3)\; j_3\,(-m_3) > \;\; exp[-S_{CP^1}(H_c')]\,.$$

EXERCISE. Prove that

$$exp[-S_{CP^1}(H_c')] = \mathcal{O}(c^k)$$

when $c \longrightarrow 0$.

Choosing $m_1 = j_1,\ m_2 = -j_2,\ m_3 = -j_3$ and noting that $D^{j_2}_{(j_1-j_3)(-j_2)}\begin{pmatrix} 0 & -1 \\ 1 & 1 \end{pmatrix} \neq 0$ as well as $< j_1\, j_1 \mid j_2\,(j_1-j_3)\; j_3\, j_3 > \neq 0$, we infer that the constant $C = 0$ unless $j_1 + j_2 + j_3 \leq k$. Thus

$$(45) \qquad j_1 = |j_2 - j_3|,\ |j_2 - j_3| + 1,\ \ldots,\ \min(j_2 + j_3,\ k - j_2 - j_3)$$

is a necessary condition for the existence of a non-zero σ satisfying (43). It is not difficult to prove that (45) is also a sufficient condition.

The reader should have noticed a similarity between the above argument and the one leading to the elimination of all but a finite-dimensional subspace of ψ_0 in (26). The first consisted of looking for the holomorphic sections of a vector bundle over the space of the gauge fields on CP^1 modulo the complex gauge transformations. The second could be reformulated as a search for the holomorphic sections of a bundle over $L\mathbf{G}^{\mathbf{C}}$ modulo left (right) multipliation by the boundary values of fields holomorphic on D (D') . The point is that the both base spaces are naturally isomorphic (they define the moduli space of the holomorphic $\mathbf{G}^{\mathbf{C}}$-bundles over CP^1 [21]).

The selection rule $j_1 + j_2 + j_3 \leq k$ is an example of the so called fusion rules [17,5] in CFT. It forces some Green functions to vanish (we could have argued directly using (42) without assuming that $\mathcal{G}(z_i, \omega_i, A)$ is a combination of expresions (44) , although the latter follows from our analysis). In general, the space of solutions of (43) is finite dimensional. It may be parametrized by $\sigma(0)$, which takes values in a subspace of linear forms on $\bigotimes_i h_{\omega_i}$ invariant under the diagonal action of $\mathbf{G}^{\mathbf{C}}$. As a function of points (z_i) , these subspaces form a holomorphic vector bundle with a flat connection defined by the (holomorphic part) of the Knizhnik-Zamolodchikov equation (38) . The holonomy of this connection defines a representation of a central extension of the braid group on CP^1 , see [1]. The above holomorphic vector bundles can be realized as "theta function" bundles corresponding to ramified covers of CP^1 [22]. This is probably related to the Coulomb gas representation of the WZW theory [23].

The above considerations, based on the coupling of the WZW model to the external gauge field, may be repeated for the amplitudes on Riemann surfaces with boundary. They lead then to the concept of a modular functor discussed recently by G. Segal [2,24].

3. VERTEX OPERATOR REPRESENTATION

There is a special case of the WZW theory where it reduces to a toroidal compactification. This is the case of $\mathbf{G}$ being a simply-laced group $\mathbf{A}_n$ (=SU(n+1)) , $\mathbf{D}_n$ (=Spin(2n)) or $\mathbf{E}_n$ in Cartan's classification and the coupling constant k equal to 1 . An overwhelming evidence shows that in this case it is enough to limit oneself in the functional integral to fields G with values in the Cartan subgroup $\mathbf{T} \subset \mathbf{G}$. Such fields may be parametrized as

$$G \;=\; exp[2\pi i \; \Phi^j \alpha_j]$$

where α_j, $j = 1, ..., n$, are the (co)roots of Lie algebra $\mathbf{g}$ (spanning its Cartan subalgebra $\mathbf{t}$) and Φ^j are defined modulo integers.

Let $g_{ij} = \langle \alpha_i, \alpha_j \rangle$. For simply-laced groups, (g_{ij}) is a positive matrix of integer entries, even on the diagonal. We may write $g_{ij} = d_{ij} + d_{ij}$ for $d_{ij} \in \mathbf{Z}$. We shall set $b_{ij} = \frac{1}{2}(d_{ij} - d_{ji})$. For a Riemann surface Σ without boundary, the WZW action (6) (with $k = 1$) of a $\mathbf{T}$-valued field G is

$$S_\Sigma(G) \;=\; \pi i \int_\Sigma g_{ij} \, (\partial \Phi^i)(\bar{\partial} \Phi^j) \;+\; \pi i \int_\Sigma b_{ij} \; d\Phi^i d\Phi^j \quad mod \; (2\pi i) \; , \tag{46}$$

compare (5) . In order to prove (46) , it is enough to show the equality of the topological term in (6) and the second term on the right hand side of (46) . The topological term in (6) is constant under homotopies of $\mathbf{T}$-valued fields (why?) so it depends only on the winding numbers $\int_c d\Phi^j$ on cycles c , $\partial c = 0$, in Σ . Since each surface Σ may be glued from tori, see Fig. 7, it is enough to prove (46) for $\Sigma = S^1 \times S^1$ and $G(e^{i\phi}, e^{i\psi}) = exp[i(\phi n^j + \psi m^j)\alpha_j]$.

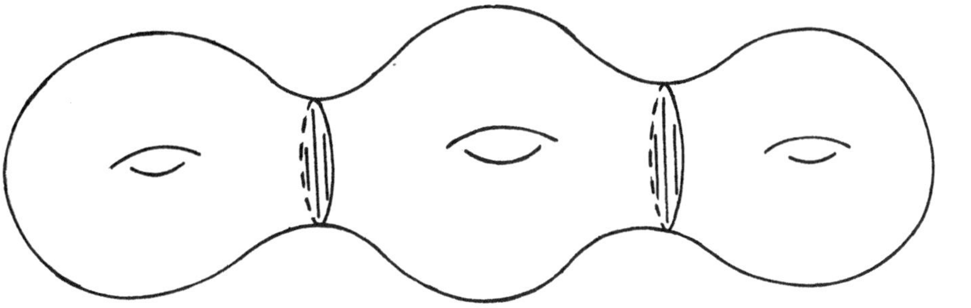

Fig. 7. Decomposition of a Riemann surface into tori

Consider an extension $\tilde{G}$ of field G to $D\times S^1$, $\tilde{G}(z,e^{i\psi}) = G^1(z)exp[i\psi m^j\alpha_j] \in \mathbf{G}$ where $G_1(e^{i\phi}) = exp[i\phi n^j\alpha_j]$. We have

$$-\frac{i}{24\pi}\int\limits_{D\times S^1}\langle\tilde{G}^{-1}d\tilde{G},\ [\tilde{G}^{-1}d\tilde{G},\tilde{G}^{-1}d\tilde{G}]\rangle = \frac{1}{8\pi}\int\limits_D\int\limits_0^{2\pi} d\psi\ \langle m^j\alpha_j,\ [G_1^{-1}dG_1, G_1^{-1}dG_1]\rangle$$

$$= -\frac{1}{2}\int\limits_D d\langle m^j\alpha_j, G_1^{-1}dG_1\rangle = -\frac{i}{2}\int\limits_0^{2\pi} d\phi\ \langle m^j\alpha_j, n^j\alpha_j\rangle = -\pi i\ g_{ij}\ m^i n^j\ .$$

On the other hand,

$$\pi i\int\limits_{S^1\times S^1} b_{ij}\ d\Phi^i d\Phi^j = \frac{i}{4\pi}\int\limits_0^{2\pi}\int\limits_0^{2\pi} b_{ij}\ (d\phi\ n^i + d\psi\ m^i)\ (d\phi\ n^j + d\psi\ m^j)$$

$$= \pi i\ b_{ij}\ (n^i m^j - m^i n^j) = -\pi i\ d_{ij}\ (m^i n^j - n^i m^j)$$

$$= -\pi i\ d_{ij}\ (m^i n^j + n^i m^j) + 2\pi i\ d_{ij}\ n^i m^j = -\pi i\ g_{ij}\ m^i n^j \quad mod\ (2\pi i)$$

as was to be shown.

It is easy to compute now the functional integrals over the $\mathbf{T}$-valued fields exactly. Choose the homology basis (marking) (a_α, b_α) , $\alpha = 1, \ldots,$ genus, on Σ , $\partial\Sigma = \emptyset$, as in Fig. 8. Let (ω_α) be the associated basis of the holomorpic forms, $\int_{a_\alpha}\omega_\beta = \delta_{\alpha\beta}$. The period matrix $(\tau_{\alpha\beta})$, $\tau_{\alpha\beta} = \int_{b_\alpha}\omega_\beta$, satisfies: $\tau_{\alpha\beta} = \tau_{\beta\alpha}$, $(Im\ \tau_{\alpha\beta}) > 0$. For the partition functions $Z(\Sigma,g) \equiv \mathbf{A}(\Sigma,g)$, we obtain

$$Z(\Sigma,g) = \int exp[-S_\Sigma(exp[2\pi i\ \Phi^j\alpha_j])]\ D\Phi \tag{47}$$

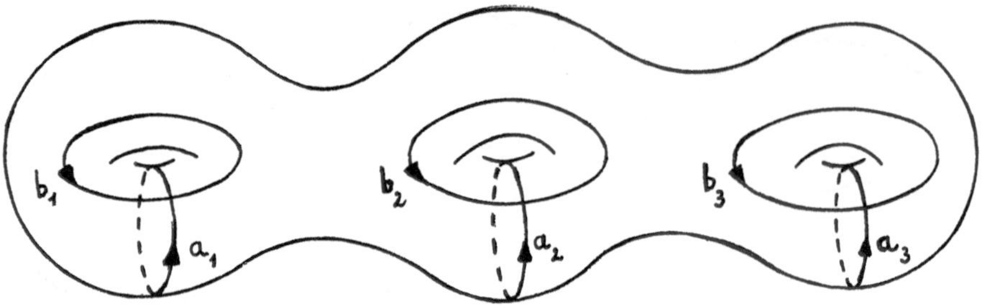

Fig. 8. Marking of a Riemann surface

$$= \sum_{\text{instantons } \Phi_{nm}} exp[-S_\Sigma(exp[2\pi i\ \Phi^j_{nm}\alpha_j])\ ] \int_{\text{fluctuations}} exp[-S_\Sigma(exp[2\pi i\ \tilde{\Phi}^j\alpha_j])\ ]\ D\tilde{\Phi}$$

where Φ_{nm} are the classical local minima of the action

$$\Phi_{nm}(\xi) = \frac{\pi}{i}\int_{\xi_0}^{\xi} [(m-\tau n)^t(Im\ \tau)^{-1}\bar{\omega} - (m-\bar{\tau}n)^t(Im\ \tau)^{-1}\omega]$$

(in a shorthand matrix notation). The fluctuations $\tilde{\Phi}$ are univalued functions (modulo an overall additive integer).

The integral in (47) is essentially Gaussian and gives $(det'(-\Delta_g)\ /\ vol_g\Sigma)^{-n/2}$ (why?) where Δ_g is the Laplacian on Σ (corresponding to metric g), $vol_g\Sigma$ denotes the volume of Σ and n is the rank of the group. det' denotes the determinant with the zero-mode contribution omitted and should be regularized using e.g. the zeta-function regularization [25]. The sum in (47) can be done explicitly.

EXERCISE. Do it. Indication: use the Poisson summation formula to sum over m_j's .

The result is:

$$Z(\Sigma,g) = \sum_{\omega\ \in\ (P/Q)^{\text{genus}}} |\Theta_{\omega+Q^{\text{genus}}}(\tau)|^2\ (\frac{det'(-\Delta_g)}{vol_g\Sigma\ det(Im\ \tau)})^{-n/2}$$

where Q is the (co)root lattice $\bigoplus_{j=1}^{n} \mathbf{Z}\alpha_j$, P is the (co)weight lattice of vectors λ in $\mathbf{t}$ such that $\langle\lambda,\alpha_j\rangle \in \mathbf{Z}$, $Q \subset P$, and

$$\Theta_{\omega+Q^{\text{genus}}}(\tau) = \sum_{\alpha\in Q^{\text{genus}}} exp[\pi i\langle\omega+\alpha,\tau(\omega+\alpha)\rangle]\ .$$

In particular, for $\mathbf{G} = \mathbf{E}_8$, $P = Q$ and

$$\mathbf{A}(\Sigma, g) = |\Theta_{Q^{\text{genus}}}(\tau)|^2 \, (\frac{det'(-\Delta_g)}{vol_g \Sigma \; det(Im\ \tau)})^{-4}$$

$$\equiv \|\Theta_{Q^{\text{genus}}}(\tau) \; (1 \otimes \omega_1 \ldots \omega_{\text{genus}})^{-4}\|^2_{\text{Quillen}}$$

where $\Theta_{Q^{\text{genus}}}(\tau) \; (1 \otimes \omega_1 \ldots \omega_{\text{genus}})^{-4}$ is treated as a holomorphic section of the "determinant bundle" [26] over the (moduli) space of Riemann surfaces whose fiber over Σ is $(\bigwedge^{\max} ker\bar{\partial})^{-4} \otimes (\bigwedge^{\max} coker\bar{\partial})^4$. $\| \cdot \|_{\text{Quillen}}$ denotes Quillen's hermitian metric on the determinant bundle defined by the above, cf. [26].

As we see, the partition functions of the $k = 1$ $\mathbf{E}_8$ WZW theory are given as hermitian squares of holomorphic sections of line bundles over the moduli spaces of complex curves. Such theories are called holomorphically factorizable and the $\mathbf{E}_8$ theory gives the simplest instance of such a case. For other groups, we have to replace line bundles by finite-dimensional vector bundles which is the situation characteristic of, more general, rational CFT's.

The computation of the functional integral over the $\mathbf{T}$-valued fields on Σ with boundary may be also done in a closed form. The space of states H becomes $\Gamma_{hol}(\mathcal{L}|_{L\mathbf{T}^{\mathbf{C}}})$ and carries a representation of $\hat{L\mathbf{T}}^{\mathbf{C}} \times \hat{L\mathbf{T}}^{\mathbf{C}}$, where $\hat{L\mathbf{T}}^{\mathbf{C}}$ is a Heisenberg group,

$$1 \longrightarrow \mathbf{C}^* \longrightarrow \hat{L\mathbf{T}}^{\mathbf{C}} \longrightarrow L\mathbf{T} \longrightarrow 1 \, .$$

H can be realized as $\sum_{\eta \, \in \, P/Q} H_\eta \otimes \bar{H}_\eta$ where the Fock space

$$H_\eta = \sum_{\alpha \in Q, \; n_I} \mathbf{C} a_{-n_I} \, |\eta + \alpha> \, .$$

Here

$$a_{-n_I} = a_{j_1(-n_1)} \cdots a_{j_{|I|}(-n_{|I|})} \, , \qquad n_j > 0 \, ,$$

$$[a_{in}, a_{jm}] = n \, \delta_{n+m,0} \, g_{ij} \, ,$$

$$a^*_{jn} = a_{j(-n)} \quad (n \neq 0) \, .$$

Let for $|z| = 1$,

$$\Phi^j(z) = \Phi^j_0 + \frac{n^j}{2\pi i} \, log\ z + \sum_{n \neq 0} \frac{1}{n} \, \Phi^j_n \, z^{-n} \, .$$

Representing $exp[2\pi i \, \Phi^j \alpha_j]$ by

$$e^{2\pi i n^j a_{j0}} \, exp[\, 2\pi i \, (\Phi^j_0 \, \langle \alpha_j, p \rangle + \sum_{n \neq 0} \frac{1}{n} \, \Phi^j_n a_{j(-n)}) \,] \, ,$$

where $[p, a_{jn}] = [e^{2\pi i n^j a_{j0}}, a_{jn}] = 0$ for $n \neq 0$ and

$$p \, |\eta + \alpha> = (\eta + \alpha) \, |\eta + \alpha>$$

and

$$e^{2\pi i n^j a_{j0}} \, |\eta + m^j \alpha_j> = (-1)^{d_{ij} m^i n^j} \, |\eta + (m^j + n^j)\alpha_j> \, ,$$

gives the left action of $\hat{LT}^{\mathbf{C}}$ in H and, by infinitesimalization, the left action of the Heisenberg algebra $\hat{Lt}^{\mathbf{C}}$.

If we put

$$\hat{\Phi}_j(z) = a_{j0} + \frac{1}{2\pi i} \log z \, \langle \alpha_j, p \rangle + \sum_{n \neq 0} \frac{i}{n} a_{jn} z^{-n}$$

then

$$: exp[\pm 2\pi i \, \hat{\Phi}_j(z)] : = \sum_n J_n^{e_{\pm \alpha j}} z^{-n-1}$$

(with the standard Fock-space normal ordering) defines the left action in H of the generators $J_n^{e_{\pm\alpha_j}}$ of $\hat{Lg}^{\mathbf{C}}$ corresponding to the step generators $e_{\pm\alpha_j}$ in $\mathbf{g}^{\mathbf{C}}$. Thus the whole action of $\hat{Lg}^{\mathbf{C}}$ in H can be explicitly described in the Fock space language. This is the celebrated vertex-operator representation of [27], see [28] for a recent treatment.

4. COSET CONSTRUCTION

The WZW theories may be used to generate new conformal models, the so called coset theories. Let $\mathbf{H} \subset \mathbf{G}$ be a (connected) subgroup. We may couple the group $\mathbf{G}$ WZW model to the gauge field A exactly as in (39) but instead of keeping A external, we shall integrate it out with the restriction that A takes values only in the Lie algebra $\mathbf{h}^{\mathbf{C}}$. The functional integrals

$$\int - \; exp[-S_\Sigma(G, A)] \; DG \; DA$$

are again computable, at least for simple geometries, by formal manipulations.

A. Parafermions

As the simplest example, consider the case $\mathbf{G}$ =SU(2) and $\mathbf{H}$ =U(1) being its diagonal subgroup. Let us compute the toroidal partition functions $Z_{\mathrm{SU(2)/U(1)}}(\tau)$ of the coset theory. This will be done by parametrizing A as

$$A = A_h - 2\pi\partial(h - i\lambda)\sigma_3 + 2\pi\bar{\partial}(h + i\lambda)\sigma_3$$

where A_h is a harmonic field

$$A_h = -\frac{1}{2}[(\theta - \bar{\tau}\phi)(Im\,\tau)^{-1}dz - (\theta - \tau\phi)(Im\tau)^{-1}d\bar{z}]\,\sigma_3 \,,$$

$$\int_0^{2\pi} A_h = -2\pi i\phi\sigma_3 \,, \qquad \int_0^{2\pi\tau} A_h = -2\pi i\theta\sigma_3 \,.$$

We shall take λ such that $e^{2\pi i\lambda}$ is a smooth function. We shall also impose the condition $\int h \, d^2z = 0$. θ and ϕ may be limited to $[0,1[$. Let us also put $x = \lambda(0)$.

$$dx \; DA = \frac{\partial(x, A)}{\partial(\lambda, h, \phi, \theta)} \, De^{2\pi i\lambda} \, D'h \; d\phi \; d\theta$$

where $D'h$ is the volume element in the subspace orthogonal to the constant mode. The Jacobian is

$$det'(-\Delta)\ /\ (Im\ \tau)^{1/2}\ .$$

Thus

$$Z_{SU(2)/U(1)}(\tau) = \int exp[-S_{T_\tau}(G,A)]\ DG\ DA = \int exp[-S_{T_\tau}(G,A)]\ DG\ DA \int_0^1 dx$$

$$= det'(-\Delta)/(Im\ \tau)^{1/2} \int exp[-S_{T_\tau}(G,A)]\ DG\ De^{2\pi i\lambda}\ D'h\ d\phi\ d\theta\ .$$

But, denoting $H = e^{2\pi(h+i\lambda)\sigma_3}$, we infer from (41) that

$$S(G,A) = S(HGH^*, A_h) + 8\pi i\ k \int (\partial h)(\bar{\partial} h)\ .$$

Hence

$$Z_{SU(2)/U(1)}(\tau) = det'(-\Delta)/(Im\ \tau)^{1/2} \int exp[-S(HGH^*, A_h) - 8\pi ik \int (\partial h)(\bar{\partial} h)]\ DG\ De^{2\pi i\lambda}\ D'h\ d\phi\ d\theta$$

$$= (Im\ \tau)^{-1/2} k^{1/2}\ det'(-\Delta)^{1/2} \int exp[-S(G,A_h)]\ DG\ d\phi\ d\theta$$

where in the last line we have used the (generalized) invariance of the DG measure, have integrated out over h (using the zeta-function regularization of the determinant) and have dropped $\int De^{2\pi i\lambda}$ i.e. the volume of the gauge group normalized to 1 .

In order to compute $\int exp[-S(G,A_h)]\ DG$, let us notice that, by applying blindly (41) with $H_1 = exp[-\frac{1}{2}(\theta - \tau\phi)(Im\ \tau)^{-1}(z-\bar{z})\sigma_3]$, we get

$$exp[-S_{T_\tau}(G,A_h)] = exp[-S_{T_\tau}(H_1 G H_1^*)]\ exp[S_{T_\tau}(H_1 H_1^*)]\ . \tag{48}$$

The problem is that $H_1 G H_1^*$ is well defined on the cylinder $C_\tau = \{z \in \mathbf{C} \mid 0 \le Imz \le 2\pi\ Im\ \tau\}\ /\ (2\pi\mathbf{Z})$ but not on T_τ . In fact

$$H_1 G H_1^*|_{Imz=0} = exp[2\pi i(\theta - \tau\phi)\sigma_3]\ H_1 G H_1^*|_{Imz=2\pi Im\tau}\ exp[-2\pi i(\theta - \bar{\tau}\phi)\sigma_3]\ .$$

It is easy to show that (48) still holds if we interprete $exp[-S_{T_\tau}(H_1 G H_1^*)]$ as

$$exp[-S_D(exp[2\pi i(\theta - \tau\phi)\sigma_3])] \bullet exp[-S_{C_\tau}(H_1 G H_1^*)] \bullet exp[-S_D(exp[-2\pi i(\theta - \bar{\tau}\phi)\sigma_3])] \tag{49}$$

where the multiplication acts on the $\mathcal{L}_{H_1 G H_1^*|_{Imz=2\pi Im\tau}}$ factor of $exp[-S_{C_\tau}(H_1 G H_1^*)]$. In particular, an easy computation gives

$$exp[-S_{T_\tau}(H_1 H_1^*)] = exp[4\pi k(Im\ \tau)\phi^2]\,.$$

We are left with the calculation of

$$\int exp[-S_{T_\tau}(H_1 G H_1^*)]\, DG = Tr\ [q^{L_0}\bar{q}^{\bar{L}_0}\ l(exp[2\pi i(\theta-\tau\phi)\sigma_3])\ r(exp[2\pi i(\theta-\tau\phi)\sigma_3])]$$

in the Schrödinger picture language. The latter expression may be immediately written in terms of the SU(2) Kac-Moody characters, see (35) , as $(c = \frac{3k}{k+2})$

$$(q\bar{q})^{-c/24}\sum_{j=0}^{k/2}|\chi_j^k(\tau, exp[2\pi i(\theta-\tau\phi)\sigma_3])|^2\,.$$

We shall need, however, a somewhat more refined expression. Expansion of the characters

$$\chi_j^k(\tau, exp[2\pi i u\sigma_3]) = \sum_{m=j\ mod\ 1}\chi_{jm}^k(\tau)\, e^{4\pi i m u}$$

gives

$$\int exp[-S_{T_\tau}(H_1 G H_1^*)]\, DG$$

$$= (q\bar{q})^{-c/24}\sum_{j=0}^{k/2}\sum_{m,n=j\ mod\ 1}\chi_{jm}(\tau)\,\overline{\chi_{jn}^k(\tau)}\, exp[4\pi i m(\theta-\tau\phi) - 4\pi i n(\theta-\bar{\tau}\phi)]\,.$$

Since

$$det'(-\Delta)^{1/2} = 2\,(Im\ \tau)\,|\eta(\tau)|^2\,,$$

where $\eta(\tau) = e^{\pi i\tau/12}\prod_{n\geq 1}(1-e^{2\pi i n\tau})$ is the Dedekind function, putting everything together, we obtain:

$$Z_{\mathrm{SU(2)/U(1)}}(\tau) = 2(Im\ \tau)^{1/2}k^{1/2}(q\bar{q})^{-c/24}\,|\eta(\tau)|^2\int_0^1 d\phi\int_0^1 d\theta\sum_{j=0}^{k/2}\sum_{m,n=j\ mod\ 1}$$
$$\chi_{jm}^k(\tau)\,\overline{\chi_{jn}(\tau)}\, exp[-4\pi k(Im\ \tau)\phi^2 + 4\pi i(m-n)\theta + 4\pi i(n\bar{\tau}-m\tau)\phi]$$

$$= 2(Im\ \tau)^{1/2}k^{1/2}(q\bar{q})^{-c/24}\,|\eta(\tau)|^2\int_0^1 d\phi\sum_{j=0}^{k/2}\sum_{m=j\ mod\ 1}$$
$$|\chi_{jm}^k(\tau)q^{-m^2/k}|^2\, exp[-4\pi k(Im\ \tau)(\phi-m/k)^2]\,.$$

$\chi_{jm}^k(\tau)q^{-m^2/k} \equiv c_{jm}^k(\tau)$ are the so called string functions [29]. They have the important property

$$c_{jm}^k = c_{j(m+k)}^k$$

which allows to extend the $d\phi$ integral to the whole line and gives finally

$$Z_{SU(2)/U(1)}(\tau) = (q\bar{q})^{-c/24} |\eta(\tau)|^2 \sum_{j=0}^{k/2} \sum_{\substack{0 \leq m < k \\ m = j \ mod\ 1}} |c^k_{jm}(\tau)|^2 .$$

This coincides with the expressions for the partition functions of the parafermionic theories of [30].

The general SU(2) WZW model can be factorized into the parafermionic theory and the free field (with values in the Cartan subgroup). Let us see how this works on the example of the diagonal WZW Green functions given formally as

$$\int \prod_i D^{j_i}_{m_i m_i}(G(z_i))\ exp[-S_{CP^1}(G)]\ DG \ \equiv\ \mathcal{E} .$$

By changing variables: $G \mapsto HGH^*$ for $H = exp[2\pi(h + i\lambda)\sigma_3]$, we obtain

$$\mathcal{E} = \int \prod_i (e^{4\pi m_i h(z_i)}\ D^{j_i}_{m_i m_i}(G(z_i)))\ \ exp[-S_{CP^1}(HGH^*)]\ DG$$

$$= \int \prod_i (e^{4\pi m_i h(z_i)}\ D^{j_i}_{m_i m_i}(G(z_i)))\ \ exp[-S_{CP^1}(G, A) + 8\pi ik \int_{CP^1} (\partial h)(\bar{\partial} h)]\ DG$$

where $A^{01} = H^{-1}\bar{\partial} H$. By averaging this expression with the normalized "measure"

$$\prod_i e^{-4\pi m_i h(z_i)}\ exp[-8\pi ik \int (\partial h)(\bar{\partial} h)]\ det'(-\Delta)\ (vol)^{-1/2}\ De^{2\pi i\lambda}\ D'h\ /\ \text{normalization} ,$$

we obtain

(50)
$$\mathcal{E} = \int \prod_i D^{j_i}_{m_i m_i}(G(z_i))\ exp[-S_{CP^1}(G, A)]\ DG\ DA$$
$$(det'(-\Delta)/vol)^{-1/2}\ \ exp[-\frac{\pi}{k} \sum_{i,j} m_i m_j (-\Delta')^{-1}(z_i, z_j)]$$
$$= \int \prod_i D^{j_i}_{m_i, m_i}(G(z_i))\ exp[-S_{CP^1}(G, A)]\ DG\ DA$$
$$\int exp[-4\pi i \sum_j m_j \Phi(z_j) - 8\pi ik \int (\partial\Phi)(\bar{\partial}\Phi)]\ D\Phi$$

where the first integral gives the contribution of the "parafermionic primaries" [30] and the last one is the contribution of the (free) field $exp[2\pi i\Phi\sigma_3]$ taking values in the Cartan subgroup of SU(2) . Both expressions are formal and require multiplicative renormalization. This of the free field is given by the normal ordering. Whenever we know the WZW Green functions (e.g. for the four-point functions), relation (50) allows to find the parafermionic correlations. For $k = 1$, the parafermionic theory is trivial (e.g. the parafermionic Green functions are constant). This is why the vertex operator representation works in this case.

B. Unitary series

Another classical example of the coset theory corresponds to **G** =SU(2)×SU(2) and **H** =SU(2) being the diagonal subgroup. The WZW model with (non simple) group **G** will be composed of two SU(2) models, one with coupling constant k and another one with $k=1$. The partition functions of the coset theory are given as

$$\int exp[-kS(G_1,A)-S(G_2,A)]\, DG_1\, DG_2\, DA$$

(with the k-dependence of S made explicit). Let us sketch their computation in the case of torus T_τ, see [31]. We may parametrize an open dense set of A's as

$$A^{01} = (H_1H)^{-1}\bar\partial(H_1H) = H^{-1}A_h^{01}H + H^{-1}\bar\partial H$$

with $H: T_\tau \longrightarrow \mathrm{SL(2,C)}$, $H_1(z)=exp[-\frac{1}{2}(\theta-\tau\phi)(Im\ \tau)^{-1}(z-\bar z)\sigma_3]$ and $A_h^{01}=H_1^{-1}\bar\partial H_1$. Consider the positive matrix $H(0)H(0)^*$ in SL(2,C). It may be written as

$$\begin{pmatrix} e^\sigma & re^{2\pi i\psi} \\ re^{-2\pi i\psi} & e^{-\sigma}(1+r^2)\end{pmatrix}.$$

The map

$$(H,\phi,\theta) \longmapsto (A,\psi,\sigma)$$

is four to one if $\phi,\ \theta\in[0,1[$. Its Jacobian is $det'(\bar\partial_A^*\bar\partial_A)\,\Delta_1$ where $\bar\partial_A$ acts on sl(2,C)-valued functions Λ by

$$\bar\partial_A\Lambda = \bar\partial\Lambda + [A^{01},\Lambda]$$

and Δ_1 is the contribution of the zero modes.
We may write

$$Z_{\mathrm{G/H}}(\tau) = \int_0^1\int \delta(\sigma)d\sigma\int exp[-kS_{T_\tau}(G_1,A)-S_{T_\tau}(G_2,A)]\, DG_1\, DG_2\, DA$$

$$= \frac{1}{4}\int_0^1 d\phi\int_0^1 d\theta\int exp[-kS_{T_\tau}(G_1,A)-S_{T_\tau}(G_2,A)]\, det'(\bar\partial_A^*\bar\partial_A^*)\,\Delta_1\,\delta(\sigma)\, DG_1\, DG_2\, DH\ .$$

By (41),

$$S(G_a,A) = S(H_1HG_aH^*H_1^*) - S(H_1HH^*H_1^*)\ .$$

Besides, by the Polyakov-Wiegmann chiral anomaly calculation [32], with a careful treatment of the zero modes,

$$det'(\bar\partial_A^*\bar\partial_A)\,\Delta_1 = const.\ (Im\tau)^{-1}\ exp[4S(H_1HH^*H_1^*)-4S(H_1H_1^*)]\ det'(\bar\partial_{A_h}\bar\partial_{A_h})\ .$$

Hence

$$Z_{\mathrm{G/H}} = const.\ (Im\ \tau)^{-1}\int_0^1 d\phi\int_0^1 d\theta\ exp[-kS_{T_\tau}(H_1HG_1H^*H_1^*)-S_{T_\tau}(H_1HG_2H^*H_1^*)$$

$$+(k+5)S_{T_\tau}(H_1HH^*H_1^*)]\ \delta(\sigma)\ exp[-4S_{T_\tau}(H_1H_1^*)]\ det'(\bar\partial_{A_h}^*\bar\partial_{A_h})\, DG_1\, DG_2\, DH$$

$$\equiv const.\ (Im\ \tau)^{-1}\int_0^1 d\phi\int_0^1 d\theta\ \ Z^k_{\mathrm{SU(2)}}(\tau,e^{2\pi i(\theta-\tau\phi)\sigma_3})\ Z^1_{\mathrm{SU(2)}}(\tau,e^{2\pi i(\theta-\tau\phi)\sigma^3})$$

$$\mathcal{Z}^k_{H_3}(\tau, e^{2\pi i(\theta-\tau\phi)\sigma_3})\ exp[-4S_{T_\tau}(H_1 H_1^*)]\ det'(\bar\partial^*_{A_h}\bar\partial_{A_h})$$

where the partition functions on the right hand side correspond to the twisted WZW models

$$Z^k_{\mathrm{SU(2)}}(\tau, e^{2\pi i(\theta-\tau\phi)\sigma_3}) = (q\bar q)^{-\frac{3k}{k+2}/24}\sum_{j=0}^{k/2}|\chi^k_j(\tau, e^{2\pi i(\theta-\tau\phi)\sigma_3})|^2$$

and to the (twisted) WZW-type sigma model with fields HH^* (which may be thought of as taking values in the three-dimensional hyperbolic space H^3 =SL(2,C)/SU(2))

$$\mathcal{Z}^{k+5}_{H_3}(\tau, e^{2\pi i(\theta-\tau\phi)\sigma_3}) = const.\ (Im\ \tau)^{1/2}\ (q\bar q)^{(k+5)\phi^2}\ det'(\bar\partial^*_{A_h}\bar\partial_{A_h})^{-1/2}\ .$$

The crucial input in the furthur argument is the observation by Goddard, Kent and Olive [33] that

$$\chi^k_j(\tau, e^{iu\sigma_3})\ \chi^1_\epsilon(\tau, e^{iu\sigma_3}) = \sum_{\substack{l=0\\ l=j+\epsilon\ mod\ 1}}^{(k+1)/2} \chi^{k+1}_l(\tau, e^{iu\sigma_3})\ \chi^{\mathrm{Vir}}_{h_{2j+1,2l+1}}(\tau) \tag{51}$$

where $\chi^{\mathrm{Vir}}_{h_{p,q}}(\tau)$ are the characters ($= Tr\ q^{L_0}$) in the irreducible unitary, HW (i.e. generated by a HW vector ψ such that $L_n\psi = 0\ ,\ \ n > 0$, and $L_0\psi = h_{p,q}\psi$) representations of the Virasoro algebra. The ones appearing in the formula correspond to the central charge $c = 1 - \frac{6}{(k+2)(k+3)}$ and $h_{p,q} = \frac{[(k+3)p-(k+2)q]^2-1}{4(k+2)(k+3)}$ and run through the complete list of the $c < 1$ irreducible unitary HW representations of the Virasoro algebra [34].

Inserting the decomposition (50) and performing the $d\phi$, $d\theta$ integrals leads finally to the expression

$$Z_{\mathbf{G}/\mathbf{H}}(\tau) = (q\bar q)^{-(1-\frac{6}{(k+2)(k+3)})/24}\sum_{j=0}^{k/2}\sum_{l=0}^{j}|\chi_{h_{2j+1,2l+1}}(\tau)|^2\ ,$$

i.e. the partition functions of the (principal series) of the unitary minimal theories [17,35].

One may also use the coset representation of the minimal unitary models to generate the Green functions of the latter. For example, the functional integrals

$$\int\prod_i(Re\sum_{m_i}D^{j_i}_{m_im_i}(G_1(z_i)))\ \ exp[-kS_{CP^1}(G_1,A) - S_{CP^1}(G_2,A)]\ DG_1\ DG_2\ DA$$

generate the Green functions of the diagonal fields $\phi_{2j_i+1,2j_i+1}$ of the minimal models [17]. By the change of variables $A^{01} = H^{-1}\bar\partial H$ they can be expressed in terms of the WZW Green functions [31]. The minimal theories are believed to describe the scaling limit of the critical $P^{2k+2}(\Phi)_2$ theories [36]. The $\phi_{2,2}$ field corresponds to the Φ field and one can compute explicitly its four-point functions in terms of the hypergeometric functions.

References

[1] J. FRÖHLICH - Contribution to this volume

[2] G. SEGAL - *The definitions of conformal field theory*, in *Links Between Geometry and Mathematical Physics*, pp. 13-17, MPI/87-58 preprint and the talk at the International Congress of Mathematical Physics, Swansea 1988

[3] C. VAFA - *Operator formalism on Riemann surface*, Phys. Lett. B 190, 47-54 (1987), *Conformal theories and punctured surfaces*, Phys. Lett. B 199, 195-202 (1987), *Conformal algebra of Riemann surfaces*, HUTP-88/A053 preprint,

L. ALVAREZ-GAUMÉ, C. GOMEZ, G. MOORE, C. VAFA - *Strings in the operator formalism*, Nucl. Phys. B 303, 455-521 (1988)

[4] E. WITTEN - *Non-abelian bosonization in two-dimensions*, Commun. Math. Phys. 92, 455-477 (1984)

[5] D. GEPNER, E. WITTEN - *String theory on group manifolds*, Nucl. Phys. B 278, 493-549 (1986)

[6] V. KNIZHNIK, A. B. ZAMOLODCHIKOV - *Current algebra and Wess-Zumino model in two dimensions*, Nucl. Phys. B 247, 83-103 (1984)

[7] G. FELDER, K. GAWĘDZKI, A. KUPIAINEN - *Spectra of Wess-Zumino-Witten models with arbitrary simple groups*, Commun. Math. Phys. 117, 127-158 (1988)

[8] A. M. POLYAKOV - *Interaction of Goldstone particles in two dimensions. Application to ferromagnets and massive Yang-Mills fields*, Phys. Lett. B 59, 79-81 (1975)

[9] A. M. POLYAKOV, P. B. WIEGMANN - *Goldstone fields in two dimensions with multivalued actions*, Phys. Lett. B 141, 223-228 (1984)

[10] A. PRESSLEY, G. SEGAL - *Loop Groups*, Clarendon Press, Oxford 1986

[11] V. G. KAC - *Infinite Dimensional Lie Algebras*, Cambridge University Press, Cambridge 1985

[12] R. GOODMAN, N. R. WALLACH - *Structure and unitary cocycle representations of loop groups and the group of diffeomorphisms of the circle*, J. Reine Angew. Math. 347, 69-133 (1984)

[13] H. SUGAWARA - *A field theory of currents*, Phys. Rev. 170, 1659-1662 (1968),

I. B. FRENKEL - *Two constructions of affine Lie algebra representations and Boson-Fermion correspondence in quantum field theory*, J. Func. Anal. 44, 259-327 (1981)

[14] V. G. KAC - *Infinite-dimensional Lie algebras and Dedekind's η-function*, Func. Anal. Appl. 8, 68-70 (1974)

[15] P. CHRISTE, R. FLUME - *The four-point correlations of all primary operators of the* d=2 *conformal invariant* SU(2) *σ- model with Wess-Zumino term*, Nucl. Phys. B 282, 466-494 (1987)

[16] A. B. ZAMOLODCHIKOV, V. A. FATEEV - *Operator algebra and correlation functions of the two-dimensional Wess-Zumino* SU(2)×SU(2) *chiral model*, Yad. Fiz. 43, 1031-1044 (1986)

[17] A. A. BELAVIN, A. M. POLYAKOV, A. B. ZAMOLODCHIKOV - *Infinite conformal symmetry in two-dimensional quantum field theory*, Nucl. Phys. B 241, 333-380 (1984)

[18] I. FRENKEL - the talk at the International Congress of Mathematical Physics, Swansea 1988,

H. SONODA - *Sewing conformal field theories I* and *II*, LBL-25140 (1988) and LBL-25316 (1988) preprints,

G. FELDER, R. SILVOTTI - *Modular covariance of minimal model correlation functions*, to appear in Commun. Math. Phys.

[19] E. WITTEN - *Quantum field theory and Jones polynomial*, IASSNS-HEP-88/33 preprint

[20] G. FELDER, K. GAWĘDZKI, A. KUPIAINEN - The spectrum of Wess-Zumino-Witten models, Nucl. Phys. B 299, 355-366 (1988)

[21] M. F. ATIYAH, R. BOTT - *The Yang-Mills equations over Riemann surfaces*, Phil. Trans. R. Soc. Lond. A 308, 523-615 (1982),

M. S. NARASIMHAN, C. S. SHESHADRI - *Stable and unitary vector bundles on a compact Riemann surface*, Ann. Math. 82, 540-567 (1965),

S. K. DONALDSON - *A new proof of a theorem of Narasimhan and Seshadri*, J. Diff. Geom. 18, 269-277 (1983) and *Infinite determinants, stable bundles and curvature*, Duke Math. J. 54, 231-247 (1987)

[22] N. HITCHIN - Private communication

[23] D. BERNARD, G. FELDER - Private communication

[24] K. GAWĘDZKI - *Conformal field theory*, IHES/P/88/56 preprint, to appear in Astérisque

[25] D. B. RAY, I. M. SINGER - *Analytic torsion for complex manifolds*, Ann. Math. 98, 154-177 (1973)

[26] D. QUILLEN - *Determinants of Cauchy-Riemann operators over Riemann surface*, Func. Anal. Appl. 19, 31-34 (1985)

[27] I. B. FRENKEL, V. G. KAC - *Basic representations of affine Lie algebras and dual resonance models*, Inv. Math. 62, 23-66 (1980),

G. SEGAL - *Unitary representations of some infinite dimensional groups*, Commun. Math. Phys. 80, 301-342 (1981)

[28] H. TSUKADA - *String path integral realization of vertex operator algebras*, University of California in San Diego thesis 1988

[29] V. G. KAC - *An elucidation of "Infinite dimensional algebras... and the very strange formula". $E_8^{(1)}$ and the cube root of the modular invariant j*, Advances in Math. 35, 264-273 (1980)

[30] A. B. ZAMOLODCHIKOV, V. A. FATEEV - *Non-local (parafermion) currents in two-dimensional conformal quantum field theory and self-dual critical points of $\mathbf{Z}_N$-symmetric statistical systems*, Zh. Eksp. Teor. Fiz. 89, 380-399 (1985),

D. GEPNER, Z. QIU - *Modular invariant partition functions for parafermionic field theories*, Nucl. Phys. B 285 [FS 19], 423-453 (1987)

[31] K. GAWĘDZKI, A. KUPIAINEN - *Coset construction from functional integrals*, IHES/P/88/45 preprint, to appear in Nucl. Phys. B

[32] A. M. POLYAKOV, P. B. WIEGMANN - *Theory of non-abelian Goldstone bosons in two dimensions*, Phys. Lett. B 131, 121-126 (1983)

[33] P. GODDARD. A. KENT, D. OLIVE - *Unitary representations of the Virasoro and super-Virasoro algebras*, Commun. Math. Phys. 103, 105-119 (1986)

[34] D. FRIEDAN, Z. QIU, S. SHENKER - *Details of the non-unitarity proof for heighest weight representations of the Virasoro algebra*, Commun. Math. Phys. 107, 535-542 (1986)

[35] A. CAPPELLI, C. ITZYKSON, J.-B. ZUBER - *The* A-D-E *classification of minimal and* $A_1^{(1)}$ *conformal invariant theories*, Commun. Math. Phys. 113, 1-26 (1987)

[36] A. B. ZAMOLODCHIKOV - *Conformal symmetry and multicritical points in two-dimensional quantum field theory*, Yad. Fiz. 44, 821-827 (1986)

TWO-DIMENSIONAL CONFORMAL FIELD THEORY AND THREE-DIMENSIONAL TOPOLOGY

J. Fröhlich [1] and C. King [2]

[1] Theoretical Physics, ETH-Hönggerberg
CH-8093 Zürich, Switzerland

[2] Department of Mathematics, Cornell University
Ithaca, NY 14853, USA

Summary

We present a survey of two-dimensional conformal field theory and show how the mathematical structures underlying conformal field theory can be used to construct invariants of links imbedded in a general class of three-dimensional manifolds. After a general introduction, we discuss chiral algebras and their representation theory. Chiral vertices are introduced as analogues of Clebsch-Gordan operators in group theory. Braiding and fusing of chiral vertices is analyzed, and it is sketched how to define conformal field theory on arbitrary Riemann surfaces by a sewing procedure. We then show how to construct link invariants from the data provided by a conformal field theory and sketch connections with quantum group theory.

1. Introduction

This paper is concerned with two subjects of recent interest: two-dimensional conformal field theory and three-dimensional topology, in particular the theory of knots and links imbedded in three-manifolds. The point of this paper is to show that there is an intimate relationship between these two subjects. This is not an entirely new idea. It has its origins roughly in 1987. Several people had studied the monodromy of conformal blocks (the holomorphic parts in a factorization of Green functions into holomorphic and anti-holomorphic factors) in a variety of models of conformal field theory, such as the minimal models.[1] It was noted that the conformal blocks in these models carry linear representations of the braid groups. Certain subsets of the conformal blocks in the minimal models and in the SU(2)-W-Z-W models gave rise to unitarizable representations of the braid groups factoring through certain Hecke algebras, the Temperley-Lieb-Jones algebras. These representations had been studied by V.F.R. Jones in connection with an investigation of the index of subfactors in von Neumann algebras.[2] This led him to the discovery of a new invariant for knots and links, the celebrated Jones polynomial. It was shown that there was a discrete series of unitarizable

Constructive Quantum Field Theory II, Edited by G. Velo and A. S. Wightman, Plenum Press, New York, 1990

representations of the braid groups factoring through Hecke algebras. Tsuchiya and Kanie showed that Jones' discrete series of braid group representations corresponds to the representations carried by the conformal blocks of the fundamental field in the level-k SU(2)-W-Z-W models, $k = 1, 2, 3, \ldots$. Friedan and Connes[a] conjectured that it also corresponds to the discrete series of unitary minimal models that had been identified by Friedan, Qiu and Shenker.

Independently, the study of the statistics of particles and fields in two- and three-dimensional quantum field theories was taken up again, and it was shown that, in such theories, statistics is described by certain representations of the braid groups generated by Yang-Baxter matrices. It was also realized that the representations of the braid groups carried by general conformal blocks of two-dimensional conformal field theory were generated by Yang-Baxter matrices, as well, the so-called braid matrices.[1]

Jones argued that the central objects needed to construct new polynomial invariants for knots and links were Yang-Baxter matrices obeying certain additional properties. This was pinned down, more precisely, by Turaev in spring 1987.[2]

Hence, it was very natural to conjecture that there should be an intimate connection between general two-dimensional conformal field theory and knot theory. One of the conjectures raised by one of us (J.F.) was that link-invariants could perhaps be constructed directly through appropriate analytic continuations of local Green functions of two-dimensional conformal field theory along curves in the complex plane prescribed by links in $\mathbb{R}^3$ (and the use of operator product expansions at places where two arguments merge, corresponding to turning points of the link). We shall prove this conjecture in Sec. 6.3, (Theorem 6.4).

A crucial element needed to understand the precise relationship between two-dimensional conformal field theory and knot theory, besides the braid group representations provided by conformal blocks, was the fusion of chiral vertices, i.e. the short distance expansions for conformal blocks. Fusion of chiral vertices was studied by several groups of people. The clearest formulation was proposed by Moore and Seiberg.[1]

Once the braid- and fusion matrices corresponding to the conformal blocks of some conformal field theory and the equations between them are known it turns out to be quite easy to associate link invariants with them. This is carried out in Sec. 6.2. Actually, the construction of link invariants does not require very detailed knowledge of conformal field theory. The role played by conformal field theory is merely to provide suitable Yang-Baxter-(braid-) and fusion matrices. Further mathematical structures intimately tied to conformal field theory are not directly relevant for the construction of link invariants.

Besides conformal field theory, there is a second source of braid- and fusion matrices out of which link invariants can be constructed: the representation theory of *quantum groups*. In the representation theory of quantum groups, braid matrices appear as generators of the commutants of tensor product representations of a quantum group. Thus the role played by representations of the braid groups in the representation theory of quantum groups is analogous to the one played by representations of the permutation groups in the representation theory of SU(n). In the context of quantum groups, the fusion

[a] private communication of A. Connes to J. F., spring 1987.

matrices can be interpreted as matrices of generalized Clebsch-Gordan coefficients. They project from a tensor product space carrying a tensor product representation of the quantum group onto an irreducible subspace. The braid- and fusion matrices of quantum group theory satisfy algebraic equations analogous to the equations satisfied by the braid- and fusion matrices of conformal field theory. The braid- and fusion matrices of quantum group theory can be used to construct link invariants. This has been worked out in detail in basic papers by Reshetikhin.[2]

Since the braid- and fusion matrices of quantum group theory play a basic role in the representation theory of quantum groups, it is natural to ask whether the braid and fusion matrices of conformal field theory have a representation theoretic meaning, too. This question was originally raised independently of the work on quantum groups carried out by the Russian and Japanese schools. In the course of the past year, it became clear that the braid- and fusion matrices of conformal field theory were intimately connected with the representation theory of certain infinite-dimensional algebras, called *chiral algebras*, which play a fundamental role in conformal field theory. In fact, there is a far-reaching parallelism between the representation theory of chiral algebras and the representation theory of quantum groups. It should have very useful applications in further work on the representation theory of infinite-dimensional algebras. For example, the theory of lowest-weight representations of current algebra $\widehat{\mathrm{su}}(n)_k$ is closely related to the theory of highest-weight representations of $U_q(\mathrm{su}(n))$, $q = \exp 2\pi i(k + n)^{-1}$. The Yang-Baxter (braid-) matrices in these two theories are related to each other by the so-called vertex-SOS transformation. It is natural to conjecture that the braid matrices associated with the representation theory of a large family of chiral algebras can be interpreted as braid matrices associated with the representation theory of some quantum group. A proof would require a suitable extension of Tannaka-Krein theory.

For the past several years, there has been quite a lot of activity in studying three-dimensional gauge theories with a Chern-Simons term in the action. It was recognized that certain particles described by such theories may have arbitrary real spin and intermediate statistics interpolating continuously between Bose- and Fermi statistics. It turned out that intermediate statistics could be described by certain representations of the braid groups. For abelian gauge theories, these representations are one-dimensional and thus not very interesting, mathematically. But the general theory leaves room for higher-dimensional, non-abelian representations. It turned out that there is a natural connection between the pure abelian Chern-Simons theory and the Gauss linking number. This led to the conjecture that the pure non-abelian Chern-Simons theory might be related to more interesting link invariants such as the Jones polynomial. This conjecture was recently proven by Witten in a beautiful paper.[3] He also showed that pure Chern-Simons gauge theory in three dimensions is basically equivalent to the holomorphic sector of a two-dimensional W-Z-W model. Thus, what the W-Z-W model of two-dimensional conformal field theory teaches us about knot theory can also be understood with the help of three-dimensional Chern-Simons gauge theory. The important aspects of this result are that Chern-Simons theory might provide a more intrinsically three-dimensional approach to knot theory and that it ties in naturally with other recent developments in three-dimensional topology connected with the Casson-invariant and Floer cohomology; see Ref. 4 for a nontechnical account of these matters. However, it is tempting to think that, in the end, it is again the relation of Chern-Simons theory to braid group representations and quantum groups[5] that make it a useful tool in knot theory and in other areas of three-dimensional topology. So, it is perhaps really quantum group theory that will become an important tool in three-dimensional topology, but quantum field theory will be a valuable guide leading to new conjectures ad heuristic ''proofs''.[b]

In fact, there is an intriguing analogy between quantum field theory, or the closely related theory of critical phenomena in statistical mechanics, and topology of manifolds: Quantum field theory tends to be noninteracting, classical Landau theory tends to correctly describe critical phenomena, in dimensions larger than four. Topology of more than four-dimensional manifolds becomes ''classical'' and comparatively ''easy''. In four dimensions, quantum corrections (statistical fluctuations) become important in quantum field theory (the theoretical description of critical phenomena, respectively), and topology becomes subtle and ''nonclassical''. In three dimensions, all degrees of freedom of a quantum field theory, or a statistical system in the thermodynamic limit, near criticality, are strongly coupled, quantum corrections are crucial; three-dimensional topology is difficult and highly ''nonclassical''. In two dimensions, critical phenomena are described by two-dimensional conformal field theory, and one might hope that (at least rational) conformal field theories can be classified; in particular those models of conformal field theory that are most immediately relevant for the theory of critical phenomena can perhaps be solved exactly. Two-dimensional topology is ''solved exactly''.

There are presumably some basic, mathematical reasons for this analogy between properties of quantum field theory and of the theory of critical phenomena, on one hand, and aspects of topology, on the other hand. Exploiting this analogy and understanding the underlying reasons for it might lead to new ideas about topology and quantum field theory. Perhaps the results in Refs. 3, 4 and 5 and in this paper represent a step in this direction.

The organization of this paper is as follows: In Sec. 2 we define chiral algebras and study their representation theory. In particular, we construct an analogue of the tensor operators of group theory, the so-called chiral vertices. We explore the structure of the algebra of chiral vertices and show how it leads to representations of the braid groups generated by Yang-Baxter (braid-) matrices. In Sec. 3, we study the fusion of chiral vertices. This corresponds, in group theory, to decomposing a product of tensor operators into a sum of irreducible tensor operators. We rederive the basic equations between braid- and fusion matrices that play a central role in our construction of link-invariants. In Sec. 4, we show how two-dimensional conformal field theories can be reconstructed from the representation theory of a pair of chiral algebras by representing local fields of some conformal field theory as sums of products of chiral vertices of the two chiral algebras. Locality yields equations between the braid matrices describing the algebra of chiral vertices of the two chiral algebras and the coefficients in the sum of products of chiral vertices representing a local field, the so-called structure constants. These equations are reviewed. In Sec. 5, it is sketched how conformal field theories over Riemann surfaces of higher genus can be constructed from conformal field theories over the Riemann sphere by

[b] It should be emphasized that link polynomials can be associated with a large class of three-dimensional quantum field theories, *not* just with pure Chern-Simons theory.

some sewing procedures, provided certain conditions related to modular invariance, involving the four-point functions on the Riemann sphere and the one-point functions on the torus, are satisfied. In Sec. 6.1, some basic definitions of knot theory are recalled. In Sec. 6.2, invariants of links imbedded in S^3 or in $S^2 \times S^1$ are constructed from the braid- and fusion matrices of Secs. 2 and 3, associated with the representation theory of some chiral algebra. By comparison with Witten's paper,[3] we find a relation between fusion matrices and the matrices representing the modular transformation $\tau \rightarrow -1/\tau$ on the space of characters of a chiral algebra.

In Sec. 6.3, it is outlined how our invariants can be obtained from an analytic continuation "along some link" and subsequent short-distance expansions of a ratio of local Green functions of conformal field theories. In the process, we recover an equation between fusion matrices and structure constants. Section 6.3 is based on the results of Secs. 2–4. In Sec. 6.4, we sketch how to construct invariants of links imbedded in a general class of three-manifolds by using the results of Secs. 2, 3 and 5. Finally, in Sec. 7, we briefly describe a (partly conjectural) connection between the representation theory of chiral algebras and the representation theory of quantum groups. This clarifies the relationship between our approach to the construction of link-invariants and the one due to N. Yu. Reshetikhin based on quantum groups.

Although this paper is preoccupied primarily with problems in pure mathematics, it is written in the style of theoretical physics ("physical mathematics"). Many arguments are heuristic and somewhat tentative, but we believe that our main conclusions can be rendered rigorous.

2. Chiral Algebras and Chiral Vertices

In this section, we recall the notion of chiral algebra and some basic concepts used in the representation theory of chiral algebras. Since this is not really a paper on two-dimensional conformal field theory, we shall not consider the most general notion of chiral algebra; but see Ref. 6. Rather, we shall focus our attention on chiral algebras generated by local fields on the unit circle.

A chiral algebra, $\mathcal{A}$, may be specified in terms of generators $\{\psi_{j,n}\}_{n=-\infty, j\in\mathcal{J}}^{\infty}$, where $\mathcal{J} \equiv \mathcal{J}_{\mathcal{A}}$ is a countable index set. It is usually assumed that $\mathcal{A}$ is a unital algebra; the identity element of $\mathcal{A}$ is given by $\psi_{0,0} \equiv 1$, with $\psi_{0,n} = 0$, for $n \neq 0$. We denote the infinite-dimensional complex vector space of formal finite linear combinations of generators, $\psi_{j,n}$, with complex coefficients by $\mathcal{G}(\mathcal{A})$. The algebra $\mathcal{A}$ is defined as the set of all words of finite length in the elements of $\mathcal{G}(\mathcal{A})$, modulo polynomial relations between elements of $\mathcal{G}(\mathcal{A})$: Two words which can be transformed into each other by applying all possible polynomial relations correspond to the same element of $\mathcal{A}$. Formally, the generators $\{\psi_{j,n}\}_{n\in\mathbb{Z}}$, $j \in \mathcal{J}$, determine a "field" $\psi_j(z)$, $z \in \mathbb{C}$, $0 < |z| < \infty$, defined by

$$\psi_j(z) = \sum_{n\in\mathbb{Z}} \psi_{j,n} z^{-n-h_j} , \tag{2.1}$$

where h_j is a real number, called the *conformal dimension of* ψ_j. In order for $\psi_j(z)$ to be single-valued, it is necessary and sufficient that

$$h_j \in \mathbb{Z}\,, \qquad j \in \mathcal{J}\,. \tag{2.2}$$

By inverting the Fourier-Laurent expansion (2.1) we obtain

$$\psi_{j,n} = \frac{1}{2\pi i}\oint_{|z|=1} z^{n+h_j-1}\psi_j(z)\,dz\,. \tag{2.3}$$

We note that $\psi_0(z) = 1$, for $h_0 = 0$.

Let $\mathcal{M}$ denote the group of Möbius transformations (fractional linear transformations) of the complex z-plane which map the unit circle into itself; $\mathcal{M} \cong PSL(2, \mathbb{R})$. We assume that $\mathcal{A}$ carries a representation of $\mathcal{M}$ as an automorphism group, $\{\tau_w : w \in \mathcal{M}\}$, of $\mathcal{A}$ given by

$$\tau_w(\psi_j(z)) = \left(\frac{dw(z)}{dz}\right)^{h_j}\psi_j(w(z))\,, \tag{2.4}$$

where $z \to w(z)$ is the Möbius transformation corresponding to w. (One sees again from (2.4) that $\psi_j(z)$ is single-valued iff $h_j \in \mathbb{Z}$) Let σ denote the transformation $z \to e^{i\sigma}z$, (rotation of the unit circle through an angle σ). Then it follows from (2.3) and (2.4) that

$$\tau_\sigma(\psi_{j,n}) = e^{-in\sigma}\,\psi_{j,n}\,,$$

for all $n \in \mathbb{Z}, j \in \mathcal{J}$. Hence $\mathcal{G}(\mathcal{A})$ is $\mathbb{Z}$-graded, with $\mathcal{G}_n(\mathcal{A})$ the vector space spanned by $\{\psi_{j,n}\}_{j\in\mathcal{J}}$.

We define $\mathcal{A}_n$ to be the linear subspace of $\mathcal{A}$ of all elements A in $\mathcal{A}$ for which

$$\tau_\sigma(A) = e^{-in\sigma}A\,. \tag{2.5}$$

Then $\mathcal{A}$ is $\mathbb{Z}$-graded, too, with

$$\mathcal{A} = \bigoplus_{n\in\mathbb{Z}} \mathcal{A}_n\,. \tag{2.6}$$

We also define

$$\mathcal{A}_> = \bigoplus_{n>0}\mathcal{A}_n\,, \qquad \mathcal{A}_0 = \mathcal{A}_{n=0}\,, \qquad \mathcal{A}_< = \bigoplus_{n<0}\mathcal{A}_n\,; \tag{2.7}$$

$\mathcal{G}_>(\mathcal{A})$, $\mathcal{G}_0(\mathcal{A})$ and $\mathcal{G}_<(\mathcal{A})$ are defined similarly. Since $\{\tau_w : w \in \mathcal{M}\}$ is an automorphism group on $\mathcal{A}$, (2.5) shows that $\mathcal{A}_>$, $\mathcal{A}_0$ and $\mathcal{A}_<$ are subalgebras of $\mathcal{A}$.

In conformal field theory one usually assumes that $\mathcal{G}(\mathcal{A})$ contains a subalgebra, *Vir*, with generators $\{L_n\}_{n\in\mathbb{Z}}$ satisfying the Virasoro algebra

$$[L_n, L_m] = (n-m)L_{n+m} + \frac{c}{12}(n^3 - n)\delta_{n+m,0}\,, \tag{2.8}$$

where c is the central charge. Then

$$T(z) = \sum_{n \in \mathbb{Z}} L_n z^{-n-2} \tag{2.9}$$

is a field of conformal dimension $h_T = 2$, called energy-momentum tensor. We denote $T(z)$ by $\psi_2(z)$.

In this situation, one assumes that infinitesimal Möbius transformations are generated by L_0, $L_1 + L_{-1}$ and $i[L_1 - L_{-1}]$; e.g.

$$\frac{d}{d\sigma} \tau_\sigma(\psi_{j,n})|_{\sigma=0} = i(L_0 \psi_{j,n} - \psi_{j,n} L_0) . \tag{2.10}$$

The generator of translations, $z \to z + a$, of the complex plane is given by L_{-1}.

From now on we shall usually assume that $\mathcal{G}(\mathcal{A})$ contains a subalgebra, *Vir*, satisfying (2.8), for some c.

Examples and remarks

(1) If there are quadratic relations of the form

$$\psi_{j,n}\psi_{i,m} - \psi_{i,m}\psi_{j,n} = \sum_{l \in \mathcal{J}} f^l_{ji}(n, m, k)\psi_{l,k} \tag{2.11}$$

then by (2.4)

$$f^l_{ji}(n, m, k) = c^l_{ji}\delta_{n+m,k} , \tag{2.12}$$

for some constants c^l_{ji} (the structure constants of $\mathcal{G}(\mathcal{A})$), and one concludes that $\mathcal{G}(\mathcal{A})$ is an infinite-dimensional Lie algebra. (An alternative to (2.11) would be to specify an operator product expansion for $\psi_j(z)\psi_i(w)$.[7]) Well-known examples of chiral algebras which are Lie algebras are the Kac-Moody algebras, $\hat{g}$, where g is a classical Lie algebra. In this case one can enlarge $\mathcal{G}(\mathcal{A})$ to include a Virasoro algebra with the help of the Sugawara construction.[8] These are typical examples which the reader ought to bear in mind throughout this paper. There are, however, important chiral algebras which are *not* Lie algebras.[9]

(2) The notion of chiral algebra can be generalized considerably by starting from fields $\psi_j(z)$, $j \in \mathcal{J}$, which may be *multi-valued* on S^1. It is then more natural to define the fields ψ_j on the real line (a "*light ray*") which is the image of S^1 under the transformation $z \to x = i(1 - z)/(1 + z)$. In this case the conformal dimensions of the fields $\psi_j(x)$ need *not* be integers, anymore. There are then quadratic relations between the fields ψ_j which typically have the form

$$\psi_i(x)\psi_j(y) = e^{2\pi i \theta_{ij}}\psi_j(y)\psi_i(x) , \tag{2.13}$$

for $x \neq y$, where θ_{ij} is a real phase, and θ_{ij} is an integer, for all i and j in $\mathcal{J}$, iff $h_j \in \mathbb{Z}$, for all $j \in \mathcal{J}$. An attempt towards analyzing such general chiral algebras is described in some detail in Ref. 10. Such algebras play a fairly important role in conformal field theory, see e.g. Ref. 9, but, for simplicity, we shall only consider chiral algebras of *local* fields ($\theta_{ij} \in \mathbb{Z}$) which have the structure described above. (Our analysis could be extended to the present case.)

(3) Chiral algebras can be *algebras if there exists an involution $*$ on $\mathcal{J}$, ($j \mapsto j^*$, for $j \in \mathcal{J}$), with $0^* = 0$, $2^* = 2$, $h_{j^*} = h_j$, for all $j \in \mathcal{J}$, such that if we define

$$\psi^*_{j,n} = \psi_{j^*,-n} , \qquad (1^* = 1, L^*_n = L_{-n}) , \tag{2.14}$$

and extend $*$ anti-linearly to $\mathcal{G}(\mathcal{A})$, and then to all of $\mathcal{A}$ by requiring that $(A \cdot B)^* = B^*A^*$, for all A, B in $\mathcal{A}$, then $*$ preserves all relations on $\mathcal{A}$. If $\mathcal{A}$ is a *algebra, then it follows from (2.1) and (2.14) that

$$\psi_j(z)^* = \left(\frac{1}{z^*}\right)^{2h_j} \psi_{j^*}\left(\frac{1}{z^*}\right) , \tag{2.15}$$

where z^* is the complex conjugate of $z \in \mathbb{C}$. Since $h_{j^*} = h_j$, $(\psi_j(z)^*)^* = \psi_j(z)$, and, by (2.4), $\{\tau_w : w \in \mathcal{M}\}$ is a *automorphism group on $\mathcal{A}$. Note that $(\mathcal{A}_>)^* = \mathcal{A}_<$.

Chiral algebras known to occur in conformal field theory are *algebras.

As we shall see, conformal field theory is, in essence, nothing but the representation theory of a pair of chiral algebras.[11,12] This motivates the study of representations of chiral algebras. We shall focus our attention on *lowest-weight representations.*

A representation π of $\mathcal{A}$ on a complex vector space $\mathcal{H}_\pi$ is a lowest-weight representation if $\mathcal{H}_\pi$ contains a cyclic vector ξ_0 such that

$$\pi(A)\xi_0 = 0 , \qquad \text{for all } A \in \mathcal{A}_> . \tag{2.16}$$

(The vector ξ_0 is called cyclic iff $\mathcal{H}_\pi = \{\pi(A)\xi_0 : A \in \mathcal{A}\}$.) We also *assume* that ξ_0 is an eigenvector of $\pi(L_0)$ corresponding to an eigenvalue $h_\pi > -\infty$. It then follows from (2.5) and (2.10) that

$$\pi(L_0)\pi(A)\xi_0 = (n + h_\pi)\pi(A)\xi_0 , \tag{2.17}$$

for all $A \in \mathcal{A}_{-n}$, $n \geq 0$. Thus $\pi(\mathcal{A}_{-n})\xi_0$ is an eigenspace of $\pi(L_0)$ corresponding to the eigenvalue $n + h_\pi$, $n = 0, 1, 2, \ldots$, and, since by (2.6)

$$\bigoplus_{n=0}^{\infty} \pi(\mathcal{A}_{-n})\xi_0 = \mathcal{H}_\pi , \tag{2.18}$$

the spectrum of $\pi(L_0)$ is given by $h_\pi + \mathbb{Z}_+$.

Finally we *assume* that the automorphism group $\{\tau_w : w \in \mathcal{M}\}$ is implemented on $\mathcal{H}_\pi$ by a representation, U_π, of SL(2, $\mathbb{R}$) such that

(i)

$$\pi(\tau_w(\mathcal{A})) = U_\pi(w)\pi(A)U_\pi(w^{-1}) , \tag{2.19}$$

for $w \in SL(2, \mathbb{R})$;

(ii) the generators of $\{U_\pi(w) : w \in SL(2, \mathbb{R})\}$ are given by $\pi(L_0)$, $\pi(L_1 + L_{-1})$, $i\pi(L_1 - L_{-1})$; and

(iii) $\mathcal{H}_\pi$ consists of analytic vectors for $\pi(L_{-1})$.

Definition 2.1

(1) Representations, π, of a chiral algebra $\mathcal{A}$ with the property that $\mathcal{G}(\mathcal{A})$ contains *Vir* and all the properties specified above (see (2.16)–(2.18), (i)–(iii)) are called *covariant lowest-weight representations*.

(2) Let $\mathcal{H}'_\pi \subseteq \mathcal{H}_\pi$ be a subspace of $\mathcal{H}_\pi$ invariant under $\pi(\mathcal{A})$. A vector $\xi \in \mathcal{H}'_\pi$ with the properties that $\pi(A)\xi = 0$, for all $A \in \mathcal{A}_>$, $\{\pi(A)\xi : A \in \mathcal{A}\} = \mathcal{H}'_\pi$, is called an *invariant* vector for $\mathcal{A}$.

(3) Let $\mathcal{H}^*_\pi$ be the space of linear functionals on $\mathcal{H}_\pi$. An element η of $\mathcal{H}^*_\pi$ is called *invariant* if

$$\langle \eta | \pi(A)\xi_0 \rangle = 0 , \qquad \text{for all } A \in \mathcal{A}_< . \tag{2.20}$$

(4) An invariant vector $\xi \in \mathcal{H}_\pi$ is called *strongly invariant* iff

$$\pi(\psi_{j,k})\xi = 0 , \qquad \text{for } k \geq -h_j + 1 , \qquad \text{for all } j \in \mathcal{J} . \tag{2.21}$$

□

Note that, since $h_2 \equiv h_T = 2$, it follows that $\pi(L_{-1})\xi = 0$ if ξ is strongly invariant. Hence

$$\pi(L_n)\xi = 0 , \qquad \text{for } n \geq -1 ,$$

and

$$h_\pi = 0 . \tag{2.22}$$

An invariant linear functional $\eta \in \mathcal{H}^*_\pi$ is called *strongly invariant* iff it is invariant and

$$\langle \eta | \pi(\psi_{j,k})\xi \rangle = 0 , \qquad \text{for all } \xi \in \mathcal{H}_\pi , \tag{2.23}$$

for all $k \leq h_j - 1$.

It follows that if η is a strongly invariant vector in $\mathcal{H}^*_\pi$ then

$$\langle \eta | \pi(L_n)\xi \rangle = 0 , \qquad \text{for all } n \leq 1 , \tag{2.24}$$

and all $\xi \in \mathcal{H}_\pi$.

In the following we shall only consider chiral algebras $\mathcal{A}$ with the property that there is precisely one representation, π_1, called *vacuum representation*, such that $\mathcal{H}_{\pi_1}$ contains a *unique* strongly invariant vector, Ω, and $\mathcal{H}^*_{\pi_1}$ contains a *unique* strongly invariant linear functional, Ω^*, with

$$\langle \Omega^* | \Omega \rangle = 1 \ .$$

The vector Ω and the linear functional Ω^* are called vacuum. It follows from the strong invariance of Ω and Ω^* that, for $w \in \mathcal{M}$,

$$\langle \Omega^* | \pi_1(\tau_w(A))\Omega \rangle = \langle \Omega^* | \pi_1(A)\Omega \rangle \ , \tag{2.25}$$

for all $A \in \mathcal{A}$.

A representation π of $\mathcal{A}$ is *irreducible* iff $\mathcal{H}_\pi$ does not contain any subspaces $\mathcal{H}'_\pi \neq \{0\}$, $\mathcal{H}_\pi$ invariant under $\pi(\mathcal{A})$. Then every invariant vector ξ is an eigenvector of $\pi(L_0)$ with eigenvalue h_π, and the space $\mathcal{H}^I_\pi$, of invariant vectors is spanned by $\{\pi(A)\xi_0 : A \in \mathcal{A}_0\}$. In most examples, $\mathcal{H}^I_\pi$ is finite-dimensional.

Remarks

If $\mathcal{A}$ is a * algebra then, with a representation π of $\mathcal{A}$ on $\mathcal{H}_\pi$, we can associate a representation π^* of $\mathcal{A}$ on $\mathcal{H}^*_\pi$ by setting

$$\langle \pi^*(A)\eta | \xi \rangle = \langle \eta | \pi(A^*)\xi \rangle \ , \tag{2.26}$$

for all $A \in \mathcal{A}$, $\eta \in \mathcal{H}^*_\pi$, $\xi \in \mathcal{H}_\pi$.

A representation π of a chiral algebra $\mathcal{A}$ which is a * algebra is *unitary* if $\mathcal{H}_\pi$ is equipped with a scalar product and $\pi(A^*) = \pi(A)^*$, for all $A \in \mathcal{A}$, where $\pi(A)^*$ is the adjoint of the operator $\pi(A)$ on the Hilbert space, $\overline{\mathcal{H}}_\pi$, obtained by taking the closure of $\mathcal{H}_\pi$ in the norm induced by the scalar product. Unitary reprecentations of chiral algebras appear in the study of unitary conformal field theories.[13] In the following, we shall usually ignore convergence and domain questions. As a rule, these questions can be settled for unitary representations.

Let $\mathcal{A}$ be a chiral algebra with generators $\{\psi_{j,n}\}_{n=-\infty, j\in\mathcal{J}}^{\infty}$. We define a *linear deformation map*, δ_z, depending on a complex number z, with $0 < |z| < \infty$, on the linear space $\mathcal{G}(\mathcal{A})$ spanned by the generators $\psi_{j,n}$ of $\mathcal{A}$ by setting, for every $n \in \mathbb{Z}$,

$$\delta_z(\psi_{j,n}) = \sum_{k=-h_j+1}^{\infty} \binom{n + h_j - 1}{k + h_j - 1} z^{n-k} \psi_{j,k} \ , \tag{2.27}$$

where $\binom{n}{m}$ are the usual binomial coefficients, with $\binom{n}{m} = 0$, for $m > n \geq 0$, $\binom{0}{0} = 1$, and with the obvious definition for $n < 0$. Our definition of δ_z is motivated by the contour integral formalism of Ref. 7 and has been proposed in Refs. 11 and 12. Note that

$$\delta_z(L_{-1}) = L_{-1} \ , \qquad \delta_z(1) = 0 \ . \tag{2.28}$$

For $n > 1 - h_j$, $\delta_z(\psi_{j,n}) = \tau_z(\psi_{j,n})$, where τ_z denotes the Möbius automorphism corresponding to translation by z. However, δ_z *cannot* be extended to an automorphism of $\mathcal{A}$, as is clear from the fact that $\delta_z(1) = 0$. Moreover, if $\mathcal{A}$ is a *algebra $\delta_z(\psi^*_{j,n}) \neq \delta_z(\psi_{j,n})^*$.

The map δ_z can be used to define a tensor product $\otimes_z$, on $\mathcal{G}(\mathcal{A})$ which takes $\psi_{j,n}$ to $\delta_z(\psi_{j,n}) \otimes 1 + 1 \otimes \psi_{j,n}$. For all examples of chiral algebras that appear in local conformal field theory, the definition of $\otimes_z$ is consistent with the algebraic structure of $\mathcal{A}$.

Let $\bar{L} = \bar{L}_{\mathcal{A}}$ be the list of all irreducible lowest-weight representations of $\mathcal{A}$. A representation π_i of $\mathcal{A}$, with $i \in \bar{L}$, will henceforth be usually denoted by i. For i and l in $\bar{L}$, we define a tensor product representation, $i \otimes_z l$, by setting

$$i \underset{z}{\otimes} l(\psi_{j,n}) = i(\delta_z(\psi_{j,n})) \otimes 1 + 1 \otimes l(\psi_{j,n}) , \tag{2.29}$$

for all generators, $\psi_{j,n}$, of $\mathcal{A}$.

Definition 2.2

A chiral algebra $\mathcal{A}$ is called *rational* iff $\mathcal{A}$ is a * algebra, $\mathcal{G}(\mathcal{A})$ contains *Vir*, and $\bar{L}_{\mathcal{A}}$ contains a *finite* set, $L_{\mathcal{A}}$, with the properties that

(1) $L_{\mathcal{A}}$ contains unique vacuum representations 1 and 1^*, with $1 \simeq 1^*$;

(2) if $i \in L_{\mathcal{A}}$ then $i^* \in L_{\mathcal{A}}$, and $i \otimes_z i^* \simeq i^* \otimes_z i$ contains 1 once;

(3) if i and l belong to $L_{\mathcal{A}}$ then, for $0 < |z| < \infty$, every irreducible lowest-weight subrepresentation of $i \otimes_z l$ is equivalent to a representation $k \in L_{\mathcal{A}}$, and the multiplicity, N_{kil}, of k in $i \otimes_z l$ is *finite* and *independent* of z, for $0 < |z| < \infty$. (2.30)

A set $L_{\mathcal{A}}$ with properties (1)–(3), above, is called a $\otimes$-invariant set of lowest-weight representations of $\mathcal{A}$.

Examples

The Kac-Moody algebra $\widehat{su}(2)$ at level k is easily seen to be a rational chiral algebra with card $L_{\widehat{su}(2)} = k + 1$.[14] The Virasoro algebra, *Vir*, with central charge $c = 1 - (6/m(m+1))$, $m = 3, 4, \ldots$, is known to be rational.[7,15]

Rational chiral algebras, $\mathcal{A}$, along with a $\otimes$-invariant set, $L_{\mathcal{A}}$, of irreducible lowest-weight representations are the basic building blocks for the theory developed in this paper.

Next, we introduce an analogue of the notion of tensor operators in group theory, the so called *chiral vertices*. For $i \in L_{\mathcal{A}}$, let $\mathcal{H}_i$ denote the representation space of i. Let i and l be in $L_{\mathcal{A}}$ and let $k \in L_{\mathcal{A}}$ be a representation of $\mathcal{A}$ appearing in $i \otimes_z l$ with multiplicity $N_{kil} > 0$.

Let $F^k_\alpha(i, l)$ denote the projection of $\mathcal{H}_i \otimes_1 \mathcal{H}_l$ onto $\mathcal{H}_{k^{(\alpha)}}$, $\alpha = 1, \ldots, N_{kil}$, with the property that

$$k^{(\alpha)}(A)F^k_\alpha(i, l)\xi \otimes \eta = F^k_\alpha(i, l)[i(\delta_1(A))\xi \otimes \eta + \xi \otimes l(A)\eta] , \tag{2.31}$$

for every $A \in \mathcal{G}(\mathcal{A})$. We note that even if the representations $i \in L_{\mathcal{A}}$ are unitary the projections $F^k_\alpha(i, l)$ are not bounded operators. We define the operator $V^{kl}_\alpha(\xi, 1)$ to be the linear map from $\mathcal{H}_l$ to $\mathcal{H}_k$ given by

$$\langle \zeta | V^{kl}_\alpha(\xi, 1)\eta \rangle = \langle \zeta^{(\alpha)} | F^k_\alpha(i, l)\xi \otimes \eta \rangle , \tag{2.32}$$

for every $\zeta \in \mathcal{H}^*_k$ and $\zeta^{(\alpha)}$ the corresponding linear functional on $\mathcal{H}_{k^{(\alpha)}}$. Since $\delta_z(L_{-1}) = L_{-1}$, it follows from (2.31) and (2.32) that, for $|w|$ small enough,

$$k(e^{wL_{-1}})V^{kl}_\alpha(\xi, 1)l(e^{-wL_{-1}}) = V^{kl}_\alpha(i(e^{wL_{-1}})\xi, 1) . \tag{2.33}$$

For $z \neq 0, \infty$, we define

$$V^{kl}_\alpha(\xi, z) = V^{kl}_\alpha(i(e^{(z-1)L_{-1}})\xi, 1) \tag{2.34}$$

$\alpha = 1, \ldots, N_{kil}$.

Theorem 2.1[11]

The operators $V^{kl}_\alpha(\xi, z)$ defined in (2.34) have the following properties:
(a) $V^{kl}_\alpha(\xi, z)$ is linear in ξ.
(b) $k(A)V^{kl}_\alpha(\xi, z) = V^{kl}_\alpha(i(\delta_z(A))\xi, z) + V^{kl}_\alpha(\xi, z)l(A)$, for all $A \in \mathcal{G}(\mathcal{A})$.
(c) $\dfrac{d}{dz} V^{kl}_\alpha(\xi, z) = V^{kl}(i(L_{-1})\xi, z)$.

Proof

(a) follows directly from (2.31) and (2.32). In order to prove (b), we use (2.34) and (2.31) to obtain

$$\begin{aligned} k(A)V^{kl}_\alpha(\xi, z) &= k(A)V^{kl}_\alpha(i(e^{(z-1)L_{-1}})\xi, 1) \\ &= V^{kl}_\alpha(i(\delta_1(A))i(e^{(z-1)L_{-1}})\xi, 1) + V^{kl}_\alpha(i(e^{(z-1)L_{-1}})\xi, 1)l(A) , \end{aligned} \tag{2.35}$$

for $A \in \mathcal{G}(\mathcal{A})$. Next, we note that

$$i(\delta_1(A))i(e^{(z-1)L_{-1}})\xi = i(e^{(z-1)L_{-1}})i(\tau_{z-1}(\delta_1(A)))\xi , \tag{2.36}$$

since i is a covariant lowest-weight representation; see Definition 2.1, (1). By (2.27) and the remark after (2.28)

$$\tau_{z-1}(\delta_1(A)) = \delta_{z-1}(\delta_1(A)) \tag{2.37}$$

and by (2.27)

$$\delta_{z-1}(\delta_1(A)) = \delta_z(A) , \tag{2.38}$$

$z \neq 0, \infty$. Inserting (2.36)–(2.38) into (2.35), we obtain

$$k(A)V_\alpha^{kl}(\xi, z) = V^{kl}(i(e^{(z-1)L_{-1}})i(\delta_z(A))\xi, 1) + V^{kl}(i(e^{(z-1)L_{-1}})\xi, 1)l(A)$$

which, by (2.34), yields (b).

Finally, (c) follows directly from (2.34)

□

For more details see Ref. 11. The operators $V_\alpha^{kl}(\xi, z)$ are called *chiral vertices*. Among all chiral vertices there are the so-called *primary vertices*.

Definition 2.3.

A chiral vertex $V_\alpha^{kl}(\xi, z)$ is called primary iff ξ is an *invariant* vector in $\mathcal{H}_i$. (Invariant vectors are defined in Definition 2.1, (2).)

Proposition 2.2.[11]

If $V_\alpha^{kl}(\xi, z)$ is a primary vertex then, for all $n \in \mathbb{Z}$,

$$k(L_n)V_\alpha^{kl}(\xi, z) - V^{kl}(\xi, z)l(L_n) = \left[z^{n+1}\frac{d}{dz} + z^n(n+1)h_\xi\right]V^{kl}(\xi, z)\,, \quad (2.39)$$

where h_ξ is the eigenvalue of $i(L_0)$ corresponding to the eigenvector ξ.

Proof

Since $L_n \in \mathcal{G}(\mathcal{A})$, for all $n \in \mathbb{Z}$,

$$k(L_n)V_\alpha^{kl}(\xi, z) - V_\alpha^{kl}(\xi, z)l(L_n) = V_\alpha^{kl}(i(\delta_z(L_n))\xi, z)\,, \quad (2.40)$$

by Theorem 2.1, (a). By (2.27),

$$i(\delta_z(L_n))\xi = \sum_{k=-1}^{\infty}\binom{n+1}{k+1}z^{n-k}i(L_k)\xi\,, \quad (2.41)$$

where we have used that the conformal dimension of $\psi_2 \equiv T$ (the energy-momentum tensor) is $h_2 = 2$. Since ξ is an invariant vector, (see Definition 2.1, (2)),

$$i(L_k)\xi = 0\,, \qquad \text{for all } k > 0\,.$$

Hence

$$i(\delta_z(L_n))\xi = [z^{n+1}i(L_{-1}) + z^n(n+1)i(L_0)]\xi\,.$$

Now we use that $i(L_0)\xi = h_\xi \xi$ and Theorem 2.1, (c) to conclude that

$$V_\alpha^{kl}(i(\delta_z(L_n))\xi, z) = z^{n+1} \frac{d}{dz} V_\alpha^{kl}(\xi, z) + z^n(n+1)h_\xi V_\alpha^{kl}(\xi, z)$$

and the proof of (2.39) is complete.

□

Remarks

(1) All we use in the proof of Proposition 2.2 is that ξ is an invariant vector for the subalgebra *Vir* of $\mathcal{G}(\mathcal{A})$. (A representation space $\mathcal{H}_i$ of $\mathcal{A}$ may in general be decomposed into many invariant subspaces for *Vir* each of which contains invariant vectors for *Vir* which need not be invariant vectors for $\mathcal{A}$.)

(2) Proposition 2.2 shows that primary vertices are primary fields in the sense of *BPZ*,[7] i.e. they are conformal tensors of weight h_ξ. This has rather far reaching consequences.

(3) Note that the fact that Ω, Ω^* are strongly invariant vectors and definitions (2.27) and (2.29) imply that

$$1 \underset{z}{\otimes} l \simeq 1^* \underset{z}{\otimes} l \simeq l\,, \tag{2.42}$$

for all $l \in L_{\mathcal{A}}$. Hence $F_\alpha^k(1, l) = \delta^{kl} F^l(1, l)$, and we may choose $F^l(1, l)$ such that

$$F^l(1, l)(\Omega \otimes \xi) = \xi\,, \qquad \text{for all } \xi \in \mathcal{H}_l\,. \tag{2.43}$$

Using (2.27), (2.34), Theorem 2.1, (b) and (2.43) we conclude that

$$V_\alpha^{kl}(\Omega, z) = V_\alpha^{kl}(\Omega^*, z) = \delta^{kl} 1_{\mathcal{H}_k}\,. \tag{2.44}$$

(4) From (2.27) and (2.29) one may conclude (using contour integrals) that, for i and l in $L_{\mathcal{A}}$,

$$\left.\begin{array}{l} i \otimes_z l \text{ and } l \otimes_z i \text{ contain the same irreducible} \\ \text{subrepresentation in } L_{\mathcal{A}}. \end{array}\right\} \tag{2.45}$$

This follows by studying $\langle \xi_{k^*} | k(\psi_j(\zeta)) V_\alpha^{kl}(\xi_i, z)\xi_l\rangle$, and $\langle \xi_{k^*} | k(\psi_j(\zeta)) V_\alpha^{ki}(\xi_l, z)\xi_i\rangle$, $\alpha = 1, \ldots, N_{kil}$ $(=N_{kli})$, for $\xi_i \in \mathcal{H}_i$, $\xi_l \in \mathcal{H}_l$ and $\xi_{k^*} \in \mathcal{H}_k^*$.

(5) By (2.42) and (2.45), $N_{kl1} = N_{k1l} = \delta_{kl}$; hence there is only one vertex

$$V_\alpha^{k1}(\xi, z) = \delta^{kl} V^{l1}(\xi, z)\,, \tag{2.46}$$

$\xi \in \mathcal{H}_l$. Using remark (4) we may extend the definition of $V^{l1}(\xi, z)\Omega$ to $z = 0$ by setting

$$V^{l1}(\xi, 0)\Omega = \xi\,. \tag{2.47}$$

Similarly,

$$V_\alpha^{1k}(\xi, z) = \delta^{kl} V^{1l}(\xi, z) , \tag{2.48}$$

$\xi \in \mathcal{H}_{l*} = \mathcal{H}_l^*$, and we may extend the definition of $\langle \Omega^* | V^{1l}(\xi, z)\rangle$ to $z = \infty$, by setting

$$\lim_{z \to \infty} \langle \Omega^* | V^{1l}(l^*(z^{2L_0})\xi, z) = \langle \xi | . \tag{2.49}$$

□

Using remark (2) and (2.46)–(2.49), we may now explicitly calculate $\langle \Omega^* | V^{1l}(\xi_1, z_1) V^{l1}(\xi_2, z_2)\Omega\rangle$ and $\langle \Omega^* | V^{1l}(\eta_1, w_1) V_\alpha^{lk}(\eta_2, w_2) V^{k1}(\eta_3, w_3)\Omega\rangle$ for *Virasoro-invariant* vectors ξ_1, $\eta_1 \in \mathcal{H}_{l*}$, $\xi_2 \in \mathcal{H}_l$, $\eta_2 \in \mathcal{H}_j$ and $\eta_3 \in \mathcal{H}_k$, $(z_1, z_2) \in M_2^>$, $(w_1, w_2, w_3) \in M_3^>$, where

$$M_n^> = \{(z_1, \ldots z_n) \in \mathbb{C}^n : |z_1| > \cdots > |z_n|, -\pi < \arg z_i \leq \pi\} . \tag{2.50}$$

One obtains

$$\langle \Omega^* | V^{1l}(\xi_1, z_1) V^{l1}(\xi_2, z_2)\Omega\rangle = \langle \xi_1 | \xi_2\rangle (z_1 - z_2)^{-h_{\xi_1} - h_{\xi_2}} , \tag{2.51}$$

$$\langle \Omega^* | V^{1l}(\eta_1, w_1) V_\alpha^{lk}(\eta_2, w_2) V^{k1}(\eta_3, w_3)\Omega\rangle$$
$$= \langle \eta_1^{(\alpha)} | F_\alpha^l(j, k)\eta_2 \otimes \eta_3\rangle \prod_{1 \leq i < j \leq 3} (w_i - w_j)^{-\Delta_{ij}} , \tag{2.52}$$

where $\Delta_{ij} = h_{\eta_i} + h_{\eta_j} - h_{\eta_k}$, with (i, j, k) a permutation of $(1, 2, 3)$.

Clearly, the right-hand side of (2.51) has an analytic continuation to $\tilde{M}_2$ and the right-hand side of (2.52) has an analytic continuation to $\tilde{M}_3$, where

$$M_n = \{(z_1, \ldots , z_n) \in \mathbb{C}^n : z_i \neq z_j, \text{ for } i \neq j\} , \tag{2.53}$$

and $\tilde{M}_n$ is the universal covering space of M_n.

Note that the domains M_n are not simply connected, $\pi_1(M_n) = P_n$ (the pure braid group on n strings[16]). The permutation group, S_n, of n elements acts on M_n in the obvious way. The braid group, B_n, on n strings can be defined as the fundamental group, $\pi_1(M_n/S_n)$, of M_n/S_n. Pick a point $p \in M_n$. The covering space, $\tilde{M}_n$, of M_n can be described as the set of pairs $(z^{(n)}, [\gamma])$, where $z^{(n)} \equiv (z_1, \ldots , z_n) \in M_n$ and $[\gamma]$ is a homotopy class of paths $\gamma \subset M_n$ from p to $z^{(n)}$. The braid group B_n can be described as the set of pairs $(\pi, [\gamma])$, where $\pi \in S_n$ and $[\gamma]$ is a homotopy class of paths $\gamma \subset M_n$ from p to $\pi(p)$. The multiplication law in B_n is given by

$$(\pi, [\gamma])(\pi', [\gamma']) = (\pi \circ \pi', [\gamma \circ \pi(\gamma')]) . \tag{2.54}$$

An action of B_n on $\tilde{M}_n$ is given by

$$(\pi, [\gamma])(z^{(n)}, [\gamma']) = (\pi(z^{(n)}), [\gamma \circ \pi(\gamma')]) ; \tag{2.55}$$

B_n acts freely on $\tilde{M}_n$. For $z^{(n)} \in M_n^>$, there is a unique homotopy class, $[\gamma_0]$, of paths, γ_0, from p to $z^{(n)}$, with $\gamma_0 \subset M_n^>$. We set

$$\tilde{M}_n^> = \{(z^{(n)}, [\gamma_0]) : z^{(n)} \in M_n^>\} . \tag{2.56}$$

Then $\tilde{M}_n^>$ is a fundamental domain for the action of B_n on $\tilde{M}_n$.

The group B_n is generated by elements $\tau_1, \ldots, \tau_{n-1}$, where $\tau_i = (t_i, [\gamma_i])$, t_i denotes the transposition of i and $i+1$, and, for $p = \left(1, \frac{1}{2}, \ldots, \frac{1}{n}\right) \in M_n^>$, γ_i is a path leaving $1, \ldots, \frac{1}{i-1}, \frac{1}{i+2}, \ldots, \frac{1}{n}$ constant and exchanging $\frac{1}{i}$ with $\frac{1}{i+1}$ along a positively oriented path not enclosing $1, \frac{1}{2}, \ldots, \frac{1}{i-1}, \frac{1}{i+2}, \ldots, \frac{1}{n}$. The relations between the generators $\tau_1, \ldots, \tau_{n-1}$ are as follows:

$$\tau_i \tau_{i+1} \tau_i = \tau_{i+1} \tau_i \tau_{i+1}$$

$$\tau_i \tau_j = \tau_j \tau_i , \qquad \text{for } |i - j| \geq 2 . \tag{2.57}$$

Every $b \in B_n$ can be written as a word in $\tau_1, \ldots, \tau_{n-1}$,

$$b = \tau_{i_1}^{\varepsilon_1} \ldots \tau_{i_k}^{\varepsilon_k} , \qquad \varepsilon_j = \pm 1 , \qquad j = 1, \ldots, k , \tag{2.58}$$

$k = 1, 2, \ldots$, modulo the relations (2.57).

Points in $\tilde{M}_n$ are henceforth written as $Z^{(n)} = (Z_1, \ldots, Z_n)$, and their projection onto M_n as $z^{(n)}$.

Next, we study products of chiral vertices. We must *assume* that there are no domain problems when considering a product of chiral vertices, even if the representations $i \in L_{\mathcal{A}}$ are unitary, since, in that case, $V_\alpha^{jk}(\xi_l, z)$ is *not* at bounded operator from $\bar{\mathcal{H}}_k$ to $\bar{\mathcal{H}}_j$, for any $\xi_l \neq \vec{0}$ in $\mathcal{H}_l$, with $N_{jlk} \neq 0$.

We *assume* that, for $(z_1, \ldots, z_n) \in M_n^>$,

$$V_{\underline{\alpha},\underline{j}}^{ik}(\xi_1 \otimes \cdots \otimes \xi_n, z_1, \ldots, z_n) = \prod_{l=1}^{n} V_{\alpha_l}^{j_{l-1} j_l}(\xi_l, z_l) \tag{2.59}$$

is an operator from $\bar{\mathcal{H}}_k$ to $\bar{\mathcal{H}}_i$, where $\bar{\mathcal{H}}_i$ is an appropriate extension of $\mathcal{H}_i$, $i \in L_{\mathcal{A}}$; ($\bar{\mathcal{H}}_i$ would be some dense domain in $\bar{\mathcal{H}}_i$ if i is unitary), for arbitrary $\xi_l \in \mathcal{H}_{i_l}$, $l = 1, \ldots, n$. In (2.59), $j_0 = i$ and $j_n = k$. From Theorem 2.1, (c) we get the equation

$$\frac{\partial}{\partial z_l} V^{ik}_{\underline{\alpha},\underline{j}}(\xi_1 \otimes \cdots \otimes \xi_n, z_1, \ldots, z_n)$$

$$= V^{ik}_{\underline{\alpha},\underline{j}}(\xi_1 \otimes \cdots \otimes i_l(L_{-1})\xi_l \otimes \cdots \otimes \xi_n, z_1, \ldots, z_n) \tag{2.60}$$

This equation serves to analytically continue $V^{ik}_{\underline{\alpha},\underline{j}}(\xi_1 \otimes \cdots \otimes \xi_n, z_1, \ldots, z_n)$ in $(z_1, \ldots, z_n)$ beyond the domain $M_n^{>}$. It is plausible, though only proven in special cases,[11,15,17] that one can analytically continue $V^{ik}_{\underline{\alpha},\underline{j}}(\xi_1 \otimes \cdots \otimes \xi_n, z_1, \ldots, z_n)$ to all of $\tilde{M}_n$. Of course, there is no reason why the function $V^{ik}_{\underline{\alpha},\underline{j}}$ $(\xi_1 \otimes \cdots \otimes \xi_n, Z_1, \ldots, Z_n)$, $Z^{(n)} \in \tilde{M}_n$, should have trivial monodromy, i.e. correspond to a single-valued function on M_n. It can be seen from (2.51) and (2.52) that this would mean that $\mathrm{spec}(i(L_0)) \subset \mathbb{Z}$, for all $i \in L_{\mathcal{A}}$—which is essentially never the case. The question then is how to describe the monodromy of $V^{ik}_{\underline{\alpha},\underline{j}}(\xi_1 \otimes \cdots \otimes \xi_n, Z^{(n)})$. The answer is that the monodromy of these operator-valued analytic functions is obtained from certain Yang-Baxter representations of the braid group B_{n+1} which we now describe. Our analysis is based on two basic assumptions which have not been derived completely from results on the representation theory of chiral algebras, so far; (but see Refs. 11, 12, 18 and 19).

The expectations $\langle \xi | V^{ik}_{\underline{\alpha},\underline{j}}(\xi_1 \otimes \cdots \otimes \xi_n, Z^{(n)})\eta \rangle$, $\xi \in \mathcal{H}_{i*}$, $\xi_l \in \mathcal{H}_{i_l}$, $l = 1, \ldots, n$, and $\eta \in \mathcal{H}_k$, define multi-linear functionals on $\mathcal{H}_{i*} \times \mathcal{H}_{i_1} \times \cdots \times \mathcal{H}_{i_n} \times \mathcal{H}_k$ which we denote by $\Phi_{\underline{\alpha},\underline{j}}(i, \underline{i}, k; Z^{(n)})$.

(A1) For fixed i, $\underline{i}$ and k, $\{\Phi_{\underline{\alpha},\underline{j}}(i, \underline{i}, k; Z^{(n)})\}$ forms a set of $d(i, \underline{i}, k)$ *linearly independent* multi-linear functionals *holomorphic* on $\tilde{M}_n \setminus \{0\}$, where

$$d(i, \underline{i}, k) = \sum_{j_1, \ldots, j_{n-1}} \prod_{l=1}^{n} N_{j_{l-1} i_l j_l}, \tag{2.61}$$

with $j_0 = i$, $j_n = k$, and N_{jmn} the multiplicity of j in $m \otimes_z n$.

□

The functionals $\{\Phi_{\underline{\alpha},\underline{j}}(i, \underline{i}, k; Z^{(n)})\}$ carry a representation of B_{n+1} which can be derived from the following assumption.

(A2) Let $(Z_1, Z_2, 0) \in \tilde{M}_3$, $\xi_1 \in \mathcal{H}_m$ and $\xi_2 \in \mathcal{H}_n$, m and n in $L_{\mathcal{A}}$. Let τ_1 be a generator of B_3. Then

$$V^{ik}_{\alpha_1,\alpha_2,j}(\xi_1 \otimes \xi_2, \tau_1^{\pm 1}[Z_2, Z_1])$$
$$= \sum_{\beta_1,\beta_2,l} R^{\pm}(i, m, n, k)^{l\beta_2\beta_1}_{j\alpha_1\alpha_2} V^{ik}_{\beta_2\beta_1,l}(\xi_2 \otimes \xi_1, Z_2, Z_1)\,. \tag{2.62}$$

Equation (2.62) can be rewritten as follows. We define

$$[V^{ij}_{\alpha_1}(\xi_1, z_1) V^{jk}_{\alpha_2}(\xi_2, z_2)]_{\pm}$$

by analytic continuation in (z', z'') from $(z_2, z_1) \in M_2^{>}$ to (z_1, z_2) along a path $\gamma_{\pm}$ not enclosing 0 with $(t, [\gamma_{\pm}]) = \tau_1^{\pm 1}$, where t is the transposition of 1 and 2. Then

$$[V_{\alpha_1}^{ij}(\xi_1, z_1)V_{\alpha_2}^{jk}(\xi_2, z_2)]_{\pm}$$

$$= \sum_{\beta_1, \beta_2, l} R^{\pm}(i, m, n, k)_{j\alpha_1\alpha_2}^{l\beta_2\beta_1} V_{\beta_2}^{il}(\xi_2, z_2)V_{\beta_1}^{lk}(\xi_1, z_1) \text{ ,}^{c} \tag{2.63}$$

where the matrices $R^{\pm}(i, m, n, k)$ do not depend on $\xi_1 \in \mathcal{H}_m$ and $\xi_2 \in \mathcal{H}_n$, but only on i, m, n, and k, and have the property that

$$R^{\pm}(i, m, n, k)_{j\cdots}^{l\cdots} = 0 \, , \qquad \text{unless} \begin{cases} N_{imj}N_{jnk} \neq 0 \, , \\ N_{inl}N_{lmk} \neq 0 \, . \end{cases} \tag{2.64}$$

□

Let

$$\Delta(i, m, n, k) \equiv \{j \in L_{\mathcal{A}} : N_{imj}N_{jnk} \neq 0\} \tag{2.65}$$

and define the complex vector space

$$W(i, m, n, k) = \bigoplus_{j \in \Delta(i,m,n,k)} \mathbb{C}^{N_{imj}} \otimes \mathbb{C}^{N_{jnk}} \, . \tag{2.66}$$

Then $R^{\pm}(i, m, n, k)$ are linear maps from $W(i, n, m, k)$ to $W(i, m, n, k)$. By analytic continuation of (2.62) to points $(Z_1, Z_2, 0) \in \tilde{M}_3$ along paths $\gamma_{\mp}$, with $(t, [\gamma_{\mp}]) = \tau_1^{\mp 1}$ and another application of (2.62) we find, using (A1), that

$$R^{\pm}(i, m, n, k)R^{\mp}(i, n, m, k) = 1|_{W(i,m,n,k)} \, . \tag{2.67}$$

Hence,

$$\dim W(i, m, n, k) = \dim W(i, n, m, k) \tag{2.68}$$

and, by (2.63), (2.64),

$$\sum_j N_{imj}N_{jnk} = \sum_l N_{inl}N_{lmk} \, , \tag{2.69}$$

for arbitrary i, m, n and k in $L_{\mathcal{A}}$. From (2.45) we also have that

$$N_{imj} = N_{ijm} \, , \tag{2.70}$$

and from (2.49) and (2.45),

$$N_{imj} = N_{m^*i^*j} \, , \tag{2.71}$$

[c] Formally, (2.63) follows from a version of Schur's lemma.

i, m and j in $L_{\mathcal{A}}$. Note also that from (2.42) and (2.44) it follows that

$$N_{i1j} = \delta_{ij} \ . \tag{2.72}$$

Equations (2.69)–(2.72) are important constraints on the multiplicites N_{imj} identical to the constraints on the multiplicities one finds in the representation theory of groups and quantum groups.[20] (The multiplicities for $\widehat{\mathrm{su}}(2)_k$ (level-k-SU(2) current algebra) and $U_q(sl(2))$, $q = e^{2\pi i/k+2}$, are worked out in Ref. 21 as an example.)

We now use assumptions (A1) and (A2), (2.62) to define a representation of B_{n+1} on the space of generalized vertices

$$\{V^{ik}_{\underline{\alpha},\underline{j}}(\xi_1 \otimes \cdots \otimes \xi_n, Z_1, \ldots, Z_n)\}$$

defined in (2.59):

$$\begin{aligned} &V^{ik}_{\alpha_1\ldots\alpha_n,j_1\ldots,j_{n-1}}(\xi_1 \otimes \cdots \otimes \xi_{l+1} \otimes \xi_l \otimes \cdots \otimes \xi_n, \tau_l^{\pm 1} Z^{(n)}) \\ &\quad = \sum_{j_l',\beta_l,\beta_{l+1}} R(j_{l-1}, i_{l+1}, i_l, j_{l+1})^{j_l' \beta_l \beta_{l+1}}_{j_l \alpha_{l+1} \alpha_l} \\ &\quad\quad \times V^{ik}_{\alpha_1\ldots\beta_l\beta_{l+1}\ldots\alpha_n,j_1\ldots j_l'\ldots j_n}(\xi_1 \otimes \cdots \otimes \xi_l \otimes \xi_{l+1} \otimes \cdots \otimes \xi_n, Z^{(n)}) \ . \end{aligned} \tag{2.73}$$

Equation (2.65) defines a representation of B_{n+1}, because the matrices $R^+(i, m, n, k)$ satisfy the *Yang-Baxter equation*

$$\begin{aligned} &\sum_n R^+(q, i, l, s)^n_r R^+(p, j, l, n)^k_q R^+(k, j, i, s)^m_n \\ &\quad = \sum_n R^+(p, j, i, r)^n_q R^+(n, j, l, s)^m_r R^+(p, i, l, m)^k_n \ . \end{aligned} \tag{2.74}$$

Here and in most of the following formulas, we omit the indices α_1, α_2, β_1, β_2 on $R^+(\ldots)^{n\beta_2\beta_1}_{q\alpha_1\alpha_2}$ which must be summed over. Hence in (2.74) a matrix product of certain blocks of the R-matrices is understood. Equation (2.74) follows from the associativity of multiplication and analytic continuation of chiral vertices; see e.g. Ref. 10.

One can explore the Möbius invariance of

$$\langle \Omega^* | V^{11}_{\underline{\alpha},\underline{j}}(\xi_1 \otimes \cdots \otimes \xi_n, Z^{(n)}) \Omega \rangle \ , \tag{2.75}$$

for $n = 2, 3, 4$, to compute some special R-matrices and derive a constraint.[21] In order to avoid cumbersome notation we assume that the operators $F^k_\alpha(i, l)$ introduced before (2.31) are normalized such that

$$\langle \zeta | F^k_\alpha(i, l) \xi \otimes \eta \rangle = \langle \zeta | F^k_\alpha(l, i) \eta \otimes \xi \rangle \ , \tag{2.76}$$

for arbitrary $\zeta \in \mathcal{H}_k^*$, $\xi \in \mathcal{H}_i$, $\eta \in \mathcal{H}_l$, and arbitrary i, l, k in $L_{\mathcal{A}}$. From (2.47) we then get that

$$\langle \zeta | V_\alpha^{kl}(\xi, 1) V^{l1}(\eta, 0)\Omega \rangle = \langle \zeta | V_\alpha^{ki}(\eta, 1) V^{i1}(\xi, 0)\Omega \rangle \,. \tag{2.77}$$

Using (2.49), (2.77) can be generalized to

$$\langle \zeta | V_\alpha^{kl}(\xi, 1) V^{l1}(\eta, 0)\Omega \rangle = \langle \xi | V_\alpha^{i^*l}(\zeta, 1) V^{l1}(\eta, 0)\Omega \rangle \,. \tag{2.78}$$

Using the normalizations (2.76)–(2.78) one can derive from (2.62) and the Möbius invariance of (2.75) the following equations:

(a) $$R^{\pm}(1, j, j^*, 1)_l^k = \delta^{jk}\delta_{j^*l} e^{\mp i\pi(h_j + h_{j^*})} \tag{2.79}$$

(b) $$R^{\pm}(j, i, l, 1)_n^m = \delta^{im}\delta_{ln} e^{\pm i\pi(h_j - h_i - h_l)} \tag{2.80}$$

and

(c) $$\sum_m R^{\pm}(j, i, l, k)_m^n D^{\pm}(m, i, k) R^{\pm}(j, l, i, k)_p^m = \delta_p^n D^{\pm}(j, i, p) \,, \tag{2.81}$$

where

$$D^{\pm}(j, i, k) = e^{\mp 2\pi i(h_j - h_i - h_k)} \,, \tag{2.82}$$

and

$$h_i = \inf \operatorname{spec} i(L_0) \,. \tag{2.83}$$

In the proof of (a), (b) and (c) one considers (2.75) for $n = 2$, 3 and 4, respectively. These equations were first proven in Ref. 22 and then in Ref. 21, where it is also shown that (c) reflects a relation between the generators of B_n characteristic of the braid group on the twice punctured two-dimensional sphere.

To conclude this section, we should emphasize that

(i) the analyticity of $V_{\alpha,j}^{ik}(\xi_1 \otimes \cdots \otimes \xi_n, Z_1, \ldots, Z_n)$ on all of $\tilde{M}_n \backslash \{0\}$, see assumption (A1),

(ii) assumption (A2), (2.62); and

(iii) Eqs. (2.77) and (2.78)

have not been derived purely from results on the representation theory of chiral algebras, so far. These assumptions—which are valid in several models of rational conformal field theory[18,19]—can, however, be derived from a system of "axioms" for two-dimensional, local conformal field theory.[11,12] Their status will be further studied elsewhere. They will be assumed to hold henceforth.

3. The Fusion of Chiral Vertices

It has been shown in Refs. 12 and 7 that from the assumptions described in Sec. 2 one can derive an *operator product expansion* for chiral vertices. Let $\{\xi_p^N\}_{N=0}^{\infty}$ be a basis for $\mathcal{H}_p$ and $\{\langle\xi_p^N|\}_{N=0}^{\infty}$ a dual basis for $\mathcal{H}_p^*$, (i.e. $\langle\xi_p^N|\xi_p^M\rangle = \delta^{NM}$). Then, for $|z-w| < |w| < |z| < \infty$, $\xi_r \in \mathcal{H}_r$ and $\xi_s \in \mathcal{H}_s$, in the sense of sesquilinear forms on $\mathcal{H}_j^* \times \mathcal{H}_k$,

$$V_\alpha^{jm}(\xi_r, z)V_\beta^{mk}(\xi_s, w)$$
$$= \sum_{p\in L_{\mathcal{A}}} \sum_{N=0}^{\infty} \sum_{\gamma,\delta} F(j, r, s, k)^{p\,\gamma\delta}_{m\alpha\beta} \langle\xi_p^N| V_\gamma^{ps}(\xi_r, z-w)\xi_s\rangle V_\delta^{jk}(\xi_p^N, w)\ , \tag{3.1}$$

where the coefficients $F(j, r, s, k)^{p\,\gamma\delta}_{m\alpha\beta}$ satisfy a system of linear equations with coefficients given in terms of the braid matrices $R^{\pm}$ introduced in Sec. 2. The main purpose of this section is to review these equations; see Refs. 7, 12, 21 and 23 for details.

Remarks

(1) The convergence of $\sum_{N=0}^{\infty}(\ldots)$ on the right-hand side of (3.1) can be proven (under natural domain hypotheses) if all representations in $L_{\mathcal{A}}$ are unitary. In more general situations, convergence of the right-hand side of (3.1) must be assumed.

(2) Let $\mathcal{V}_{prs} \simeq \mathbb{C}^{N_{prs}}$ be the vector space of independent vertices $V_\alpha^{ps}(\xi_r, z)$, $\alpha = 1, \ldots, N_{prs}$, for fixed p, r, s and $\xi_r \in \mathcal{H}_r$. Equation (2.62) shows that the matrices $R^{\pm}(i, m, n, k)$ define linear maps from $\oplus_j \mathcal{V}_{inj} \otimes \mathcal{V}_{jmk}$ to $\oplus_l \mathcal{V}_{iml} \otimes \mathcal{V}_{lnk}$, while we see from (3.1) that the matrices $F(i, m, n, k)$ define linear maps from $\oplus_p \mathcal{V}_{pmn} \otimes \mathcal{V}_{ipk}$ to $\oplus_j \mathcal{V}_{imj} \otimes \mathcal{V}_{jnk}$. From now on we shall omit the subscripts α, β and the superscripts γ, δ on $F(i, m, n, k)^{p\,\gamma\delta}_{j\alpha\beta}$ which refer to bases in $\mathcal{V}_{imj} \otimes \mathcal{V}_{jnk}$, $\mathcal{V}_{pmn} \otimes \mathcal{V}_{ipk}$, respectively.

(3) Note that if, in Eq. (3.1), we set $k = 1$, $m = s$ and $j = p$, and use the completeness relation

$$\sum_{N=0}^{\infty} \xi_p^N \langle\xi_p^N| = 1|_{\mathcal{H}_p}$$

then Eq. (3.1) yields

$$F(j, r, s, 1)^p_m = \delta^s_m \delta^p_j\ . \tag{3.2}$$

□

Next, we describe the equations between the R- and the F-matrices that follow from the structure described in Sec. 2. It is helpful to use a graphical notation.

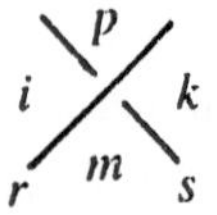

$\leftrightarrow \quad R^{+}(i, r, s, k)^{p}_{m}$ (3.3)

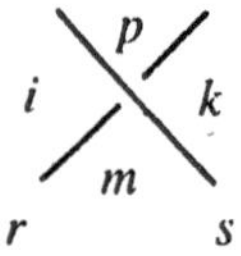

$\leftrightarrow \quad R^{-}(i, r, s, k)^{p}_{m}$ (3.4)

and

$\leftrightarrow \quad F(i, r, s, k)^{p}_{m}$. (3.5)

Then the equations between R- and F-matrices are

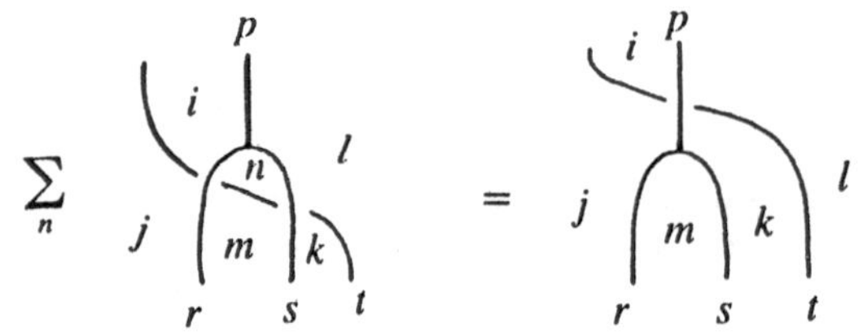

(3.6)

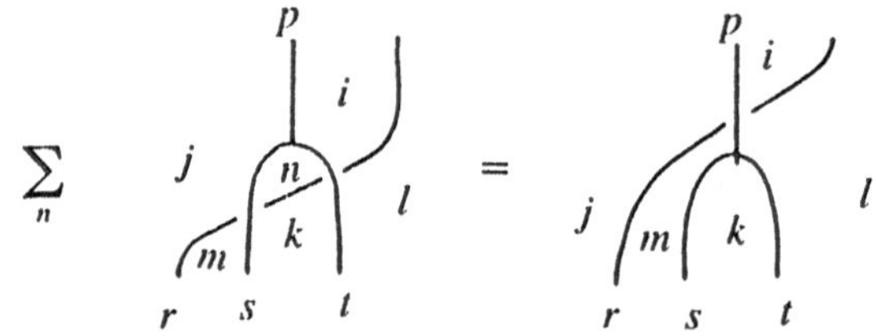

(3.7)

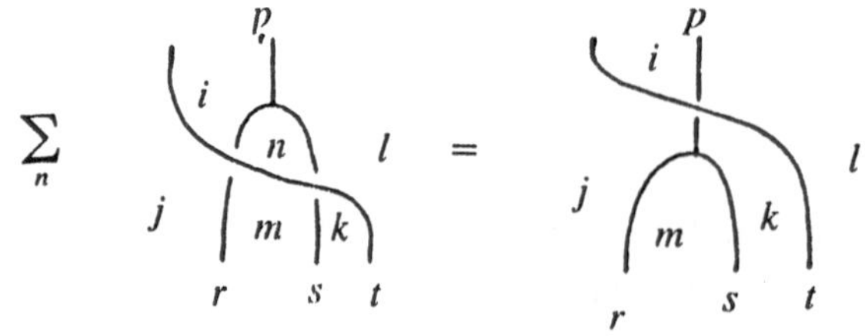

(3.8)

and

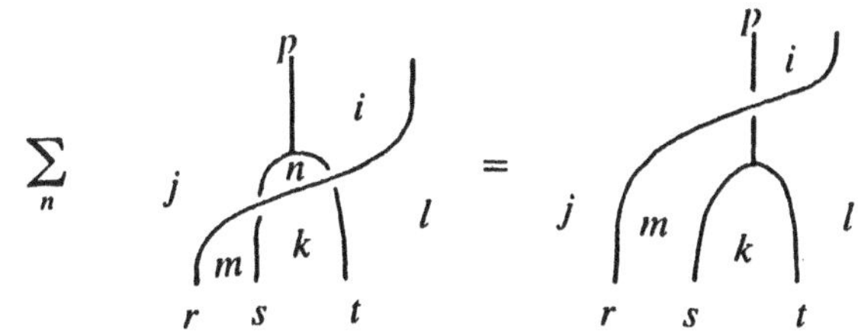

$$\sum_n \qquad = \qquad (3.9)$$

The analytical form of Eqs. (3.6), (3.8) is obviously

$$\sum_n R^\pm(m, s, t, l)_k^n R^\pm(j, r, t, n)_m^i F(i, r, s, l)_n^p = F(j, r, s, k)_m^p R^\pm(j, p, t, l)_k^i \,.$$

$$(3.10)$$

Equations (3.7)–(3.9) can be transcribed similarly; see Ref. 21.

We also have equations relating $R^\pm$- and F-matrices to the conformal dimensions of chiral vertices:

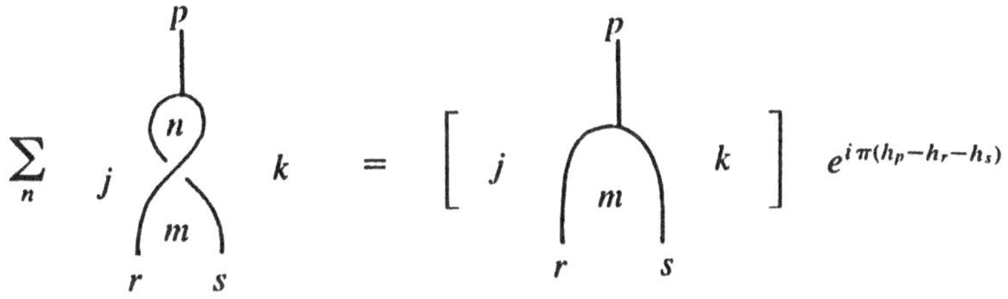

$$\sum_n \qquad = \qquad e^{i\pi(h_p - h_r - h_s)} \,, \quad (3.11)$$

and

$$\sum_n \qquad = \qquad e^{-i\pi(h_p - h_r - h_s)} \,, \quad (3.12)$$

where $h_p = \inf \operatorname{spec} p(L_0)$, for $p \in L_{\mathcal{A}}$.

Thus

$$\sum_n R^\pm(j, r, s, k)_m^n F(j, s, r, k)_n^p = e^{\pm i\pi(h_p - h_r - h_s)} F(j, r, s, k)_m^p \,. \qquad (3.13)$$

Equations (2.67) and (2.74) (the Yang-Baxter equation) are, in graphical notation,

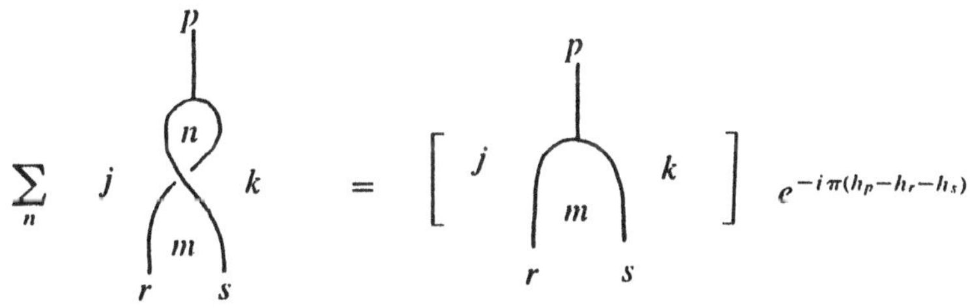

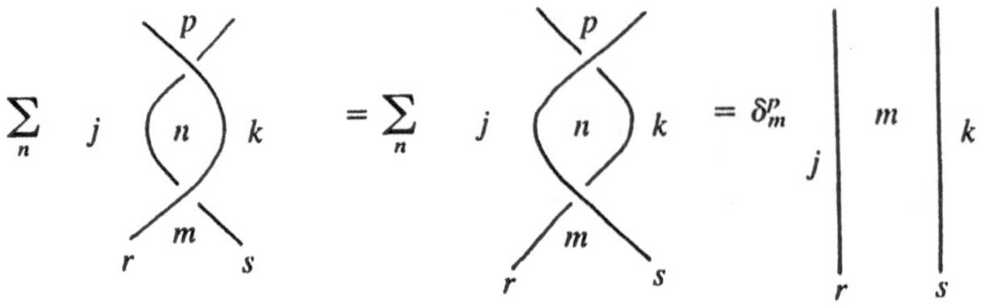

(3.14)

and

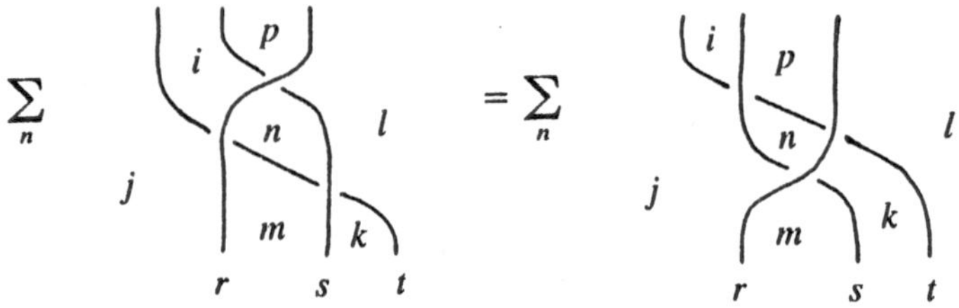

, (3.15)

respectively.

Equations (3.6)–(3.9), (3.11) and (3.12), and (3.14) and (3.15) are a complete system of equations between the $R^{\pm}$-matrices, F-matrices and the conformal dimensions h_i, $i \in L_{\mathscr{A}}$. Equations (2.79)–(2.82) may easily be derived from them; (see e.g. Ref. 21).

From (3.6) and (3.14) we derive

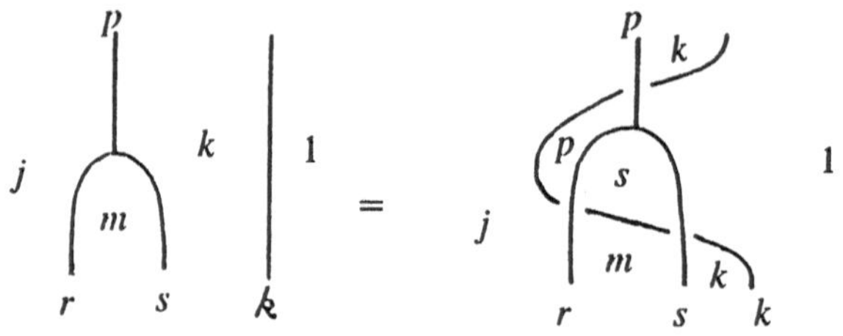

(3.16)

Hence

$$F(j, r, s, k)^p_m = R^{\pm}(m, s, k, 1)^s_k R^{\pm}(j, r, k, s)^p_m F(p, r, s, 1)^p_s R^{\mp}(j, k, p, 1)^k_p \, .$$

By (2.80)

$$R^{\pm}(m, s, k, 1)^s_k = e^{\pm i\pi(h_m - h_s - h_k)} \, ,$$

and by (3.2)

$$F(p, r, s, 1)^p_s = 1 \, .$$

Hence

$$F(j,\, r,\, s,\, k)^p_m = e^{\pm i\pi(h_m + h_p - h_j - h_s)} R^{\pm}(j,\, r,\, k,\, s)^p_m \,. \tag{3.17}$$

Equations (3.10) and (3.17) have an important consequence. A chiral vertex $V^{jk}_{\alpha}(\xi_r, z)$, with $\xi_r \in \mathcal{H}_r$, is called of "*type r*". Let us assume we know the $R^{\pm}$-matrices for vertices of type s and t, r and t, r and l, and r and k, respectively. Then in Eq. (3.10) all matrices, except possibly $R^{\pm}(j, p, t, l)^i_k$, are known, and we may view (3.10) as an equation for the braid matrices $R^{\pm}(j, p, t, l)^i_k$ between a vertex of type p obtained by fusing vertices of type r and s and a vertex of type t. This observation serves to compute all braid matrices from some basic ones. In order to explain this, we introduce the notion of fundamental representations of $\mathcal{A}$.

Definition 3.1.

The representations $i_1, \ldots, i_a$ in $L_{\mathcal{A}}$ are called a complete family of fundamental representations of $\mathcal{A}$ if every representation $k \in L_{\mathcal{A}}$ can be obtained by successively forming tensor products, $\otimes_z$, of the representations $i_1, \ldots, i_a$ and decomposing them into irreducible representations.

It follows that if the $R^{\pm}$-matrices for matrices of type i_m and i_n, $m, n = 1, \ldots, a$, are known then all remaining $R^{\pm}$-matrices can, in principle, be determined from (3.10), (3.17). This is demonstrated on an example in Ref. 18.

We conclude this section by describing a special case of Eqs. (3.6)–(3.9) and (3.11), (3.12) which will play a crucial role in Sec. 6: We set $p = 1$ (the vacuum representation) in all these equations. Because of Eq. (2.44)

$$\left.\begin{aligned} R^{\pm}(j,\, r,\, 1,\, k)^n_m &= \delta_{mk}\delta^{nj} \\ R^{\pm}(j,\, 1,\, r,\, k)^n_m &= \delta_{mj}\delta^{nk} \end{aligned}\right\} . \tag{3.18}$$

Let us denote by  a vertex of type 1. Then, using (3.18), (3.6) becomes

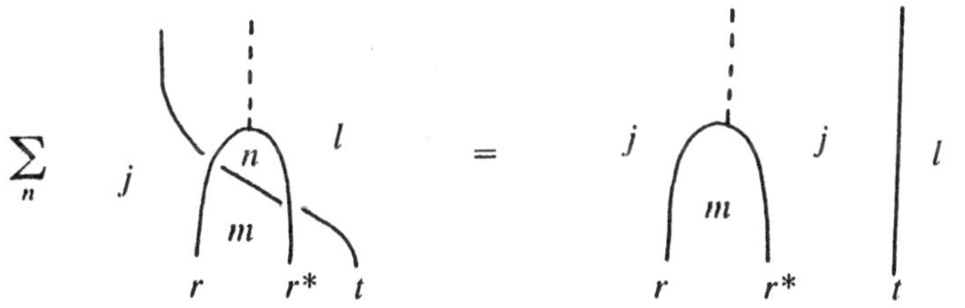

(3.19)

By (3.14) this yields

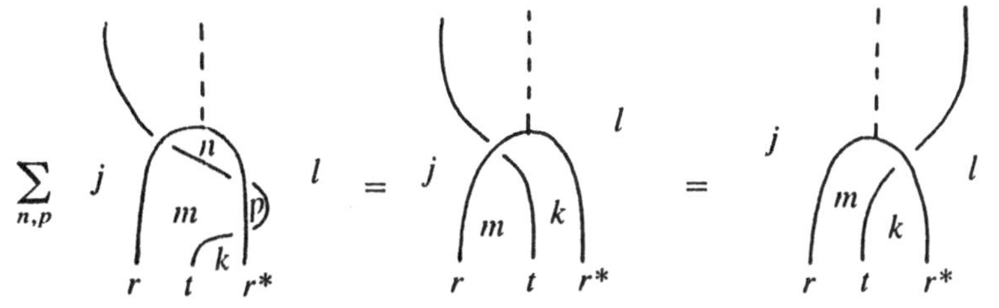

(3.20)

We shall abbreviate $F(l, r, r^*, l)^1_n$ by

$$F(l, r, r^*, l)^1_n \equiv F^l_{rn} \,. \tag{3.21}$$

Then (3.20) yields

$$R^{\pm}(j, r, t, k)^l_m F^l_{rk} = R^{\mp}(m, t, r^*, l)^j_k F^j_{rm} \,. \tag{3.22}$$

Note that by (3.1) the image of the matrices F^l_{rk} and F^j_{rm} is one-dimensional; ($\gamma = \delta = 1$ in (3.1)). However, viewed as a linear map from $\mathcal{V}_{lrk}$ to $\mathcal{V}_{kr^*l}$, F^l_{rk} is invertible, and F^j_{rm} is an invertible linear map from $\mathcal{V}_{mr^*j}$ to $\mathcal{V}_{jrm}$, provided, of course, F^l_{rk} and F^j_{rm} are nonzero. Moreover, when reinterpreted in this way, F^l_{rk} and F^j_{rm} act on different spaces and hence they commute. Thus, by multiplying (3.22) from the right by $(F^l_{rk})^{-1}(F^j_{rm})^{-1}$ we find

$$(F^j_{rm})^{-1}R^{\pm}(j, r, t, k)^l_m = (F^l_{rk})^{-1}R^{\mp}(m, t, r^*, l)^j_k \,. \tag{3.23}$$

It is useful to reinstall the indices $\alpha, \beta, \ldots$ to understand the meaning of Eqs. (3.22) and (3.23). Equation (3.22) then reads

$$\sum_{\delta} R^{\pm}(j, r, t, k)^{l\gamma\delta}_{m\alpha\beta}(F^l_{rk})^{\delta}_{\mu} = \sum_{\nu} R^{\mp}(m, t, r^*, l)^{j\nu\gamma}_{k\beta\mu}(F^j_{rm})^{\nu}_{\alpha}$$

from which it follows, using

$$\sum_{\mu} (F^l_{rk})^{\lambda}_{\mu}[(F^l_{rk})^{-1}]^{\mu}_{\varepsilon} = \delta^{\lambda}_{\varepsilon} \,,$$

that

$$\sum_{\alpha} [(F^j_{rm})^{-1}]^{\alpha}_{\lambda} R^{\pm}(j, r, t, k)^{l\gamma\nu}_{m\alpha\beta} = \sum_{\mu} [(F^l_{rk})^{-1}]^{\mu}_{\nu} R^{\mp}(m, t, r^*, l)^{j\lambda\gamma}_{k\beta\mu}$$

which is (3.23).

Equation (3.23) can be expressed, graphically, as follows:

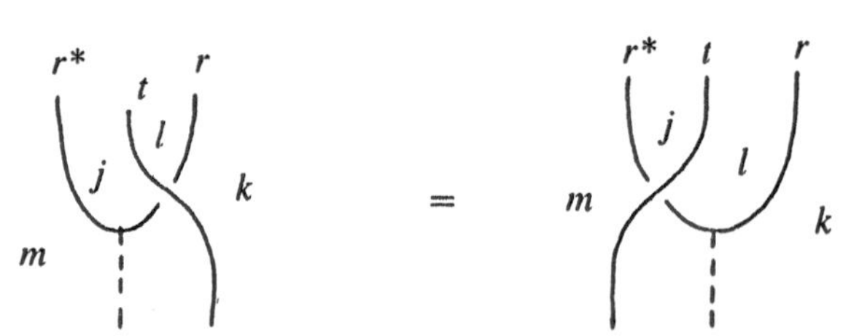

(3.24)

with the rule

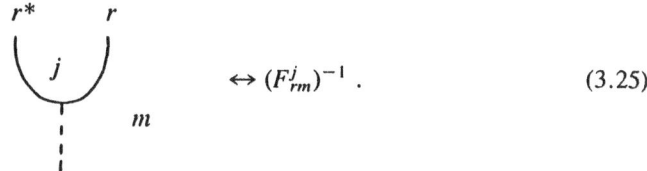

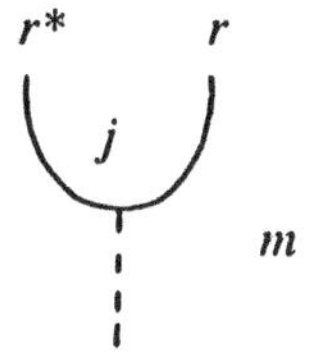

$$\leftrightarrow (F^{j}_{rm})^{-1} \,. \tag{3.25}$$

From (3.21) and (3.25) we have that

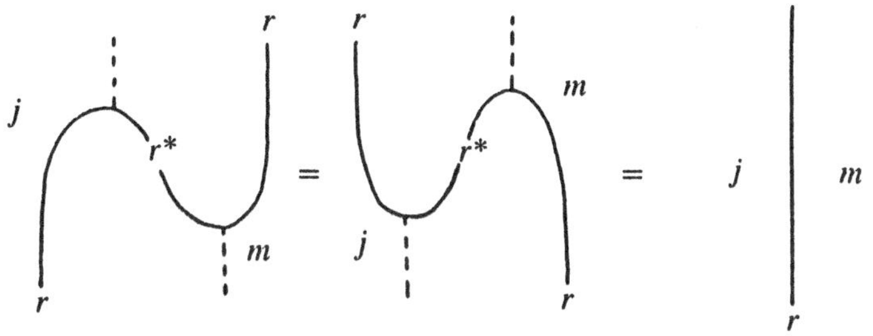

(3.26)

corresponding to

$$F^{j}_{rm}(F^{j}_{rm})^{-1} = (F^{m}_{r^*j})^{-1}F^{m}_{r^*j} = 1 \,, \tag{3.27}$$

for $F^{j}_{rm} \neq 0$, $F^{m}_{r^*j} \neq 0$. Since $N_{1rr^*} = 1$, it follows from (3.1) and (3.22) that $F^{j}_{rm} \neq 0$ whenever $V^{jm}(\xi, z) \neq 0$, for $\xi \in \mathcal{H}_r$, i.e. $N_{jrm} \neq 0$.

Equations (3.20), (3.24) and (3.26) will play an important role in Sec. 6.

4. The Reconstruction of Local Fields from Chiral Vertices

In this section we consider two rational chiral algebras, $\mathcal{A}$ and $\bar{\mathcal{A}}$. For each algebra, the basic assumptions, (A1) and (A2), (see (2.61)–(2.64)), described in Sec. 2 are assumed to hold. Hence the results of Sec. 2 and Sec. 3 are valid.

The purpose of this section is to reconstruct from the chiral vertices associated with the representation theory of $\mathcal{A}$ and $\bar{\mathcal{A}}$ local fields of some two-dimensional conformal field theory. We propose the following *ansatz*: Let $\xi \in \mathcal{H}_r$, $r \in L_{\mathcal{A}}$, $\bar{\xi} \in \mathcal{H}_{\bar{r}}$, $\bar{r} \in L_{\bar{\mathcal{A}}}$. We try to choose coefficients $D\left(\begin{smallmatrix} r \\ ij \end{smallmatrix}\middle|\begin{smallmatrix} \bar{r} \\ \bar{i}\bar{j} \end{smallmatrix}\right)$ such that

$$\phi_{\xi\otimes\bar{\xi}}(z, \bar{z}) = \sum_{i,j,\bar{i},\bar{j}} D\left(\begin{matrix} r \\ ij \end{matrix}\middle|\begin{matrix} \bar{r} \\ \bar{i}\bar{j} \end{matrix}\right) V^{ij}(\xi, z) \otimes V^{\bar{i}\bar{j}}(\bar{\xi}, \bar{z}) \tag{4.1}$$

is a *local* field in the sense that

$$\phi_{\xi\otimes\bar{\xi}}(z, \bar{z})\phi_{\eta\otimes\bar{\eta}}(w, \bar{w}) \tag{4.2}$$

is *symmetric* under exchanging the order of the two factors, (by analytic continuation in (z, w) along a path $\gamma^{(2)} \in \tilde{M}_2\backslash\{0\}$ and in $(\bar{z}, \bar{w})$ along a path, $\bar{\gamma}^{(2)}$, homotopic to the complex conjugate of $\gamma^{(2)}$. This guarantees, in particular, that, for $\bar{z} = z^* =$ complex conj. of z, $\bar{w} = w^*$, $\phi_{\xi\otimes\bar{\xi}}(z, z^*)\phi_{\eta\otimes\bar{\eta}}(w, w^*)$ is symmetric under exchanging the two factors). This requirement yields an overdetermined system of algebraic equations for the coefficients $D\left(\begin{smallmatrix} r \\ ij \end{smallmatrix}\middle|\begin{smallmatrix} \bar{r} \\ \bar{i}\bar{j} \end{smallmatrix}\right)$ involving the braid matrices $R^{\pm}$ and $\bar{R}^{\pm}$ of the algebras $\mathcal{A}$ and $\bar{\mathcal{A}}$:[d]

$$\sum_{j,\bar{j}} D\left(\begin{array}{c|c} r & \bar{r} \\ \hline ij & \bar{i}\bar{j} \end{array}\right) D\left(\begin{array}{c|c} s & \bar{s} \\ \hline jk & \bar{j}\bar{k} \end{array}\right) R^{\pm}(i, r, s, k)_j^l \bar{R}^{\mp}(\bar{i}, \bar{r}, \bar{s}, \bar{k})_{\bar{j}}^{\bar{l}}$$

$$= D\left(\begin{array}{c|c} s & \bar{s} \\ \hline il & \bar{i}\bar{l} \end{array}\right) D\left(\begin{array}{c|c} r & \bar{r} \\ \hline lk & \bar{l}\bar{k} \end{array}\right) ; \tag{4.3}$$

see e.g. Refs. 17, 21 and 22. Further equations involving the coefficients D are obtained from the simple observation that if the chiral vertices of a product of two local fields are fused (see Sec. 3), one obtains a new field which is again *local*. These equations involve the fusion matrices and the coefficients, $C\left(\begin{smallmatrix} p \\ rs \end{smallmatrix}\middle|\begin{smallmatrix} \bar{p} \\ \bar{r}\bar{s} \end{smallmatrix}\right)$, of the operator product expansion.[7,21] It is useful to introduce a graphical notation summarizing the contents of all these equations in a simple and intuitive way: We choose Cartesian coordinates in the plane, with a t-axis, called the *axis of transfer*, and an x-axis. All chiral vertices referring to the algebra $\mathcal{A}$ are represented by lines moving upwards into the $\{t \geq 0\}$ half-plane with slopes that become horizontal (i.e. with a tangent vector that is parallel to the x-axis) only at points where two chiral vertices are fused. These lines are labelled by representations r, r^*, s, . . . in $L_{\mathcal{A}}$. All chiral vertices referring to the algebra $\bar{\mathcal{A}}$ are represented by lines, labelled by $\bar{r}$, $\bar{r}^*$, $\bar{s}$, . . . in $L_{\bar{\mathcal{A}}}$, moving downwards into the $\{t < 0\}$ half-plane with slopes that become horizontal only at points where two chiral vertices of $\bar{\mathcal{A}}$ are fused. Every line is the boundary between two regions of the plane labelled by representations belonging to $L_{\mathcal{A}}$, $L_{\bar{\mathcal{A}}}$, respectively. Lines referring to vertices of $\mathcal{A}$ are hooked up to lines referring to vertices of $\bar{\mathcal{A}}$ on the x-axis, with "amplitudes" given by the D matrices. Thus

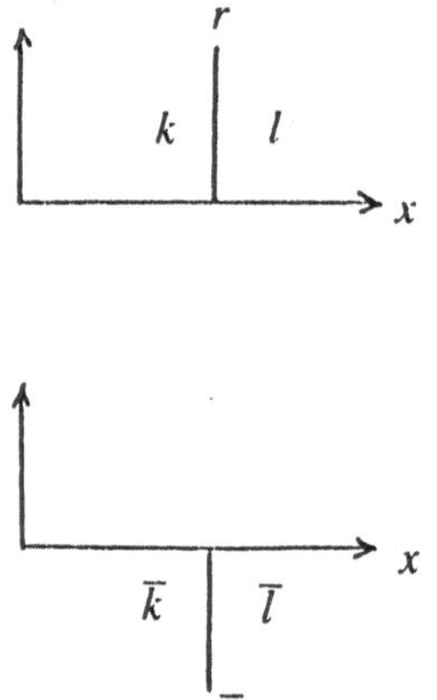

$\leftrightarrow [V^{kl}(\xi_r, z)]$

(4.4)

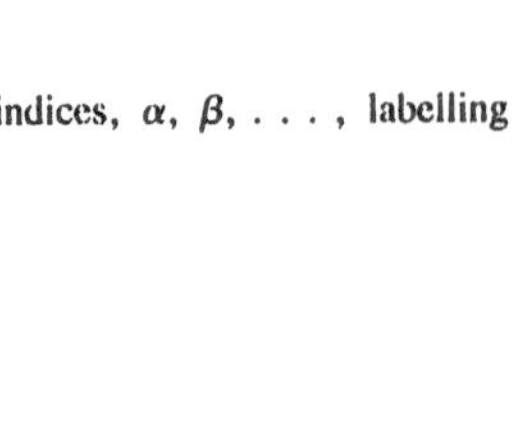

$\leftrightarrow [V^{\bar{k}\bar{l}}(\bar{\xi}_{\bar{r}}, \bar{z})]$

---

[d] We suppress the greek indices, α, β, . . . , labelling different vertices of the same type.

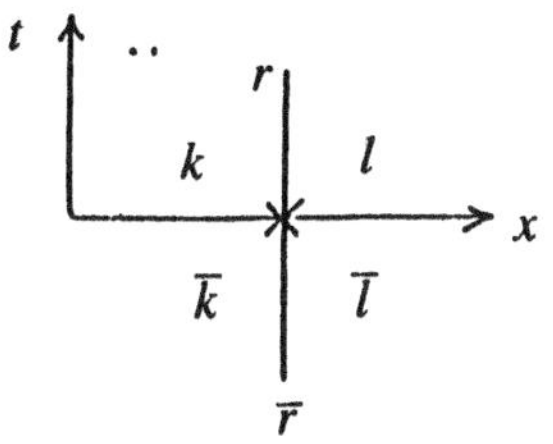

$$\leftrightarrow D\left(\begin{array}{c|c} r & \bar{r} \\ kl & \bar{k}\bar{l} \end{array}\right) \tag{4.5}$$

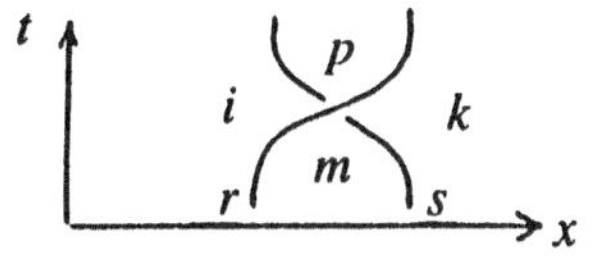

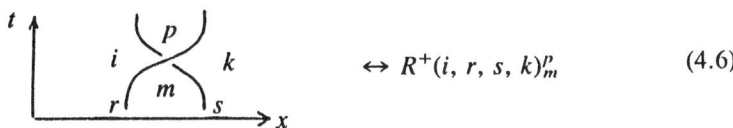

$$\leftrightarrow R^{+}(i,\, r,\, s,\, k)^{p}_{m} \tag{4.6}$$

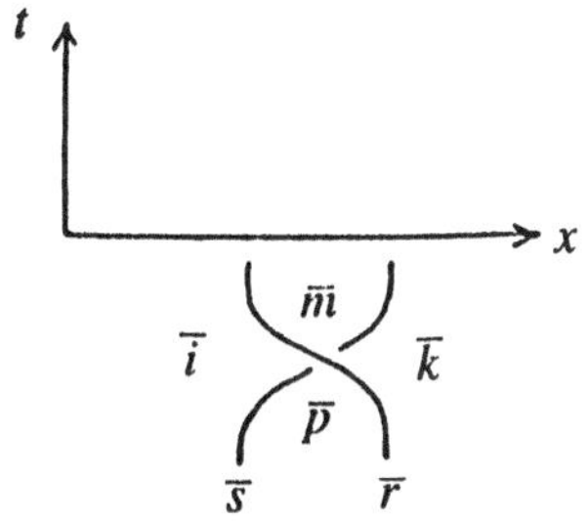

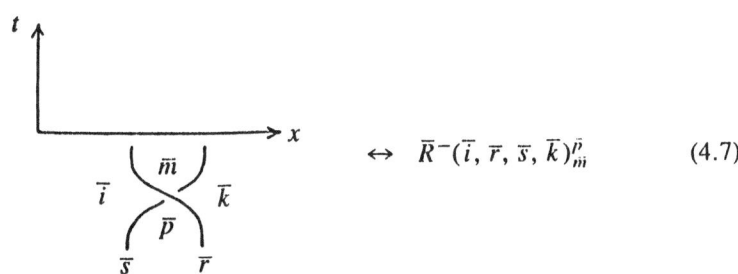

$$\leftrightarrow \quad \bar{R}^{-}(\bar{i},\, \bar{r},\, \bar{s},\, \bar{k})^{\bar{p}}_{\bar{m}} \tag{4.7}$$

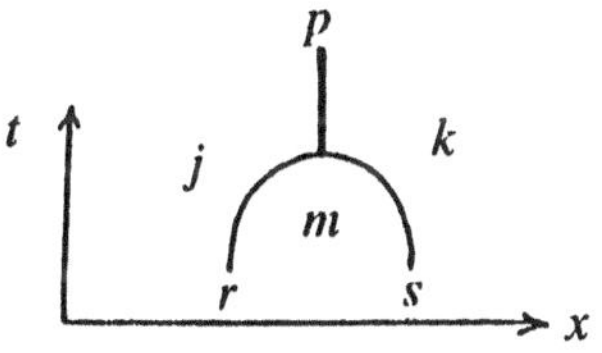

$$\leftrightarrow F(j,\, r,\, s,\, k)^{p}_{m} \tag{4.8}$$

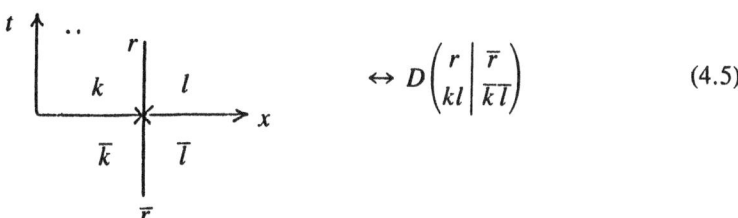

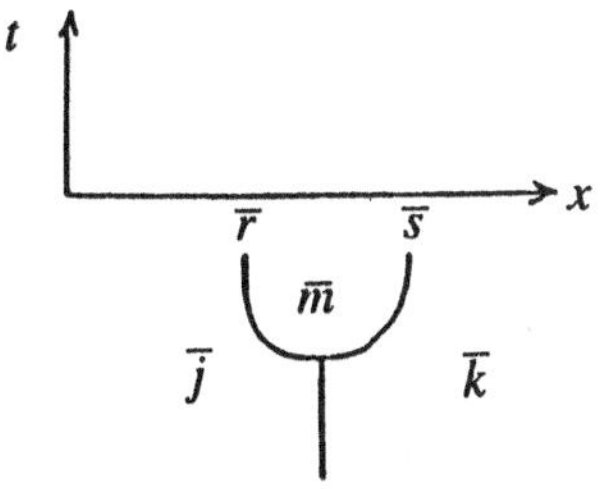

$$\leftrightarrow \bar{F}(\bar{j},\, \bar{r},\, \bar{s},\, \bar{k})^{\bar{p}}_{\bar{m}}\ . \tag{4.9}$$

The graphical representations of R^- and $\bar{R}^+$ are chosen in the obvious way.

Equation (4.3) reads, in graphical notation, as follows:

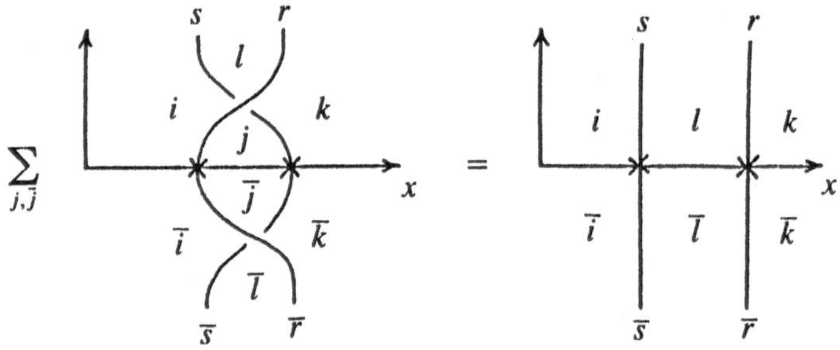

(4.10)

We also have that

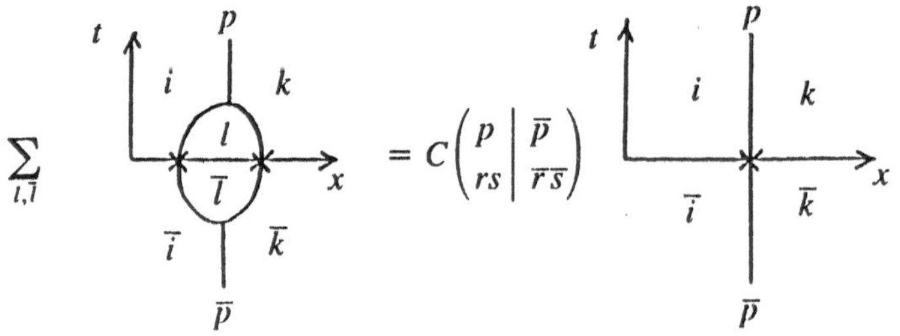

(4.11)

If the matrices D are determined from (4.10) then the coefficients, $C\left(\begin{smallmatrix} p \\ rs \end{smallmatrix} \middle| \begin{smallmatrix} \bar{p} \\ \bar{r}\bar{s} \end{smallmatrix}\right)$, of the operator product expansions for products of two local fields can be obtained from (4.11); see Ref. 21.

Clearly, all the equations in Secs. 2 and 3 hold separately for the $R^{\pm}$- and F-matrices, the $\bar{R}^{\pm}$- and $\bar{F}$-matrices, respectively.

A special case of the theory developed here concerns the so-called left-right $(L - R)$ symmetric conformal field theories: In such theories $\mathscr{A} = \bar{\mathscr{A}}$, $L_{\mathscr{A}} = L_{\bar{\mathscr{A}}}$, and one can then make the ansatz that

$$D\left(\begin{matrix} r \\ ij \end{matrix} \middle| \begin{matrix} \bar{r} \\ \bar{i}\bar{j} \end{matrix}\right) = D_{irj}\delta^{r\bar{r}}\delta_{i\bar{i}}\delta_{j\bar{j}} \,. \tag{4.12}$$

Then Eqs. (4.3) reduce to

$$\sum_j R^{\pm}(i, r, s, k)^l_j R^{\mp}(i, r, s, k)^m_{\bar{j}} D_{irj} D_{jsk} = \delta^{lm} D_{isl} D_{lrk} \,. \tag{4.13}$$

Note that Eqs. (4.3) and (4.13) are *overdetermined* systems of algebraic equations for the D-coefficients; (there are more equations than unknowns, as an easy count shows). It is therefore not obvious that these equations have non-trivial solutions, let alone that they may have different series of solutions. However, this has been proven to be the case in several classes of local conformal field theories[24]; (the A-series of minimal models, Wess-Zumino-Witten models, . . .). We shall find a method for calculating the D-coefficients in Sec. 6. For the remainder of this section we shall assume that Eqs. (4.3) (or (4.13)) have nontrivial solutions. We then show how to construct the Green functions of a local conformal field theory from these data.

First, we introduce the so-called *conformal blocks*. Let $\xi_{r_i} \in \mathcal{H}_{r_i}$, $r_i \in L_{\mathcal{A}}$, for $i = 1, \ldots, n$. Let $j_l \in L_{\mathcal{A}}$, $l = 1, \ldots, n$, with $j_0 = 1$ and $j_n = 1$. Let $r \equiv (r_1, \ldots, r_n)$, $j \equiv (j_0, \ldots, j_n)$. Let Ω, Ω^* denote the vacuum vectors in $\mathcal{H}_1$, $\mathcal{H}_1^*$, respectively; (recall that Ω, Ω^* are strongly invariant vectors in the sense of Definition 2.1, (4); in particular they are Möbius-invariant). Then we define the conformal blocks, f, by

$$f^{j}_{\xi_{r_1}\cdots\xi_{r_n}}(z_1, \ldots, z_n) \equiv \left\langle \Omega^* \middle| \prod_{i=1}^{n} V^{j_{i-1}j_i}(\xi_{r_i}, z_i)\Omega \right\rangle ; \tag{4.14}$$

(again, we suppress the label $\alpha = 1, \ldots, N_{jrk}$ labelling the different types of vertices $V^{jk}_{\alpha}(\xi_r, z)$). The anti-conformal blocks, $f^{\bar{j}}_{\bar{\xi}_{\bar{r}_1}\cdots\bar{\xi}_{\bar{r}_n}}(\bar{z}_1, \ldots, \bar{z}_n)$, derived from the representation theory of $\bar{\mathcal{A}}$, are defined similarly; the vacua are denoted by $\bar{\Omega}$, $\bar{\Omega}^*$.

Next, we define a "metric", K, on the space of conformal blocks:

$$K^{r\bar{r}}_{j\bar{j}} = \prod_{i=1}^{n} D\begin{pmatrix} r_i & \bar{r}_i \\ j_{i-1}j_i & \bar{j}_{i-1}\bar{j}_i \end{pmatrix} . \tag{4.15}$$

The Green functions of a local conformal field theory are now defined by

$$\begin{aligned} &H_{\xi_{r_1}\otimes\bar{\xi}_{\bar{r}_1}\cdots\xi_{r_n}\otimes\bar{\xi}_{\bar{r}_n}}(z_1, \bar{z}_1, \ldots, z_n, \bar{z}_n) \\ &\quad = \sum_{j,\bar{j}} K^{r\bar{r}}_{j\bar{j}} f^{j}_{\xi_{r_1}\cdots\xi_{r_n}}(z_1, \ldots, z_n) f^{\bar{j}}_{\bar{\xi}_{\bar{r}_1}\cdots\bar{\xi}_{\bar{r}_n}}(\bar{z}_1, \ldots, \bar{z}_n) . \end{aligned} \tag{4.16}$$

In view of Eqs. (4.1), (4.14) and (4.15), we have that

$$\begin{aligned} &H_{\xi_{r_1}\otimes\bar{\xi}_{\bar{r}_1}\cdots\xi_{r_n}\otimes\bar{\xi}_{\bar{r}_n}}(z_1, \bar{z}_1, \ldots, z_n, \bar{z}_n) \\ &\quad = \left\langle \Omega^* \otimes \bar{\Omega}^* \middle| \prod_{i=1}^{n} \phi_{\xi_{r_i}\otimes\bar{\xi}_{\bar{r}_i}}(z_i, \bar{z}_i)\Omega \otimes \bar{\Omega} \right\rangle . \end{aligned} \tag{4.17}$$

We shall sometimes abbreviate $H_{\xi_{r_1}\otimes\bar{\xi}_{\bar{r}_1}\cdots\xi_{r_n}\otimes\bar{\xi}_{\bar{r}_n}}$ by $H_{r\bar{r}}$. It should be emphasized that, due to Eqs. (4.1) and (2.63), (2.73), the monodromy of the Green functions $H_{r\bar{r}}(z_1, \bar{z}_1, \ldots, z_n, \bar{z}_n)$, for $(z_1, \ldots, z_n) \in M_n$, $(\bar{z}_1, \ldots, \bar{z}_n) \in M_n$ (with M_n given by (2.53)) is completely determined by the braid matrices $R^{\pm}$ and $\bar{R}^{\pm}$. Moreover, by (3.1) and (2.51), (2.52), the singularities of $H_{r\bar{r}}(z_1, \bar{z}_1, \ldots, z_n, \bar{z}_n)$, as $|z_i - z_j| \to 0$, or $|\bar{z}_i - \bar{z}_j| \to 0$, are independent of n, with coefficients proportional to the F-matrices.

If the algebras $\mathcal{A}$ and $\bar{\mathcal{A}}$ are * algebras and the representations $L_{\mathcal{A}}$, $L_{\bar{\mathcal{A}}}$ are unitary the constructions of Secs. 3 and 4 yield unitary conformal field theories from which relativistic conformal field theories can be reconstructed; see e.g. Refs. 11 and 13.

We conclude this section with a brief discussion of a simple example, the massless free field in two dimensions. In spite of its simplicity, this example will play a fairly important role, later on. In this example, $\mathcal{A}$ is chosen to be U(1)-current algebra. This algebra is generated by a chiral current $(\partial_z \varphi)(z)$, where

$$\langle \Omega^* | \varphi(z)\varphi(w)\Omega \rangle = \frac{1}{4\pi} \ln\left(\frac{1}{z-w}\right) . \tag{4.18}$$

It can be extended to a larger algebra, $\tilde{\mathcal{A}}$, with the property that $\mathcal{G}(\tilde{\mathcal{A}})$ contains a Virasoro algebra with central charge $c = 1$ by setting

$$T(z) \propto \frac{1}{2} : (\partial_z \varphi)^2 : (z) , \tag{4.19}$$

: — : denotes standard Wick ordering. The algebra $\tilde{\mathcal{A}}$ is a * algebra and has unitary representations, $\varepsilon(\varepsilon \in \mathbb{R})$, with the property that $\mathcal{H}_\varepsilon$ contains an invariant vector, ξ_ε, such that

$$\varepsilon(L_0)\xi_\varepsilon = h_\varepsilon \xi_\varepsilon , \tag{4.20}$$

for

$$h_\varepsilon = \frac{\varepsilon^2}{8\pi} \geq 0 . \tag{4.21}$$

The corresponding primary chiral vertices (see Definition 2.3) can be expressed as Wick-ordered exponentials of $\varphi(z)$:

$$V^{\alpha+\varepsilon,\alpha}(\xi_\varepsilon, z) = : \exp i\varepsilon\varphi(z) : P_\alpha , \tag{4.22}$$

where $P_\alpha \xi = \xi$, for $\xi \in \mathcal{H}_\alpha$, and $P_\alpha \xi = 0$, for $\xi \in \mathcal{H}_\beta$, $\beta \neq \alpha$. The Wick order, : — :, is chosen so that

$$\langle \Omega^* | V^{1\varepsilon}(\xi_{-\varepsilon}, z) V^{\varepsilon 1}(\xi_\varepsilon, w)\Omega \rangle = (z-w)^{-(\varepsilon^2/4\pi)} . \tag{4.23}$$

By (4.22) we can define

$$V_\varepsilon(z) \equiv \sum_\alpha V^{\alpha+\varepsilon,\alpha}(\xi_\varepsilon, z) = : \exp i\varepsilon\varphi(z) : . \tag{4.24}$$

Wick's theorem tells us that

$$V_\varepsilon(z)V_{\varepsilon'}(w) = (z-w)^{\varepsilon\varepsilon'/4\pi} : \exp i(\varepsilon\varphi(z) + \varepsilon'\varphi(w)):$$

$$= (z-w)^{\varepsilon\varepsilon'/4\pi} V_{\varepsilon+\varepsilon'}(w)[1 + 0(z-w)] \,. \tag{4.25}$$

Clearly

$$\langle \Omega^* | V_\varepsilon(z)\Omega\rangle = 0 \,, \qquad \text{for } \varepsilon \neq 0 \,. \tag{4.26}$$

From (4.25) and (4.26) we get

$$\left\langle \Omega^* \middle| \prod_{i=1}^{n} V_{\varepsilon_i}(z_i)\Omega \right\rangle = \begin{cases} 0 \,, & \text{for } \sum_{i=1}^{n} \varepsilon_i \neq 0, \\ \prod_{1\leq i<j\leq n} (z_i - z_j)^{\varepsilon_i\varepsilon_j/4\pi} \,, & \text{for } \sum_{i=1}^{n} \varepsilon_i = 0 \end{cases} . \tag{4.27}$$

This formula is ancient; see e.g. Ref. 25. If, in (4.27), we set $n = 3$, let $z_1 \to \infty$, $z_2 = z - w$, $z_3 \to 0$, we obtain, using (2.47) and (2.49), that

$$(z-w)^{\varepsilon\varepsilon'/4\pi} = \langle \xi_{\varepsilon+\varepsilon'} | V_\varepsilon(z-w)\xi_{\varepsilon'}\rangle \,. \tag{4.28}$$

Comparing (4.25) and (4.28) with formula (3.1), we conclude that

$$F(\alpha+\varepsilon, \varepsilon, \varepsilon', \alpha-\varepsilon')^\gamma_\beta = \delta^\alpha_\beta \delta^\gamma_{\varepsilon+\varepsilon'} \,. \tag{4.29}$$

The braid matrices of this simple theory are easily calculated from (4.24), (4.27) and (4.28), as in (2.80). We get

$$R^\pm(\alpha+\varepsilon', \varepsilon', \varepsilon, \alpha-\varepsilon)^\gamma_\beta = \delta^\alpha_\beta \delta^\gamma_{\alpha-\varepsilon+\varepsilon'} \exp\left(\pm\frac{i}{4}\varepsilon\varepsilon'\right) . \tag{4.30}$$

In this model, the representation, ε^*, conjugate to ε is $\varepsilon^* = -\varepsilon$. Hence, the coefficients $F^\alpha_{\varepsilon\beta}$ analogous to the coefficients F^l_{rn} defined in (3.21) are given by

$$F^\alpha_{\varepsilon\beta} = F(\alpha, \varepsilon, \varepsilon^*, \alpha)^1_\beta = \delta^\alpha_{\beta+\varepsilon} \,. \tag{4.31}$$

Local fields can be constructed from the vertex operators $V_\varepsilon(z)$ as follows:

(1) If $\varepsilon^2/8\pi \in \mathbb{Z}$ then $V_\varepsilon(z)$ and $V_\varepsilon(\bar{z})$ are local.

(2) If $\varepsilon^2/8\pi - \bar{\varepsilon}^2/8\pi \in \mathbb{Z}$ then $V_\varepsilon(z)V_{\bar{\varepsilon}}(\bar{z})$ is local.

This yields an infinity of solutions of Eqs. (4.3), (4.13). Solutions, C, of Eq. (4.11) are trivially obtained from Wick's theorem. Finally, the conformal blocks of the theory are given by formula (4.27).

Note that Eqs. (3.6)–(3.9), (3.11), (3.12), (3.14) and (3.15), and (3.20), (3.24) all hold for the free field, as one easily checks using (4.29), (4.30)

Nonunitary representations of *Vir* with $c < 1$, based on the free, massless field are systematically discussed in Ref. 15 and references given there.

5. Conformal Field Theories on Higher-Genus Riemann Surfaces

As we shall see in the next section, it will be important for us to know how to extend a local conformal field theory defined over the Riemann sphere, as studied in Secs. 3 and 4, to a conformal field theory defined over a Riemann surface of arbitrary genus g. This problem has been studied in Refs. 26, 27, 28 and 29. It is a nontrivial proposition that a conformal field theory defined over the Riemann sphere determines theories defined over arbitrary Riemann surfaces. But, in Ref. 27, simple sufficient conditions have been formulated which guarantee that the proposition is valid.

Two ways of sewing Riemann surfaces will be convenient for our purposes:

(1) Let M be a Riemann surface with puncture P and a local coordinate w vanishing at P. Let M' be a Riemann surface with puncture P' and a local coordinate w' vanishing at P'. We assume that w and w' are well defined inside the discs $\{w : |w| < \rho\theta\}$ and $\{w' : |w'| < \rho\theta\}$, respectively, for some positive numbers $\rho > 1$, $\theta \in (0, 1]$. We excise the disk $\{w : |w| < \rho^{-1}\theta\}$ from M and the disk $\{w' : |w'| < \rho^{-1}\theta\}$ from M' and glue M to M' along the annuli $\{w : \frac{\theta}{\rho} < |w| < \theta\rho\}$ and $\{w' : \frac{\theta}{\rho} < |w'| < \theta\rho\}$ by imposing the condition

$$ww' = \theta^2 e^{i\alpha}, \tag{5.1}$$

for some twisting angle $\alpha \in [0, 2\pi)$.

The resulting surface is denoted by $M \infty_{\theta,\alpha} M'$ and is independent of ρ.

(2) The second procedure for sewing Riemann surfaces is related to the Schottky parametrization. We pick $2n$ points $w_1, w_1', \ldots, w_n, w_n'$ on the Riemann sphere and choose some $\rho > 1$. We also choose numbers $\theta_i \in (0, 1]$, $i = 1, \ldots, n$. We excise disks of radii $\theta_i/\rho < 1$ around each of the points w_i, w_i', $i = 1, \ldots, n$. We then glue the annular region

$$\left\{w : \frac{\theta_i}{\rho} < |w - w_i| < \theta_i\rho\right\}$$

to the annular region

$$\left\{w' : \frac{\theta_i}{\rho} < |w' - w_i'| < \theta_i\rho\right\}$$

by setting

$$(w - w_i)(w' - w_i') = \theta_i^2 e^{i\alpha_i} \tag{5.2}$$

for some angles $\alpha_i \in [0, 2\pi)$, $i = 1, \ldots, n$. The resulting Riemann surface is denoted by $M_n(\underline{w}, \underline{w}', \underline{\theta}, \underline{\alpha})$; (it does not depend on ρ).

We realize that our description of procedures (1) and (2) is a bit sketchy, but it is all that is needed in the following.

The point is that the sewing procedures (1) and (2) can be represented as sewing procedures of Green functions in conformal field theory. These procedures will permit us to define Green functions on Riemann surfaces of arbitrary genus in terms of Green functions on the Riemann sphere. We define $\mathcal{R}$ to be the index set

$$\mathcal{R} = \left\{ (r, \bar{r}) : r \in L_{\mathcal{A}}, \bar{r} \in L_{\bar{\mathcal{A}}}, D\begin{pmatrix} r & \bar{r} \\ ij & \bar{i}\bar{j} \end{pmatrix} \not\equiv 0 \right\} . \tag{5.3}$$

We let Φ_i be one of the operators $\psi_{j_i}(z)$, $j_i \in \mathcal{J}_{\mathcal{A}}$ (see (2.1)), $\psi_{\bar{j}_i}(\bar{z})$, $\bar{j}_i \in \mathcal{J}_{\bar{\mathcal{A}}}$, or $\phi_{\xi_{s_i} \otimes \bar{\xi}_{\bar{s}_i}}(z_i, \bar{z}_i)$, with $(s_i, \bar{s}_i) \in \mathcal{R}$, $(\xi_{s_i} \in \mathcal{H}_{s_i}, \bar{\xi}_{\bar{s}_i} \in \mathcal{H}_{\bar{s}_i})$, $i = 1, \ldots, m$.

We pick a basis $\{\xi_r^N\}_{N=0}^{\infty}$ in $\mathcal{H}_r$ and a dual basis $\{\xi_{r*}^N\}_{N=0}^{\infty}$ in $\mathcal{H}_{r*}$, i.e.

$$\langle \xi_{r*}^N | \xi_r^K \rangle = \delta^{NK} , \tag{5.4}$$

for all $r \in L_{\mathcal{A}}$, ($r \in L_{\bar{\mathcal{A}}}$, respectively).

A basic *completeness assumption* in conformal field theory is expressed in the following condition. Let Ω, Ω^* be the vacua for $\mathcal{A}$ and $\bar{\Omega}$, $\bar{\Omega}^*$ the vacua for $\bar{\mathcal{A}}$; see Eq. (2.25). We set

$$\langle 1 | \equiv \langle \Omega^* \otimes \bar{\Omega}^* | , \qquad | 1 \rangle \equiv \Omega \otimes \bar{\Omega} . \tag{5.5}$$

We set $q = \theta e^{i\alpha}$, $\bar{q} = \theta e^{-i\alpha}$, $q^* = \bar{q}^* = \theta$,

$$\begin{aligned} \xi_r(q) &= q^{r(L_0)} \xi_r , \qquad \bar{\xi}_{\bar{r}}(\bar{q}) = \bar{q}^{\bar{r}(L_0)} \bar{\xi}_{\bar{r}} , \\ \Phi_i(q, \bar{q}) &= q^{L_0} \bar{q}^{\bar{L}_0} \Phi_i q^{-L_0} \bar{q}^{-\bar{L}_0} , \end{aligned} \tag{5.6}$$

where $L_0 \equiv \oplus_{r \in L_{\mathcal{A}}} r(L_0)$, $\bar{L}_0 \equiv \oplus_{\bar{r} \in \bar{L}_{\bar{\mathcal{A}}}} \bar{r}(L_0)$.

The completeness condition is that

(C1)

$$\begin{aligned} \langle 1 | \Phi_1 \ldots \Phi_m | 1 \rangle = \sum_{(r,\bar{r}) \in \mathcal{R}} \sum_{N,M=0}^{\infty} \langle 1 | \Phi_1(q, \bar{q}) \ldots \Phi_k(q^*, \bar{q}) \xi_r^N(q) \otimes \bar{\xi}_{\bar{r}}^M(\bar{q}) \rangle \\ \times \langle \xi_{r*}^N(q^*) \otimes \bar{\xi}_{\bar{r}*}^M(\bar{q}^*) | \Phi_{k+1}(q^*, \bar{q}^*) \ldots \Phi_m(q^*, \bar{q}^*) | 1 \rangle , \end{aligned} \tag{5.7}$$

for arbitrary $k = 0, \ldots, m$ and arbitrary operators $\Phi_1, \ldots, \Phi_m$, $m = 0, 1, 2, \ldots$. It is nontrivial to show that the series on the right-hand side of (5.7) converges, with a result that does not depend on our choice of bases $\{\xi_r^N\}_{N=0}^{\infty}$, $\{\xi_{r*}^N\}_{N=0}^{\infty}$. In a class of unitary conformal theories, this has presumably been proven. In the following, (5.7) will be assumed to hold.

Note that (5.7) is an example of how one represents the sewing procedure (1) as sewing of Green functions, where M and M' are the Riemann sphere.

Next, we show how to represent the sewing procedure (2) as sewing of Green functions. We define the Green functions $\langle\Phi_1 \ldots \Phi_m\rangle_{M_n(\underline{w},\underline{w}',\underline{\theta},\underline{\alpha})}$ on the Riemann surface $M_n(\underline{w}, \underline{w}', \underline{\theta}, \underline{\alpha})$ defined above by the following formula. We define

$$\Psi_{r_j,\bar{r}_j}(N_j, M_j) \equiv \phi_{\xi_{r_j}^{N_j}(q_j)\otimes\bar{\xi}_{\bar{r}_j}^{M_j}(\bar{q}_j)}(w_j, \bar{w}_j) \,, \tag{5.8}$$

$q_j = \theta_j e^{i\alpha_j}$, $\bar{q}_j = \theta_j e^{-i\alpha_j}$; and

$$\Psi_{r_j^*,\bar{r}_j^*}(N_j, M_j) \equiv \phi_{\xi_{r_j}^{N_j}(q_j^*)\otimes\xi_{\bar{r}_j}^{M_j}(\bar{q}_j^*)}(w_j', \bar{w}_j') \,. \tag{5.9}$$

where $q_j^* = \bar{q}_j^* = \theta_j$. Then

$$Z\langle\Phi_1 \ldots \Phi_m\rangle_{M_n(\underline{w},\underline{w}',\underline{\theta},\underline{\alpha})}$$

$$= \sum_{\substack{(r_j,\bar{r}_j)\in\mathfrak{R} \\ j=1,\ldots,n}} \sum_{\substack{N_j,M_j=0 \\ j=1,\ldots,n}}^{\infty} \left\langle 1 \left| \Phi_1 \ldots \Phi_m \prod_{j=1}^{n} \Psi_{r_j,\bar{r}_j}(N_j, M_j)\Psi_{r_j^*,\bar{r}_j^*}(N_j, M_j) \right| 1 \right\rangle , \tag{5.10}$$

where Z may be chosen so that

$$\langle 1\rangle_{M_n(\underline{w},\underline{w}',\underline{\theta},\underline{\alpha})} = 1 \,.$$

Again, it is a nontrivial assumption on the theory that the series on the right-hand side (5.10) converge and are independent of the bases $\{\xi_r^N\}$, $\{\xi_{r*}^N\}$, ... We believe that this assumption can be proven for a class of unitary, rational conformal field theories.

It is now clear how to express the sewing procedure (1) in the general case. Let $M = M_n(\underline{w}, \underline{w}', \underline{\theta}, \underline{\alpha})$, $M' = M_l(\underline{u}, \underline{u}', \underline{\theta}', \underline{\alpha}')$, and set

$$N = M \underset{\theta,\alpha}{\infty} M' \,.$$

Then (5.7) can be generalized to

$$\langle\Phi_1 \ldots \Phi_m\rangle = \sum_{(r,\bar{r})\in\mathfrak{R}} \sum_{N,M=0}^{\infty} \langle\Phi_1 \ldots \Phi_k \phi_{\xi_r^N(q)\otimes\bar{\xi}_{\bar{r}}^M(\bar{q})}(P, \bar{P})\rangle_M$$

$$\times \langle\phi_{\xi_{r^*}^N(q^*)\otimes\bar{\xi}_{\bar{r}^*}^M(\bar{q}^*)}(Q, \bar{Q})\Phi_{k+1} \ldots \Phi_m\rangle_{M'} \,. \tag{5.11}$$

Equations (5.10) and (5.11) can be used to define $\langle\Phi_1 \ldots \Phi_m\rangle_N$, for an arbitrary Riemann surface N, in terms of local Green functions on the Riemann sphere. It is nontrivial to prove that $\langle\Phi_1 \ldots \Phi_m\rangle_N$ only depends on the complex structure on N, but *not* on the particular way $\langle\Phi_1 \ldots \Phi_m\rangle_N$ has been constructed by sewing Green functions on the Riemann sphere, using Eqs. (5.10) and (5.11).

It has been shown in Ref. 27 that, assuming the series in (5.10) and (5.11) converge and are basis-independent, it suffices to check the following two conditions, in order to ensure that $\langle \Phi_1 \ldots \Phi_m \rangle_N$ only depend on N and its complex structure:

(C1) Condition (C1) holds; see (5.7).

(C2) The one-point Green function $\langle \Phi_1 \rangle_{T(\tau)}$ on the torus $T(\tau)$, (τ, Im $\tau > 0$, is the modular parameter) is modular-invariant, for arbitrary Φ_1. The arguments in Ref. 27 are formal, because the analysis part of the problem is not dealt with. Again, we believe that the analysis part of the problem can be settled for unitary, rational conformal field theories; see e.g. Refs. 29 and 15.

The reader may wonder why we have chosen to describe how to sew local Green functions rather than the more fundamental conformal blocks. The reasons are simple: First, explicit sewing formulars for conformal blocks are even more complicated to write out explicitly than Eqs. (5.10) and (5.11); second, when one sews conformal blocks some care is needed about multiplying them by suitable powers of determinant line bundles.[e] Fortunately, and this is our third reason, only some crude aspects of sewing which can be expressed graphically will be relevant for the theory developed in Sec. 6. The appropriate graphical representation of (5.10) can be obtained by writing each term on the right-hand side of (5.10) as a sum of products of conformal blocks of the theory over the Riemann sphere and then applying to the resulting expressions the graphical rules developed in Secs. 3 and 4. Thus, let $\Phi_i = \phi_{\xi_{s_i} \otimes \bar{\xi}_{\bar{s}_i}}(z_i, \bar{z}_i)$, $i = 1, \ldots, m$. Then the right-hand side of (5.10) can be expressed, graphically, as follows:

Right-hand side of (5.10) $\leftrightarrow$ 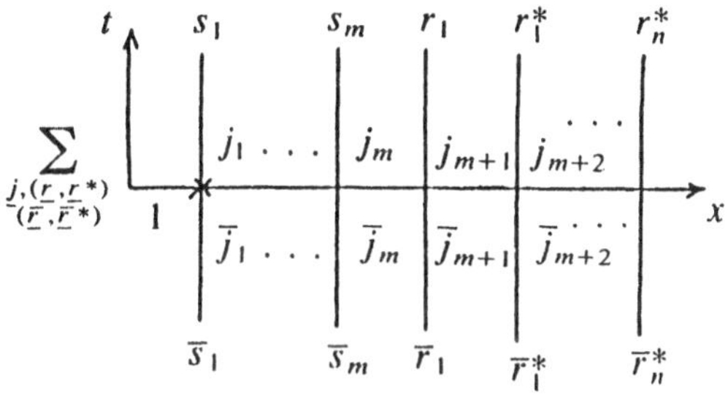

(5.12)

The meaning of the lines and crosses has been explained in Sec. 4. Dropping the crosses and the lines in the half-space $\{t \leq 0\}$ in (5.12), and restricting the sum over $\bar{j}$ to a sum over $j_{m+1}, \ldots, j_{m+2n-1}$, we obtain an appropriate graphical expression for the sewing of conformal blocks; (we shall not need the exact analytical expressions for the sewing of conformal blocks).

Given (5.12), the right-hand side of (5.11) can be expressed, graphically, by

Right-hand side of (5.11) $\leftrightarrow$ 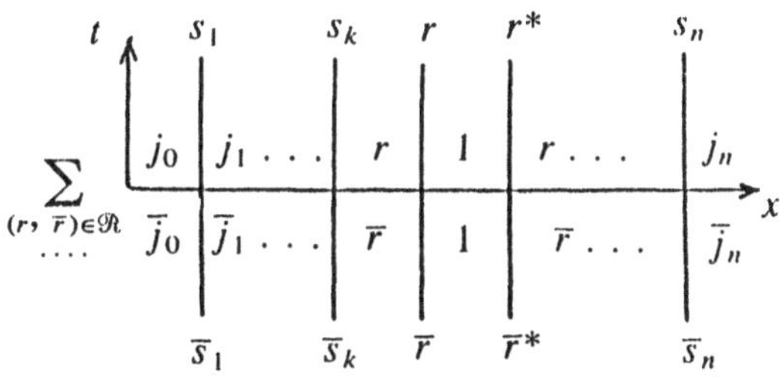

(5.13)

[e] The appropriate powers of det ∂, det $\bar{\partial}$ can be determined, for example, by representing conformal field theory in terms of three-dimensional Chern-Simons theory.[3]

In the next section, we shall apply the results of Secs. 3–5 to construct link invariants for a rather general class of three-dimensional manifolds. This will be the main result of this paper.

6. Link Invariants for a Class of Three-Dimensional Manifolds

In this section we establish our main results: We show how the braid- and fusion matrices of conformal field theory can be used to construct link invariants for links imbedded in three-dimensional manifolds of rather general type. Our analysis has a pay-off for conformal field theory. We find simple ways to calculate the "structure constants" D introduced in Sec. 4 and the matrix elements, S_{ij}, of the representation of the modular transformation $\tau \to 1/\tau$ on the space of characters of a chiral algebra $\mathcal{A}$: D and S can be expressed in terms of fusion matrices, F, which, as was shown in Sec. 3, can be expressed in terms of the braid matrices. Connections between our construction of link invariants and other related constructions will be discussed in Sec. 7.

6.1. *Some preparatory remarks and definitions*

Let $\mathcal{M}$ be some three-dimensional manifold. A *knot*, K, in $\mathcal{M}$ is defined as follows: let

$$f: S^1 \to \mathcal{M}$$

be a continuous function on the circle with values in $\mathcal{M}$, with the property that

$$f(\alpha) \neq f(\beta)\,, \qquad \text{for } \alpha,\ \beta \text{ in } S^1 \text{ with } \alpha \neq \beta\,. \tag{6.1}$$

Two such functions, f and g, are ambient isotopic ("equivalent") if there is a homeomorphism h: $\dot{M} \times [0, 1] \to \dot{M} \times [0, 1]$ with $h(x, t) = (h_t(x), t)$, such that h_0 is the identity, and

$$h_1 \circ f = g\,. \tag{6.2}$$

A knot, K, is defined as (the image of) an arbitrary function in the equivalence class, $[f]$, represented by the function f. An example of a knot in $\mathbb{R}^3$ (or S^3) is

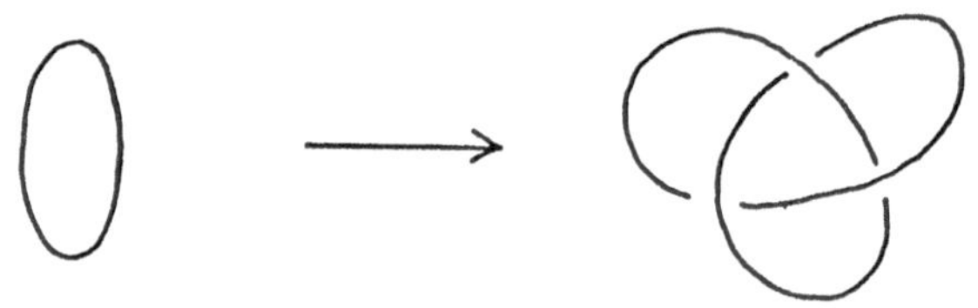

Fig. 1

A knot is *oriented* if S^1 and the image of S^1 under $f \in [f]$ are oriented. Example:

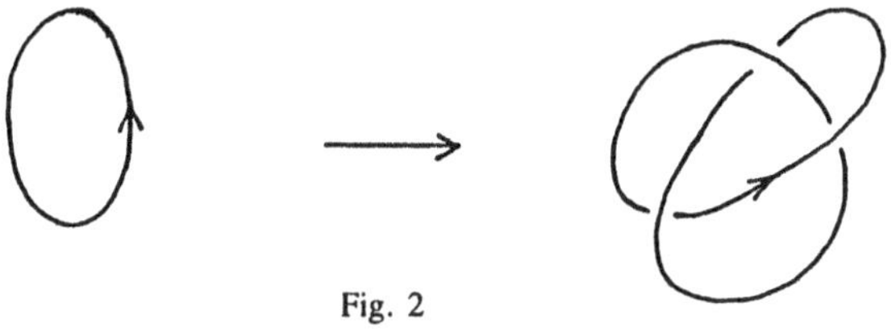

Fig. 2

A *link*, L, with n components in $\mathcal{M}$ is defined as follows: We consider n functions, f_i,

$$f_i: S^1 \to \mathcal{M}\,, \qquad i = 1, \ldots, n \tag{6.3}$$

with the properties that, for each i,

$$f_i(\alpha) \neq f_i(\beta)\,, \qquad \text{for } \alpha,\ \beta \text{ in } S_i^1 \simeq S^1 \text{ with } \alpha \neq \beta\,, \tag{6.4}$$

and, for $i \neq j$,

$$f_i(\alpha) \neq f_j(\beta)\,, \qquad \text{for all } \alpha \in S_i^1 \text{ and all } \beta \in S_j^1\,. \tag{6.5}$$

We say that $(f_1, \ldots, f_n)$ and $(g_1, \ldots, g_n)$, both satisfying (6.3)–(6.5), are equivalent if there is a permutation $\pi \in S_n$ and there is a homeomorphism $h: M \times [0, 1] \to M \times [0, 1]$, with $h(x, t) = (h_t(x), t)$, such that $h_0 = id.$, and

$$h_1 \circ f_i = g_{\pi(i)}\,, \qquad \text{for all } i\,. \tag{6.6}$$

A link, L, with n components is defined as an equivalence class $[f_1, \ldots, f_n]$ of (images of) functions, $f_1, \ldots, f_n$, under ambient isotopy. A link, L, is partially oriented if a nonempty subset of the circles $S_1, \ldots, S_n$, and their images are oriented.

Example of a partially oriented link.

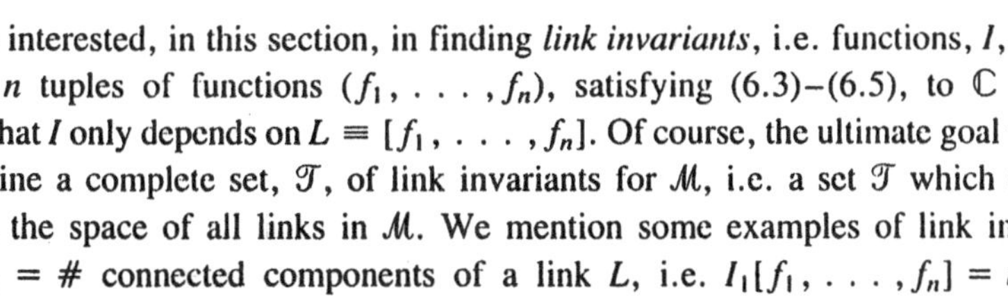

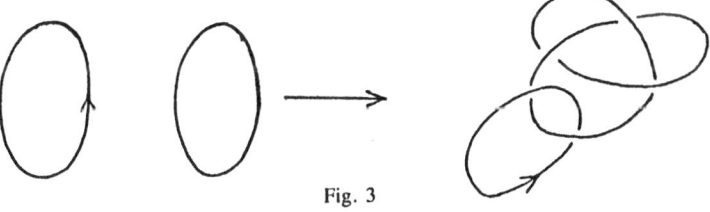

Fig. 3

We are interested, in this section, in finding *link invariants*, i.e. functions, I, from the space of n tuples of functions $(f_1, \ldots, f_n)$, satisfying (6.3)–(6.5), to $\mathbb{C}$ with the property that I only depends on $L \equiv [f_1, \ldots, f_n]$. Of course, the ultimate goal would be to determine a complete set, $\mathcal{T}$, of link invariants for $\mathcal{M}$, i.e. a set $\mathcal{T}$ which separates points on the space of all links in $\mathcal{M}$. We mention some examples of link invariants.

(E1) I_1 = # connected components of a link L, i.e. $I_1[f_1, \ldots, f_n] = n$.

(E2) I_2 = Gauss linking number.

I_2 () $= 0$

I_2 () $= 1$

I_2 () $= -1$.

The invariants I_1 and I_2 have been known, for a long time. They are easy to compute, but not very effective.

(E3) I_3 = fundamental group of the complement of a link in $\mathcal{M}$. I_3 is a strong invariant, but is very difficult to compute. It is a less ambitious goal to attempt to compute representations of the fundamental group of a link complement with values in some representation of a Lie group, although this is usually hard, too. See also Refs. 3, 4 and 5 and Sec. 7.

(E4) I_4 = Jones polynomial, $V_L(t)$.[2] In this example, each link $L \subset S^3$ is mapped to a Laurent polynomial in $\sqrt{t} \in \mathbb{C}$, $V_L(t)$, depending on L. The Jones polynomial is a very powerful invariant which can be computed relatively easily from the so-called *skein relations*.[2] As an example we mention that, for L as in Fig. 2,

$$V_L(t) = t + t^3 - t^4 .$$

(E5) I_5 = two-variable polynomial $P_L(t, x)$.[2] This is a generalization of the Jones polynomial. The classical Alexander polynomial may be viewed as a specialization of $P_L(t, x)$. For recent results on link invariants see Ref. 2. These invariants are usually defined and computed *only* for links in S^3.

It is the purpose of this section to show that the $R^{\pm}$- and F-matrices of conformal field theory can be used to recover the invariants mentioned above and construct other invariants. Moreover, we show how these invariants can be generalized to obtain link invariants for a fairly general class of three-manifolds; (for an alternative approach see Witten's basic paper[3]). Finally, our invariants have a three-dimensional interpretation and can be computed from the conformal blocks, or the local Green functions, of conformal field theory.

Our construction of link invariants employs the graphical formalism developed in Secs. 3–5 which we now propose to refine slightly.

Let $\mathcal{A}$ be a rational chiral algebra and $L_{\mathcal{A}}$ a $\otimes$-invariant set of lowest-weight representations of $\mathcal{A}$; see Definition 2.2, Sec. 2. We decompose $L_{\mathcal{A}}$ into three subsets $i_{\mathcal{A}}$, $l_{\mathcal{A}}$ and $l^*_{\mathcal{A}}$, such that

(i) $L_{\mathscr{A}} = l_{\mathscr{A}} \cup l^*_{\mathscr{A}} \cup i_{\mathscr{A}}$;

(ii) the vacuum representation $1 \simeq 1^*$ belongs to $i_{\mathscr{A}}$; for every $r \in i_{\mathscr{A}}$, $r^* \simeq r$, and $r \otimes_1 r$ contains 1 precisely once;

(iii) given any $r \in l_{\mathscr{A}}$, r^* belongs to $l^*_{\mathscr{A}}$ and $r \otimes_1 r^*$, $r^* \otimes_1 r$ contain 1 precisely once. We now represent a class of vertices $[V^{ik}_{\alpha}(\xi_r, z)]$, $r \in l_{\mathscr{A}}$, by upward oriented lines

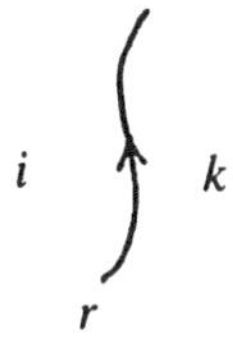

(6.7)

and $[V^{ik}_{\alpha}(\xi_{r^*}, z)]$, $r^* \in l^*_{\mathscr{A}}$, by downward oriented lines

(6.8)

Vertices $[V^{ik}_{\alpha}(\xi_r, z)]$, $r \in i_{\mathscr{A}}$, are represented by unoriented lines

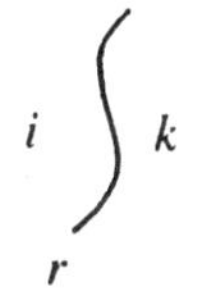

(6.9)

If the label α is important to record it will be set next to r. All other conventions introduced in Secs. 3–5 will remain valid.

We now propose to define a map between the braid- and fusion matrices, $R^{\pm}$ and F, discussed in Secs. 3–5 and elements of what one calls the *diagram* of a link in S^3; (we think of S^3 as being the one-point compactification, $\dot{\mathbb{R}}^3 \simeq S^3$, of $\mathbb{R}^3$). We choose Cartesian coordinates, t, x and y, in $\dot{\mathbb{R}}^3$. Consider a link $L = [f_1, \ldots, f_n]$. By $\underline{\lambda} = (\lambda_1, \ldots, \lambda_n)$ we denote the image of $S^1_1 \cup \cdots \cup S^1_n$ under $(f_1, \ldots, f_n)$ in $\dot{\mathbb{R}}^3$. Let π denote the (t, x)-plane in $\dot{\mathbb{R}}^3$; by "projection onto π" we mean orthogonal projection. For $p \in \underline{\lambda}$, let $t(p)$, $x(p)$ and $y(p)$ denote the t-, x- and y-coordinates of p.

We shall consider the projection of some $\underline{\lambda} \in L$ onto π. The following statements are simple consequences of the definition of a link L:

(1) One can choose $\underline{\lambda} \in L$ such that $t(p)$, $p \in \underline{\lambda}$, has only a finite, discrete set of local maxima and minima. (It might be noted that, by choosing $\underline{\lambda} \in L$ so as to minimize the number of local maxima and minima of $t(p)$, $p \in \underline{\lambda}$, one can construct a link invariant.[30]

(2) One can always choose $\underline{\lambda} \in L$ such that the projection of $\underline{\lambda}$ onto π has only a discrete set of double points, but no n^{ple} points, for $n \geq 3$.

(3) One can choose $\underline{\lambda} \in L$ such that if $p \in \underline{\lambda}$ and $p' \in \underline{\lambda}$ are projected onto the same point of π, (i.e. $t(p) = t(p')$, $x(p) = x(p')$), then the derivatives, $t'(p)$ and $t'(p')$, of t at the points $p \in \underline{\lambda}$ and $p' \in \underline{\lambda}$ are nonzero.

If two points $p \in \underline{\lambda}$ and $p' \in \underline{\lambda}$ are projected onto the same point of π then the piece of $\underline{\lambda}$ passing through p may be drawn as an *undercrossing* of the piece of $\underline{\lambda}$ passing through p' if $y(p) > y(p')$. If $y(p) < y(p')$ then the piece of $\underline{\lambda}$ passing through p is drawn as an *overcrossing* of the piece of $\underline{\lambda}$ passing through p'.

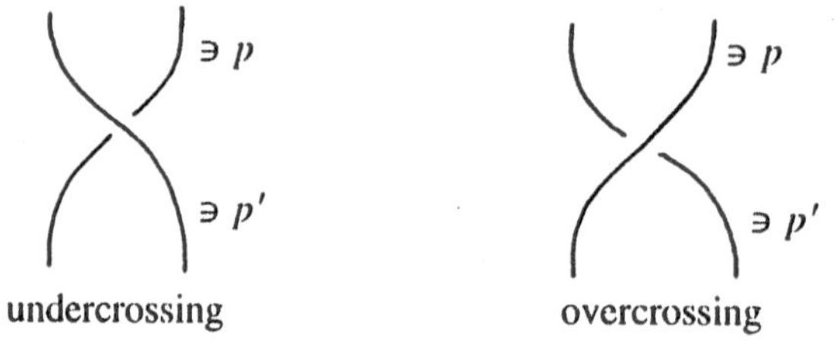

(6.10)

Let $\underline{\lambda} \in L$ have properties (1)–(3) above. Then the projection of $\underline{\lambda}$ onto π, with undercrossings and overcrossings recorded, is called a *diagram*, $D(L)$, of L. If undercrossings and overcrossings are not recorded one calls the projection of $\underline{\lambda}$ onto π a *shadow*, $S(L)$, of L.

It is intuitively easy to see that a link, L, can be reconstructed from a diagram $D(L)$, but *not* from a shadow $S(L)$; see also Ref. 31.

We note that a shadow, $S(L)$, of L decomposes π into a finite number of disjoint regions, $\Omega_1(S), \ldots, \Omega_N(S)$. We shall always suppose that $\Omega_1(S)$ contains the point at infinity in π; it is called the *outer region* of $S(L)$ or $D(L)$, whereas $\Omega_2(S), \ldots \Omega_N(S)$ are called *inner regions* of $S(L)$ or $D(L)$.

It is important to have a criterion that permits one to decide whether two different diagrams, D, and D', represent the same link L. Such a criterion is obtained by considering the following three types of so-called *Reidemeister moves*:

Reidemeister-I moves:

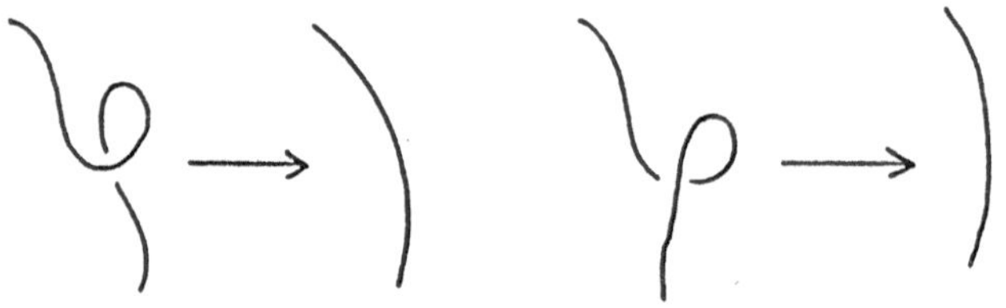

(6.11)

Reidemeister-II moves:

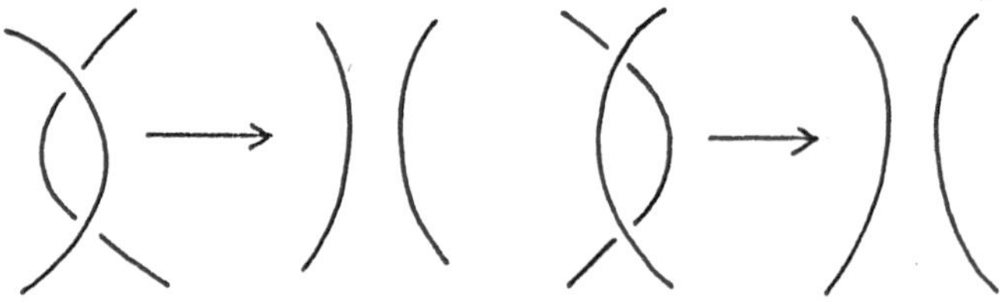

(6.12)

Reidemeister-III moves:

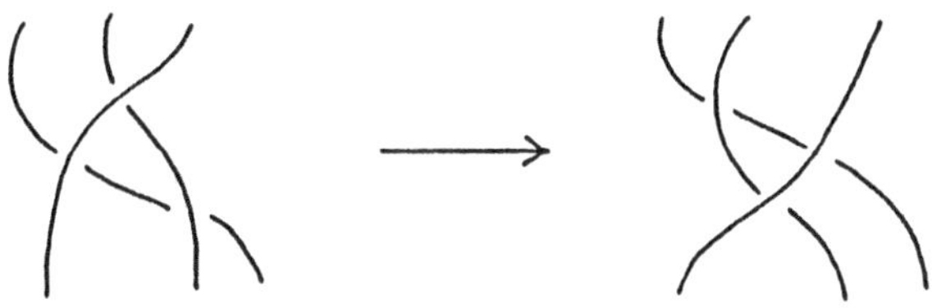

(6.13)

The orientation of the lines in (6.11)–(6.13) is omitted; these moves are defined for any assignment of orientation.

It is known that two diagrams, D and D', represent the same link, L, in S^3 if D can be transformed to D' by a sequence of Reidemeister moves of types I, II and III.[31]

Consider some representative $\underline{\lambda} = (\lambda_1, \ldots, \lambda_n)$ of a partially oriented link L. The image of λ_i under the projection onto the (t, x)-plane, π, is called ω_i and inherits the orientation of λ_i if λ_i is oriented, for $i = 1, \ldots, n$.

We are now prepared to define the map between the braid- and fusion matrices, $R^{\pm}$ and F, described in Secs. 2, 3, and their graphical representations, as modified in (6.7)–(6.9), and "elements" of a diagram, $D(L)$, of a link L. In order to do that we must identify the "elements" of $D(L)$. Recall that $S(L)$ decomposes π into disjoint regions $\Omega_1(S(L)), \ldots, \Omega_N(S(L))$. To each inner region $\Omega_i(S(L))$, $i = 2, \ldots, N$, we attach a representation $k_i \in L_{\mathcal{A}}$; to the outer region $\Omega_1(S(L))$ we attach the vacuum representation 1. To each *oriented* curve $\omega_i \subset D(L)$ we attach a representation $r_i \in l_{\mathcal{A}}$, while to each *unoriented* curve $\omega_j \subset D(L)$ we attach a representation $r_j \in i_{\mathcal{A}}$. The resulting labelled figure consisting of labelled loops ω_i and regions $\Omega_j(S)$ in the plane π is called a *decorated diagram*, $\mathcal{D}(k_2, \ldots, k_N | r_1, \ldots, r_n | L)$, of L. We now define *elements of decorated link diagrams* as follows:

(a) Let γ be a piece of a loop ω_i with the property that $t'(p)$, $p \in \gamma$, has a single zero, at some point $p_0 \in \gamma$, such that $t''(p_0) < 0$

(6.14)

This is called an element of type F.

(b) Let γ be a piece of a loop ω_i with the property that $t'(p)$, $p \in \gamma$, has a single zero, at some point $p_0 \in \gamma$, such that $t''(p_0) > 0$.

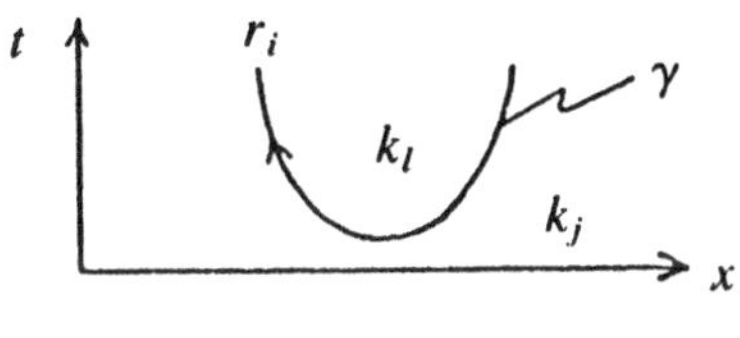

(6.15)

This is called an element of type F^{-1}.

(c) Let γ and γ' be pieces of loops ω_i and ω_j (ω_i may be equal to ω_j) which have a single common point:

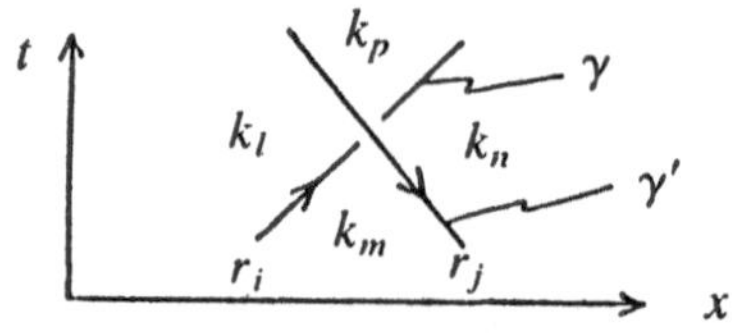

(6.16)

57

This is called an element of type R.

It is now quite clear how to associate the braid matrices, $R^{\pm}$, and the fusion matrices, F^{j}_{rk} and $(F^{j}_{rk})^{-1}$, introduced in Secs. 2 and 3 with elements of a decorated link diagram of types R, F and F^{-1}, respectively:

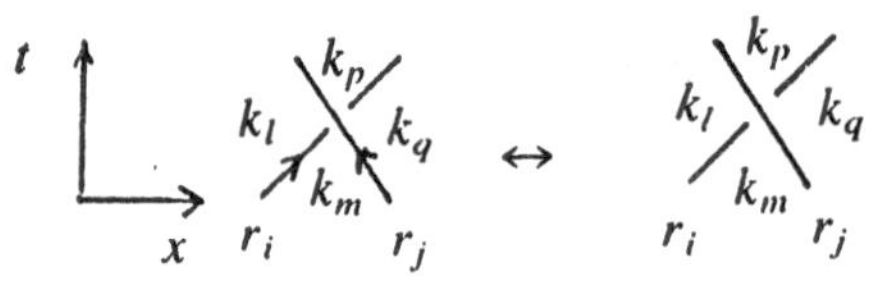

$\leftrightarrow R^{-}(k_l, r_i, r_j, k_q)^{k_r}_{k_m}$, (6.17)

(see (3.3) and (6.7))

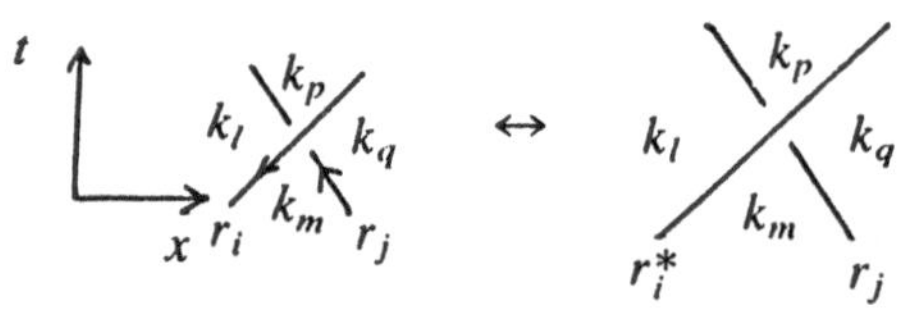

$\leftrightarrow R^{+}(k_l, r_i^*, r_j, k_q)^{k_p}_{k_m}$, (6.18)

(see (3.3) and (6.8))

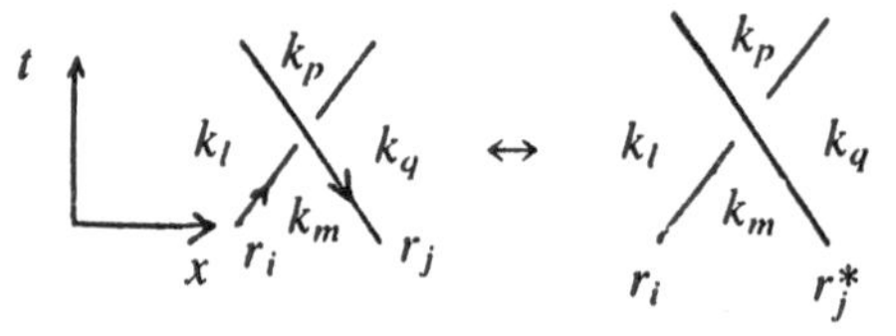

$\leftrightarrow R^{-}(k_l, r_i, r_j^*, k_q)^{k_p}_{k_m}$, (6.19)

(see (3.4) and (6.8)) .

It is obvious how the remaining assignments of crossings of a decorated link diagram to $R^{\pm}$-matrices must be defined. If a line is unoriented, e.g.

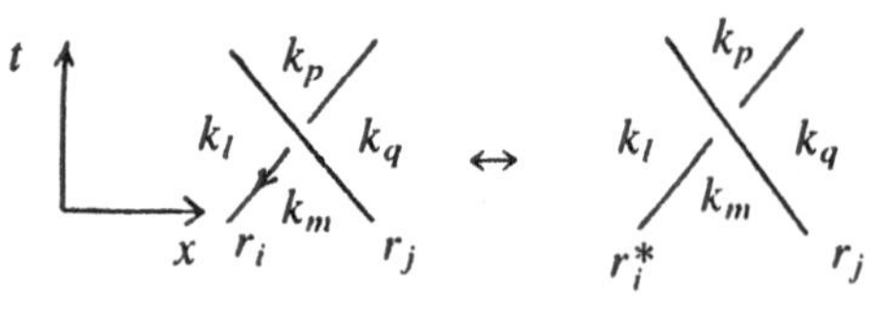

$\leftrightarrow R^{-}(k_l, r_i^*, r_j, k_q)^{k_p}_{k_m}$, (6.20)

(see (3.4) and (6.9))

then the corresponding label r_j belongs to $i_{\mathcal{A}}$, while labels of oriented lines, e.g. r_i in (6.20), belong to $l_{\mathcal{A}}$; see (6.7)–(6.9).

Next, we make the assignments

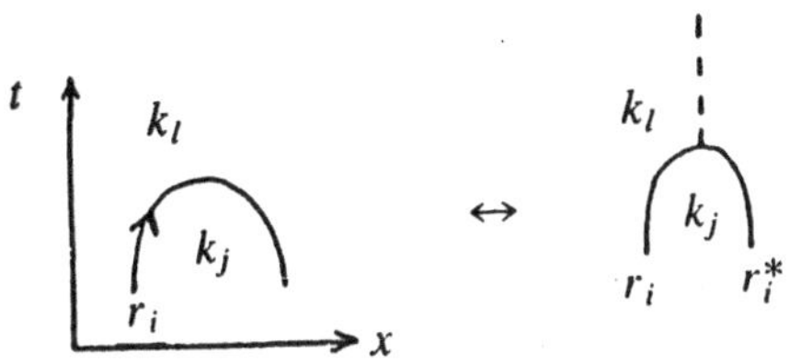

$$\leftrightarrow F^{k_l}_{r_i k_j}, \tag{6.21}$$

(see (3.20), (3.21), (6.7) and (6.8))

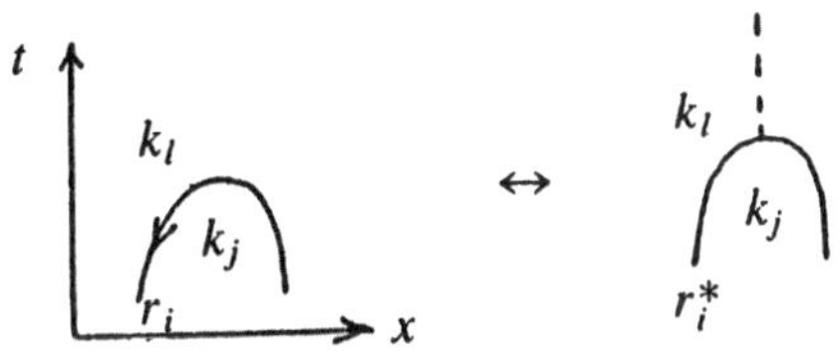

$$\leftrightarrow F^{k_l}_{r_i^\dagger k_j}. \tag{6.22}$$

If the lines in the diagrams on the left of (6.21) and (6.22) are oriented they are labelled by representations $r_i \in l_{\mathcal{A}}$; if they are unoriented then $r_i \simeq r_i^* \in i_{\mathcal{A}}$.

Finally, we have that, for $r_i \in l_{\mathcal{A}}$,

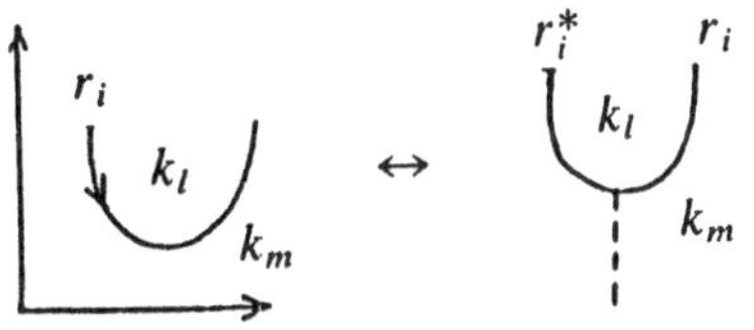

$$\leftrightarrow (F^{k_l}_{r_i k_m})^{-1}, \tag{6.23}$$

(see (3.25), (6.7) and (6.8))

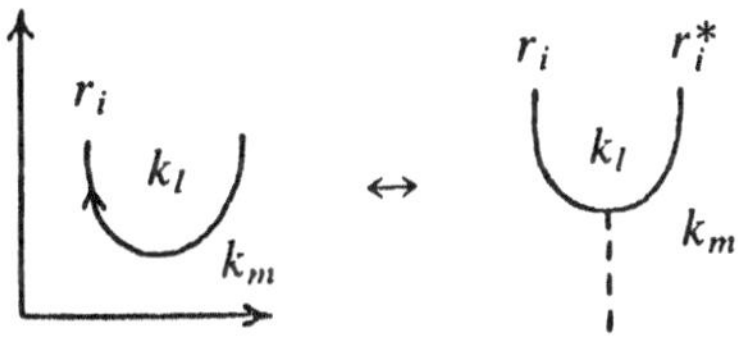

$$\leftrightarrow (F^{k_l}_{r_i^\dagger k_m})^{-1}, \tag{6.24}$$

while, for $r_i \in i_{\mathcal{A}}$,

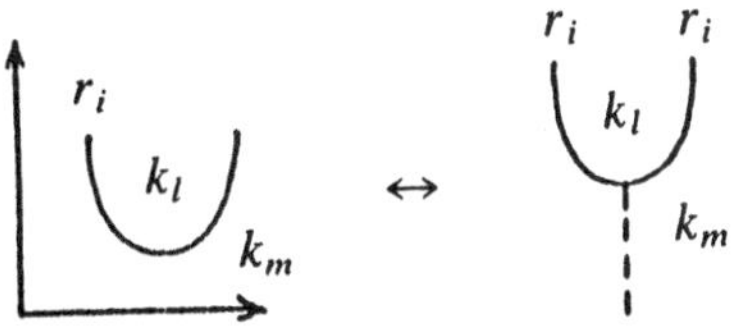

$$\leftrightarrow (F^{k_l}_{r_i k_m})^{-1} \tag{6.25}$$

Next, we define analogous assignments for the example of the free, massless chiral field discussed in Sec. 4, (4.18)–(4.30). Given a representation $r \in L_{\mathcal{A}}$, let

$h_r = \inf \operatorname{spec} r(L_0)$ be the conformal dimension of r, (i.e. the eigenvalue of $r(L_0)$ corresponding to an invariant, cyclic vector in $\mathcal{H}_r$; see Definition 2.1, (2), Sec. 2). We define

$$\varepsilon_r = \sqrt{8\pi h_r}\,, \tag{6.26}$$

see formula (4.21). We also define

$$\varepsilon_{r*} = -\sqrt{8\pi h_r}\,, \tag{6.27}$$

no matter whether $r^* \simeq r$ or $r^* \not\simeq r$. Note that if r^* is the representation contragredient to r in the sense of Eq. (2.26) then $h_r = h_{r*}$, as follows from (2.26), for $A = L_0$.

Given a decorated link diagram, $\mathcal{D}(k_2, \ldots, k_N | r_1, \ldots, r_n | L)$, we shall choose an orientation for *each* loop ω_i, $i = 1, \ldots, n$, and assign a "charge" ε_{r_i} to each upward oriented segment of ω_i and a "charge" $\varepsilon_{r_i^*} = -\varepsilon_{r_i}$ to each downward oriented segment of ω_i. The regions $\Omega_1(S(L)), \ldots, \Omega_N(S(L))$ in which $S(L)$ partitions the (t, x)-plane remain unlabelled. The resulting partially decorated link diagram is denoted by $\mathcal{D}_0(r_1, \ldots, r_n | L)$. The "elements" of $\mathcal{D}_0(r_1, \ldots, r_n | L)$ are chosen as follows:

(a.)

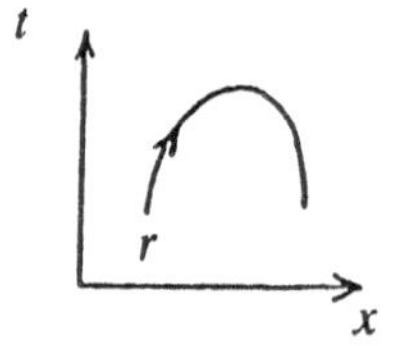

$$\leftrightarrow F_r^{(0)} = 1 \tag{6.28}$$

(b.)

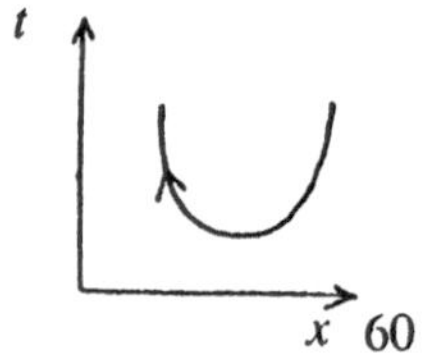

$$\leftrightarrow F_r^{(0)-1} = 1 \tag{6.29}$$

Assignments (6.28) and (6.29) are made in accordance with Eq. (4.29).

(c.)

$$\leftrightarrow R^{(0)}(r_i, r_j) = \exp\left(\frac{i}{4}\,\varepsilon_{r_i}\varepsilon_{r_j}\right) \tag{6.30}$$

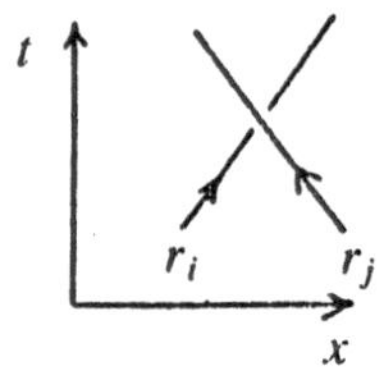

$$\leftrightarrow R^{(0)}(r_i, r_j)^{-1} = \exp\left(-\frac{i}{4}\,\varepsilon_{r_i}\varepsilon_{r_j}\right) \tag{6.31}$$

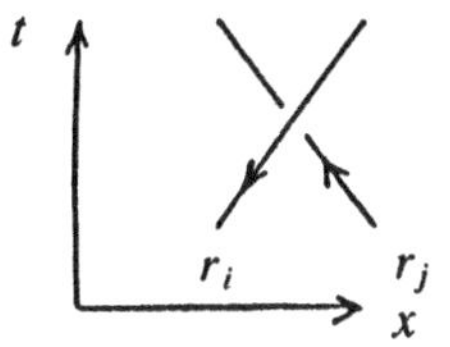

$$\leftrightarrow R^{(0)}(r_i^*, r_j) = \exp\left(\frac{i}{4}\varepsilon_{r_i^*}\varepsilon_{r_j}\right)$$

$$= \exp\left(-\frac{i}{4}\varepsilon_{r_i}\varepsilon_{r_j}\right). \quad (6.32)$$

with $\varepsilon_r = -\varepsilon_{r*} = \sqrt{8\pi h_r}$, as in (6.26), (6.27) etc.

The assignments (6.30)–(6.32) are made in accordance with Eq. (4.30) for the braid matrices of the massless field. We recall that, with these assignments, Eqs. (3.6)–(3.12), (3.14) and (3.15) and (3.20), (3.24) all hold for the free massless field. They will play an impotant role in the following.

We are now prepared to prove the basic results of this paper.

6.2. *Link-invariants for $\mathcal{M} = S^3$, $S^2 \times S^1$ from conformal field theory*

In this section we show how the results of Secs. 2 and 3 permit us to associate invariants for links, L, in the 3-sphere, or in $S^2 \times S^1$, with every rational chiral algebra. The manifold $\mathcal{M} = S^2 \times S^1$ serves to illustrate some intricacies related to the circumstance that $S^2 \times S^1$ is not simply connected. More general 3-manifolds can be dealt with by using the results of Secs. 4 and 5; see Sec. 6.4 for a sketch.

We start by studying links, L, in S^3. Let $\mathfrak{D} \equiv \mathfrak{D}(k_2, \ldots, k_N | r_1, \ldots, r_n | L)$ be a decorated diagram of L. The elements of $\mathfrak{D}$ are defined as in (6.14)–(6.16), Sec. 6.1. With every element of type F of $\mathfrak{D}$ we associate a matrix element of $F^{k_l}_{r_i k_j}$, as explained in (6.21), (6.22); $F^{k_l}_{r_i k_j} = 0$, unless $N_{k_l r_i k_j}, N_{k_j r_i k_l}, N_{k_j r_i^* k_l} \neq 0$. (Note that $F^{k_l}_{r_i k_j}$ is a 1×1 matrix, i.e. a number, only if the multiplicities $N_{k_l r_i k_j}$ and $N_{k_j r_i^* k_l}$, see Definition 2.2, (2.30) and (2.32), are $= 1$.) With every element of type F^{-1} of $\mathfrak{D}$ we associate a matrix element of $(F^{k_l}_{r_i k_m})^{-1}$, as explained in (6.23)–(6.25). We set $(F^{k_l}_{r_i k_m})^{-1} = 0$ if $N_{k_m r_i^* k_l} = 0$, or $N_{k_l r_i k_m} = 0$.

With every element of type R we associate a matrix element of $R^{\pm}(k_l, r_i, r_j, k_q)^{k_p}_{k_m}$, provided $N_{k_l r_i k_m} \neq 0, N_{k_m r_j k_q} \neq 0, N_{k_l r_j k_p} \neq 0$ and $N_{k_p r_i k_q} \neq 0$; see (6.17)–(6.20). If any one of these multiplicities vanishes we set $R^{\pm}(k_l, r_i, r_j, k_q)^{k_p}_{k_m} = 0$, etc.

We then take the sum of the products of all these matrix elements, with $r_1, \ldots, r_n$ and $k_1 = 1, k_2, \ldots, k_N$, all in $L_{\mathcal{A}}$, fixed, and call it $\nu(k_2, \ldots, k_N | r_1, \ldots, r_n | L)$. Thus we have defined a function ν,

$$\nu : \mathfrak{D}(k_2, \ldots, k_N | r_1, \ldots, r_n | L) \mapsto \nu(k_2, \ldots, k_N | r_1, \ldots, r_n | L) \quad (6.33)$$

on the set of all decorated link diagrams, $\mathfrak{D}$, of links $L \subset S^3$, with values in $\mathbb{C}$.

We define

$$\nu(r_1, \ldots, r_n | L) = \sum_{k_2, \ldots, k_N \in L_{\mathcal{A}}} \nu(k_2, \ldots, k_N | r_1, \ldots, r_n | L). \quad (6.34)$$

This is called the "*numerator of a decorated link*". (The link L is called *decorated*, since we have assigned representations $r_i \in L_{\mathcal{A}}$ to all its connected components λ_i (projected onto ω_i), $i = 1, \ldots, n$.)

Given L, let a partially decorated diagram, $\mathfrak{D}_0(r_1, \ldots, r_n | L)$, be defined as in Sec. 6.1, above (6.28). We define the "*denominator of a decorated link*" L, $\delta(r_1, \ldots, r_n | L)$, by associating with each element of type F or F^{-1} the number 1, see (6.28), (6.29), and with every element of type R the phase factor $R^{(0)}(.,.)^{\pm 1}$, satisfying the rules (6.30)–(6.32).

We then have the following theorem.

Theorem 6.1.

Let L be a (partially) oriented link in S^3 with the property that each connected component, λ_i, is assigned a representation $r_i \in L_{\mathcal{A}}$, $i = 1, \ldots, n$, according to the rules explained in Sec. 6.1. Then the quotient

$$I(r_1, \ldots, r_n | L) \equiv \frac{\nu(r_1, \ldots, r_n | L)}{\delta(r_1, \ldots, r_n | L)}, \tag{6.35}$$

with ν and δ as defined above, is an invariant of (partially) oriented links.

Proof.

The proof is surprizingly easy, given the results in Secs. 2 and 3. As recalled in Sec. 6.1, it is enough to verify that $I(r_1, \ldots, r_n | L)$ is invariant under Reidemeister moves of types I–III; see (6.11)–(6.13). This is certainly not quite obvious, given the definition of I.

The first problem to cope with is the following one: In a link diagram, in the sense of Reidemeister, lines which differ from each other only locally, (i.e. which can be deformed into each other *without* using Reidemeister moves of Types I–III), are identified, while they may represent a different number of elements of type F and F^{-1}, in our formalism, and hence might, a priori, change the value of $\nu(r_1, \ldots, r_n | L)$. Thus, the problem is to show that

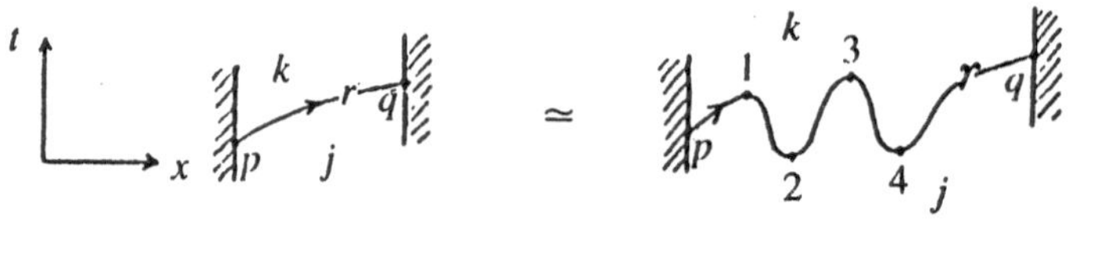

(6.36)

In (6.36), it is assumed that $t'(p)$ and $t'(q)$ (see Sec. 6.1, after (6.9)) have the same sign on both sides of the equation.

Now note, that the line on the left-hand side of (6.36) does not exhibit any elements of types F or F^{-1}, whereas the line on the right-hand side exhibits two elements of type F (1 and 3) and two elements of type F^{-1} (2 and 4). Fortunately, by (6.21)–(6.25), the product of these elements is

$$F^k_{rj}(F^k_{rj})^{-1}F^k_{rj}(F^k_{rj})^{-1} = 1 \; .$$

It is easy to prove, using these ideas, a lemma which says that if the *signs* of $t'(p)$ and $t'(q)$ are *fixed* then the contribution of

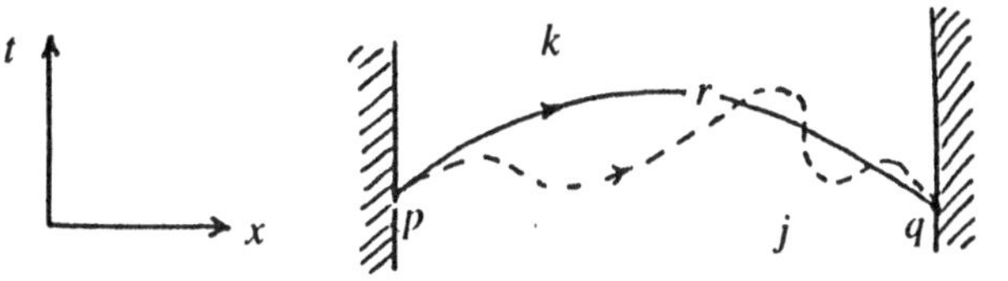

(6.37)

to $\nu(k_2, \ldots, k_N | r_1, \ldots, r_n | L)$ and to $\delta(r_1, \ldots, r_n | L)$ is *independent* of the behavior of the line between p and q.

Next, we must show that

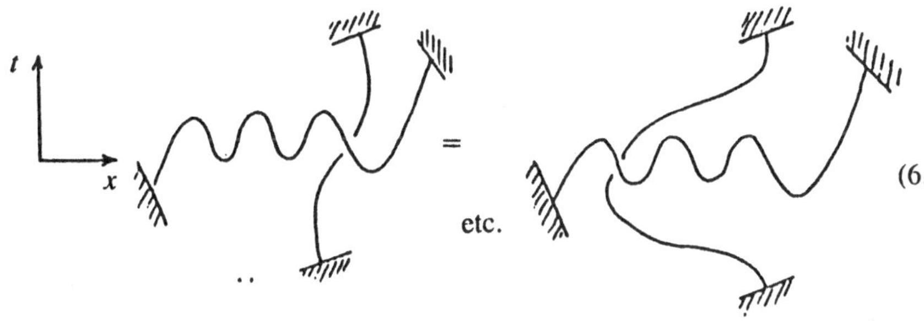

(6.38)

To show (6.38) we use (6.17)–(6.25) and Eqs. (3.20) and (3.24). Thus

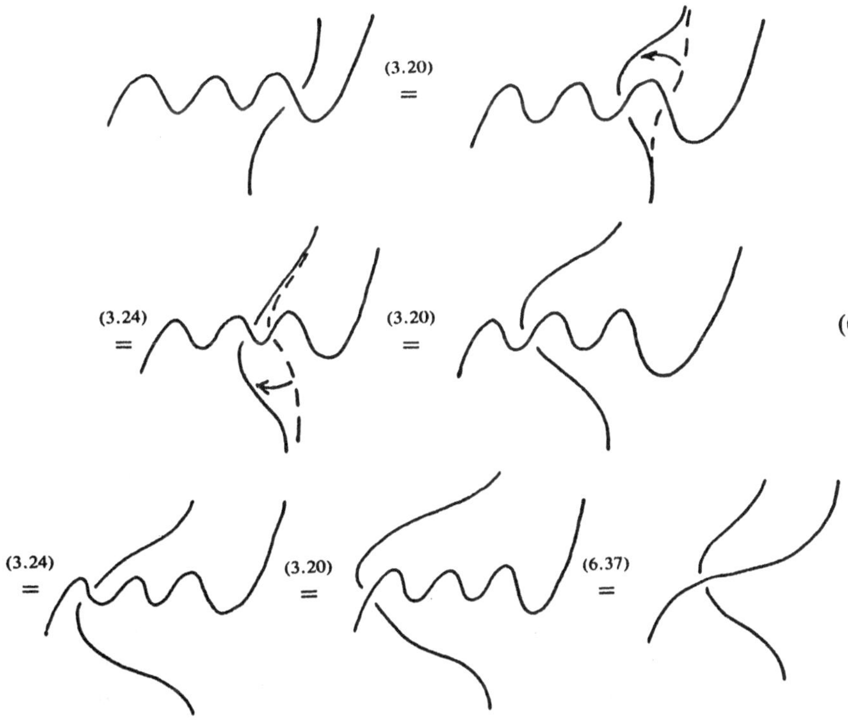

(6.39)

It follows, for example, that

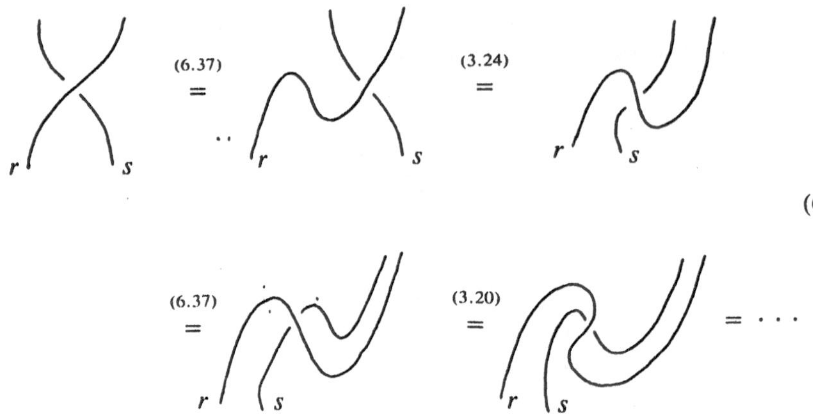

(6.40)

In conclusion, we may deform all lines locally *even when they cross other lines*, as long as no Reidemeister moves of types I–III are involved. (We have omitted drawing the t- and x-axis and the markings of the regions $\Omega_1(S(L)), \ldots, \Omega_N(S(L))$ in (6.39), (6.40); but this does not create any confusion.) As an example of the use of (6.37)–(6.40), note that

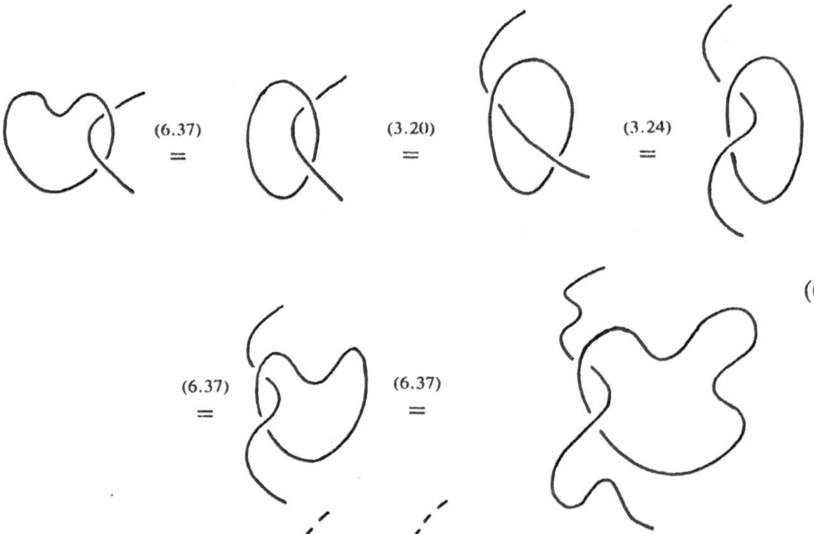

(6.41)

The passage from  to  is trivial, since it

does not introduce new elements into the link diagram. In conclusion, we have shown that a piece of a loop ω_i can be passed through any number of elements of type F and type F^{-1} of a piece of a loop ω_j, using (3.20) and (3.24). Using also that any local deformation of a loop ω_i always introduces the same number of elements of type F and of type F^{-1}, we have proven that $I(r_1, \ldots, r_n | L)$ is *invariant* under *all* deformations of the loops $\omega_1, \ldots, \omega_n$ not involving Reidemeister moves of type I–III.

Next, we attack the problem of invariance of $I(r_1, \ldots, r_n | L)$ under Reidemeister moves of type I; see (6.11). We must distinguish three cases.

Case 1. $I(r_1, \ldots, r_n | L)$ does not change if

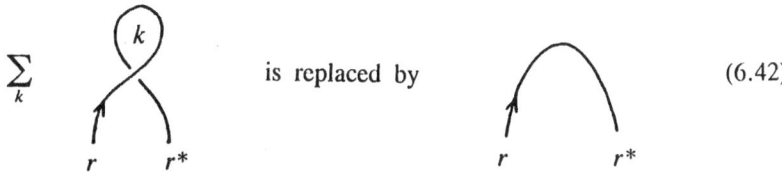

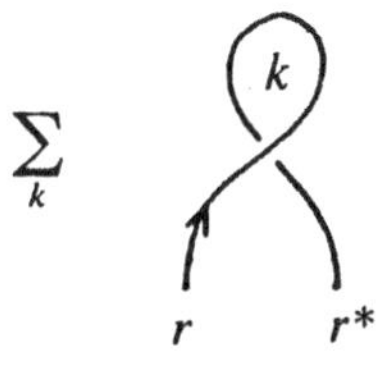

is replaced by
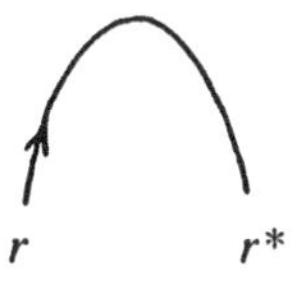

(6.42)

By Eq. (3.11), and using that $h_r = h_{r^*}$, we have that

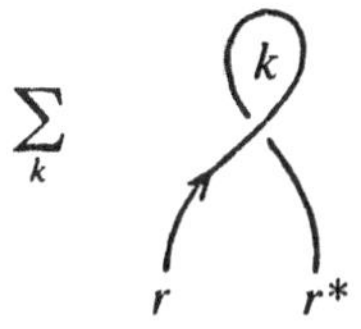

$= e^{-2i\pi h_r}$

(6.43)

Hence, under the move (6.42), $\nu(r_1, \ldots, r_n | L)$ changes by the phase factor $e^{-2i\pi h_r}$.

By (4.29) and (4.30), $\delta(r_1, \ldots, r_n | L)$ changes, under the move (6.42), by the phase factor

$$R^{(0)}(r, r^*) = \exp\left(\frac{i}{4}\, \varepsilon_r \varepsilon_{r^*}\right) = e^{-2i\pi h_r}\,, \tag{6.44}$$

where we have used (4.30), (6.30) and (6.26), (6.27), (i.e. $\varepsilon_r = -\varepsilon_{r^*} = \sqrt{8\pi h_r}$).

Hence the phase changes in $\nu(r_1, \ldots, r_n | L)$ and $\delta(r_1, \ldots, r_n | L)$ under the move (6.42) are the same, so $I(r_1, \ldots, r_n | L)$ is invariant under (6.42).

Using (3.12), (4.29), (4.30), (6.31), one also sees that

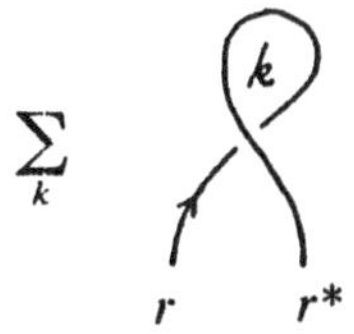

$\rightarrow$

(6.45)

leaves $I(r_1, \ldots, r_n | L)$ invariant.

Case 2.

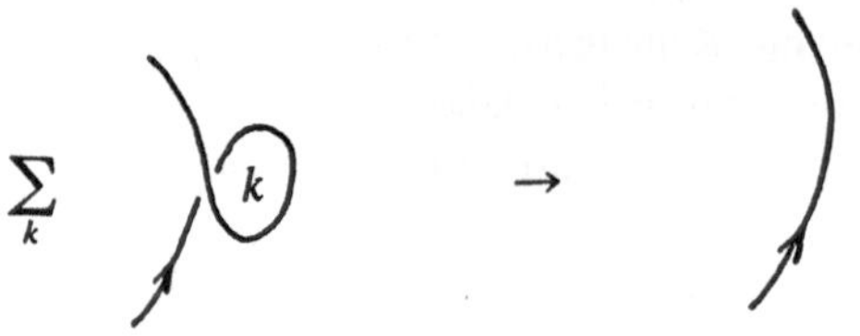

(6.46)

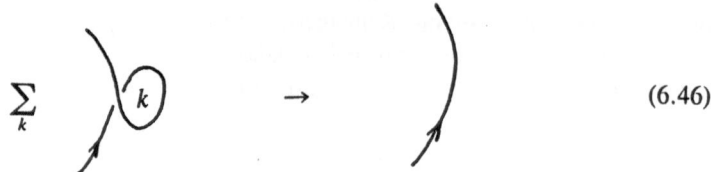

leaves $I(r_1, \ldots, r_n|L)$ invariant.
By (3.24),

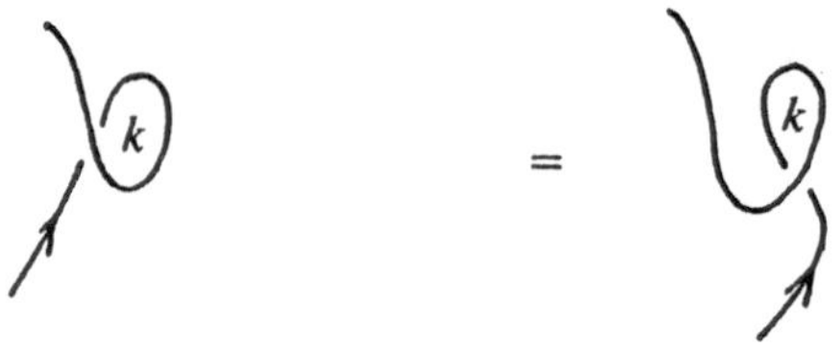

(6.47)

By (6.45)

(6.48)

leaves $I(r_1, \ldots, r_n|L)$ invariant.

The right-hand side of (6.48) exhibits an element of type F and an element of type F^{-1} which cancel; see (6.36), (6.37). Hence

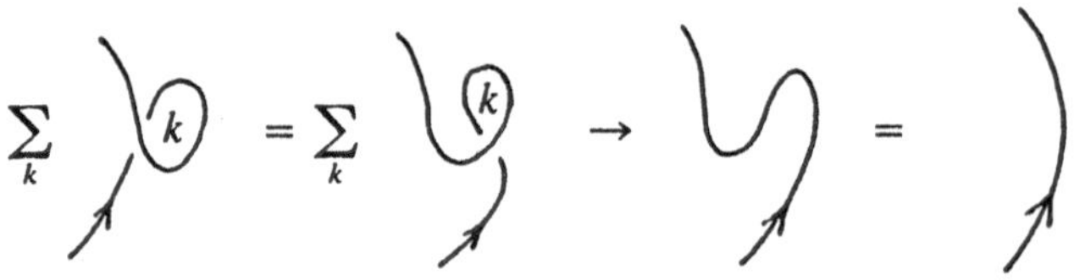

(6.49)

leaves $I(r_1, \ldots, r_n|L)$ invariant. The cases

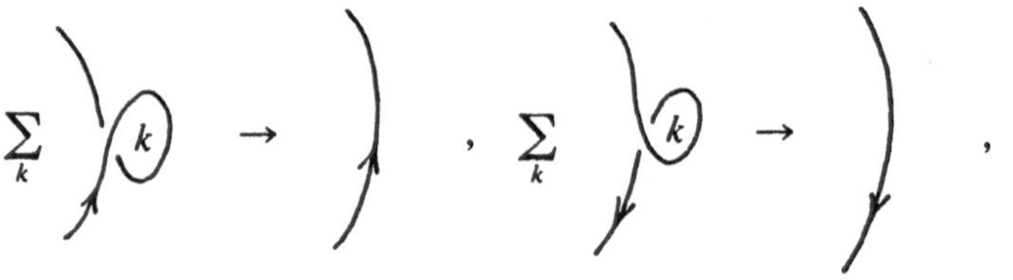

(6.50)

, etc. are treated similarly.

Case 3. Finally, we establish invariance of $I(r_1, \ldots, r_n | L)$ under the moves

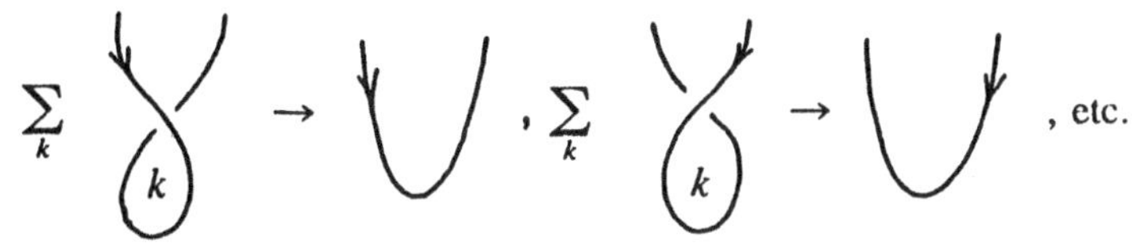

By (6.35)–(6.37) and (3.20),

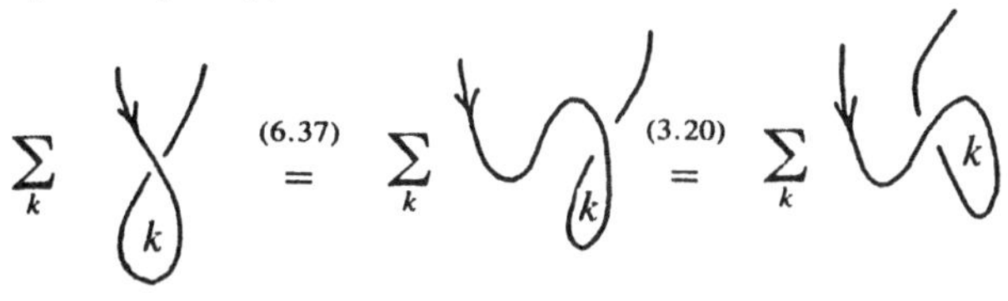

(6.51)

By (6.50), the move

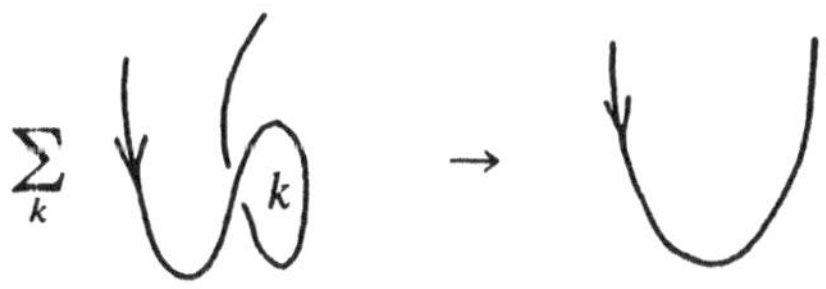

(6.52)

leaves $I(r_1, \ldots, r_n | L)$ invariant. The other moves in case 3 are dealt with similarly.

Thus we have established invariance of $I(r_1, \ldots, r_n | L)$ under Reidemeister moves of type I.

Next, we must establish invariance of $I(r_1, \ldots, r_n | L)$ under Reidemeister moves of type II; see (6.12). There are two cases to be studied:

Case 1.

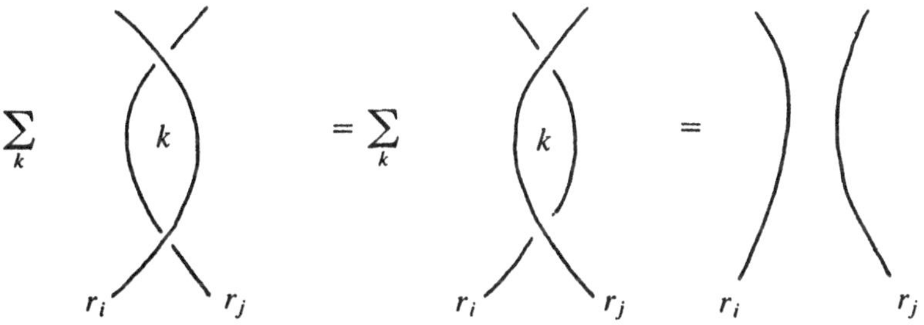

(6.53)

This is an immediate consequence of Eq. (3.14).

Case 2.

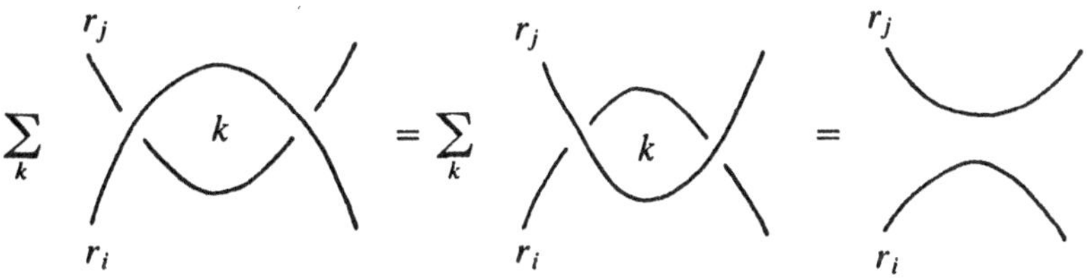

(6.54)

By (3.20) and (3.24)

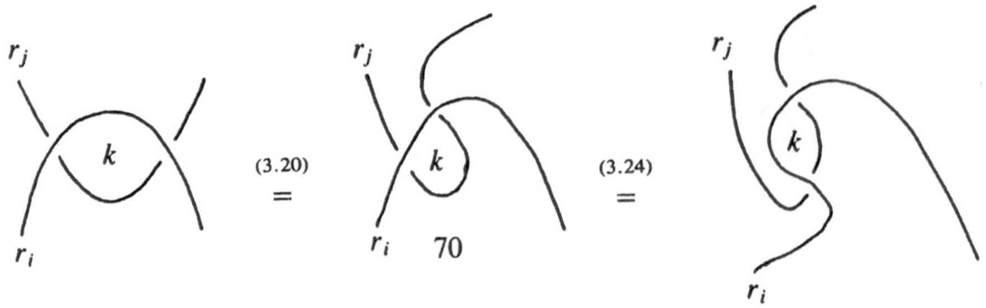

Next, by (6.53),

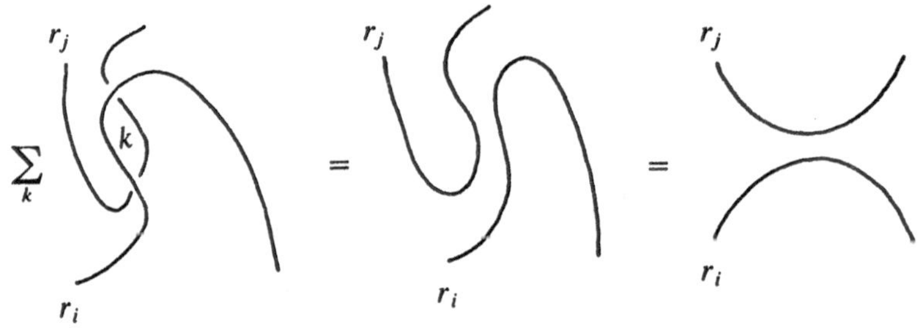

(6.55)

The other identities that fall into case 2 are proven similarly. This proves invariance of $I(r_1, \ldots, r_n|L)$ under Reidemeister moves of type II.

It remains to establish invariance of $I(r_1, \ldots, r_n|L)$ under Reidemeister moves of type III, see (6.13). Hence we must show that

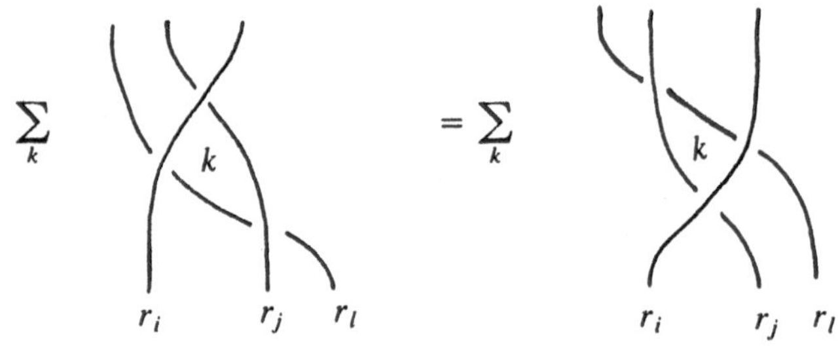

(6.56)

But this follows from the Yang-Baxter equation (3.15); see also (2.70). We note that variants of (6.56), like

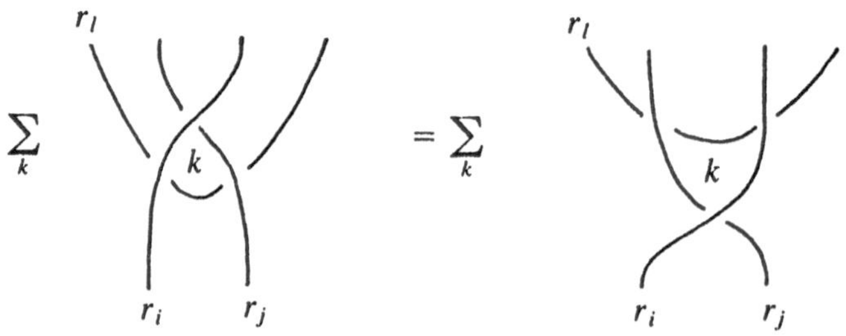

(6.57)

are easily dealt with, using (3.20) and (3.24). In this example,

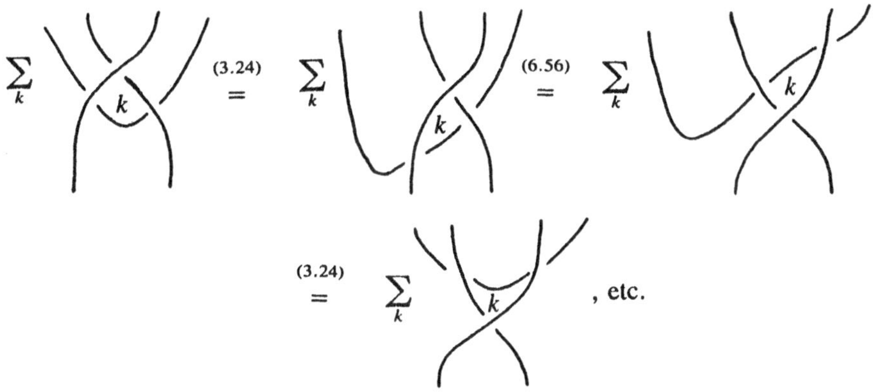

This completes our proof of invariance of $I(r_1, \ldots, r_n | L)$ under Reidemeister moves of types I–III, and hence of Theorem 6.1.

□

Remark

The pure mathematician may forgive us that, as physicists, we have not formalized our proof beyond the level reached above.

Example

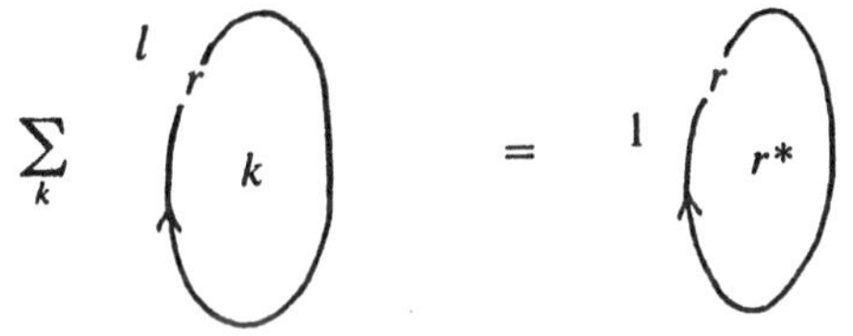

$$= \mathrm{tr}(F^{1}_{rr*}(F^{r*}_{r*1})^{-1}) \tag{6.58}$$

Proof of (6.58) We consider the sequence of identities

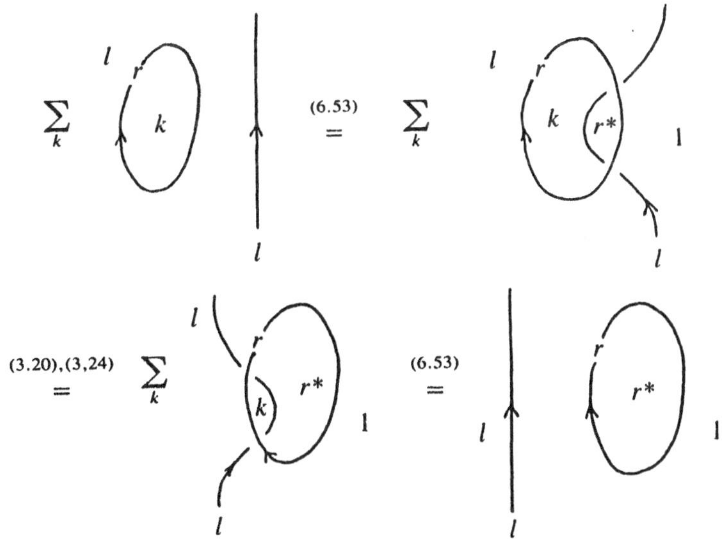

These identities can be generalized to show that the labelling of the region $\Omega_1(S(L))$ containing ∞ is irrelevant.

Next, we extend the formalism developed, so far, to links in the nonsimply connected manifold $S_2 \times S_1$. This is quite straightforward and will yield a possibly interesting relation between the matrix elements of the matrix, S, representing the modular transformation $\tau \to -1/\tau$ on the space of characters of the chiral algebra $\mathcal{A}$ and the fusion matrices, F and F^{-1}; (τ is the modular parameter).

To get started, we must define suitable link diagrams for links, L, in $S^2 \times S^1$. We shall represent S^2 by a disk with boundary points identified. We may then embed $S^2 \times S^1$ in $\mathbb{R}^3$: Choose polar coordinates in the (t, x)-plane, π, with polar angle $\varphi \in [0, 2\pi)$; the angle φ is a coordinate for S^1. We identify the sphere, S_0^2, at $\varphi = 0$ with the disk

$$\{(x, y) : (x - 2)^2 + y^2 \leq 1\} \tag{6.59}$$

in the (x, y)-plane, with boundary points identified. Then S_φ^2 is obtained by rotating S_0^2 around the y-axis through an angle φ. In this parametrization, $S^2 \times S^1$ is represented as a full torus, T, in $\mathbb{R}^3$ whose symmetry axis is the y-axis. Its projection onto π is the annulus

$$A = \{(x, t) : 1 \leq x^2 + t^2 \leq 9\} . \tag{6.60}$$

By our construction, a link, L, in $S^2 \times S^1$ is imbedded in T avoiding ∂T. That imbedding can be projected onto A, and the projection is a shadow, $S(L)$, of L. If over- and undercrossings, see (6.10), are recorded then one gets a diagram, $D(L)$, of L; $S(L)$ and $D(L)$ are contained in A. The shadow, $S(L)$, of L decomposes π into disjoint regions, $\Omega_1(S(L)), \ldots, \Omega_N(S(L))$. We agree on the convention that $\Omega_1(S(L))$ will contain the outer boundary

$$\partial_+ A = \{(x, t) : x^2 + t^2 = 9\} \tag{6.61}$$

of A, while $\Omega_N(S(L))$ contains the inner boundary

$$\partial_- A = \{(x, t) : x^2 + t^2 = 1\} \tag{6.62}$$

of A.

Next, we must explain how to decorate $D(L)$. In accordance with our parametrization of $S^2 \times S^1$, the points $(r = 1, \varphi)$, $(r = 3, \varphi)$ on $\partial_- A$, $\partial_+ A$, respectively correspond to the same point, P, (the point at infinity, for example), of S^2; here $r = \sqrt{x^2 + t^2}$, $\varphi = \arctan t/x$. A neighborhood of P will be labelled 1, the label of the vacuum representation of $\mathcal{A}$. Hence $\Omega_1(S(L))$ and $\Omega_N(S(L))$ must bear the label 1. The labels k_j of $\Omega_j(S(L))$, $j = 2, \ldots, N - 1$, range over $L_{\mathcal{A}}$, (the $\otimes$-invariant set of irreducible lowest-weight representations of $\mathcal{A}$). Every loop $\omega_j \in D(L)$ bears a label $r_j \in L_{\mathcal{A}}$, $j = 1, \ldots, n$, ($n = \#$ components of L). These rules define a decoated link diagram, $\mathfrak{D}(k_2, \ldots, k_{N-1} | r_1, \ldots, r_n | L)$, of the link $L \subset S^2 \times S^1$. Further definitions, e.g. the definition of elements of $\mathfrak{D}(k_2, \ldots, k_{N-1} | r_1, \ldots, r_n | L)$, are taken over from

Sec. 6.1, (6.14)–(6.32) and from (6.33), (6.34). In order to record the dependence of the numerator, $\nu(r_1, \ldots, r_n|L)$, on the manifold $\mathcal{M}$, in which L is imbedded, we denote it henceforth by $\nu_{\mathcal{M}}(r_1, \ldots, r_n|L)$. The definition of the denominator $\delta_{\mathcal{M}}(r_1, \ldots, r_n|L)$ of a decorated link L is somewhat more subtle when $\mathcal{M} = S^2 \times S^1$ than the definition for $\mathcal{M} = S^3$: Some of the loops $\lambda_1, \ldots, \lambda_n$ may be non-contractible in $S^2 \times S^1$; suppose $\lambda_1 \ldots, \lambda_l$, $0 \leq l \leq n$, are non-contractible. Then their projections, $\omega_1, \ldots, \omega_l$, onto the (t, x)-plane, π, are non-contractible loops in A. We suppose that the loops ω_i, $i = 1, \ldots, l$, are oriented and let $r_1, \ldots, r_l$ be the representations labelling $\omega_1, \ldots, \omega_l$. We define

$$\varepsilon_i = \pm\sqrt{8\pi h_{r_i}} \tag{6.63}$$

to be the "charge" corresponding to ω_i in the definition of the denominator, $\delta_{S^2\times S^1}$, of L; see (6.26). The + sign is chosen if ω_i is positively oriented, the − sign if ω_i is negatively oriented. It is easy to see that the numerator $\nu_{S^2\times S^1}(r_1, \ldots, r_n|L)$ vanishes unless

$$r_1 \underset{z_1}{\otimes} r_2 \underset{z_2}{\otimes} \cdots \underset{z_{l-1}}{\otimes} r_l \text{ contains } 1 , \tag{6.64}$$

the vacuum representation. Unfortunately, (6.64) does not imply, in general, that $\sum_{j=1}^{l} \varepsilon_j = 0$. If $\sum_{j=1}^{l} \varepsilon_j \neq 0$ then, in accordance with (4.27), $\delta_{S^2\times S^1}(r_1, \ldots, r_n|L)$ should be set $= 0$. Thus we must introduce one further noncontractible loop λ_0, whose projection onto π, ω_0, we suppose to be everywhere closer to $\partial_- A$ than $\omega_1, \ldots, \omega_n$. We choose ω_0 to be positively oriented and assign the "charge"

$$\varepsilon_0 = -\sum_{j=1}^{l} \varepsilon_j \tag{6.65}$$

to ω_0. We emphasize that, in general, there will not exist any representation $r_0 \in L_{\mathcal{A}}$ such that $\varepsilon_0 = \pm\sqrt{8\pi h_{r_0}}$, but that turns out to be irrelevant. One might imagine that the loops $\lambda_1, \ldots, \lambda_l$ would wind around λ_0. That would change the numerical value of the invariants to be calculated below. (This ambiguity is not unrelated to the ambiguity in choosing a "framing" for the manifold $S^2 \times S^1$ in Witten's approach to link invariants.[3]) For concreteness, we have chosen λ_0 so that it is not linked to any one of the loops $\lambda_1, \ldots, \lambda_n$.

With these modifications in mind, we calculate the denominator, $\delta_{S^2\times S^1}(r_1, \ldots, r_n|\lambda_0, L)$, of a decorated link, by carrying over the assignments (6.28)–(6.32) to the present case. We then have the following theorem.

Theorem 6.2

The numbers

$$I_{S^2\times S^1}(r_1, \ldots, r_n|\lambda_0, L) = \frac{\nu_{S^2\times S^1}(r_1, \ldots, r_n|L)}{\delta_{S^2\times S^1}(r_1, \ldots, r_n|\lambda_0, L)} , \tag{6.66}$$

with $r_1, \ldots, r_n$ *in* $L_{\mathcal{A}}$, *are invariants for partially oriented links in* $S^2 \times S^1$, *(with* $S^2 \times S^1$ *"framed" by* λ_0*).*

Proof. In addition to invariance under Reidemeister moves of types I–III, we must verify some further invariance property of $I_{S^2\times S^1}$ reflecting the fact that the generators of the braid group on the sphere, S^2, satisfy an additional relation

$$\tau_1\tau_2 \ldots \tau_{n-1}^2\tau_{n-2} \ldots \tau_2\tau_1 = id\ . \tag{6.67}$$

Invariance under Reidemeister moves of types I–III is proven in the same way as for $\mathcal{M} = S^3$. So the real issue is to find out what kind of invariance is implied by (6.67) and then to check that it holds. Fortunately, this problem can be discussed graphically. We consider some part of a link diagram and claim that the following identity holds for $I_{S^2\times S^1}$:

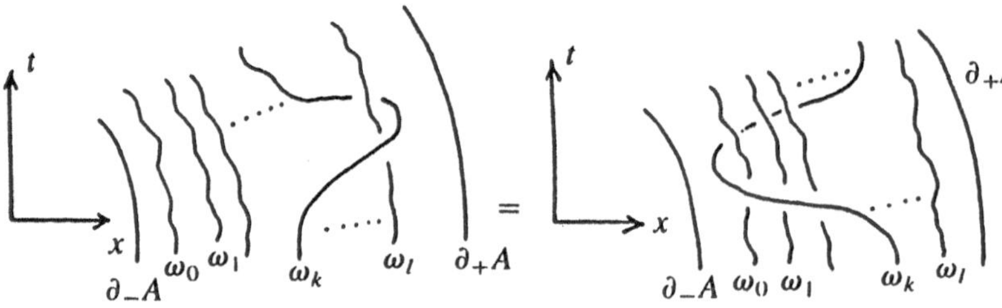

(6.68)

More abstractly, (6.68) expresses invariance of $I_{S^2\times S^1}$ under the move

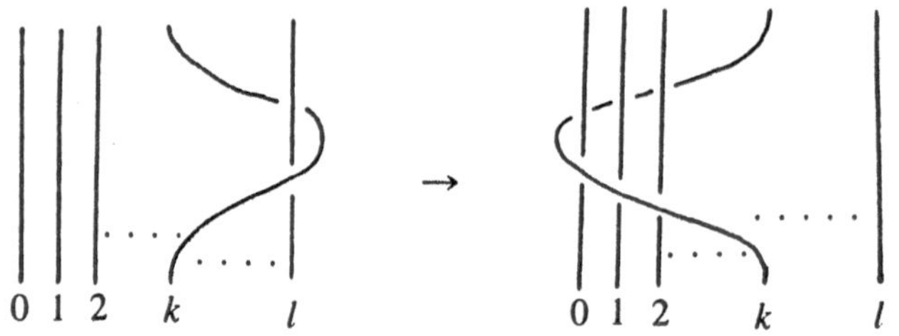

(6.69)

We first determine how $\nu_{S^2\times S^1}(r_1, \ldots, r_n|L)$ changes under the move (6.69): By (2.80),

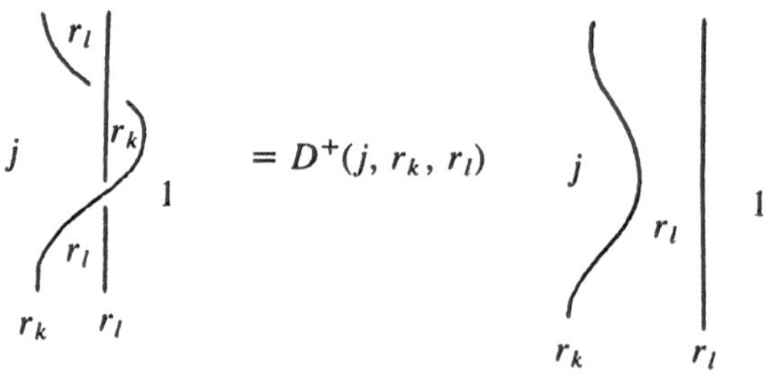

(6.70)

where

$$D^+(j, r_k, r_l) = \exp 2i\pi(h_{r_k} + h_{r_l} - h_j) \tag{6.71}$$

see (2.82). Furthermore, by (2.81)

$$\sum_j D^+(j, r_k, r_l)$$ 

$$= \delta^p_m D^+(n, r_k, p)$$ 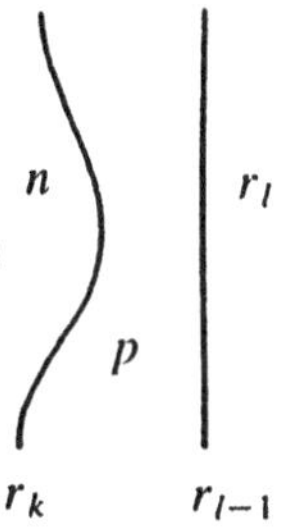

(6.72)

This equation can be iterated. When r_k has been pulled through $r_{k+1}, \ldots, r_l$, using (6.72) repeatedly, we use that

$$\sum_j D^+(n, r_k, j)$$ 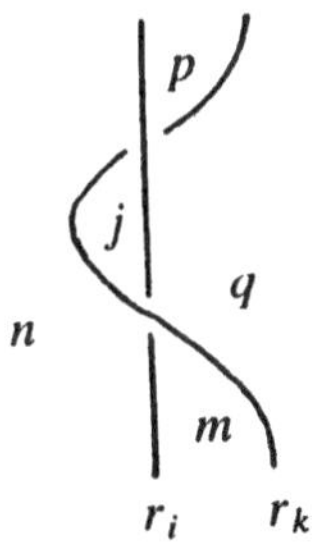

$$= \delta^p_m D^+(p, r_k, q)$$ 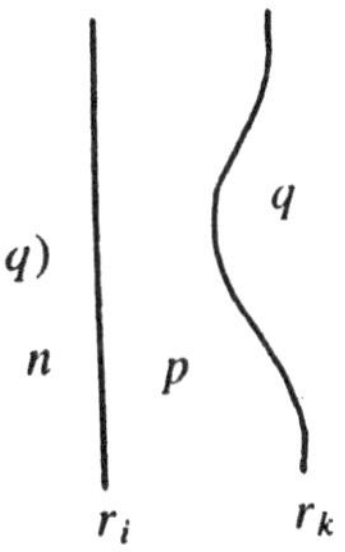

(6.73)

to pull r_k successively through $r_{k-1}, \ldots, r_1$. At the end of this process we are left with

$$D^+(1, r_k, r_k^*)$$ 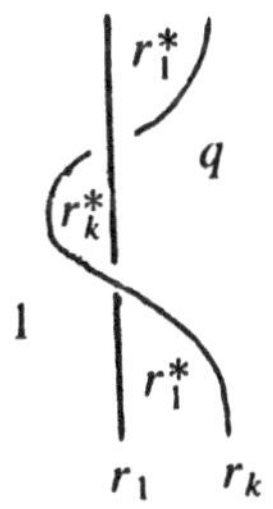

$$= D^+(r_1^*, r_k, q)$$ 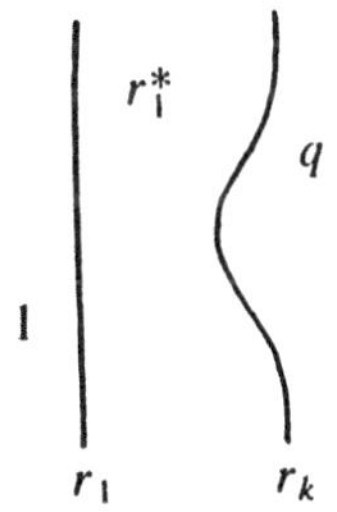

(6.74)

Hence, the move (6.69) changes $\nu_{S^2 \times S^1}(r_1, \ldots, r_n | L)$ by a phase factor

$$D^+(1, r_k, r_k^*) = \exp(4i\pi h_{r_k}) \tag{6.75}$$

This can also be verified, more quickly, by fusing $r_1, \ldots, r_{k-1}$ and $r_{k+1}, \ldots, r_l$, using (3.9) and (3.6) repeatedly, and, finally, using (2.80).

The change of $\delta_{S^2 \times S^1}(r_1, \ldots, r_n | \lambda_0, L)$ under the move (6.69) is calculated easily from (6.30)–(6.32). We find, using (6.65),

$$\exp\left(-\frac{i}{2}\varepsilon_k\left(\sum_{\substack{j=0\\ j\neq k}}^{l}\varepsilon_j\right)\right) = \exp\left(\frac{i}{2}\varepsilon_k^2\right) = \exp(4i\pi h_{r_k}) , \tag{6.76}$$

where, in the last step, we have used (6.63).

By (6.75) and (6.76), $I_{S^2 \times S^1}(r_1, \ldots, r_n | \lambda_0, L)$ is invariant under the move (6.69).

□

Next, we propose to establish a relation between the link invariants I_{S^3} and $I_{S^2 \times S^1}$ found in Theorems 6.1 and 6.2, respectively. Given a link L in $S^2 \times S^1$, we define a link $\bar{L} \subset S^2 \times S^1$ by adding to L one further noncontractible component, λ, not linked to any component, λ_i, of L, $i = 1, \ldots, n$. We may choose λ such that its projection, ω, onto π coincides with $\partial_- A = \{(x, t) : x^2 + t^2 = 1\}$. We label λ with a representation $r \in L_{\mathscr{A}}$. The loop λ_0 introduced after (6.64) may be chosen so that ω_0 encloses ω. It is almost trivial to prove the following proposition.

Proposition 6.3.

$$I_{S^3}(r_1, \ldots, r_n | L) = \sum_{r \in L_{\mathscr{A}}} I_{S^2 \times S^1}(r, r_1, \ldots, r_n | \lambda_0, \bar{L}) \operatorname{tr}(F^{r}_{r1}(F^{1}_{r*r})^{-1})^{-1} \tag{6.77}$$

Proof. Clearly, we may choose the representative $\underline{\lambda} = (\lambda_1, \ldots, \lambda_n)$ of the link $L \subset S^3 \simeq \dot{\mathbb{R}}^3$ to lie inside the full torus, $T \subset \dot{\mathbb{R}}^3$, representing $S^2 \times S^1$. Hence all the loops $\omega_1, \ldots, \omega_n$ are contained in the annulus A, (but some or all of these loops may not be contractible in A). Since λ is chosen such that $\omega \cap \omega_i = \emptyset$, for all $i = 0, \ldots, n$, and since we may choose λ_0 such that $\omega_0 \cap \omega_j = \emptyset$, $j = 1, \ldots, n$, $I_{S^2 \times S^1}(r, r_1, \ldots, r_n | \lambda_0, \bar{L})$ can be factorized as follows:

$$I_{S^2 \times S^1}(r, r_1, \ldots, r_n | \lambda_0, \bar{L}) = I(r|\lambda) I^{(r)}_{S^3}(r_1, \ldots, r_n | L) . \tag{6.78}$$

Here $I(r, \lambda)$ denotes the numerical contribution of the diagram

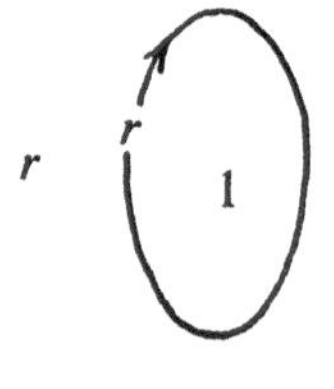

which, according to (6.21) and (6.24), is given by

$$\operatorname{tr}(F^{r}_{r1}(F^{1}_{r*r})^{-1}) \tag{6.79}$$

The factor $I^{(r)}_{S^3}(r_1, \ldots, r_n | L)$ is computed as in (6.33)–(6.35), with the condition that the label of the region $\Omega_{i_0}(S(L))$ (see above (6.11)) which contains the origin of π is given by $k_{i_0} = r \in L_{\mathcal{A}}$. Then (6.77) follows from (6.78), (6.79) and (6.34).

□

Suppose that L_0 consists of a single loop in T which is labelled by the vacuum representation, 1. Our conventions imply that

$$I_{S^3}(1 | L_0) = 1 .$$

Furthermore

$$I_{S^2 \times S^1}(r, 1 | \lambda_0, \bar{L}_0) = \delta_{r1} ,$$

and

$$F^1_{11} = 1 . \qquad (6.80)$$

Labelling L_0 by s, we have from (6.77) that

$$I_{S^3}(s | L_0) = \sum_{r \in L_{\mathcal{A}}} I_{S^2 \times S^1}(r, s | \lambda_0, \bar{L}_0) \, \mathrm{tr}(F^r_{r1}(F^1_{r*r})^{-1})^{-1} . \qquad (6.81)$$

Let $S = (S_{rt})_{r,t \in L_{\mathcal{A}}}$ be the matrix representing the modular transformation $\tau \mapsto -1/\tau$ on the space of characters of $\mathcal{A}$; (the character $\chi_r(q)$ is given by $\sum_{N=0}^{\infty} \langle \xi^N_r | q^{r(L_0)} \xi^N_r \rangle$, where $\{\xi^N_r\}_{N=0}^{\infty}$ is a basis in $\mathcal{H}_r$ and $\{\langle \xi^N_r |\}_{N=0}^{\infty}$ a dual basis in $\mathcal{H}^*_r$, $q \equiv e^{i\tau}$). Comparing (6.78)–(6.81) with Witten's formula (4.42) in Ref. 3, we conclude that

$$\mathrm{tr}(F^r_{r1}(F^1_{r*r})^{-1})^{-1} = S_{1r}/S_{11} . \qquad (6.82)$$

(Witten uses the label "0" instead of our label "1", and "i" replaces "r" in his formula.) An expression for S_{rs}/S_{11} is derived similarly.

6.3. *A three-dimensional interpretation of the invariants* $I_{S^3}(r_1, \ldots, r_n | L)$

In this section we use the results in Secs. 3 and 4 to give a three-dimensional interpretation of the invariants $I_{S^3}(r_1, \ldots, r_n | L)$. Our approach is based on using objects from a left-right symmetric conformal field theory to calculate $I_{S^3}(r_1, \ldots, r_n | L)$ in a new way. Our method works for general conformal field theories constructed from the representation theory of two rational chiral algebras, $\mathcal{A}$ and $\bar{\mathcal{A}}$, but, for the sake of notational symplicity, we confine our analysis to left-right symmetric theories; see (4.12)–(4.13); (the reader will easily see how to treat the general case—no new invariants emerge from it. However the analysis yields conditions on pairs of chiral algebras, $\mathcal{A}$, $\bar{\mathcal{A}}$, that can give rise to a *local* conformal field theory. They will be discussed elsewhere).

Let $D(L)$ be a diagram of the link L, as defined in Sec. 6.1, after (6.10). Using Reidemeister moves of types II and III, we can deform the diagram $D(L)$ into a new diagram $\bar{D}(L)$ with the following properties:

(1) $\bar{D}(L)$ is a diagram of L.

(2) All elements of type F of $\bar{D}(L)$ are contained in the half-plane.

$$\pi_+ \equiv \{(t, x) : t > 0\} \; , \tag{6.83}$$

and all elements of type F^{-1} of $\bar{D}(L)$ are contained in $\pi_- \equiv \{(t, x) : t < 0\}$. Let $\bar{\mathscr{D}}(k_2, \ldots, k_N | r_1, \ldots, r_n | L)$ be a decoration of $\bar{D}(L)$. We introduce new "elements" for $\bar{D}(L)$, $\bar{\mathscr{D}}(k_2, \ldots | \ldots, r_n | L)$ as follows:

(a) Elements of type F and R of $\bar{\mathscr{D}}(k_2, \ldots | \ldots, r_n | L)$ in the half-plane π_+ are defined and assigned matrix elements of fusion- and braid matrices as in (6.17)–(6.22).

(b) The intersection of a line in $\bar{\mathscr{D}}(k_2, \ldots | \ldots, r_n | L)$ with the x-axis is called an element of type D. To an element of type D we assign the matrix $D_{k_i r_j k_l}$, defined in (4.1), (4.12), according to the rules

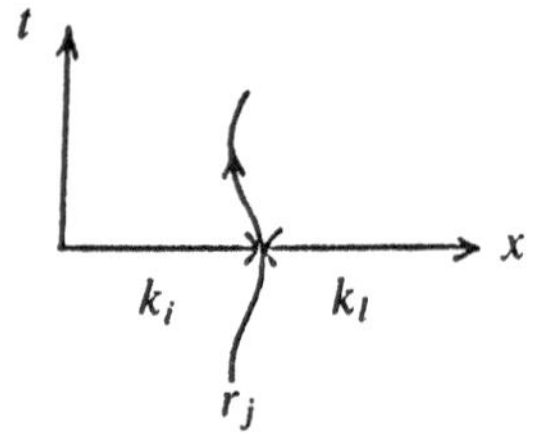

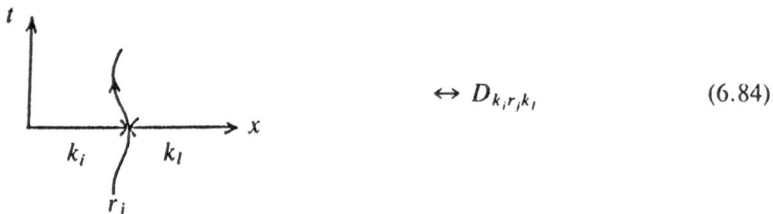

$$\leftrightarrow D_{k_i r_j k_l} \tag{6.84}$$

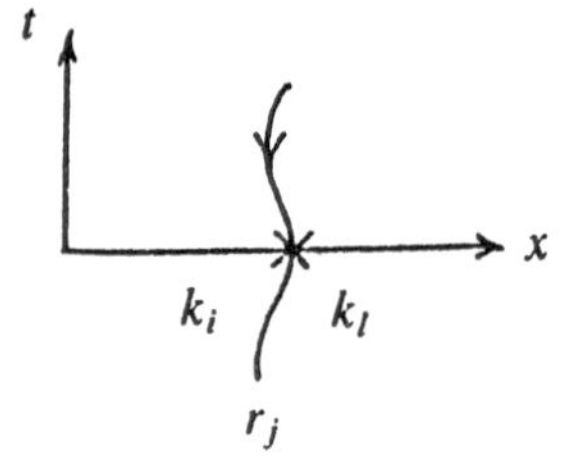

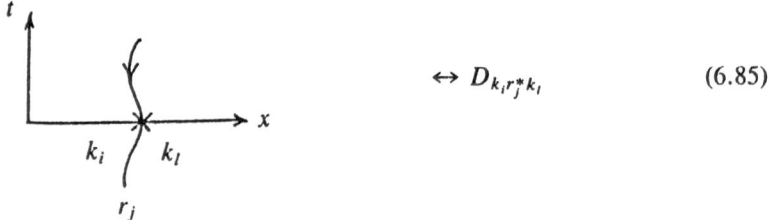

$$\leftrightarrow D_{k_i r_j^* k_l} \tag{6.85}$$

(c) Elements of type F^{-1} of $\mathscr{D}(k_2, \ldots | \ldots, r_n | L)$ are henceforth called "*elements of type* $\bar{F}$" and are assigned the following matrix elements:

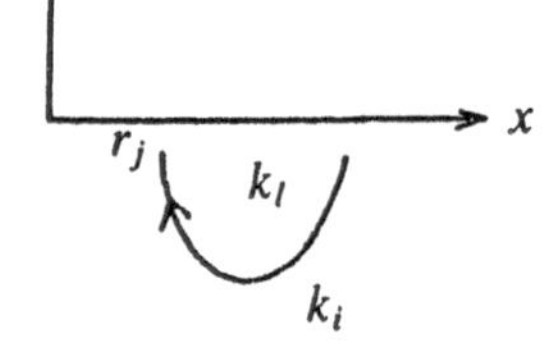

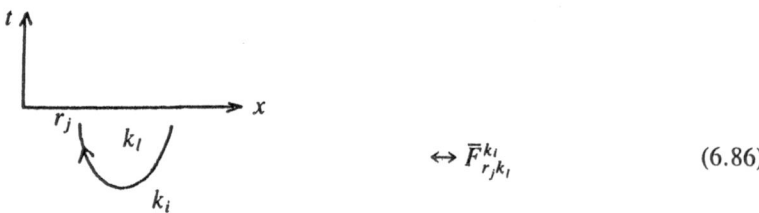

$$\leftrightarrow \bar{F}^{k_l}_{r_j k_l} \tag{6.86}$$

where

$$\bar{F}^{p}_{rq} \equiv \bar{F}(p, r, r^*, p)^{1}_{q} ,$$

and $\bar{F}(p, r, s, k)^{m}_{q}$ has been defined in (4.9).

Similarly

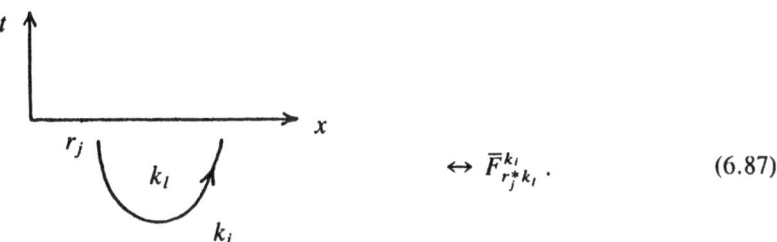

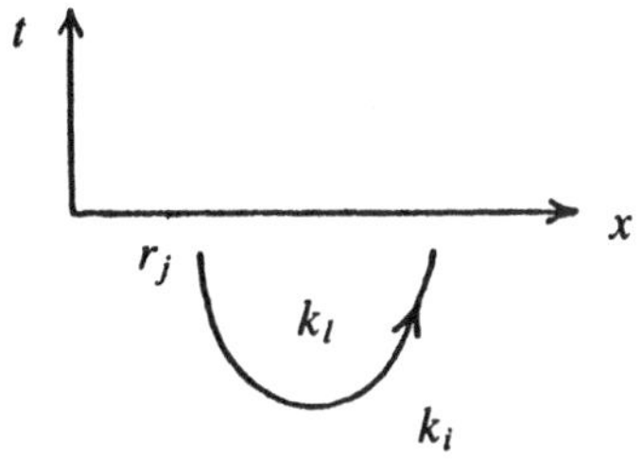

$$\leftrightarrow \bar{F}^{k_l}_{r_j^* k_l} . \qquad (6.87)$$

(d) Elements of type R of $\mathfrak{D}(k_2, \ldots | \ldots, r_n | L)$ contained in the half-plane π^- are now called "elements of type $\bar{R}$" and are assigned the following matrix elements of braid matrices:

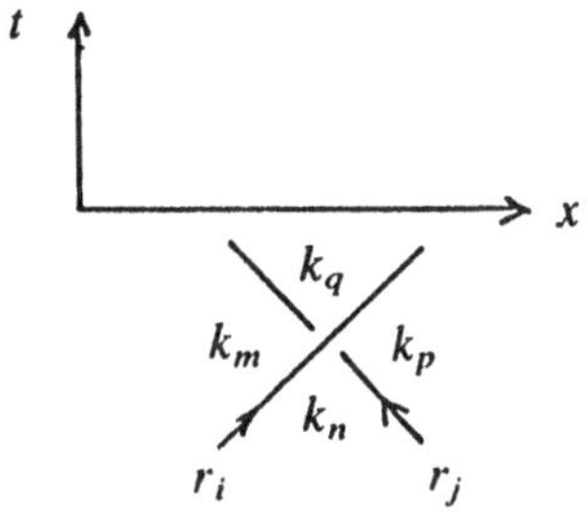

$$\leftrightarrow \bar{R}^{+}(k_m, r_j, r_i, k_p)^{k_n}_{k_q} \qquad (6.88)$$

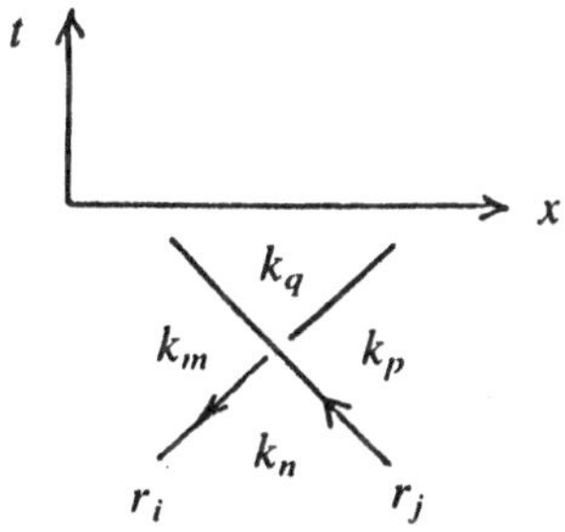

$$\leftrightarrow \bar{R}^{-}(k_m, r_j, r_i^*, k_p)^{k_n}_{k_q} \qquad (6.89)$$

etc.

These rules are chosen in accordance with (4.7). With these assignments, one has that

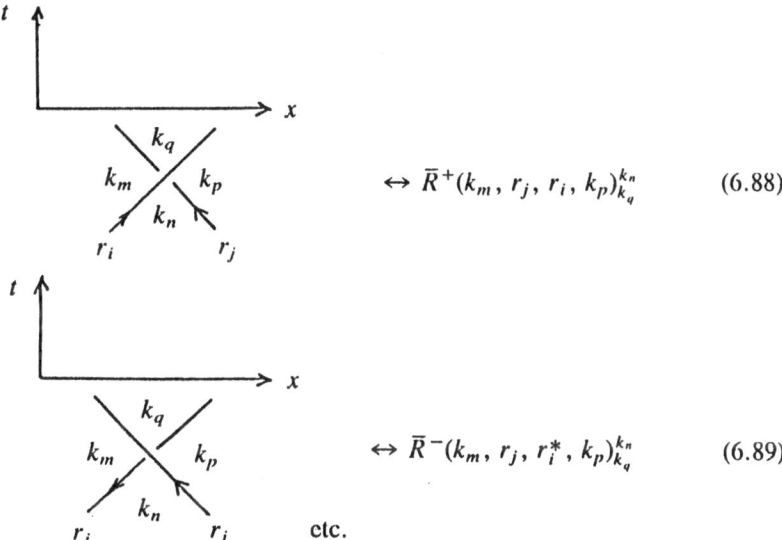

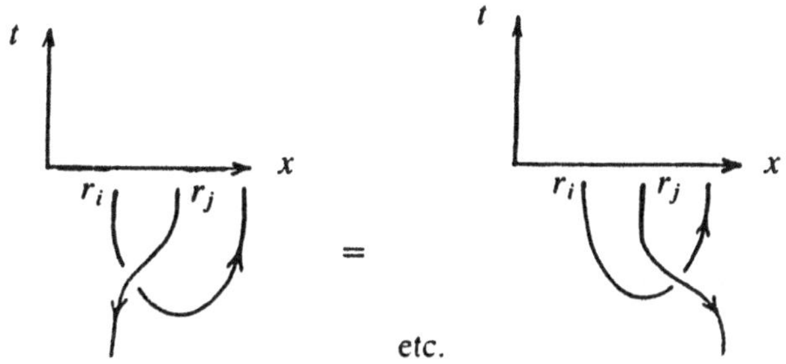

(6.90)

etc.

These identities follow from Eqs. (3.6)–(3.10) if one replaces $R^{\pm}$ by $\bar{R}^{\pm}$ and F by $\bar{F}$. From Eqs. (4.10) and (3.14) one deduces that

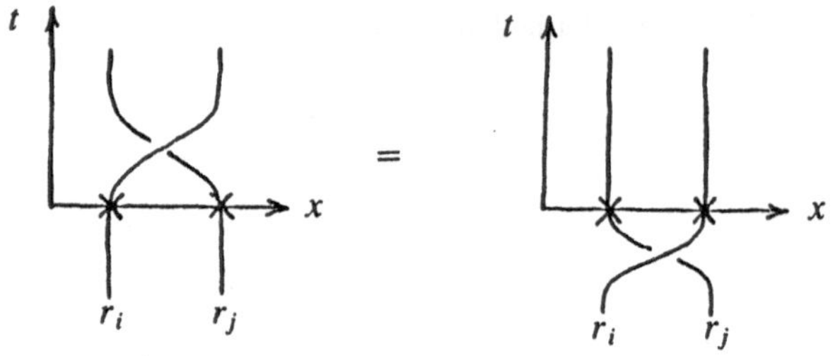

(6.91)

In order to construct link invariants from the present formalism, we require the identity

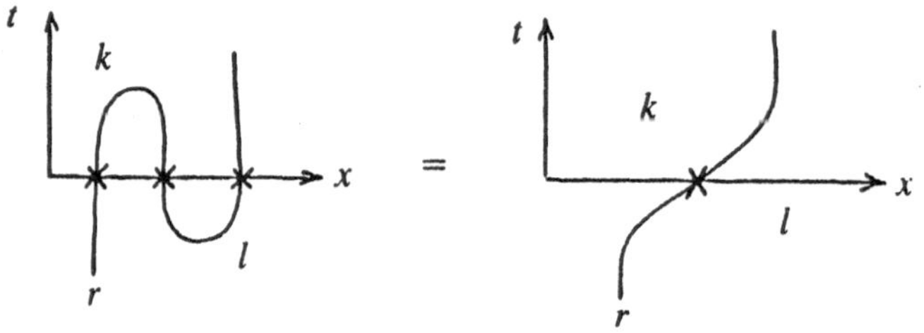

(6.92)

In formulas,

$$D_{krl} F^{k}_{rl} D_{lr^*k} \bar{F}^{l}_{r^*k} D_{krl} = D_{krl} . \tag{6.93}$$

Thus we must normalize the local fields $\phi_{\xi\otimes\bar{\xi}}(z, \bar{z})$ introduced in (4.1) such that

$$F^{k}_{rl} D_{lr^*k} \bar{F}^{l}_{r^*k} D_{krl} = 1 . \tag{6.94}$$

One should verify that this is possible. To do this, we note that, by choosing an appropriate overall normalization for the local field $\phi_{\xi\otimes\bar{\xi}}(z, \bar{z})$, $\xi \in \mathcal{H}_r$, $\bar{\xi} \in \mathcal{H}_r$, $(\phi_{\xi\otimes\bar{\xi}}(z, \bar{z}), \mapsto \phi'_{\xi\otimes\bar{\xi}}(z, \bar{z}) = \lambda_r \phi_{\xi\otimes\bar{\xi}}(z, \bar{z})$, for some $\lambda_r \neq 0)$, we can achieve that

$$F^{r}_{r1} D_{1r^*r} \bar{F}^{1}_{r^*r} D_{rr1} = 1 , \qquad \text{i.e.}$$

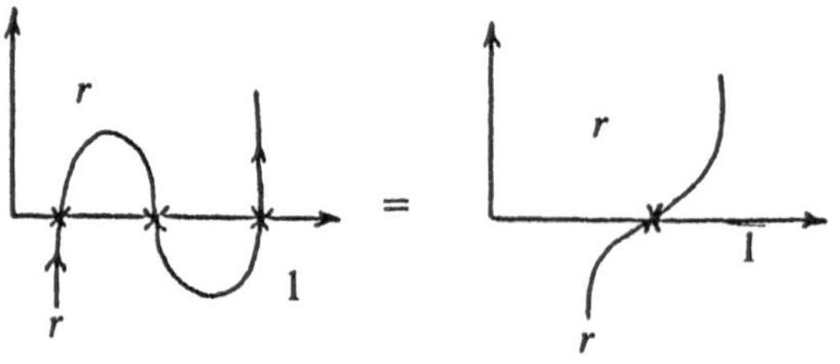

(6.95)

Now we note that

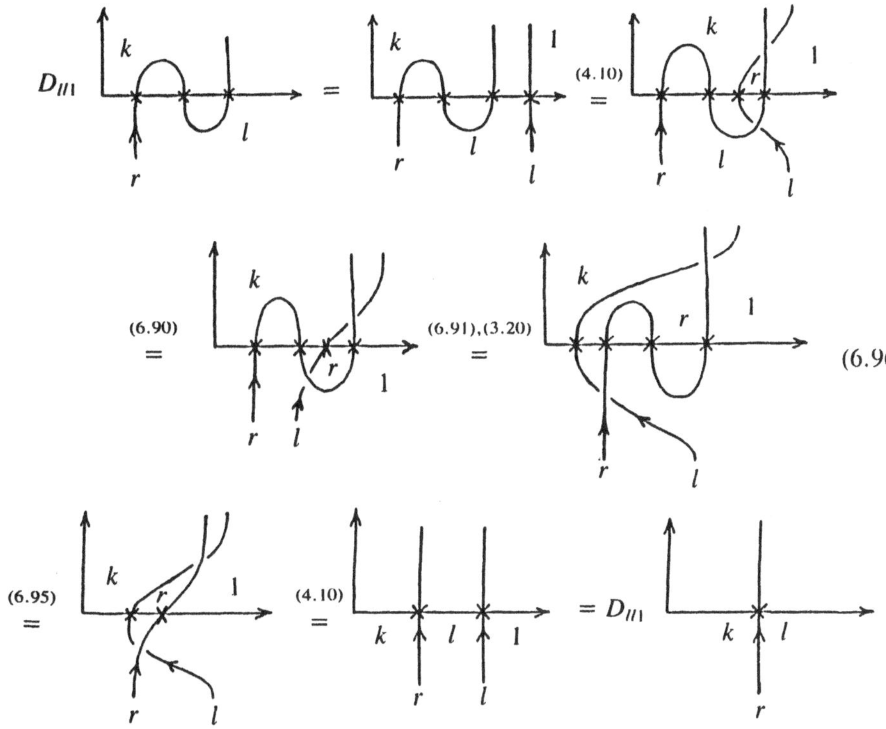

(6.96)

Thus (6.95) implies (6.94), and we conclude that if the local fields $\phi_{\xi\otimes\bar{\xi}}$ are normalized in accordance with (6.95)

$$D_{lr*k}\bar{F}^{l}_{r*k}D_{krl} = (F^{k}_{rl})^{-1} \ . \tag{6.97}$$

Equation (6.97) can be used to calculate the coefficients D_{krl} from the fusion matrices F^k_{rl} and their adjoints and inverses.[21]

Using the modified rules (a)–(d), above, we may define a function $\bar{\nu}$,

$$\bar{\nu}\colon \mathfrak{D}(k_2, \ldots, k_N|r_1, \ldots, r_n|L) \mapsto \bar{\nu}(k_2, \ldots, k_N|r_2, \ldots, r_n|L)$$

as in (6.33), and define the numerator of a decorated link L, $\bar{\nu}(r_1, \ldots, r_n|L)$, as in (6.34). Using (6.91) and (6.97) repeatedly, it is very straightforward to show that

$$\bar{\nu}(r_1, \ldots, r_n|L) = \nu(r_1, \ldots, r_n|L) \ ; \tag{6.98}$$

(convert all elements of type $\bar{R}$ into elements of type R, using (6.91), then apply (6.97)). Thus, we shall not obtain new invariants by our new procedure. Nevertheless, there is a pay-off: The new formalism permits us to interpret the invariants, $I_{S^3}(r_1, \ldots, r_n|L)$ in terms of monodromy of the Green functions

$$G_{\underline{r}}(\underline{z}, \underline{\bar{z}}) \equiv H_{\xi_{s_1}\otimes\bar{\xi}_{s_1}\cdots\xi_{s_m}\otimes\bar{\xi}_{s_m}}(z_1, \bar{z}_1, \ldots, z_m, \bar{z}_m), \tag{6.99}$$

introduced in Sec. 4, (4.16), where s_i or s_i^* belongs to $\underline{r} \equiv \{r_1, \ldots, r_n\}$, and $m = 2n + 2k$, $k = 0, 1, 2, \ldots$.

To see this, consider a representative $\underline{\lambda} = (\lambda_1, \ldots, \lambda_n)$ of a link $L \subset \mathbb{R}^3$ such that its diagram, $\bar{D}(L)$, has properties (1) and (2) required at the beginning of this section; (see (6.83)). Let

$$\underline{z}(t_+) = \underline{\lambda} \cap \{(t, x, y) : t = t_+\},\ t_+ > 0 \tag{6.100}$$

and

$$\underline{\bar{z}}(t_-) = \underline{\lambda} \cap \{(t, x, y) : t = t_-\},\ t_- < 0\,. \tag{6.101}$$

For $\underline{z}(t_+) \equiv (z_1, \ldots, z_m)$, choose $s_1, \ldots, s_m$ such as specified by the labellings, $r_1, \ldots, r_n$, and the orientations of the components, $\lambda_1, \ldots, \lambda_n$, of $\underline{\lambda}$, with $m = 2n + 2k$, $k = 0, 1, 2, \ldots$. Suppose that z_i, z_l are contained in $\lambda_j \cap \{(t, x, y) : t = t_+\}$, with $s_i = r_j$ and $s_l = r_j^*$. We choose ξ_{s_i} and ξ_{s_l} to be invariant vectors in $\mathcal{H}_{r_j}$, $\mathcal{H}_{r_j^*}$, respectively, with $\langle \xi_{s_l} | \xi_{s_i} \rangle = 1$. Similar conventions are imposed on the choice of the vectors $\bar{\xi}_{s_i}$, $i = 1, \ldots, m$.

Define the spaces $M_m^>$ and M_m, $\tilde{M}_m$ as in (2.46), (2.49), respectively. We require that $\underline{z}(0) \in M_m^>$ and $\underline{\bar{z}}(0) = \underline{z}^*(0)$ (the complex conjugate of $\underline{z}(0)$). At the first maximum of the "time"-coordinate, $t(p)$, $p \in \lambda_j$, $j = 1, \ldots, n$, $\underline{z}(t)$ reaches the boundary of M_m. At this point, two chiral vertices, $V_\alpha^{kl}(\xi_{s_i}, z_i)$ and $V_\beta^{pq}(\xi_{s_l}, z_l)$, $\xi_{s_i} \in \mathcal{H}_{r_j}$, $\xi_{s_l} \in \mathcal{H}_{r_j^*}$, $\langle \xi_{s_l} | \xi_{s_i} \rangle = 1$, should be fused according to Eq. (3.1), Sec. 3. But, instead of letting $\underline{z}(t)$ approach ∂M_m and fusing the two chiral vertices into the identity, we keep z_i at a distance $\varepsilon > 0$ from z_l and omit fusion. At later "times", z_i and z_l are kept constant. We introduce this regularization at all subsequent maxima of some $t(p)$, $p \in \lambda_k$, $k = 1, \ldots, n$. Then the regularized path, $\{\underline{z}^{(\varepsilon)}(t) : 0 \leq t \leq t_+\}$, determines a point $\underset{\sim}{Z}^{(\varepsilon)}(t_+) \in \tilde{M}_m$, and we may evaluate the Green function $G_{\underline{r}}$ at $(\underset{\sim}{Z}^{(\varepsilon)}(t_+), \underset{\sim}{\bar{Z}}^{(\varepsilon)}(t_-))$.

Next, we define

$$G_{\underline{r}}^0(\underline{z}, \underline{\bar{z}}) = \left\langle \Omega^* \otimes \Omega^* \middle| \prod_{i=1}^{m} V_{\varepsilon_{s_i}}(z_i) \otimes V_{\varepsilon_{s_i}}(\bar{z}_i)\Omega \otimes \Omega \right\rangle, \tag{6.102}$$

where $m, s_1, \ldots, s_m$ are chosen as in (6.99), $\varepsilon_{s_i} = \pm\sqrt{8\pi h_{s_i}}$, and the sign is chosen according to the orientations of $\lambda_1, \ldots, \lambda_n$. The function $G_{\underline{r}}^0(\underset{\sim}{Z}^{(\varepsilon)}(t_+), \underset{\sim}{\bar{Z}}^{(\varepsilon)}(t_-))$ is defined by analytic continuation of $G_r^0(\underline{z}, \underline{\bar{z}})$ along the paths $\{\underline{z}^{(\varepsilon)}(t) : 0 \leq t \leq t_+\}$ and $\{\underline{\bar{z}}^{(\varepsilon)}(t) : t_- \leq t \leq 0\}$. (We use the *same* ε-regularization to keep $\underline{z}(t)$ inside M_m as above.)

We now define

$$Y_{\underline{r}}^{(\varepsilon)}(t_+, t_-) \equiv \frac{G_{\underline{r}}(\underset{\sim}{Z}^{(\varepsilon)}(t_+), \underset{\sim}{\bar{Z}}^{(\varepsilon)}(t_-))}{G_r^0(\underset{\sim}{Z}^{(\varepsilon)}(t_+), \underset{\sim}{\bar{Z}}^{(\varepsilon)}(t_-))}\,. \tag{6.103}$$

Then one can prove the following result.

Theorem 6.4.

$$I_{S^3}(r_1, \ldots, r_n | L) = \lim_{\varepsilon \downarrow 0} \lim_{\substack{t_+ \to \infty \\ t_- \to -\infty}} Y_{\underline{r}}^{(\varepsilon)}(t_+, t_-) . \qquad (6.104)$$

The *proof* of Theorem 6.4 is an easy consequence of our results in Secs. 2–4. We think that the details are not sufficiently interesting to be dwelled upon, at this point, but see (4.17), (4.27), (2.69), (3.1).

The formalism developed in this section has another pay-off: On the basis of Eqs. (6.91) and (3.6)–(3.15), one can construct invariants for *framed*, closed networks (networks of "ribbons") with vertices of order *three*, embedded in S^3. Although we do not know how useful such invariants are, we believe that our results are new. They will be reported elsewhere. Suffice it to say that Eq. (6.91), i.e.

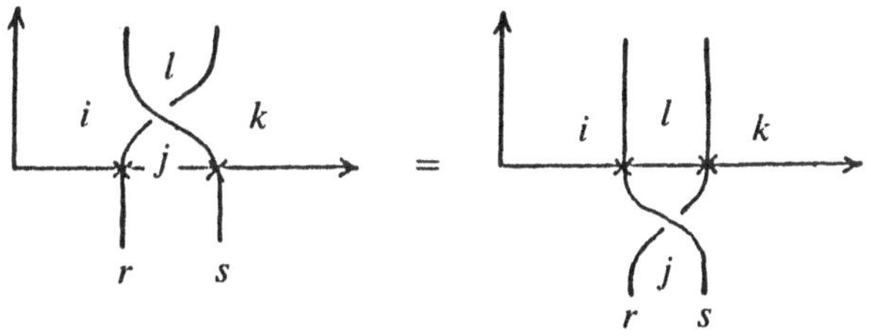

or

$$D_{irj} D_{jsk} R^{\pm}(i, r, s, k)_j^l = D_{lrk} D_{isl} \bar{R}^{\pm}(i, s, r, k)_l^j$$

implies

$$D_{isl}^{-1} D_{lrk}^{-1} R^{\pm}(i, r, s, k)_j^l = D_{jsk}^{-1} D_{irj}^{-1} \bar{R}^{\pm}(i, s, r, k)_l^j$$

which will play an important role in our construction.

6.4. *Link invariants for general three-manifolds*

In this section we sketch how to construct link invariants, $I_{\mathcal{M}}$, for a general class of three-manifolds. Details are deferred to a future publication. Our construction is based on a combination of the results of Secs. 6.1–6.3 with those reviewed in Secs. 3–5, in particular with Eqs. (5.8)–(5.11) and the graphical notations (5.12) and (5.13).

In order to explain the basic ideas, we start with a simple example:

$$\mathcal{M} = \Sigma \times S^1 \qquad (6.105)$$

where $\Sigma = M_n(\underline{w}, \underline{w}', \underline{\theta}, \underline{\alpha})$ is a Riemann surface parametrized as in Sec. 5, (2); see (5.2). For the purpose of calculating invariants for links embedded in $\mathcal{M}$ we may proceed in two ways:

(i) We can follow the approach of Secs. 6.1 and 6.2 and introduce diagrams for links $L \subset \mathcal{M}$, decorated diagrams, and rules for how to associate numbers with decorated link diagrams.

(ii) We can follow the approach of Sec. 6.3 and calculate invariants from the monodromy of conformal blocks or local Green functions of conformal field theories on Σ.

While the first approach is very useful in practical evaluations of link invariants, the second approach is more three-dimensional and might lead to elegant proofs.

We first follow (i). Thus, we must define diagrams for links $L \subset \mathcal{M}$. This can be done as follows: We view $\mathcal{M}$ as arising from $S^2 \times S^1$, where S^2 is the Riemann sphere, by the sewing procedure (2) of Sec. 5. We represent S^2 by a disk with boundary points identified, so that $S^2 \times S^1$ is represented as a full torus in $\mathbb{R}^3$, whose symmetry axis is the y-axis, as explained in Sec. 6.2. Inside the torus, we draw $2n$ loops $l_1, l_1', \ldots, l_n, l_n'$ contained in planes parallel to the (t, x)-plane and at constant distance from the y-axis. They intersect S^2 at the points $w_i, w_i', i = 1, \ldots, n$, introduced in Sec. 5, (2). Arcs on the loop l_i intersecting S^2 in w_i are labelled by a representation $s_i \in L_{\mathcal{A}}$, corresponding arcs on l_i' by s_i^*, (with $s_i = s_i^*$, for $s_i \in i_{\mathcal{A}}$). On every sphere S^2 embedded in the full torus as a disk, the relations (5.2), Sec. 5, (2), are imposed; the representations s_i will be summed over, eventually; (this is recorded by putting s_i and s_i^* in between round parentheses.)

An (oriented or partially oriented) link L in $\mathcal{M}$ can now be represented as a family, $\mathcal{L}$, of (oriented) loops and lines inside the torus, not intersecting $l_1 \cup l_1' \cdots \cup l_n \cup l_n'$. The lines of $\mathcal{L}$ are not necessarily all closed, but may have endpoints inside the annular regions introduced in Sec. 5, (2), at some fixed angle φ, just to reemerge in an annular region equivalent to the former by (5.2), at the same value of φ. In other words, the lines in $\mathcal{L}$ must be closed (oriented) curves inside the torus, *modulo the relations* (5.2).

The lines $l_1, l_1', \ldots, l_n, l_n'$ and those in $\mathcal{M}$ may now be projected onto the (t, x)-plane. Their projection is a shadow, $S(L)$, of $L \subset \Sigma \times S^1$. If under- and overcrossings are recorded we obtain a diagram, $D(L)$, of L. Such a diagram may be decorated according to the rules described in Secs. 6.1 and 6.2, with some modifications that we shall sketch.

Remark. In our drawings we may represent the full torus as a cylinder with top and bottom identified. Hence, the line, $l_1, l_1', \ldots, l_n, l_n'$ will be represented as straight lines parallel to the t-axis.

Consider two lines, l_i and l_i', decorated by representations $s_i \equiv s$ and s^*, respectively. These lines represent the ith handle of Σ. Let $\{a_i, b_i\}_{i=1}^n$ be the standard basis of the homology of Σ, as indicated in the following figure:

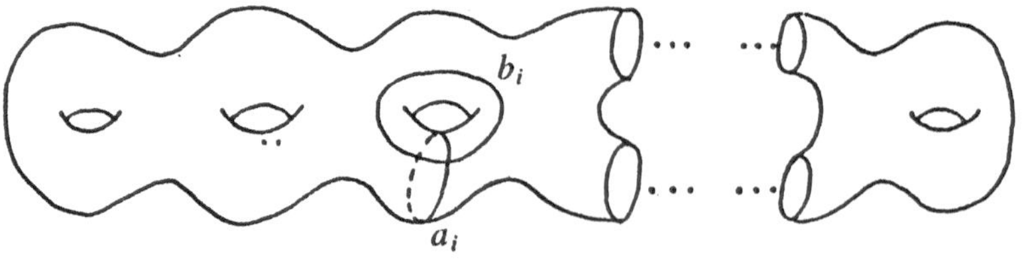

(6.106)

We imagine that the ith handle of Σ arises from the following sewing operation:

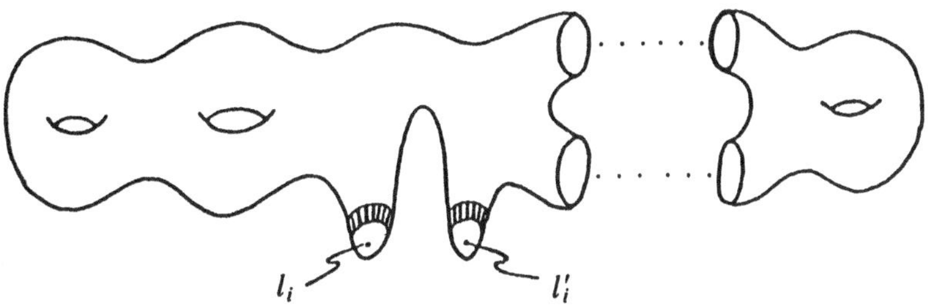

(6.107)

The two shaded annular regions are identified, as prescribed by (5.2). The task is to identify those lines in a link diagram that correspond to loops homologous to $n_i a_i$, $m_i b_i$, n_i, $m_i \in \mathbb{Z}$. (We set $n_i = m_i = \pm 1$ in the following examples.)

A line in a link diagram corresponding to a loop homologous to a_i is represented, graphically, as follows:

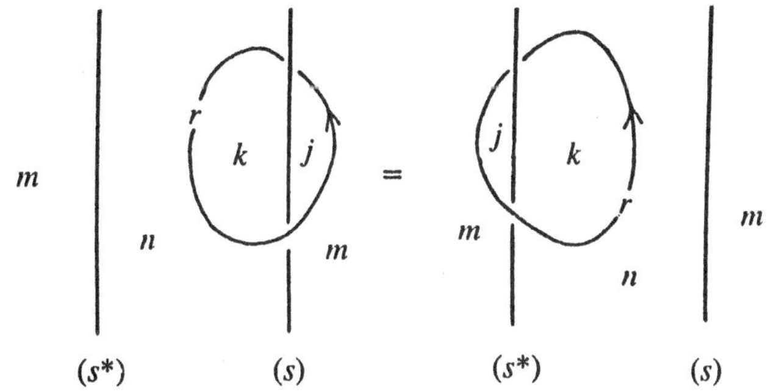

(6.108)

The proof of identity (6.108) is based on some simple algebraic identities between multiplicities and braid matrices. For example,

$$N_{krn} = N_{r^*k^*n} = N_{r^*nk^*} = N_{n^*rk^*} = N_{nr^*k} , \tag{6.109}$$

and

$$R^+(k, r, s, m)^j_n = R^-(m, s^*, r^*, k)^j_n \tag{6.110}$$

which follow easily from the results in Secs. 2 and 3; see e.g. Refs. 18 and 21.

The loop in (6.108) is also equivalent to

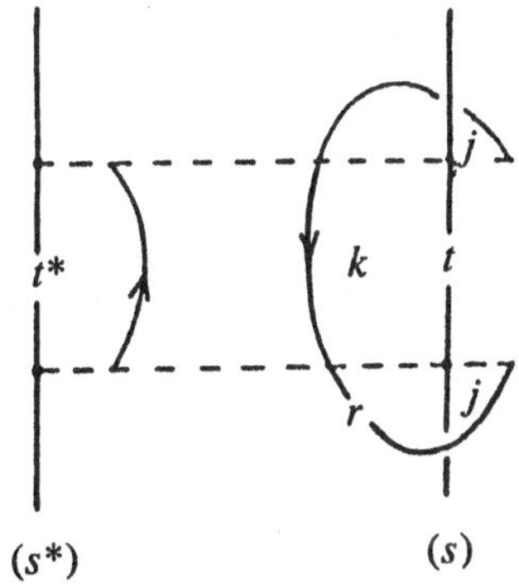

(6.111)

but we shall only use the representation (6.106). Similarly,

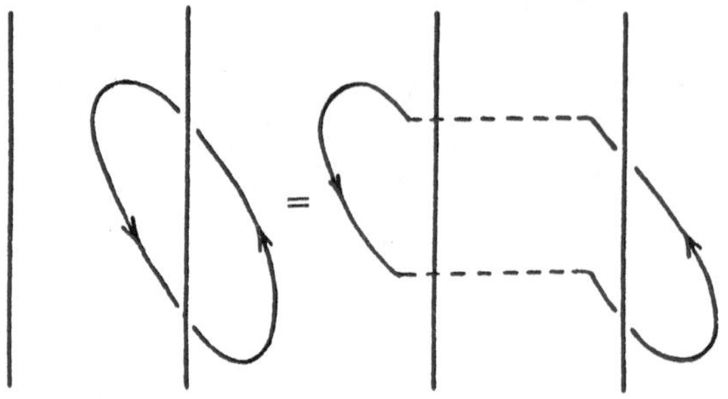

(6.112)

but we shall use the representation corresponding to the left-hand side of (6.112). (Of course, this loop is contractible.)

In these identities it is quite unimportant whether we use the formalism of Sec. 6.2 or the one of Sec. 6.3, but the one of Sec. 6.2 is more natural. (We realize that, in Sec. 5, we have not discussed any details of how to sew conformal blocks, rather than local Green functions. This is not important here.)

A loop of a link, L, tracing out a cycle on Σ homologous to b_i is represented, graphically, by

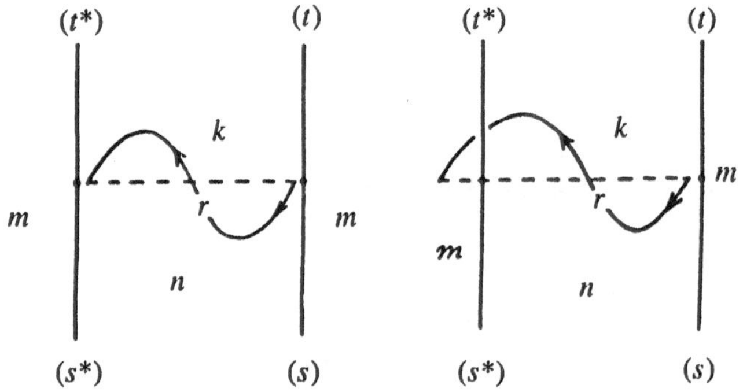

(6.113)

Whether the line bearing the label r is braided with one or both of the vertical lines depends on the position of the line bearing the label r inside the full torus and on the choice of the parameters θ_i and α_i in relation (5.2). Of course, after summing over labels of vertical lines and regions and dividing by the denominator of the link, the dependence of $\underline{\theta}$ and $\underline{\alpha}$ must disappear.

For $i = n = 1$, i.e. when Σ is a torus, loops represented by (6.108) and loops represented by (6.113) are interchanged by the modular transformations S. This yields some useful identities.

More general noncontractible loops in $\mathcal{M}$ are represented by composing loops of type (6.108) and loops of type (6.113).

With the above rules in hand, one can, in principle, calculate the numerator, $\nu_{\mathcal{M}}(r_1, \ldots, r_n | L)$, of a link $L \subset \mathcal{M} \equiv \Sigma \times S^1$, as in Sec. 6.2. (Of course, the representations assigned to segments of the vertical lines $l_1, \ldots, l_n$ must be summed over.) From what we learned in Sec. 6.2 and above it is not very difficult to infer how to calculate the denominator of a link in this more general context. We must bear in mind, though, that certain additional noncontractible loops must be introduced in the calculation of the denominator, δ, as explained in Sec. 6.2, (6.63)–(6.66). The value of the link invariants will depend on the topology of these additional loops relative to the loops in L, as explained in Sec. 6.2. Details will have to be discussed elsewhere.

Next, suppose that the surface Σ arises from sewing two surfaces, M and M':

$$\Sigma = M \underset{\theta,\alpha}{\infty} M' \tag{6.114}$$

According to (5.11), the sewing operation (5.1) can be represented, graphically, as follows:

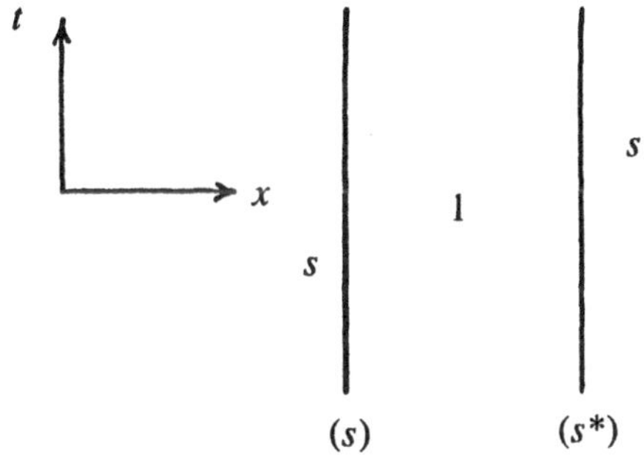

(6.115)

and we have identities, like

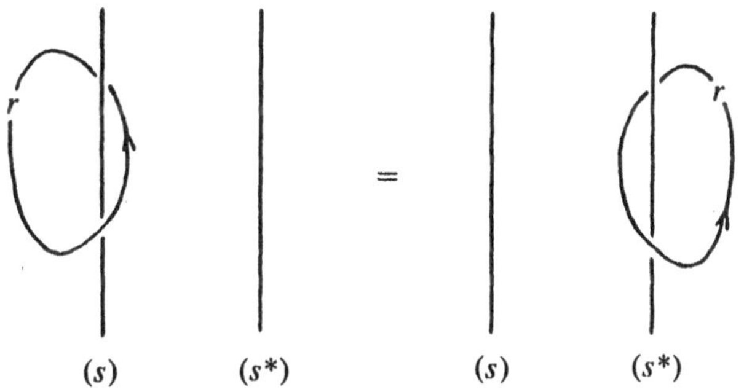

(6.116)

The following diagram represents a loop:

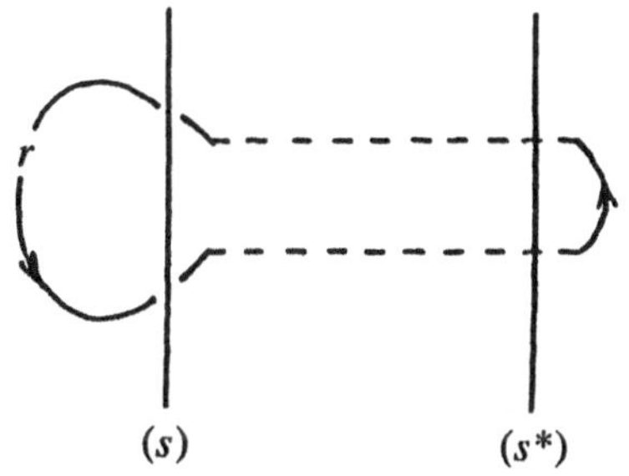

(6.117)

It is quite clear, what the rules are to calculate numerator and denominator of links, in this situation. We shall not enter into details.

Next, we turn our attention to the problem of constructing link invariants for links embedded in general connected, compact three-dimensional manifolds. We choose such a manifold, $\mathcal{M}$. Let f be a (real-valued) Morse function on $\mathcal{M}$. Let $I \subset \mathbb{R}$ be the interval of values of f. For $a \in I$, we define

$$\mathcal{M}_a = \{p \in \mathcal{M} : f(p) \leq a\} \; . \tag{6.118}$$

If a is a point in I not corresponding to a critical point of f, $\mathcal{M}_a$ is a manifold with boundary, $\partial\mathcal{M}_a$, consisting of finitely many, nonsingular Riemann surfaces $\Sigma_a^{(1)}, \ldots, \Sigma_a^{(\nu_a)}$. At values, a, of f corresponding to critical points, some surface $\Sigma_a^{(l)}$ may grow or loose a handle, or two surfaces, $\Sigma_a^{(\alpha)}$ *and* $\Sigma_a^{(\beta)}$ are sewed, as prescribed in (5.1), (with $\theta \to 0$), or a surface $\Sigma_a^{(\alpha)}$ develops a node to split into two surfaces, $\Sigma_a^{(\alpha,1)}$ and $\Sigma_a^{(\alpha,2)}$, at larger values of a. The results sketched in Sec. 5 and the constructions explained above for the example, where $\mathcal{M} = \Sigma \times S^1$, tell us how to construct $\partial\mathcal{M}_a$ by sewing $\nu = \max_{a \in I^{\nu_a}}$ Riemann spheres, using (5.1), and attaching handles, using (5.2). When a is close to $\min_{p\in\mathcal{M}} f(p)$ or to $\max_{p\in\mathcal{M}} f(p)$, $\partial\mathcal{M}_a$ is a sphere.

Diagrams for links embedded in $\mathcal{M}$ are constructed as follows: We consider the strip $\sigma_I = \{(t, x) : t \in I\}$ in the (t, x)-plane. We decompose σ_I into ν strips, $\sigma_1, \ldots, \sigma_\nu$, parallel to the t-axis, separated by $\nu - 1$ strips and two outer regions, all bearing the label 1. During sub-intervals of I, the Riemann sphere corresponding to σ_α may be sewed to the Riemann sphere corresponding to σ_β. This is represented by adding, in that t-interval, a line on the boundary of σ_α bearing a label $s \in L_{\mathcal{A}}$ and a line on $\partial\sigma_\beta$ bearing a label s^*, and summing over s, as explained in (6.115)–(6.117). It is clear that the addition of a handle to a sphere represented by the αth strip, during a certain t-interval $\subseteq I$, corresponds to adding two lines, l and l', inside σ_α, assigning conjugate representations in $L_{\mathcal{A}}$ to segments of l and l', and to sum over these representations, as explained for the example where $\mathcal{M} = \Sigma \times S^1$.

The manifold $\mathcal{M} = S^2 \times S^1$ can, in this general setting, be represented as follows:

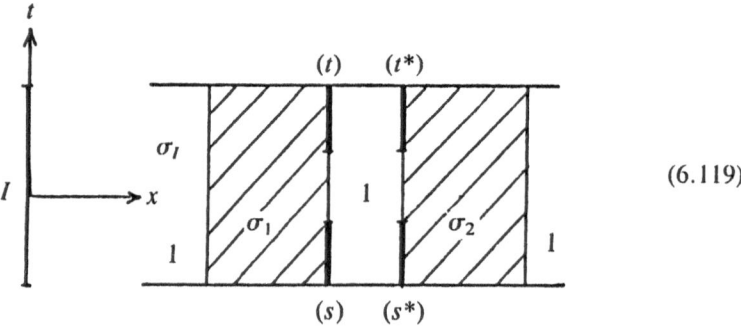

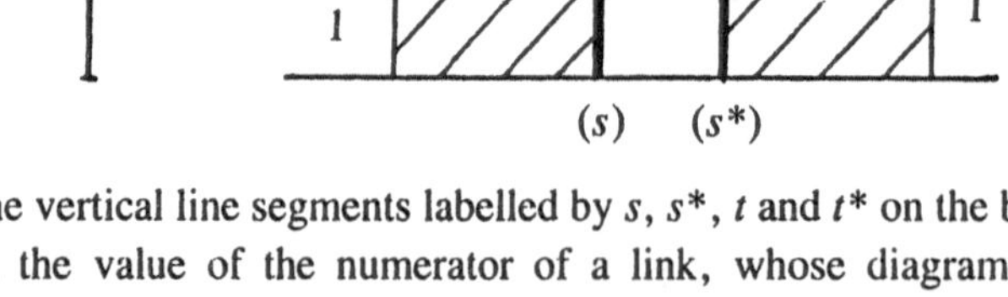

(6.119)

The effect of the vertical line segments labelled by s, s^*, t and t^* on the boundaries of σ_1 and σ_2 is that the value of the numerator of a link, whose diagram is inscribed in $\sigma_1 \cup \sigma_2$, calculated according to the rules (6.115)–(6.117), is identical to the one calculated in Sec. 6.2.

If, during an interval $I_1 \subset I$, a handle is formed the link diagrams must be inscribed in

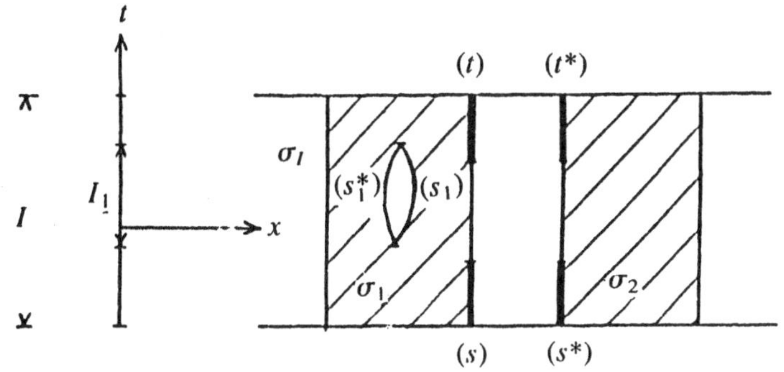

(6.120)

The details of this method to construct link invariants for links embedded in general connected, compact three-manifolds are bound to be lengthy and must be described elsewhere.

Route (ii) mentioned at the beginning of this subsection consists in proving an analogue of Theorem 6.4, Sec. 6.3. This requires constructing conformal blocks and calculating their monodromy on general Riemann surfaces. Although it is known how to deform the conformal structure of a Riemann surface by inserting powers of the energy-momentum tensor T into conformal blocks,[27] the details of this procedure are unimportant, since the monodromy of conformal blocks is not affected by small deformations of the conformal structure. On the basis of (a slight extension of) the results in Sec. 5, one can probably establish a generalization of Theorem 6.4, Sec. 6.3.

7. Relations to Quantum Groups, Outlook

The purpose of this last section is to outline the relations between the approach to constructing link invariants developed in this paper which is based on the representation theory of chiral algebras and an approach based on the representation theory of quantum groups, explored in papers by Reshetikhin.[32] These relations have already been clarified in examples: If the chiral algebra, $\mathscr{A}$, is chosen to be the Kac-Moody algebra, $\widehat{\mathrm{su}}(n)_k$, at level k and the quantum group is $U_q(\mathrm{su}(n))$, for some $q = q(k)$ which is a root of unity if k is an integer then our invariants agree with those constructed by Reshetikhin.[32] One way of showing this is to compare the approach developed in Sec. 6.3 with the work of Knizhnik and Zamolodchikov,[33] as refined by Tsuchiya and Kanie,[34] which is based on analyzing the so-called Knizhnik-Zamolodchikov differential equations for the conformal blocks of the Wess-Zumino-Witten models. The braid (R-) matrices and the fusion (F-) matrices for the conformal blocks of the W-Z-W models can be calculated from the Knizhnik-Zamolodchikov equations; see e.g. Ref. 18, 19 and 34. From the work of Belavin and Drinfel'd and others[35] it is known that the R- and F-matrices obtained from the Knizhnik-Zamolodchikov equations for SU(n) W-Z-W models are precisely those that appear in the representation theory of the corresponding q-deformations of su(n) if $q = q(k)$ is chosen correctly. Of course this is not an accident. We thus propose to briefly sketch some tentative connections between the representation theory of Kac-Moody algebras (or, more generally, chiral algebras) and the representation theory of quantum groups.

Let $U_q \equiv U_q(g)$ be the q-deformation of a classical Lie algebra g. The finite-dimensional, irreducible representations of U_q can be labelled by some set of highest weights of g. (If q is a root of unity then there are usually only finitely many finite-dimensional, irreducible representations of U_q.) Let L_{U_q} be a list of all finite-dimensional, irreducible representations, with $1 \in L_{U_q}$ denoting the trivial representation of U_q. Let $\Delta: U_q \to U_q \otimes U_q$ denote comultiplication on U_q. Given i and j in L_{U_q}, we define a tensor product representation, $i \otimes j$, of U_q by setting, for each generator A of U_q,

$$A \to (i \otimes j)(\Delta(A)) . \tag{7.1}$$

See Refs. 32, 36 and 37. If q is not a root of unity the tensor product representation $i \otimes j$ can be decomposed into a direct sum of finite-dimensional irreducible representations,[f]

$$i \otimes j \simeq \bigoplus_{k \in L_{U_q}} \bigoplus_{\alpha=1}^{N_{kij}} k^{(\alpha)} , \tag{7.2}$$

where $k^{(\alpha)} \simeq k \in L_{U_q}$, $\alpha = 1, \ldots, N_{kij}$, and N_{kij} is the multiplicity of k in $i \otimes j$. The multiplicities N_{kij} have the same properties, (2.65)–(2.68), as the multiplicities appearing in the representation theory of a chiral algebra, $\mathcal{A}$.

Let V^i be the representation space of $i \in L_{U_q}$. Given i and j, let $K_k^{ij}(q, \alpha)$ be the projection from $V^i \otimes V^j$ onto $V^{k(\alpha)}$, $\alpha = 1, \ldots, N_{kij}$; K is called the matrix of Clebsch-Gordan coefficients.[32] Let $x_i \in V^i$, $x_j \in V^j$ and $x_k \in (V^k)^*$, the dual space of V^k. We define a tensor operator, $v_\alpha^{kj}(x_i)$ by setting

$$\langle x_k | v_\alpha^{kj}(x_i) x_j \rangle \equiv \langle x_k | K_k^{ij}(q, \alpha) x_i \otimes x_j \rangle . \tag{7.3}$$

Clearly

$$\begin{aligned} v_{m,\alpha,\beta}^{kj}(x_r \otimes x_s) x_j &\equiv v_\alpha^{km}(x_r) v_\beta^{mj}(x_s) x_j \\ &= K_k^{rm}(q, \alpha) x_r \otimes (K_m^{sj}(q, \beta) x_s \otimes x_j) . \end{aligned} \tag{7.4}$$

For i and j in L_{U_q}, let

$$P^{ij}: V^i \otimes V^j \to V^j \otimes V^i$$

denote the permutation operator. Let $R \in U_q \otimes U_q$ be the universal R-matrix of U_q; (note that R is a Yang-Baxter operator[32]). We set

[f] The representation theory when q is a root of unity does not appear to be worked out completely, yet. Of course, this is the case of main interest to us.

$$R^{ij} = P^{ij}(i \otimes j)(R) ; \tag{7.5}$$

R^{ij} is a linear map from $V^i \otimes V^j$ to $V^j \otimes V^i$. The R-matrices $\{R^{ij}\}_{i,j \in L_{U_q}}$ satisfy the Yang-Baxter equation (with spectral parameter at ∞), so that they generate representations of the colored braid groups, and have the property that

$$j \otimes i(\Delta(A))R^{ij} = R^{ij} i \otimes j(\Delta(A)) . \tag{7.6}$$

The matrices R^{ij} and $K_j^{mn}(q, \alpha)$ satisfy algebraic equations analogous to (3.10) and (3.13)[32]:

$$R^{ij} K_j^{mn}(q, \alpha) = K_j^{mn}(q, \alpha) R^{ln} R^{lm} ; \tag{7.7}$$

$$K_j^{mn}(q, \alpha) R^{nm} = (-1)^{\bar{j}} q^{(c(j)-c(m)-c(n))/4} K_j^{nm}(q, \alpha) , \tag{7.8}$$

where $\bar{j}$ is the parity of j in $m \otimes n$, and $c(j)$ is the eigenvalue of the Casimir operator of g on $\mathcal{V}^j$, the representation space of the irreducible representation j of g.

From Eqs. (7.3), (7.7) and (7.8) one can, in principle, determine the quadratic relations in the algebra of tensor operators $v_\alpha^{kj}(x_r)$, $x_r \in V^r$. They must have the form

$$v^{kj}_{m,\alpha,\beta}([R^{rs}]^{\pm 1} x_r \otimes x_s) = \sum_{n,\gamma,\delta} R^{\pm}(k, r, s, j)^{n\gamma\delta}_{m\alpha\beta} v^{kj}_{n,\gamma,\delta}(x_s \otimes x_r) . \tag{7.9}$$

Clearly, $v^{kj}_{m,\alpha,\beta}(x_r \otimes x_s)$ is a linear map from V^j to V^k. It can be expanded in a sum of operators $v_\delta^{kj}(x_p)$, $x_p \in V^p$. This expansion has the form

$$v^{kj}_{m,\alpha,\beta}(x_r \otimes x_s) = \sum_p F(k, r, s, j)^{p\gamma\delta}_{m\alpha\beta} v_\delta^{kj}(K_p^{rs}(q, \gamma) x_r \otimes x_s) . \tag{7.10}$$

For $g = \mathrm{su}(2)$, the connection between the matrices $[R^{rs}]^{\pm 1}$ and $R^{\pm}(k, r, s, j)^{n\gamma\delta}_{m\alpha\beta}$, and between $K_p^{rs}(q, \gamma)$ and $F(k, r, s, j)^{p\gamma\delta}_{m\alpha\beta}$, is given by the so-called vertex-SOS transformation.[38] For more general quantum groups, this transformation remains to be worked out in more detail.

Formulas (7.9) and (7.10) suggest a precise connection between the representation theory of quantum groups briefly reviewed above and the representation theory of chiral algebras, as developed in Secs. 2 and 3: It is tempting to imagine that the matrices $R^{\pm}(k, r, s, j)^{n\gamma\delta}_{m\alpha\beta}$ and $F(k, r, s, j)^{p\gamma\delta}_{m\alpha\beta}$ are in one-one correspondence with the braid matrices $R^{\pm}$ introduced in (2.58), (2.59) and the fusion matrices F introduced in (3.1), respectively; i.e. given a chiral algebra, $\mathcal{A}$, and a $\otimes$-invariant set, $L_{\mathcal{A}}$, of lowest-weight representations of $\mathcal{A}$, one can find a quantum group, U_q, with a family, L_{U_q}, of finite-dimensional, irreducible highest-weight representations, and a map $\wedge$

$$\begin{aligned} \wedge : L_{\mathcal{A}} &\to L_{U_q} \\ j &\mapsto \hat{j} \\ 1 &\mapsto 1 \end{aligned} \tag{7.11}$$

such that

$$N_{kil} = N_{\hat{k}\hat{i}\hat{l}} \qquad (7.12)$$

where N_{kil} are the multiplicities introduced in Definition 2.2, (3), and $N_{\hat{k}\hat{i}\hat{l}}$ are the multiplicities appearing in Eq. (7.2).

Furthermore, the braid matrices of Eq. (2.58) and (2.59) are related to those in Eq. (7.9) by

$$R^{\pm}(k, r, s, j)^{n\gamma\delta}_{m\alpha\beta} = R^{\pm}(\hat{k}, \hat{r}, \hat{s}, \hat{j})^{\hat{n}\gamma\delta}_{\hat{m}\alpha\beta} ; \qquad (7.13)$$

and, similarly, the fusion matrices in Eq. (3.1) are related to those in Eq. (7.10) by

$$F(k, r, s, j)^{p\gamma\delta}_{m\alpha\beta} = F(\hat{k}, \hat{r}, \hat{s}, \hat{j})^{\hat{p}\gamma\delta}_{\hat{m}\alpha\beta} . \qquad (7.14)$$

While this connection appears to be realized in the examples of Kac-Moody algebras $\widehat{\mathrm{su}}(n)_k$ and the corresponding q-deformations of su(n), it is conceivable that, in general, the braid matrices of a chiral algebra $\mathcal{A}$ may differ from those of a corresponding quantum group by phase factors corresponding to *abelian representations* of the braid groups. In this case, the F-matrices would also differ by phase factors. A general solution to the problem of associating some quantum group to a pair, $(\mathcal{A}, L_{\mathcal{A}})$, of a chiral algebra $\mathcal{A}$ and a $\otimes$-invariant set of irreducible, lowest-weight representations of $\mathcal{A}$ (obeying certain conditions that remain, as yet, to be spelled out precisely) would require an extension of Tannaka-Krein theory perhaps somewhat analogous to the one discovered by Doplicher and Roberts.[39] This would elucidate the role played by quantum groups in conformal field theory (and other quantum field theories in two and three space-time dimensions with braid group statistics[10]): Quantum groups would play the role of "internal symmetries", in analogy to the role played by internal symmetry groups ("gauge groups of the first kind") in conventional local quantum field theory.[29] An idea like this was presumably first proposed by Woronowicz.[40]

The action of the quantum group, U_q, associated to a pair $(\mathcal{A}, L_{\mathcal{A}})$ on the elements of $\mathcal{A}$ would be trivial, while the chiral vertices constructed from $(\mathcal{A}, L_{\mathcal{A}})$ would transform nontrivially under U_q; (one would explicitly introduce a dependence of chiral vertices associated with $(\mathcal{A}, L_{\mathcal{A}})$ on vectors in representation spaces of U_q).

If this theory can be developed it might have an important application: One could use the representation theory of the quantum group U_q associated to $(\mathcal{A}, L_{\mathcal{A}})$ to determine the braid- and fusion matrices of the chiral vertices associated with $(\mathcal{A}, L_{\mathcal{A}})$ and then attempt to calculate the conformal blocks, see (4.14), by solving a generalized Riemann-Hilbert problem for the given R- and F-matrices.

Acknowledgments

The interest of J. F. in the subject of this paper was triggered by discussions with A. Connes and, especially, V. F. R. Jones during stay at I.H.É.S., in spring 1987. He wishes to thank them for inspiring remarks and encouraging interest in our efforts. He also thanks the I.H.É.S. for hospitality during various stages of the work that has led to this paper. We are much indebted to G. Felder and G. Keller for their interest and their invaluable help in learning conformal field theory. Sections 2–5 are based on joint work with them. J. F. thanks E. Trubowitz for some useful comments on the use of Morse functions in topology.

Note added in proof

(1) Some of the material in this paper (parts of Secs. 2–4) was presented by one of us (J.F.) at the Erice school on *Constructive Quantum Field Theory II*, Summer 1988.

(2) The relations between S-matrix and fusion matrices, see (6.82), had also been established by G. Moore and N. Seiberg.

(3) After submitting this paper for publication we learnt of the work by Reshetikhin and Turaev on invariants for networks of ribbons with vertices of order three and invariants for general three-manifolds that appears to overlap with the work sketched at the end of Sec. 6.3 and in Sec. 6.4. In order to construct invariants for general three-manifolds they propose to make use of Kirby calculus. This was also suggested to us by Louis Kauffman. L. Crane has proposed an alternative approach, involving the use of Hegaard splittings.

All this work is inspired by Witten's paper.[3]

It would be interesting to clarify the relations between Witten's invariants and the Floer cohomology groups.

(4) Ideas related to those in Sec. 7 have independently been described by Alvarez-Gaumé *et al.*; Moore and Reshetikhin; Buchholz, Mack and Todorov; and Ganchev and Petkova.

References

1. Basic references on conformal field theory:
   A. A. Belavin, A. M. Polyakov and A. B. Zamolodchikov, *Nucl. Phys.* **B241** (1984) 333;
   D. Friedan, Z. Qiu and S. Shenker, *Phys. Rev. Lett.* **52** (1984) 455; *Commun. Math. Phys.* **107** (1986) 535.
   Monodromy of conformal blocks and braid group representations:
   Vl.S. Dotsenko and V. A. Fateev, *Nucl. Phys.* **240 [FS12]** (1984) 312; *Nucl. Phys.* **B251 [FS13]** (1985) 691; *Phys. Lett.* **154B** (1985) 291;
   V. G. Knizhnik and A. B. Zamolodchikov, *Nucl. Phys.* **B247** (1984) 83;
   A. Tsuchiya and Y. Kanie, *Lett. Math. Phys.* **13** (1987) 303;
   J. Fröhlich, "Statistics of fields, the Yang-Baxter equation, and the theory of knots and links", 1987 Cargèse lectures, in *Non-Perturbative Quantum Field Theory*, eds. G't Hooft *et al.* (Plenum Press, New York, 1988);
   G. Moore and N. Seiberg, *Phys. Lett.* **B212** (1988) 451.
2. The revolution in knot theory was initiated in:
   V.F.R. Jones, *Bull. AMS* **12** (1985) 103; *Ann. Math.* **126** (1987) 335.
   Background from the theory of operator algebras can be found in:
   V. F. R. Jones, *Invent. Math.* **72** (1983) 1; *Braid Groups, Hecke Algebras and type II_1 Factors*, Proceedings of Japan-US Conf. on Operator Algebras, 1983.

General texts on knot theory:
D. Rolfsen, *Knots and Links*, Publish or Perish, Math. Lecture Series, 1976;
G. Burde and H. Zieschang, *Knots*, de Gruyter Studies in Mathematics 5.
Recent papers on link-invariants:
P. Freyd, D. Yetter, J. Hoste, W. B. R. Lickorish, K. Millet and A. Ocneanu, *Bull. AMS* **12** (1985) 239;
L. H. Kauffman, *Topology* **26** (1987) 395.
Y. Akutsu and M. Wadati, *J. Phys. Soc. Japan* **56** (1987) 839, 3039;
Y. Akutsu, T. Deguchi and M. Wadati, *J. Phys. Soc. Japan* **56** (1987) 3464; **57** (1988) 757, 1905;
V. F. R. Jones, "On knot invariants related to some statistical mechanical models," Berkeley Preprint, 1988;
V. G. Turaev, "The Yang-Baxter equation and invariants of links", LOMI Preprint E-3-87;
N. Yu. Reshetikhin, "Quantized universal enveloping algebras, the Yang-Baxter equation and invariants of links" I, II, LOMI Preprints E-4-87, E-17-87.

3. E. Witten, "Quantum field theory and the Jones polynomial", Proc. IAMP Conf. 1988, to appear in *Int'l J. Mod. Phys. A*.
4. M. F. Atiyah, "New invariants of three- and four-dimensional manifolds" in: "The Mathematical Heritage of Hermann Weyl, ed. R. Wells, Providence R. I.: (AMS Publ., 1988).
5. J. Fröhlich and C. King, "The Chern-Simons theory and knot polynomials", Preprint ETH-TH/89-10, to appear in *Commun. Math. Phys.*
6. G. Felder, J. Fröhlich and G. Keller, *Commun. Math. Phys.*, in press; G. Moore and N. Seiberg, "Naturality in conformal field theory", Preprint IAS SNS-HEP-88/31; J. Fröhlich, unpublished.
7. A. A. Belavin, A. M. Polyakov and A. B. Zamolodchikov, *Nucl. Phys.* **B241** (1984) 333.
8. V. G. Knizhnik and A. B. Zamolodchikov, *Nucl. Phys.* **B247** (1984) 83; P. Goddard, A. Kent and D. Olive, *Phys. Lett.* **152B** (1985) 88; *Commun. Math. Phys.* **103** (1986) 105.
9. A. B. Zamolodchikov, *Theor. Math. Phys.* **65** (1985) 1205; H. Eichenherr, *Phys. Lett.* **151B** (1985) 26; A. B. Zamolodchikov and V. A. Fateev, *Sov. Phys. JETP* **62** (1985) 215; **63** (1985) 913.
10. J. Fröhlich, "Statistics of fields, the Yang-Baxter equation and the theory of knots and links", 1987 Cargèse lectures, in: *Non-Perturbative Quantum Field Theory*, eds. G. 't Hooft *et al.* (Plenum Press, New York, 1988).
11. G. Felder, J. Fröhlich and G. Keller, "On the structure of unitary conformal field theory I: existence of conformal blocks", *Commun. Math. Phys.*, in press, and "On the Structure . . . II: Representation Theoretic Approach", to appear in *Commun. Math. Phys.*
12. G. Moore and N. Seiberg, "Classical and quantum conformal field theory", Preprint IAS SNS-HEP-88/39, to appear in *Commun. Math. Phys.*
13. D. Friedan, Z. Qiu, and S. Shenker, *Phys. Rev. Lett.* **52** (1984) 455; see also Ref. 11.
14. G. Felder, K. Gawędzki, and A. Kupiainen, *Nucl. Phys.* **B299** (1988) 355; *Commun. Math. Phys.* **117** (1988) 127; and references given there.
15. G. Felder, "BRST approach to minimal models", *Nucl. Phys. B*, in press.
16. J. Birman, "Braids, links and mapping class groups", *Ann. Math. Studies* **82**, (Princeton University Press, Princeton; 1974).
17. Vl. S. Dotsenko and V. A. Fateev, *Nucl. Phys. B* **240** [**FS12**] (1984) 312; **B251** [**FS13**] (1985) 691; *Phys. Lett.* **154B** (1985) 291.
18. G. Felder, J. Fröhlich and G. Keller, "Braid Matrices and structure constants for minimal models", to appear in *Commun. Math. Phys.*
19. D. Bernard and G. Felder, ETH-preprint 1989.

20. M. Jimbo, *Lett. Math. Phys.* **10** (1985) 63; *Lett Math. Phys.* **11** (1986) 247; V. G. Drinfel'd, "Quantum Groups", in *Proceedings of ICM*, Berkeley 1986, ed. A. M. Gleason, Providence R.I.: (AMS Publ., 1987); M. Rosso, *Commun. Math. Phys.* **117** (1988) 581; N. Yu. Reshetikhin, "Quantized universal enveloping algebras, the Yang-Baxter equation and invariants of links, I, II", LOMI Preprints E-4-87, E-17-87.
21. G. Felder, J. Fröhlich and G. Keller, "On the structure . . . , II"; see Ref. 11.
22. K.-H. Rehren and B. Schroer, "Einstein causality and Artin braids", Berlin Preprint FU-88-0439.
23. G. Moore and N. Seiberg, *Phys. Lett.* **B212** (1988) 451.
24. Ref. 17; and
P. Di Francesco, "Structure constants for rational conformal field theories"; Preprint, Saclay PhT-88/139; Ref. 18.
25. J. Fröhlich, *Commun. Math. Phys.* **47** (1976) 233.
26. D. Friedan and S. Shenker, *Nucl. Phys.* **B281** (1987) 509; T. Eguchi and H. Ooguri, *Nucl. Phys.* **B282** (1987) 308; C. Vafa, *Phys. Lett.* **B190** (1987) 47; *Phys. Lett.* **B199** (1987) 195; L. Alvarez-Gaumé, C. Gomez, G. Moore and C. Vafa, *Nucl. Phys.* **B303** (1988) 455; P. West, "A review of duality, string vertices, overlap identities and the group theoretical approach to string theory, Preprint CERN Th 4819/87.
27. H. Sonoda, "Sewing conformal field theories I, II," Preprints LBL-25140, LBL-25316, 1988.
28. J. Fröhlich, Notes for seminars at I.H.É.S. and at École Polytechnique (Palaiseau), 1988.
29. G. Felder and R. Silvotti, "Modular covariance of minimal model correlation functions", Preprint IAS SM 1988.
30. N. Kuiper, *Math. Ann.* **278** (1987) 193–209.
31. D. Rolfsen, *Knots and Links*, Publish or Perish, Math. Lecture Series, 1976; G. Burde and H. Zieschang, *Knots*, de Gruyter Studies in Mathematics **5**.
32. N. Yu. Reshetikhin, see Ref. 20.
33. V. G. Knizhnik and A. B. Zamolodchikov, see Ref. 8.
34. A. Tsuchiya and Y. Kanie, *Lett. Math. Phys.* **13** (1987) 303.
35. A. A. Belavin and V. G. Drinfel'd, *Funct. Anal. Appl.* **16** (1982) 1; **17** (1983) 69.
36. M. Jimbo, see Ref. 20.
37. M. Rosso, see Ref. 20.
38. V. Pasquier, *Commun. Math. Phys.* **118** (1988) 355.
39. S. Doplicher and J. Roberts, "C*-algebras and duality for compact groups . . .", in *Proceedings of VIIIth International Congress on Mathematical Physics*, ed. M. Mebkhout and R. Sénéor (World Scientific, 1989); see also: S. Doplicher, R. Haag and J. E. Roberts, *Commun. Math. Phys.* **23** (1971) 199; **35** (1974) 49.
40. S. Woronowicz, private communication (to J. F.), 1987.

RENORMALIZATION THEORY AND THE TREE EXPANSION

Lon Rosen

Mathematics Department
University of British Columbia
Vancouver, B.C., Canada

Perturbative renormalization theory may not be a topic of CQFT, but it does have a long and rich history and is the basis of every textbook account of QFT. Still, few students really understand it. This is hardly surprising since no text really explains it. I believe that it <u>is</u> possible to give a simple and complete account of perturbative renormalization, and it is fitting that I try to do so in a school on CQFT; for the approach I shall advocate shares a number of ideas with current work in CQFT, and an understanding of the cancellation of ∞'s in perturbation theory is important for an understanding of what happens in the "small field region" of CQFT.

What has rendered renormalization theory comprehensible (for me) is the tree expansion developed by Gallavotti and Nicolò[1,2] on the basis of renormalization group ideas going back to Wilson. The GN tree expansion will be the starting point for these lectures. The main reference for my talks is the book[3] by Joel Feldman, Tom Hurd, Jill Wright and myself, in which we extend the GN method to general models involving massless particles and gauge symmetries (at least for the abelian case of QED).

Our renormalization scheme can be carried out in a variety of ways and situations. I shall work in Euclidean x-space where there are the fewest technical difficulties, and I'll use Feynman α-parameters and Wick ordering since I think there's a gain in clarity. The proof of ultraviolet- (UV-) renormalizability given in Theorem 6 will be essentially complete. As for the complications of massless particles and gauge invariance, I shall content myself with a few remarks at the end and urge you to consult Ref. 3 for full details.

Constructive Quantum Field Theory II, Edited by G. Velo and A. S. Wightman, Plenum Press, New York, 1990

STATEMENT OF THE PROBLEM

Consider a Euclidean quantum field theory (EQFT) in $\mathbf{R}^d$, d > 2, with Lagrangian density $\mathcal{L}(\Phi(x))$, a local polynomial in the fields $\Phi = (\Phi_1,\ldots,\Phi_N)$ and their derivatives:

$$\mathcal{L} = \mathcal{L}_0 + \lambda\mathcal{L}_{int}$$

where $\mathcal{L}_0$ is quadratic, $\mathcal{L}_{int}$ has degree ≥ 3 and λ is the coupling constant. $\mathcal{L}_0$ also depends linearly on physical parameters $\vec{Z}$ such as masses and field strengths.

<u>Examples</u>

1. ϕ_d^{2n} model $\quad \Phi = \phi : \mathbf{R}^d \to \mathbf{R}$

$$\mathcal{L}_0 = \frac{1}{2}\,(Z_0(\partial\phi)^2 + m^2\phi^2), \qquad \mathcal{L}_{int} = :\phi^{2n}: \tag{1}$$

where $Z_0 > 0$ is the field strength and m ≥ 0 the mass. (Wick monomials such as $:\phi^{2n}:$ are defined in (8) below.)

2. QED_4 $\quad \Phi = (\Phi_1,\Phi_2,\Phi_3) = (A,\psi,\bar{\psi})$

where the vector field $A = (A_\mu)_{\mu=1,\ldots,4}$ is the photon field and the anticommuting spinor fields $(\psi_\nu)_{\nu=1,\ldots,4}$ and $(\bar{\psi}_\nu)$ are the electron-positron fields. (I shall generally suppress vector and spinor indices.) In Feynman gauge,

$$\mathcal{L}_0 = \frac{1}{2}\, Z_0(\partial A)^2 + \bar{\psi}(-Z_1 i\not{\partial}+m)\psi, \qquad \mathcal{L}_{int} = :\bar{\psi}\not{A}\psi: \tag{2}$$

where $(\partial A)^2 = \sum_{\mu,\nu} (\partial_\nu A_\mu)^2$, $\not{\partial} = \partial_\nu\gamma^\nu$, the 4 x 4 Euclidean Dirac matrices satisfy

$$\gamma^\mu\gamma^\nu + \gamma^\nu\gamma^\mu = -2\delta^{\mu\nu},$$

and m > 0 is the electron mass.

The central problem of CQFT is to construct a measure on the space of fields corresponding to the formal expression

$$d\nu(\Phi) = \text{const.}\ e^{-\int\mathcal{L}dx}\mathcal{D}\Phi \tag{3}$$

where $\mathcal{D}\Phi = \prod_{i,x} d\Phi_i(x)$ and const. $= (\int e^{-\int\mathcal{L}dx}\mathcal{D}\Phi)^{-1}$ to that dν is normalized. Let E(J) be the generating functional for dν:

$$E(J) = \int e^{(\Phi,J)}d\nu(\Phi) \tag{4}$$

where $(\Phi,J) = \sum_i \int \Phi_i(x)J_i(x)dx$ and the source functions $J_i(x)$ are in Schwartz space and have the same nature as the corresponding field Φ_i; in particular, J_i is fermionic if Φ_i is. Unlike the other braver lecturers at this school I won't analyze E(J) in the constructive sense, but rather in the perturbative sense: my goal will be to make sense of (4) as a formal power series (fps) in the coupling constant, i.e., to show that the coefficients of the power series are well-defined without regard to the question of whether the series converges (it surely doesn't!).

As you all know, even this more modest goal is beset with serious difficulties, namely the infamous divergent integrals of QFT. The formal remedy is to introduce counterterms $\delta\mathcal{L}$ having the same form as terms in $\mathcal{L}$ by "adjusting" the parameters in $\mathcal{L}$:

$$\mathcal{L} + \delta\mathcal{L} = \mathcal{L}_0(\vec{z}(\lambda)) + \tilde{\lambda}(\lambda)\mathcal{L}_{int} \tag{5a}$$

where $\vec{z}(\lambda)$ and $\tilde{\lambda}(\lambda)$ are fps in λ; e.g.,

$$\tilde{\lambda}(\lambda) \sim \lambda + \sum_{n=2}^{\infty} a_n\lambda^n. \tag{5b}$$

The coefficients a_n are infinite and are supposed to cancel the infinities in E(J). Letting

$$\begin{aligned} E_{ren}(J) &= \int e^{(\Phi,J)}d\nu_{ren} \\ &\equiv \int e^{(\Phi,J)}e^{-\int(\mathcal{L}+\delta\mathcal{L})dx}\mathcal{D}\Phi \Big/ \int e^{-\int(\mathcal{L}+\delta\mathcal{L})dx}\mathcal{D}\Phi \end{aligned} \tag{6}$$

the basic problem becomes:

Problem. Show that $\delta\mathcal{L}$ can be chosen so that E_{ren} has a well-defined fps in λ.

As it stands, this problem reads: show that

$$\infty - \infty = \text{finite}.$$

A rigorous interpretation (see below) consists of introducing regularizations to render all the ∞'s finite, choosing the counterterms $\delta\mathcal{L}$ to cancel the would-be ∞'s, and then proving that the coefficients of E_{ren} are finite, uniformly as the regularizations are removed.

GAUSSIAN INTEGRALS AND GRAPHS

When $\lambda = 0$, the free measure

$$d\nu_0(\Phi) = \text{const. } e^{-\int\mathcal{L}_0 dx}\mathcal{D}\Phi$$

is well-understood. If we write (using integration by parts)

$$\int\mathcal{L}_0 dx = \frac{1}{2}(\Phi, C^{-1}\Phi)$$

where C^{-1}_{ij} is a first or second order differential operator, then $d\nu_0$ has generating functional

$$E_0(J) \equiv \int e^{(\Phi,J)} d\nu_0(\Phi) = e^{(CJ,J)/2}. \tag{7}$$

To compute the integral w.r.t. $d\nu_0$ of any monomial in the fields we simply take the appropriate functional derivative of $E_0(J)$. (When fermi fields are involved these derivatives are Grassmannian and the resulting ± signs are a nuisance, but the computations are straightforward.)

A Wick monomial $W = :\Phi_{i_1}(x_1) \ldots \Phi_{i_n}(x_n):$ is defined as

$$W = \frac{\delta^n}{\delta J_{i_1}(x_1) \ldots \delta J_{i_n}(x_n)} e^{\Phi(J) - (CJ,J)/2}\Bigg|_{J=0} . \tag{8}$$

For example,

$$:\Phi_i(x)\Phi_j(y): = \Phi_i(x)\Phi_j(y) - C_{ij}(x,y).$$

In a perturbation theory analysis the only functional integrals we need to compute are Gaussian, i.e. of the form

$$\mathcal{E}(W_1 W_2 \ldots W_p) = \int W_1 \ldots W_p \, d\nu_0(\Phi) \tag{9}$$

where $W_1, \ldots, W_p$ are Wick monomials. There is an elegant expression for the answer as a sum of graphs ("Euclidean Feynman diagrams") the prototypical calculation being

$$\int \Phi_i(x)\Phi_j(y) d\nu_0(\Phi) = C_{ij}(x,y) = \underset{x}{\bullet}\!\xrightarrow{\quad ij \quad}\!\underset{y}{\bullet}.$$

In this graphical picture, we represent the argument x of $\Phi_i(x)$ as a point in $\mathbf{R}^d$ and the "covariance" $C_{ij}(x,y)$ as an ij-line joining x to y. We speak of the fields $\Phi_i(x)$ and $\Phi_j(y)$ "contracting" to form the line $C_{ij}(x,y)$. If Φ_i and Φ_j are fermi fields, then

$$C_{ij}(x,y) = - C_{ji}(y,x)$$

and so it is necessary to associate a direction with the line going from the vertex corresponding to the field on the left to that on the right. A general graph G contributing to (9) has vertices given by the arguments of the fields in $W_1 \ldots W_p$ and lines formed by contracting the fields in pairs, and the value V(G) of G is the product of its lines. From (7) and (8) we then obtain:

<u>Gaussian Integration Rule 1</u>

$$\mathcal{E}(W_1 \ldots W_p) = \sum_G (-1)^{\pi(G)} V(G) \tag{10}$$

where the sum over G is over all graphs formed by contracting the fields of $W_1 \ldots W_p$ in pairs in all possible ways except that no two fields within the same Wick monomial may contract. $\pi(G)$ is the number of commutations of fermi fields required to move each field in $W_1 \ldots W_p$ next to the field with which it contracts.

There are a couple of variations on this rule that we'll need. We define the "connected" or "truncated" expectation by

$$\mathcal{E}_T(W_1,\ldots,W_p) = \frac{\partial^p}{\partial\lambda_1 \ldots \partial\lambda_p} \log \mathcal{E}(e^{\lambda_1 W_1+\ldots+\lambda_p W_p})\Big|_{\lambda = 0} . \quad (11)$$

For example,

$$\mathcal{E}_T(W_1) = \mathcal{E}(W_1), \quad \mathcal{E}_T(W_1,W_2) = \mathcal{E}(W_1W_2) - \mathcal{E}(W_1)\mathcal{E}(W_2).$$

We regard the arguments $x_1^j,\ldots,x_{n_j}^j$ of each W_j as forming a "cluster" $\vec{x}^j$ in $\mathbf{R}^{dn_j}$. Then:

<u>Gaussian Integration Rule 2</u>

$$\mathcal{E}_T(W_1,\ldots,W_p) = \sum_G (-1)^{\pi(G)} V(G) \quad (12)$$

where the sum over G is over those G's in (10) whose lines connect the clusters $\vec{x}^1,\ldots,\vec{x}^p$.

Rules 1 and 2 are statements about "vacuum expectations". Suppose that we decompose the fields as a sum of two independent fields: $C = C^{(h)} + C^{(s)}$ and $\Phi = \Phi^{(h)} + \Phi^{(s)}$ where $\Phi^{(h)}$ has covariance $C^{(h)}$ and $\Phi^{(s)}$ has covariance $C^{(s)}$. Then we can integrate out $\Phi^{(h)}$ treating $\Phi^{(s)}$ as an "external" field:

$$\mathcal{E}^{(h)}(F) \equiv \int F(\Phi^{(h)}+\Phi^{(s)})\, d\nu_0(\Phi^{(h)}). \quad (13)$$

$\mathcal{E}_T^{(h)}$ is defined as in (11). The graphical expression for (13) involves a sum over graphs with lines $C^{(h)}$ and with "legs" representing the unintegrated fields $\Phi^{(s)}$. If F is a product of Wick monomials then the resulting monomial in $\Phi^{(s)}$ will not be a Wick monomial; however, using the definition (8) we can rewrite this monomial as a sum over Wick monomials in $\Phi^{(s)}$ with coefficients involving $C^{(s)}$'s:

<u>Gaussian Integration Rule 3.</u>

$$\mathcal{E}_T^{(h)}(W_1,\ldots,W_p) = \sum_G (-1)^{\pi(G)} V(G) \quad (14)$$

where the sum over G is over graphs whose vertices $\mathcal{V}(G)$ consist of the arguments of the fields in $W_1,\ldots,W_p$, whose lines $\ell \in \mathcal{L}(G)$ join a vertex x to a vertex y in a different cluster and are either h-lines formed by contracting $\Phi_i^{(h)}(x)$ and $\Phi_j^{(h)}(y)$ (in which case ℓ has the value $C_\ell = C_{ij}^{(h)}(x,y)$) or s-lines formed by contracting $\Phi_i^{(s)}(x)$ and $\Phi_j^{(s)}(y)$ (in which case $C_\ell = C_{ij}^{(s)}(x,y)$), and whose legs $\lambda \in \Lambda(G)$ are half-lines emanating from each vertex x whose field $\Phi_\lambda = \Phi_i^{(s)}(x)$ does not contract. The h-lines of G must connect the clusters $\vec{x}^1,\ldots,\vec{x}^p$. The value of G is

$$V(G) = \prod_{\ell\in\mathcal{L}(G)} C_\ell \; : \prod_{\lambda\in\Lambda(G)} \Phi_\lambda : . \quad (15)$$

(The fermi fields in the product occur in the same order as in $W_1 \ldots W_p$.)

Example. In the ϕ_d^4 model, let $W_j = :\phi(x_j)^4:$ so that the cluster $\vec{x}^j = (x_j, x_j, x_j, x_j)$ is represented as a vertex with 4 attached legs. (With only one bose field the covariance matrix C is 1 x 1.) Then one graph contributing to $\mathcal{E}_T^{(h)}(W_1,W_2,W_3,W_4)$ is

G = 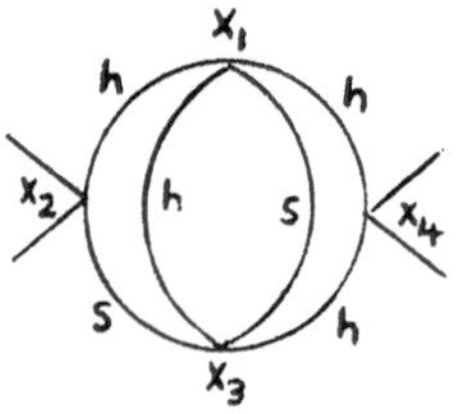

(16)

with

$$V(G) = C^{(h)}(x_1,x_2)C^{(s)}(x_2,x_3)C^{(h)}(x_3,x_4)C^{(h)}(x_4,x_1)$$

$$C^{(h)}(x_1,x_3)C^{(s)}(x_1,x_3) \ :\phi^{(s)}(x_2)^2 \ \phi^{(s)}(x_4)^2:.$$

In fact, this same graph occurs $(4!)^2(4\cdot 3)^2$ times in the sum (14), as a result of the ways in which the 4 legs from each vertex can be assigned to the lines meeting that vertex. In general, we shall include this combinatoric factor in V(G) and list the graph only once in the sum (14).

The covariance $C = (-\Delta+m^2)^{-1}$ has the UV or short-distance singularity

$$C(x,y) \sim \frac{1}{|x-y|^{d-2}} \qquad \text{as } x-y \to 0. \tag{17}$$

As a result, for the graph G of (16) with all lines given the value C, V(G) has a non-integrable singularity when $d \geq 4$. The same is true for the subgraphs

G_1 =  G_2 =  G_3 =  . (18)

The renormalization program undertakes to cancel the singularities of G and of its subgraphs with counterterms, starting with the smallest divergent subgraphs and working successively up to larger ones. This is basically the right strategy except for the notorious problem of "overlapping divergences," i.e., two divergent subgraphs which have lines in common but which are not included in one another (such as the above G_2 and G_3). It is far from clear that the renormalization cancellation attempted on one subgraph won't interfere with that attempted on the other. In fact there is interference and it is necessary to decompose each graph as a sum of topologically similar graphs which differ in the "strengths" of their lines and which do not require renormalization of overlapping subgraphs. This is the idea behind Hepp sectors[4] and Zimmermann forests[5] and leads us to:

We first transform (4) slightly by setting $\Phi^e = CJ$ and factoring out the free contribution (7):

$$E(J)/E_0(J) = e^{-(\Phi^e, C^{-1}\Phi^e)/2} \int e^{(\Phi, C^{-1}\Phi^e)} d\nu(\Phi)$$

$$= \text{const.} \int e^{V(\Phi)} e^{-(\Phi-\Phi^e, C^{-1}(\Phi-\Phi^e))/2} \mathcal{D}\Phi \tag{19}$$

where

$$V(\Phi) = -\lambda \int \mathcal{L}_{int}(\Phi(x))dx.$$

Making the change of variables $\Phi \to \Phi + \Phi^e$, we define the effective potential V_e as the log of (19):

$$V_e(\Phi^e) \equiv \log \frac{\int e^{V(\Phi+\Phi^e)} d\nu_0(\Phi)}{\int e^{V(\Phi)} d\nu_0(\Phi)}$$

$$= [\log \mathcal{E}(e^{V(\Phi+\Phi^e)})]_0 \tag{20}$$

where the notation $[\ldots]_0$ means "drop constant terms independent of Φ^e ". Formally, V_e is the generator of connected Green's functions, amputated by C^{-1}. In terms of V_e, the basic problem becomes:

<u>Problem</u>. Show that the counterterms

$$\delta V(\Phi) = - \int \delta\mathcal{L}(\Phi(x))dx$$

can be chosen so that

$$V_{e,ren}(\Phi^e) = [\log \mathcal{E}(e^{(V+\delta V)(\Phi+\Phi^e)})]_0 \tag{21}$$

has a well-defined fps in λ.

<u>Scale decomposition</u>.

Fix $M > 1$, e.g., $M = 4$. We decompose the covariance of a line ℓ as

$$C_\ell = \sum_{h=-\infty}^{\infty} C_\ell^{(h)} \tag{22a}$$

where $C_\ell^{(h)}$ has length scale M^{-h}, i.e.,

$$|C_\ell^{(h)}(x,y)| \le c\, M^{d_\ell h} e^{-M^h|x-y|} \tag{22b}$$

where d_ℓ is the "dimension" of the line ℓ:

$$d_\ell = \begin{cases} d-2 & \text{if } \ell \text{ is a bose line} \\ d-1 & \text{if } \ell \text{ is a fermi line.} \end{cases} \tag{23}$$

Typically, C_ℓ has the form (we have set the field strength equal to 1)

$$C_\ell = D_\ell(-\Delta+m_\ell^2)^{-1} \tag{24a}$$

where for fermi fields D_ℓ is the first order operator

$$D_\ell = i\partial\!\!\!/+m_\ell \tag{24b}$$

and for bose fields D_ℓ is a zero order operator, possibly depending on vector indices. For C_ℓ given by (24) we define the decomposition (22) using α-parameters:

$$C_\ell = \int_0^\infty d\alpha_\ell D_\ell e^{-\alpha_\ell(-\Delta+m_\ell^2)}$$
$$= \sum_h \int d\alpha_\ell \chi_h(\alpha_\ell) D_\ell e^{-\alpha_\ell(-\Delta+m_\ell^2)} \equiv \sum_h C_\ell^{(h)} \qquad (25)$$

where $\{\chi_h\}$ is a suitable partition of $[0,\infty)$. If $m_\ell = 0$ we choose

$$\chi_h = \text{characteristic function of } [M^{-2h}, M^{-2h+2}]. \qquad (26a)$$

If $m_\ell \neq 0$, we may suppose $m_\ell \geq 1$ (by scaling) and we choose

$$\chi_h = \begin{cases} (26a) & \text{if } h > 0 \\ \text{characteristic fn. of } [1,\infty) & \text{if } h = 0 \\ 0 & \text{if } h < 0. \end{cases} \qquad (26b)$$

Using the formula

$$e^{\alpha_\ell \Delta}(x,y) = (4\pi\alpha_\ell)^{-d/2} e^{-(x-y)^2/4\alpha_\ell} \qquad (27)$$

for the heat kernel, it is easy to check the bound (22b). Until the final lecture I shall assume that all particles are massive so that χ_h is given by (26b) and the sum (22a) over scales starts at $h = 0$.

The effect of the decomposition (22) is to resolve the singularities of C_ℓ; in particular, as $x-y \to 0$,

$$C_\ell(x,y) \sim |x-y|^{-d_\ell} \sim \sum_{h=0}^\infty M^{d_\ell h} e^{-M^h|x-y|}.$$

Corresponding to the decomposition (22a) we represent the fields as $\Phi = \sum_h \Phi^{(h)}$ where the $\Phi^{(h)}$'s are independent and $\Phi^{(h)}$ has covariance $C^{(h)}$. We write $\mathcal{E}^{(h)}$ for the expectation with regard to $\Phi^{(h)}$. To regularize the theory we truncate the infinite sums over h: the covariance with an UV-cutoff $U > 0$ is $C^{(\leq U)} = \sum_{h=0}^U C^{(h)}$. We also write

$$\mathcal{E}^{(\leq U)} = \prod_{h=0}^U \mathcal{E}^{(h)}, \quad \Phi^{(\leq k)} = \Phi^e + \sum_{h=0}^k \Phi^{(h)}, \quad \Phi^{(>k)} = \sum_{h=k+1}^U \Phi^{(h)}, \text{ etc.}$$

The regularized version of (20) is

$$V_e^U(\Phi^e) = [\log \mathcal{E}^{(\leq U)}(e^{V(\Phi^{(\leq U)})})]_0. \qquad (28a)$$

With Φ^e in Schwartz space (as we shall assume throughout), it is easy to verify that V_e^U has a well-defined fps in λ (i.e., the divergences have been regularized).

Iterated Cumulant Expansion

We also introduce the intermediate effective potential at scale k

$$V_k^U(\Phi^{(\le k)}) = [\log \mathcal{E}^{(>k)}(e^{V(\Phi^{(\le U)})})]_0. \tag{28b}$$

Suppressing the superscript U, we have $V_{-1} = V_e$, and, letting $V_U = V$, we have the recursion relation

$$V_{k-1} = [\log \mathcal{E}^{(k)}(e^{V_k})]_0 \qquad k = U, U-1, \ldots, 0$$

$$= \sum_{p=1}^{\infty} \frac{1}{p!} [\partial_s^p \log \mathcal{E}^{(k)}(e^{sV_k})|_{s=0}]_0$$

$$= \sum_{p=1}^{\infty} \frac{1}{p!} [\mathcal{E}_T^{(k)}(\underbrace{V_k, \ldots, V_k}_{p \text{ arguments}})]_0. \tag{29}$$

We have already seen how to compute the connected expectation $\mathcal{E}_T^{(k)}$ in (14). In the notation of (14), $\Phi^{(s)} = \Phi^{(<k)}$ and $\Phi^{(h)} = \Phi^{(k)}$. We represent the expectation $1/p![\mathcal{E}_T^{(k)}(W_1, \ldots, W_p)$ by the "fork"

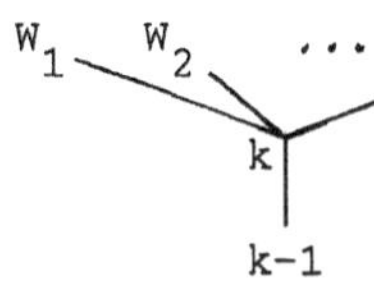

⟵ functions of input field $\Phi^{(\le k)}$

⟵ fork ≡ connected expectation $\mathcal{E}_T^{(k)}$

⟵ highest scale of output field Φ.

In this pictorial representation, we suppress the "trivial forks" (i.e., those with p=1 corresponding to ordinary expectations) so that a straight line represents a sequence of $\mathcal{E}^{(h)}$'s:

⟷ $\mathcal{E}^{(>k)}(W)$.

If we now iterate (29) we start growing "trees", e.g.,

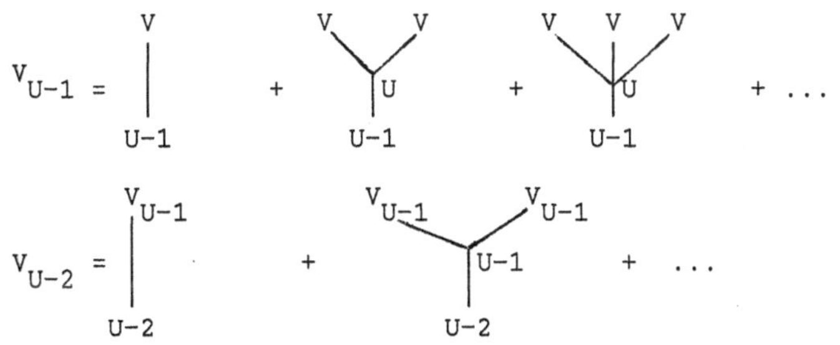

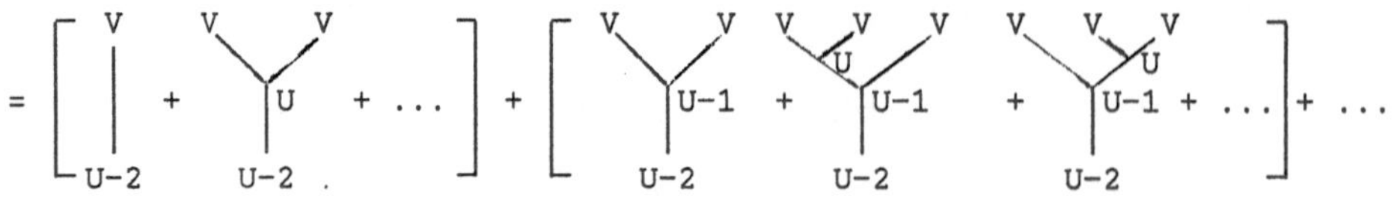

Some arboreal terminology

A tree τ is a tree graph (i.e. no closed loops) with a distinguished end-vertex at the bottom (the root), the other end-vertices at the top (called endpoints), and each remaining vertex (called a fork) having one branch down and the other $p_f \geq 2$ branches going up. If $f_1, f_2 \in \mathfrak{F}(\tau)$ (the forks of τ), we write $f_1 > f_2$ if f_1 lies above f_2 on the tree. Each $f \in \mathfrak{F}(\tau)$ bears a scale label h_f such that the scales $h = (h_f)_{f \in \mathfrak{F}}$ belong to the set

$$\mathcal{H}_r(\tau) = \{h | r < h_f \leq U;\ h_{f_1} > h_{f_2} \text{ if } f_1 > f_2\}$$

where r is the root scale, $-1 \leq r \leq U$. Each endpoint of τ is assigned the label or value V. Given $f \in \mathfrak{F}(\tau)$ we let τ_f be the subtree of τ with lowest fork f and root scale $h_{\pi(f)}$. Here $\pi(f)$ denotes the fork immediately below f.

The value $V(\tau, h)$ of such a tree is most easily described inductively: if $f_1, \ldots, f_p$ are the forks immediately above f (including endpoints) then

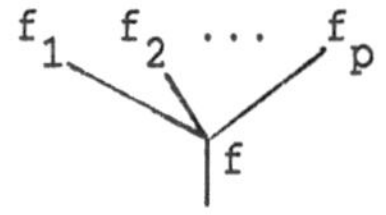

$$V(\tau_f, h) = \frac{1}{p!} [\mathcal{E}_T^{(h_f)} (\mathcal{E}^{(>h_f)} V(\tau_{f_1}, h), \ldots, \mathcal{E}^{(>h_f)} V(\tau_{f_p}, h))]_0 \tag{30}$$

where $V(\tau_f, h)$ depends only on the scales $h_{f'}$, $f' \geq f$.

Upon iterating (29), as illustrated above, we obtain:

Theorem 1. (Unrenormalized Tree Expansion) For $-1 \leq r \leq U$

$$V_r = \sum_\tau \sum_{h \in \mathcal{H}_r(\tau)} V(\tau, h). \tag{31}$$

We next give a more explicit description of $V(\tau, h)$ as a sum over the set of labelled graphs $\mathcal{G}(\tau)$ associated with τ. If τ has v endpoints then each $G \in \mathcal{G}(\tau)$ is a connected graph with v vertices, each corresponding to a monomial in V. Each leg λ of G corresponds to an uncontracted field $\Phi^{(\leq r)}(x_\lambda)$ and is attached to the vertex x_λ. Let G_f be the subgraph of G whose vertices $\mathcal{V}(G_f)$ correspond to endpoints of τ_f, whose lines $\mathcal{L}(G_f)$ are the lines of G which join two of its vertices, and whose legs $\Lambda(G_f)$ are the legs of G attached to a vertex of G_f together with the lines of G that join a vertex of G_f to a vertex not in G_f (actually the leg is the "half" of each such line that touches G_f). The external vertices of G_f are those vertices to which a leg is attached.

Example 3 (ϕ_d^4) Let τ be the tree ($h_j = h_{f_j}$)

τ = 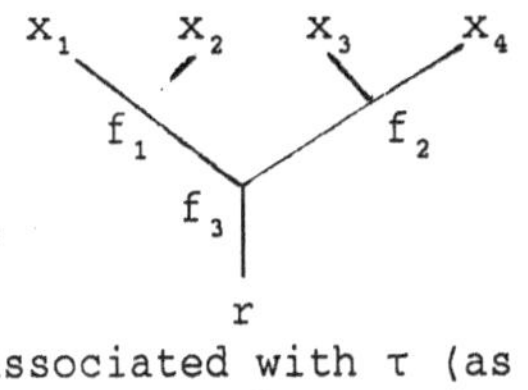

$$r < h_3 < h_1 \le U$$
$$h_3 < h_2 \le U. \qquad (32a)$$

A graph associated with τ (as yet unlabelled) is the graph of (15):

G = 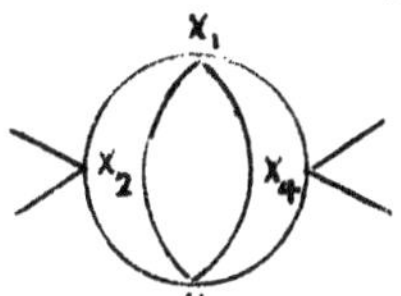

. (32b)

Then

G_{f_1} = 

and G_{f_2} = 

. (32c)

If $\pi(f') = f$ we view $G_{f'}$ as a generalized vertex for the graph G_f and we consider the reduced graph

$$g_f = G_f/\{G_{f'} \mid \pi(f') = f\}$$

formed by contracting each $G_{f'}$ to a point. In Example 3,

$$g_{f_3} = G/\{G_{f_1}, G_{f_2}\} =$$  .

Each line ℓ of G is formed at a fork $f = f(\ell)$ by the joining of two legs attached to different vertices of g_f, and is designated as being "hard" or "soft". It is hard if its value is $C^{(h_f)}$ (an "h-line" in the language of Gaussian Integration Rule 3) and is soft if its value is $C^{(<h_f)}$ (an "s-line"). According to Rule 3, g_f must be connected by its hard lines. We let $\mathcal{M}_f$ be any minimal subset of the hard lines of g_f that connects g_f, and we define

$$\mathcal{M}_{\ge f} = \bigcup_{f' \ge f} \mathcal{M}_{f'}, \qquad \mathcal{M} = \bigcup_f \mathcal{M}_f .$$

Now we can say what the labelled graphs in $\mathcal{G}(\tau)$ are: they are graphs G constructed as above with each line ℓ labelled by $h_{f(\ell)}$ (hard) or by $<h_{f(\ell)}$ (soft) in such a way that each g_f is connected by its hard lines.

<u>Example 3</u> (cont'd). A possible labelling for G compatible with τ is

G = 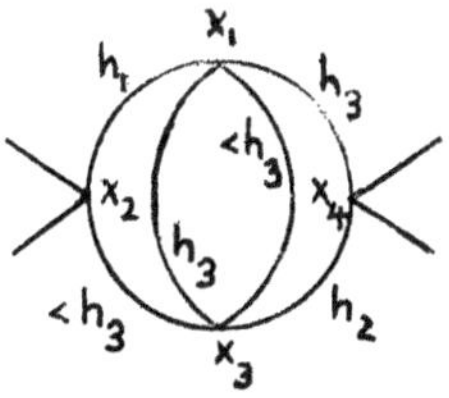

(32d)

The combinatoric factor c_G associated with G is $(4!)^2(4\cdot 3)^2$. $\mathcal{M}_{f_1}$ is the h_1-line, $\mathcal{M}_{f_2}$ the h_2-line and $\mathcal{M}_{f_3}$ may be chosen to be either one of the h_3-lines.

Finally, we can say explicitly what $V(\tau,h)$ is:

$$V(\tau,h) = \frac{1}{n(\tau)} \sum_{G\in\mathcal{G}(\tau)} V^h(G) \tag{33}$$

where

$$n(\tau) = \prod_{f\in\mathcal{F}(\tau)} p_f!$$

and

$$V^h(G) = (-1)^{\pi(G)} c_G \lambda^v \int dx_1 \ldots dx_v \prod_{\ell\in\mathcal{L}(G)} C_\ell^h : \prod_{\lambda\in\Lambda(G)} \Phi^{(\le r)}(x_\lambda):. \tag{34}$$

Here the combinatoric factor c_G is the number of ways of assigning the legs from each vertex to its lines, and

$$C_\ell^h = \begin{cases} C_\ell^{(h_{f(\ell)})}(x_\ell, y_\ell) & \text{if } \ell \text{ is hard} \\ C_\ell^{(<h_{f(\ell)})}(x_\ell, y_\ell) & \text{if } \ell \text{ is soft} \end{cases}$$

where x_ℓ and y_ℓ are the coordinates of the endpoints of ℓ.

What have we accomplished with the expansions (31) and (33)? Take $r = -1$ in (31). Then we have an expansion for V_e in which each topological graph that contributes to V_e has been decomposed as a sum over labelled graphs associated with different trees and different labellings. The expansion organizes the way in which the graphs are estimated and renormalized, the procedure being quite different for the labelled graphs associated with different trees. Each $V^h(G)$ is naturally estimated and renormalized inductively in terms of its nested subgraphs G_f. In particular, we never see any overlapping divergences: any two subgraphs G_f and $G_{f'}$ of a labelled graph G are either contained in one another (if $f < f'$ or $f' < f$) or are disjoint.

The sum over τ in (31) is infinite (and doubtless divergent), but to a fixed order of perturbation theory there is a finite number of τ's in (31) and of G's in (33). The convergence question is: does the sum over h converge uniformly in U? Focussing on a particular τ and $G\in\mathcal{G}(\tau)$ the question becomes: does

$$\sum_{h\in\mathcal{H}_{-1}(\tau)} \int dx_1 \ldots dx_v \prod_{\ell\in\mathcal{L}(G)} C_\ell^h \prod_{\lambda\in\Lambda(G)} \Phi^e(x_\lambda) \tag{35}$$

converge uniformly in U?

Let $\tilde{\Psi}(x_1,\ldots,x_v) = \prod_{\lambda\in\Lambda(G)} \Phi^e(x_\lambda)$. Using the translation invariance of $\prod_\ell C_\ell^h$, we can rewrite (35) in terms of

$$K^h(x_1,\ldots,x_{v-1}) = \prod_\ell C_\ell^h \Big|_{x_v=0} \tag{36a}$$

as

$$(35) = \sum_h \int dx\, K^h(x)\, \Psi(x), \tag{36b}$$

where $x = (x_1,\ldots,x_{v-1})$ and

$$\Psi(x) = \int dy\ \tilde{\Psi}(x_1+y,\ldots,x_{v-1}+y,\ y). \qquad (36c)$$

From now on it is understood that $x_v = 0$.

From (25), (26b) and (27),

$$K^h = \prod_\ell \int d\alpha_\ell \chi_\ell(\alpha_\ell, h_{f(\ell)})(4\pi\alpha_\ell)^{-d/2} D_\ell\ e^{-\Delta_\ell^2/4\alpha_\ell - \alpha_\ell m_\ell^2} \qquad (37)$$

where $\Delta_\ell = x_\ell - y_\ell$ and $\chi_\ell(\alpha_\ell, h)$ is the characteristic function

$$\chi_\ell(\alpha_\ell,h) = \text{char. fn. of} \begin{cases} [M^{-2h},\ M^{-2h+2}] & \text{if } \ell \text{ if hard and } h > 0 \\ [1,\infty] & \text{if } \ell \text{ is hard and } h = 0 \\ [M^{-2h+2},\infty) & \text{if } \ell \text{ is soft } (h > 0). \end{cases} \qquad (38)$$

The operators D_ℓ (for $\ell \in \mathcal{L}_F$, the set of fermi lines) produce factors $-i\not{\Delta}_\ell/2\alpha_\ell + m_\ell$ which participate in products and traces according to what chains and loops of fermi lines occur in G. As far as estimates on K^h are concerned, this structure and the order of the factors are unimportant and will be ignored. In the following, c will stand for (different) constants which possibly depend on G but are bounded by $c_0^{\ell(G)}$ where c_0 is independent of G and $\ell(G)$ is the number of lines of G. We let D(G) be the <u>UV degree of divergence</u> of a graph G:

$$D(G) = \sum_{\ell \in \mathcal{L}(G)} d_\ell - d(v(G) - 1) \qquad (39)$$

where the dimension d_ℓ is given in (23) and v(G) is the number of vertices of G.

<u>Lemma 2</u>.

$$\|K^h\|_1 \le c \prod_f M^{D(g_f)h_f} \qquad (40a)$$

$$= c \prod_f M^{D(G_f)(h_f - h_{\pi(f)})} \qquad (40b)$$

where, for the lowest fork F, $h_{\pi(F)} = 0$.

<u>Proof</u>. If $\ell \in \mathcal{L}_F$ (the set of fermi lines)

$$|D_\ell e^{-\Delta_\ell^2/4\alpha_\ell}| = |-i\not{\Delta}_\ell/2\alpha_\ell + m_\ell| e^{-\Delta_\ell^2/4\alpha_\ell} \le (\alpha_\ell^{-1/2} + m_\ell) e^{-\Delta_\ell^2/8\alpha_\ell}$$

$$\le c\ M^{h_{f(\ell)}} e^{-\Delta_\ell^2/8\alpha_\ell}$$

by (38). Therefore, dropping the lines in $\mathcal{M}^c = \mathcal{L}(G)\backslash\mathcal{M}$,

$$\prod_\ell |D_\ell e^{-\Delta_\ell^2/4\alpha_\ell}| \le c \prod_{\ell \in \mathcal{L}_F} M^{h_{f(\ell)}} \prod_{\ell \in \mathcal{M}} e^{-\Delta_\ell^2/8\alpha_\ell}. \qquad (41)$$

But

$$\int dx \prod_{\ell\in\mathcal{M}} e^{-\Delta_\ell^2/8\alpha_\ell} = c \prod_{\ell\in\mathcal{M}} \alpha_\ell^{d/2} \tag{42a}$$

and so

$$\begin{aligned}\|K^h\|_1 &\le c \prod_{\ell\in\mathcal{L}_F} M^{h_{f(\ell)}} \prod_{\ell\in\mathcal{M}} \int d\alpha_\ell \chi_\ell e^{-\alpha_\ell m_\ell^2} \prod_{\ell\in\mathcal{M}^c} \int d\alpha_\ell \chi_\ell \alpha_\ell^{-d/2} e^{-\alpha_\ell m_\ell^2} \\ &\le c \prod_{\ell\in\mathcal{L}_F} M^{h_{f(\ell)}} \prod_{\ell\in\mathcal{M}} M^{-2h_{f(\ell)}} \prod_{\ell\in\mathcal{M}^c} M^{(d-2)h_{f(\ell)}}\end{aligned} \tag{42b}$$

by (38). Now

$$\begin{aligned}&\sum_{\ell\in\mathcal{L}_F} h_{f(\ell)} - 2\sum_{\ell\in\mathcal{M}} h_{f(\ell)} + (d-2)\sum_{\ell\in\mathcal{M}^c} h_{f(\ell)} \\ &= \sum_\ell d_\ell h_{f(\ell)} - d\sum_{\ell\in\mathcal{M}} h_{f(\ell)} = \sum_f h_f\Big[\sum_{\ell:f(\ell)=f} d_\ell - d(p_f-1)\Big] \\ &= \sum_f h_f D(g_f).\end{aligned} \tag{42c}$$

This establishes (40a). (40b) follows from the following elementary lemma with $a_f = h_f$, $b_f = D(g_f)$ and f_1 the root of τ. ∎

Lemma 3. (Summation by parts) Let a_f and b_f be functions defined on the forks of a tree τ. For f_1 a fixed fork or the root of τ,

$$\sum_{f>f_1} (a_f - a_{f_1}) b_f = \sum_{f>f_1} (a_f - a_{\pi(f)}) B_f \tag{43}$$

where $B_f = \sum_{f'\ge f} b_{f'}$.

On the basis of (40b) it is now easy to determine whether $\sum_h \|K^h\|_1$ converges, uniformly in U. Since each h_f is summed from $h_{\pi(f)}+1$ to U, we obtain:

Theorem 4. (Dyson-Weinberg Power Counting Theorem)
If $D(G_f) < 0$ for every f, then

$$\sum_{h\in\mathcal{H}_{-1}(\tau)} \|K^h\|_1 \le c$$

where c is independent of U.

On the other hand, if $D(G_f) \ge 0$ for even one f, then the sum is not bounded uniformly in U. We must renormalize!

Renormalization = Taylor Subtraction

Given a subgraph G_f with $D(G_f) \ge 0$ we wish to renormalize G_f by subracting a suitable "local part" of G_f that makes the degree of divergence negative. (By a "local" monomial we mean one in which all the fields have the same argument.) We also want to arrange that these subtractions are implemented via an adjustment δV of the original interaction V. This is another advantage of the GN tree expansion: there is a clear connection between the renormalization of individual (sub-)graphs and the choice of Lagrangian counterterms.

Suppose that as the output from a fork f we have (we simply write G_f instead of $V(G_f)$ for the value of G_f)

$$G_f = \int K_f(x)\Pi(x)\,dx \tag{44}$$

where $x = (x_1,\ldots,x_n)$ and

$$\Pi(x) = :\partial^{q_1}\Phi_{i_1}^{(\le k)}(x_1)\ldots\partial^{q_n}\Phi_{i_n}^{(\le k)}(x_n):,$$

where the kernel K_f contains δ-functions if some of the x's coincide. Here k is the scale of $\pi(f)$ and the x-derivatives ∂^{q_j} arise from the renormalization operations that I'm about to define. We let $N_f = \sum|q_j|$ be the total number of such derivatives on the legs of G_f. The <u>dimension of Π</u> counts these derivatives:

$$\dim \Pi \equiv \sum_{j=1}^{n} \dim \Phi_{i_j} + N_f \tag{45a}$$

where $\delta_i \equiv \dim \Phi_i$ is $\frac{d-2}{2}$ for a bose field and $\frac{d-1}{2}$ for a fermi field.

($\delta_i = d_\ell/2$ where ℓ is a line formed by contracting Φ_i with another field.)

The <u>external degree of G_f</u> is

$$\delta(G_f) \equiv d - \dim \Pi. \tag{45b}$$

The <u>criterion for UV-renormalizability</u> is that every vertex v in V has degree

$$\delta(v) \equiv d - \sum_{\lambda\in\Lambda(v)} \delta_\lambda \ge 0. \tag{46a}$$

Here $\Lambda(v)$ is the set of legs or fields attached to v. For simplicity we shall assume that V is <u>dimensionless</u>, i.e.,

$$\delta(v) = 0 \text{ for every } v. \tag{46b}$$

Then

$$\begin{aligned} D(G_f)-N_f &= \sum_{\ell\in\mathcal{L}(G_f)} d_\ell - d(v(G_f)-1) - N_f \\ &= \sum_{v\in\mathcal{V}(G_f)} \Big(\sum_{\lambda\in\Lambda(v)} \delta_\lambda - d\Big) - \sum_{\lambda\in\Lambda(G_f)} \delta_\lambda + d - N_f \\ &= \delta(G_f). \end{aligned} \tag{47}$$

That is, the "internal" degree $D(G_f)-N_f$ and the external degree $\delta(G_f)$ agree.

<u>Definitions</u>. Select one of $x_1,\ldots,x_n$ as the <u>localization vertex</u> x_f for G_f, and let

$$x(t_f) = (x_1(t_f),\dots,x_n(t_f)) \quad \text{where } x_j(t_f) = x_f + t_f(x_j - x_f) \tag{48}$$

for $0 \le t_f \le 1$. Let $\mu_f = \delta(G_f)$. The <u>local parts</u> of Π and G_f are

$$L\Pi(x) = \begin{cases} \sum_{m_f=0}^{\mu_f} \frac{1}{m_f!} \partial_{t_f}^{m_f} \Pi(x(t_f))\Big|_{t_f=0} & \text{if } \mu_f \ge 0 \\ 0 & \text{if } \mu_f < 0 \end{cases} \tag{49a}$$

and

$$LG_f = \int K_f(x) L\Pi(x)dx. \tag{49b}$$

The <u>renormalization</u> of G_f is

$$RG_f = (1-L)G_f. \tag{49c}$$

Note that LG_f is the integral of a local Wick polynomial in $\Phi^{(\le k)}$. By Taylor's Theorem, if $\mu_f \ge 0$,

$$RG_f = \frac{1}{(\mu_f+1)!} \int_0^1 dt_f (1-t_f)^{\mu_f} \partial_{t_f}^{\mu_f+1} \int K_f(x)\Pi(x(t_f))dx$$

$$= \frac{1}{(\mu_f+1)!} \int_0^1 dt_f (1-t_f)^{\mu_f} \int K_f(x) (\Delta\cdot\partial)^{\mu_f+1} \Pi(x(t_f))dx \tag{50}$$

where $\Delta\cdot\partial = \sum_j (x_j - x_f)\cdot\partial_{x_j}$ and the ∂'s in (50) do not act on the Δ's.

The intuitive reason that the definition (49) works is this. Suppose that

$$\mu_f = D(G_f) - N_f \ge 0,$$

where N_f counts the derivatives on the legs of G_f from renormalization operations at forks $f' > f$. If we apply L of (49a) at f then we introduce an additional m_f derivatives on the legs of G_f, where $m_f = 0,1,\dots,$ or μ_f. If we apply R of (50), then we introduce an additional $m_f = \mu_f+1$ derivatives. The new external degree of G_f,

$$\delta_f = \mu_f - m_f, \tag{51a}$$

thus satisfies

$$0 \le \delta_f < d \qquad \text{if L is applied at f} \tag{51b}$$

$$\delta_f \le -1 \qquad \text{if R is applied at f.} \tag{51c}$$

Associated with the additional m_f derivatives there is an extra Δ^{m_f} in the kernel K_f of G_f. Since $K_f(x)$ has exponential decay between its arguments with "mass" M^{h_f}, this extra Δ^{m_f} gives rise to an extra $M^{-m_f h_f}$ in the power counting. Of course, something this good doesn't come for free. The price we have to pay is the action of ∂^{m_f} on the fields Π. Typically, these contract at scale $k = h_{\pi(f)}$ and so the price is $M^{m_f k}$ for a net gain of

$$M^{-m_f(h_f - h_{\pi(f)})}. \tag{51d}$$

As in (40b) the bound on K_f contains the factor $M^{\mu_f(h_f-h_{\pi(f)})}$, the term $-N_f$ in μ_f coming from the N_f Δ's already present in K_f before renormalization at f. Combining this "internal" factor with the new renormalization factor (51d), we obtain the net power counting factor

$$M^{(\mu_f-m_f)(h_f-h_{\pi(f)})} = M^{\delta_f(h_f-h_{\pi(f)})} \quad . \tag{51e}$$

By (51c), $\delta_f \leq -1$ and so (51e) allows us to sum over $h_f > h_{\pi(f)}$, uniformly in U. The actual argument is a little more complicated than what I've just outlined, but not by much (see Theorem 6).

THE RENORMALIZED TREE EXPANSION

How do we introduce these renormalization cancellations? We modify the tree notation (in particular (30)) as follows: the label R attached to a fork f means

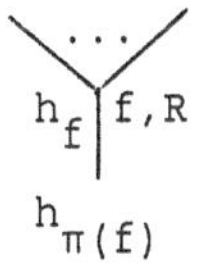

$= \chi(h_f > h_{\pi(f)})$ (1−L)

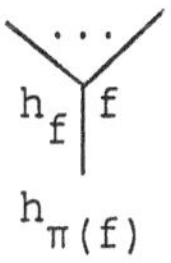

. (52a)

In order to introduce these R-operations via legitimate counterterms in the potential we have to include the local parts for all h_f, not just $h_f > h_{\pi(f)}$. So we must also include the following "useless counterterms":

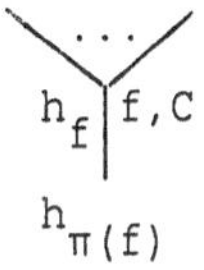

$= -\chi(h_f \leq h_{\pi(f)})$ L

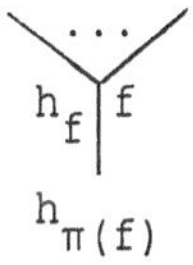

. (52b)

A <u>renormalized tree</u> is a tree τ with a label ρ_f = R or C at each fork f and the label V at each endpoint. Its value $V(\tau,\rho,h)$ is defined as in (30) but with the modifications (52). The appropriate set of scales for a tree τ with root scale r is

$$\mathcal{H}_r(\tau,\rho) = \{h \mid h_{\pi(f)} < h_f \leq U \text{ of } \rho_f = R;\ 0 \leq h_f \leq h_{\pi(f)} \text{ if } \rho_f = C\} \tag{53}$$

where, here, $h_{\pi(F)} = r$ for the lowest fork F.

We define the renormalized potential to be

$$V_{ren} = \sum_\tau \sum_\rho \sum_{h\in\mathcal{H}_U(\tau,\rho)} V(\tau,\rho,h) \equiv V+\delta V. \tag{54a}$$

Note that the trivial tree in the above sum contributes the unrenormalized potential $V(\Phi^{(\leq U)})$ and that every non-trivial tree contributes local counterterms since the root scale U forces $\rho_F = C$.

We introduce the notations

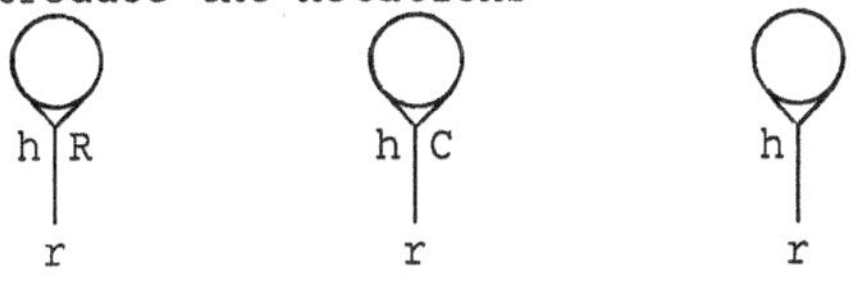

etc.

to denote the sum of the values of all non-trivial trees whose lowest fork F and whose root have the labels shown. If the scale $h = h_F$ is omitted then it is understood to be summed over its allowed range. With this notation

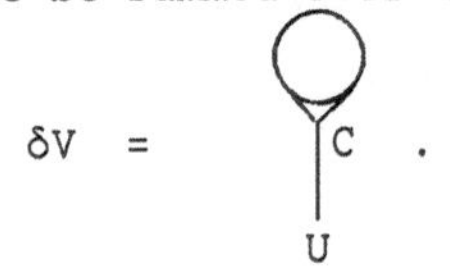

$$\delta V = \quad \bigcirc\!\!\!\mid C \quad . \tag{54b}$$

From the definitions (52) we have the basic identity

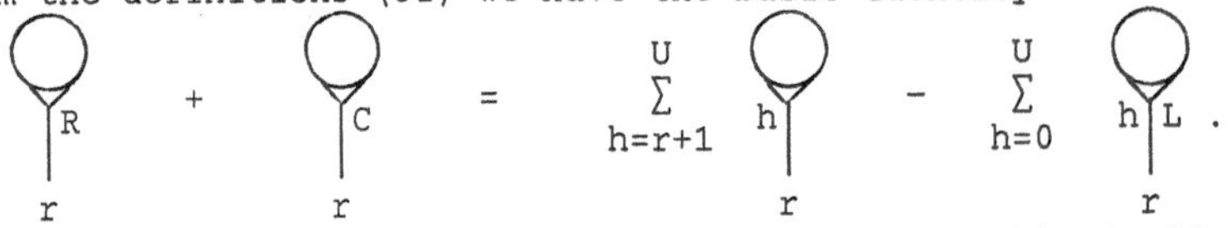

$$\bigcirc\!\!\!\mid_r R \;+\; \bigcirc\!\!\!\mid_r C \;=\; \sum_{h=r+1}^{U} h \bigcirc\!\!\!\mid_r \;-\; \sum_{h=0}^{U} h \bigcirc\!\!\!\mid_r L \;. \tag{54c}$$

As in (28b) we define the intermediate renormalized effective potential by

$$V_{ren,r} = [\log \mathcal{E}^{(>r)} (e^{V_{ren}})]_0,$$

and as in Theorem 1 we have:

Theorem 5. (Renormalized Tree Expansion)   For $-1 \leq r \leq U$,

$$V_{ren,r} = \sum_{\tau} \sum_{\rho} \sum_{h \in \mathcal{H}_r(\tau,p)} V(\tau,\rho,h). \tag{55}$$

Proof. Assuming that (55) holds for r, we establish it for $r-1$. (It holds for $r=U$ by the definition (54a).) By (54c) we have, dropping the subscript ren,

$$V_r = \Big|_r^V \;+\; \sum_{h=r+1}^{U} h \bigcirc\!\!\!\mid_r \;-\; \sum_{h=0}^{U} h \bigcirc\!\!\!\mid_r L \quad .$$

Integrating out $\Phi^{(r)}$ gives

$$V_{r-1} = \Big|_{r-1}^{V\,|_r} \;+\; \{ \tfrac{1}{2!} \; {}^{V|_r}\!\!\bigvee_{r-1}^{\,V|_r} \;+\; \ldots\}$$

$$= \Big|_{r-1}^{V} \;+\; \sum_{h=r+1}^{U} h \bigcirc\!\!\!\mid_{r-1} \;-\; \sum_{h=0}^{U} h \bigcirc\!\!\!\mid_{r-1} L \;+\; \{ r \bigcirc\!\!\!\mid_{r-1} \}$$

$$= \Big|_{r-1}^{V} \;+\; \sum_{h=r}^{U} h \bigcirc\!\!\!\mid_{r-1} \;-\; \sum_{h=0}^{U} h \bigcirc\!\!\!\mid_{r-1} L$$

$$= \Big|_{r-1}^{V} \;+\; \bigcirc\!\!\!\mid_{r-1} R \;+\; \bigcirc\!\!\!\mid_{r-1} C \qquad \text{(by (54c)).} \quad \blacksquare$$

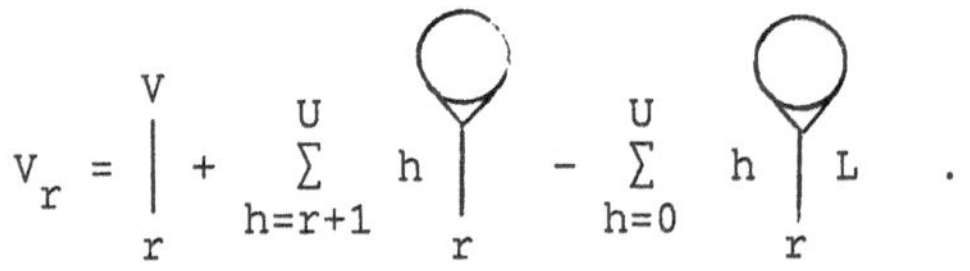

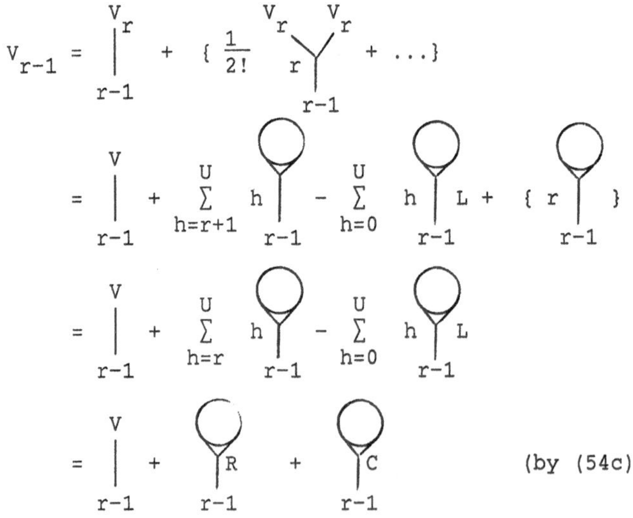

Thus the initial counterterms δV (defined in (54b)) are distributed throughout each tree to produce renormalization cancellations at R-forks and "unused" or "useless" counterterms at C-forks. These useless counterterms contribute to graphs just like interaction vertices except that they come with a scale-dependent coupling constant. In Gallavotti's lectures, a slightly different tree expansion is developed in which there are no C-forks; there, the useless counterterms are incorporated into the running coupling constants. In (55) $V(\tau,\rho,h)$ collects together many labelled graphs, all of which are renormalized and estimated in the same way. To reiterate: there are no overlapping divergences and no combinatorial mess, and the connection between the Lagrangian counterterms δV and the subtractions on individual subgraphs is clear.

It remains to fill in the proof of convergence of (55) that we sketched above:

<u>Theorem 6</u>. Let G^h_{ren} be a graph contributing to $V(\tau,\rho,h)$ in (55) with $r = -1$. Then

$$|G^h_{ren}| \le c \prod_f M^{\delta_f(h_f - h_{\pi(f)})} \tag{56}$$

where for the bottom fork $h_{\pi(F)} = 0$.

First we explain in more detail the structure of a renormalized graph G^h_{ren}. Returning to the formula (36) for the kernel of the unrenormalized graph G^h, we rewrite G^h as

$$G^h = c \int d\alpha\, \zeta(\alpha,h) \int dx\, De^{-\mathcal{B}/4}\, \Psi(x) \tag{57a}$$

where $\alpha = (\alpha_\ell)_{\ell\in\mathcal{L}(G)}$,

$$\zeta(\alpha,h) = \prod_\ell \chi_\ell(\alpha_\ell, h_{f(\ell)})\alpha_\ell^{-d/2} e^{-\alpha_\ell m_\ell^2} \tag{57b}$$

$$\mathcal{B} = \sum_\ell \beta_\ell \Delta_\ell^2 \qquad (\beta_\ell = \alpha_\ell^{-1}) \tag{57c}$$

and

$$D = \prod_{\ell\in\mathcal{L}_F} (-i\beta_\ell \not{\Delta}_\ell/2 + m_\ell). \tag{57d}$$

We obtain the renormalized graphs corresponding to G^h by moving down the tree, applying the operations (48) - (50) at each fork f in succession. First we apply the interpolation operation (48) to each external argument x_j of G_f wherever x_j occurs as the argument of a leg of G_f. Then we take t_f-derivatives, as in (49a) at a C-fork and as in (50) at an R-fork. We regard a single renormalized graph G^h_{ren} as arising from a particular choice of m_f in (49a) and a particular choice of which external leg each ∂_{t_f} acts on. Thus to each unrenormalized graph G^h there will correspond a sum of $O(c_0^{|\mathcal{F}|})$ renormalized graphs G^h_{ren}. As we move down the tree, an x_j may continue to acquire further t-dependence, so long as it remains an external vertex. In the end a line difference Δ_ℓ will depend on $t_{>f(\ell)}$, where $t_{>f} = (t_{f'})_{f'>f}$; I'll simply write $\Delta_\ell(t)$ instead of $\Delta_\ell(t_{>f(\ell)})$.

When a derivative ∂_{t_f} acts on an external vertex $x_j(t_{\geq f})$ of G_f, x_j either occurs as the argument of an external field of G or as the endpoint of a line ℓ whose other endpoint is outside G_f. In the first case we obtain

$$\partial_{t_f}\Psi(x(t_{\geq f})) = (x_j(t_{>f}) - x_f(t_{>f}))\cdot\partial_{x_j}\Psi(x(t_{\geq f})), \tag{58a}$$

and in the second

$$\partial_{t_f}\Delta_\ell(t_{\geq f}) = x_j(t_{>f}) - x_f(t_{>f}) \tag{58b}$$

where Δ_ℓ may occur in D or in $\mathcal{B}$. In the latter occurrence the t_f-derivative brings down a factor

$$-\frac{1}{2}\,\beta_\ell(\partial_{t_f}\Delta_\ell)\cdot\Delta_\ell = -\frac{1}{2}\,\beta_\ell^{1/2}(x_j-x_f)\cdot\beta_\ell^{1/2}\Delta_\ell \tag{59a}$$

from $e^{-\mathcal{B}/4}$; in the former occurrence we obtain the factor

$$-\frac{i}{2}\,\beta_\ell^{1/2}(x_j-x_f)\cdot\gamma\,\beta_\ell^{1/2}. \tag{59b}$$

Indexing the t-derivatives by ν, and letting $x_{j(\nu)}$ be the vertex of the line or external field $\ell(\nu)$ on which $\partial_{t_{f(\nu)}}$ acts, we see that when we apply all the t-derivatives to $De^{\mathcal{B}/4}\Psi$ in (57a) we obtain a sum of terms of the form (we suppress the γ-matrices in D)

$$\prod_{\ell\in\tilde{\mathcal{L}}_F}\beta_\ell^{1/2}\;W\;\prod_\nu\beta_{\ell(\nu)}^{1/2}\,\Delta_\nu\;e^{-\mathcal{B}/4}\;\Pi \tag{60}$$

where $\tilde{\mathcal{L}}_F \subset \mathcal{L}_F$, W is a monomial in the quantities $\beta_\ell^{1/2}\Delta_\ell$, $\Delta_\nu = x_{j(\nu)} - x_{f(\nu)}$, $\beta_{\ell(\nu)} = 1$ if $\ell(\nu)$ is an external field, and $\Pi = \partial_x^q\,\Psi$. The quantities W, Δ_ν, and Π in (60) are all t-dependent. Since L = 0 and R = 1 at a fork with $\delta_f < 0$, W is at most of degree 3 in any particular $\beta_\ell^{1/2}\Delta_\ell$.

<u>Example 3</u> (cont'd) Take d = 4 in (32). Suppose that $\rho_f = R$ at each of the 3 forks of τ. Since $\delta_{f_1} = \delta_{f_2} = d - 6\,\frac{d-2}{2} = -2$, R = 1 at f_1 and f_2. But $\delta_{f_3} = 0$ so that at the lowest fork the R-operation (50) involves one $t_3 = t_{f_3}$ derivative. With $x_4 = 0$ the x-integral in (57a) is

$$\int dx_1dx_2dx_3e^{-\mathcal{B}/4}\Psi(x_2) \tag{61}$$

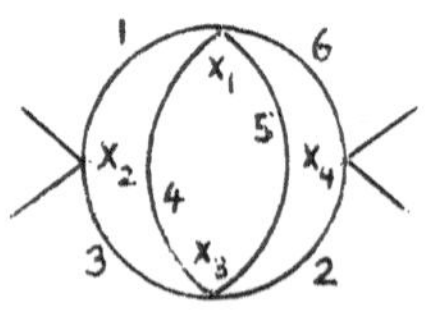

where (see (36c))

$$\Psi(x_2) = \int dy\;\Phi^e(x_2+y)^2\;\Phi^e(y)^2$$

and

$$\mathcal{B} = \sum\beta_\ell\Delta_\ell^2 = \beta_1(x_1-x_2)^2 + \beta_2 x_3^2 + \beta_3(x_2-x_3)^2 + (\beta_4+\beta_5)(x_1-x_3)^2 + \beta_6 x_1^2.$$

Choosing $x_{f_3} = x_4 = 0$ so that $x_2(t_3) = t_3 x_2$, we have the interpolated version of (61):

$$\int dx\, e^{-\mathcal{B}/4}\Psi(t_3 x_2). \tag{62}$$

The t_3-derivative of the integrand is

$$e^{-\mathcal{B}/4} x_2 \cdot \partial_{x_2}\Psi(t_3 x_2) \tag{63}$$

which is of the form (60) ($\Delta_\nu = x_2$).

If we want to see the subgraph G_2 of (18) undergo renormalization, we consider the tree τ:

τ =

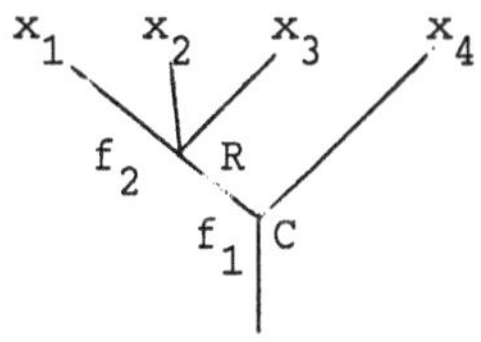

Choosing $x_{f_2} = x_2$ so that for $j = 1,3$ $x_j(t_2) = x_2 + t_2(x_j - x_2)$, we obtain the t_2-interpolated version of (61) by replacing $\mathcal{B}$ by

$$\mathcal{B}(t_2) = \beta_1(x_1-x_2)^2 + \beta_2 x_3(t_2)^2 + \beta_3(x_2-x_3)^2 + (\beta_4+\beta_5)(x_1-x_3)^2 + \beta_6 x_1(t_2)^2.$$

The t_2-derivative of the R-operation yields

$$\partial_{t_2} e^{-\mathcal{B}(t_2)/4} = -\frac{1}{2}[\beta_2(x_3-x_2)\cdot\Delta_2(t_2) + \beta_6(x_1-x_2)\cdot\Delta_6(t_2)]e^{-\mathcal{B}(t_2)/4}.$$

Finally, at the C-fork f_1 we choose $x_{f_1} = x_4 = 0$ so that $x_2(t_1) = t_1 x_2$. Setting $t_1 = 0$, we obtain the renormalized version of (61):

$$-\frac{1}{2}\int_0^1 dt_2 \int dx[\beta_2(x_3-x_2)\cdot\Delta_2(t_2) + \beta_6(x_1-x_2)\cdot\Delta_6(t_2)]e^{-\mathcal{B}(t_2)/4}\int dy\,\Phi^e(y)^4.$$

There are a couple of noteworthy points in the above example. First, in passing from G^h to G^h_{ren} not every occurrence of a variable x_j is interpolated, only those instances where x_j is the argument of an external leg (= field) of a subgraph G_f. Secondly, each occurrence of a difference Δ_ν arising from a t-derivative may be bounded by lines in $\mathcal{B}$ belonging to $\mathcal{M}_{\geq f(\nu)}$. For instance, with G labelled as in (32d) and with $\mathcal{M}_{f_3}$ chosen to be the h_3-line x_1x_3, then $\Delta_\nu = x_2$ in (63) is bounded by lines in $\mathcal{M}_{\geq f_3}$,

$$|x_2| \leq |x_2-x_1| + |x_1-x_3| + |x_3|, \tag{64a}$$

so that

$$|\Delta_\nu|e^{-\mathcal{B}/16} \leq c(\alpha_1^{1/2} + \alpha_4^{1/2} + \alpha_2^{1/2})$$

$$\le c\, M^{-h_3} = c\, M^{-h_{f(\nu)}} . \tag{64b}$$

In general, we wish to extract a factor $M^{-h_{f(\nu)}}$ from each Δ_ν in (60). In doing so we must control the percolation of coordinate differences up the tree (as illustrated in (64a)). The danger is that the constants in the bounds will grow with the number of t-derivatives; for example, if 2n factors of Δ_ℓ have collected, then

$$\Delta_\ell^{2n}\, e^{-\beta_\ell \Delta_\ell^2/16} \sim n!\,(16\alpha_\ell)^n .$$

However, this number singularity is controlled by the strict increase of scales, $h_f > h_{\pi(f)}$, at an R-fork, and we have the following bound:

Lemma 7.

$$|\prod_\nu \Delta_\nu|\; e^{-(\mathcal{B} + \sum \alpha_\ell m_\ell^2)/16} \le c \prod_f M^{-m_f h_f} . \tag{65}$$

Proof. Let $L_f = \sum_{\ell\in \mathcal{M}_f} |\Delta_\ell|$ be the length of the tree $\mathcal{M}_f$. Now a coordinate difference Δ_ν is bounded by the L_f's above $f(\nu)$ such that f and all intervening forks belong to the set $\mathcal{F}_R$ of R-forks:

$$|\Delta_\nu| \le L_{\ge f(\nu)} \equiv \sum_{\substack{f\ge f(\nu)\\ (f(\nu),f]\subset \mathcal{F}_R}} L_f .$$

Thus

$$|\prod_\nu \Delta_\nu| \le \prod_f L_{\ge f}^{m_f} . \tag{66a}$$

For $\ell \in \mathcal{M}_f$

$$\Delta_\ell^2/\alpha_\ell + \alpha_\ell \ge 2M^{h_f}\, |\Delta_\ell| - M^2$$

since $\alpha_\ell \le M^{-2h_f+2}$ if $h_f > 0$. Therefore

$$e^{-(\mathcal{B} + \sum \alpha_\ell m_\ell^2)/16} \le c\, e^{-\sum_f M^{h_f} L_f/8} \tag{66b}$$

Now

$$\sum_f M^{h_f} L_{\ge f} = \sum_{f'} L_{f'} \sum_{\substack{f\le f'\\ (f,f']\subset \mathcal{F}_R}} M^{h_f} \le 2 \sum_{f'} M^{h_{f'}} L_{f'} . \tag{66c}$$

It follows from (66) that

$$|\prod_\nu \Delta_\nu|\; e^{-(\mathcal{B} + \sum \alpha_\ell m_\ell^2)/16} \le c \prod_f L_{\ge f}^{m_f}\, e^{-M^{h_f} L_{\ge f}/16}$$
$$\le c \prod_f M^{-m_f h_f} . \qquad \blacksquare$$

Proof of Theorem 6. We wish to show that when (60) is inserted into the α- and x-integrals of (57a) the result is bounded as in (56). The first factor in (60) is bounded as in Lemma 2:

$$\prod_{\ell\in\mathcal{L}_F} \beta_\ell^{1/2} \le \prod_{\ell\in\mathcal{L}_F} M^{h_{f(\ell)}} . \tag{67a}$$

Since each $\beta_\ell^{1/2}\Delta_\ell$ occurs at most 3 times in W,

$$|W| e^{-\mathcal{B}/16} \le c. \tag{67b}$$

By (38)

$$\prod_\nu \beta_{\ell(\nu)}^{1/2} \le \prod_\nu M^{h_{f(\ell(\nu))}} \tag{67c}$$

where $h_{f(\ell(\nu))} = 0$ if $\ell(\nu)$ is an external field. By (65) and (67),

$$\int d\alpha\, \zeta(\alpha,h) \int dx\, |(60)| \le c \prod_{\ell\in\mathcal{L}_F} M^{h_{f(\ell)}} \prod_\nu M^{h_{f(\ell(\nu))}-h_{f(\nu)}} \|\pi\|_\infty$$

$$\int d\alpha\, \zeta\, e^{\sum\alpha_\ell m_\ell^2/16} \int dx\, e^{-\mathcal{B}/8}. \tag{68}$$

The factors

$$\prod_\nu M^{h_{f(\ell(\nu))}-h_{f(\nu)}} = \prod_f M^{-N_f(h_f-h_{\pi(f)})} \tag{69}$$

represent the power-counting contributions of the renormalization derivatives, the equality (69) following from

$$\sum_\nu (h_{f(\nu)}-h_{f(\ell(\nu))}) = \sum_\nu \sum_{f(\ell(\nu))<f\le f(\nu)} (h_f-h_{\pi(f)})$$
$$= \sum_f N_f(h_f-h_{\pi(f)}).$$

As for $\int dx\, e^{-\mathcal{B}/8}$, it is bounded as in (42), in the t-independent case. We drop the factors $e^{-\beta_\ell\Delta_\ell(t)^2/8}$ for $\ell \in \mathcal{M}^c$ and we change from x to "hard-line" variables $\xi = (\xi_\ell)_{\ell\in\mathcal{M}}$ where $\xi_\ell = \Delta_\ell(t)$. In spite of the t-dependence the Jacobian of the change of variables $x \to \xi$ is ± 1 (see Lemma B.2 of Ref. 3), and so

$$\int dx\, e^{-\mathcal{B}/8} \le \prod_{\ell\in\mathcal{M}} \int d\xi_\ell\, e^{-\beta_\ell\xi_\ell^2/8} = c \prod_{\ell\in\mathcal{M}} \alpha_\ell^{d/2}. \tag{70}$$

By (57b) and (70),

$$\int d\alpha\, \zeta\, e^{\sum\alpha_\ell m_\ell^2/16} \int dx\, e^{-\mathcal{B}/8} \le c \prod_{\ell\in\mathcal{M}} \int d\alpha_\ell \chi_\ell e^{-\alpha_\ell m_\ell^2/2}$$

$$\prod_{\ell\in\mathcal{M}^c} \int d\alpha_\ell \chi_\ell \alpha_\ell^{-d/2} e^{-\alpha_\ell m_\ell^2/2}.$$

The argument now continues as in Lemma 2: by (42) and (69),

$$(68) \le c\|\Pi\|_\infty \prod_f M^{(D(G_f)-N_f)(h_f-h_{\pi(f)})}$$

$$= c\|\Pi\|_\infty \prod_f M^{\delta_f(h_f-h_{\pi(f)})}$$

by (47). ∎

We call the bound (56) the "Spring-Loaded Bound" because (see (51)) at an R-fork

$$\delta_f \le -1 \quad \text{and} \quad h_f-h_{\pi(f)} > 0, \tag{71a}$$

and at a C-fork

$$0 \le \delta_f < d \quad \text{and} \quad h_f-h_{\pi(f)} \le 0. \tag{71b}$$

Thus, (56) supplies a stiff exponential spring between forks except in the case of a marginal C-fork for which $\delta_f = 0$. This enables us to sum over scales, the resulting geometric series converging uniformly in U. Powers of scales do accumulate because of marginal C-forks. For example, if there are κ marginal C-forks above the bottom R-fork F, the last sum over h_F is bounded by

$$c \sum_{h_F>r} h_F^\kappa M^{\delta_F h_F} \cong c\, \kappa!\, (r+2)^\kappa M^{\delta_F r}.$$

Corollary 8. (UV-Renormalizability) Consider a EQFT with dimensionless interaction and positive masses. Let G^h_{ren} be a graph contributing to $V(\tau,\rho,h)$ in the renormalized tree expansion (55). Then

$$\sum_{h\in \mathcal{H}_r(\tau,\rho)} |G^h_{ren}| \le c_0^{\ell(G)} \kappa!\,(r+2)^\kappa M^{\delta_F r} \tag{72}$$

where c_0 is a constant independent of U, G, and r, κ is the number of marginal C-forks in τ, and $\delta_F = \delta(G_{ren})$ satisfies (71).

IR-RENORMALIZABILITY

Suppose that the model involves massless fields. Then the covariance C_ℓ of a zero mass line ℓ has power law rather than exponential decay at ∞. For example, the covariance $C_\ell = (-\Delta)^{-1}$ for massless scalar bosons falls off like

$$C_\ell(x,y) \sim |x-y|^{2-d} \qquad \text{as } |x-y| \to \infty. \tag{73}$$

As a result, the kernel of a graph can have a non-integrable infrared singularity at ∞ as well as UV-singularities at coinciding arguments. As in the UV case, we resolve the IR-singularity (73) by means of the scale decomposition (22), where $C_\ell^{(h)}$ and χ_h are given by (25) and (26a). To regularize the theory, we now impose both a UV-cutoff $U > 0$ and an IR-cutoff $I \le 0$ and truncate sums like (22a) at both ends; e.g.,

$$C^{[I,U]} = \sum_{h=I}^{U} C^{(h)}, \quad \mathcal{E}^{[I,U]} = \prod_{h=I}^{U} \mathcal{E}^{(h)}, \quad \Phi^{(\leq k)} = \Phi^e + \sum_{h=I}^{k} \Phi^{(h)}, \quad \text{etc.}$$

Just as in the UV regime, our goal is to introduce R- and C-operations at the forks of τ, to develop a renormalized tree expansion, and to prove that the resulting graphs are finite, uniformly as $U \to \infty$ and as $r, I \to -\infty$. But our UV analysis falls short in the IR regime in two ways:

i) at a marginal C-fork ($\delta_f = 0$), the sum over h_f behaves like

$$\sum_{h_f=I}^{h_{\pi(f)}} M^{\delta_f(h_f - h_{\pi(f)})} = h_{\pi(f)} - I + 1$$

which diverges as $I \to -\infty$;

ii) if $\rho_F = R$ then the factor $M^{\delta_F r}$ in (72) diverges as $r \to -\infty$.

To overcome difficulty i) we simply do not introduce marginal counterterms in δV in the IR regime. For many models, these marginal counterterms are not required for the cancellations at R-forks. However, if we do require or want them at R-forks, we can insert them during the course of the expansion. In more detail, we decompose the localization operator L of (49a) as $L = L^0 + L^+$ where L^0 produces marginal counterterms (i.e. $m_f = \mu_f$ in (49a) so that by (51a) $\delta_f = 0$) and L^+ produces counterterms with $\delta_f > 0$. We generalize the UV-definitions (52) of the R- and C-operations at a fork f, namely,

$$R = \chi(h_f > h_{\pi(f)}) \, (1-L),$$

$$C = -\chi(h_f \leq h_{\pi(f)}) \, L,$$

as follows: at $f > F$

$$R = \chi(h_f > h_{\pi(f)}) \, (1-L) \tag{74a}$$

and

$$C = -\chi(h_f \leq h_{\pi(f)}) \, [L^+ + \chi(h_f \geq 0) L^0] + \chi(h_{\pi(f)} < h_f < 0) L^0 \equiv C_- + C_+ \tag{74b}$$

whereas at F (r is the root scale)

$$R = \chi(h_F > r) \, [1 - L^+ - \chi(h_F \geq 0) L^0] \tag{74c}$$

and

$$C = -\chi(h_F \leq r) \, [L^+ + \chi(h_F \geq 0) L^0] = C_- . \tag{74d}$$

With the definition (74d) the initial counterterms δV, given by (54b), do not contain marginal counterterms with $h_F < 0$. However, at forks $f > F$ we have inserted marginal counterterms by adding them at C-forks (see 74b)) and subtracting them at R-forks. The R- and C-operations satisfy the basic identity

$$R + C = \sum_{h>k} h - \sum_{h} [L^+ + \chi(h \geq 0) L^0] \, h \tag{75}$$

where the bottom fork in (75) is $f > F$.

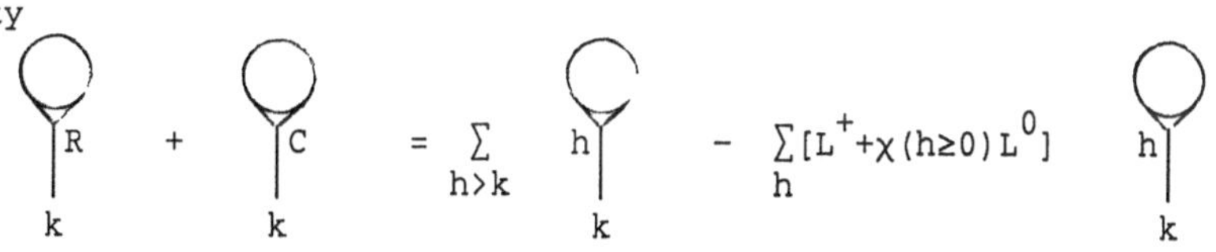

It is now straightforward to establish the renormalized true expansion for the IR regime. Each fork in the tree now bears the label $\rho_f = R, C_-$ or C_+ (the label C denotes the sum $C_- + C_+$) and the scales h run over the set $\mathcal{H}_r(\tau,\rho) = \{h \mid h_{\pi(f)} < h_f \le U$ if $\rho_f = R$ or C_+; $I \le h_f \le h_{\pi(f)}$ if $\rho_f = C_-\}$ where, here, $h_{\pi(F)} = r$. The intermediate effective potential is defined as in (55):

$$V_{ren,r} = [\log \mathcal{E}^{(>r)}(e^{V_{ren}})]_0 \ . \tag{76}$$

With (75) replacing (54c), the proof of the tree expansion is identical to that of Theorem 5:

<u>Theorem 9</u>. (<u>Renormalized Tree Expansion in the IR Regime</u>)

For $I - 1 \le r \le U$

$$V_{ren,r} = \sum_\tau \sum_\rho \sum_{h \in \mathcal{H}_r(\tau,\rho)} V(\tau,\rho,h)$$

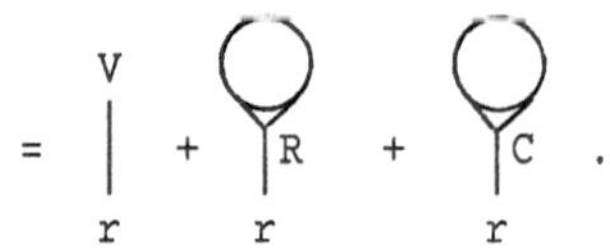

(77)

Difficulty ii) arises for two reasons. First, the L^1-norm is too crude in the IR regime, in the sense that it permits too many integrations of x_j's over all of $\mathbf{R}^d$; each such integration produces a factor $\alpha_\ell^{d/2}$ which is good in the UV- but bad in the IR-regime:

$$|\int dx e^{-\mathcal{E}/8}\, \Pi| \le \|\Pi\|_\infty \prod_{\ell \in} \int d\xi_\ell e^{-\beta_\ell \xi_\ell^2/8}$$
$$= c\|\Pi\|_\infty \prod_{\ell \in} \alpha_\ell^{d/2} \ .$$

If an x_j is external, i.e. if it is the argument of Π, then we need not and should not integrate it over $\mathbf{R}^d$ in the IR regime. Suppose that g_f has $v^e(g_f)>1$ external vertices (a generalized vertex G_f' of g_f is external if it contains an external x_j). Then, if $h_f<0$, we remove $v^e(g_f)-1$ lines from $\mathcal{M}_f$ to give a set $\tilde{\mathcal{M}}_f \subset \mathcal{M}_f$ of lines which connect the vertices of g_f into $v^e(g_f)$ separate connected components, each having exactly one external vertex which pins down that component. For example (external x_j's are marked with an x):

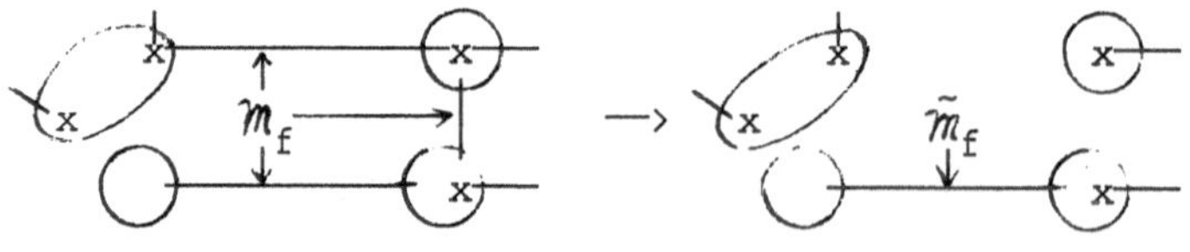

where $v^e(g_f) = 3$. If $h_f \geq 0$, we remove no lines: $\tilde{\mathfrak{M}}_f = \mathfrak{M}_f$. Let $\tilde{\mathfrak{M}} = \bigcup_f \tilde{\mathfrak{M}}_f$. We replace the estimate (78a) by

$$|\int dx e^{-\mathcal{B}/8} \Pi| \leq c(\Pi) \prod_{\ell \in \tilde{\mathfrak{M}}} \alpha_\ell^{d/2}, \tag{78b}$$

where the constant $c(\Pi)$ includes the integration over some of the x_j's. The improvement of (78b) over (78a) is

$$\prod_{f \in \mathfrak{F}_{en}} M^{dh_f(v^e(g_f)-1)} \tag{79}$$

where $\mathfrak{F}_{en}$ is the set of forks f with $v^e(g_f)>0$ and $h_f<0$.

There is one catch in the above reasoning. It works fine in the absence of renormalization. However, if we have interpolated at f so that an external variable x_j has acquired a t_f-dependence then integration over x_j as an argument of Π produces a factor t_f^{-d} in $c(\Pi)$. This is related to the second reason for difficulty ii): the renormalization operation (58a) is harmful in the IR regime. The factor $\Delta_\nu = x_j - x_f$ in (58a) gives a factor $\alpha_{f(\nu)}^{1/2} > 1$ which is not compensated for by a factor $\beta_{\ell(\nu)}^{1/2} < \alpha_{f(\nu)}^{-1/2}$. To get around this difficulty we pull apart the operation R=1-L at a fork $f \in \mathfrak{F}_{en}$ and at the "1-fork" we extract the factor in (79) instead of a renormalization factor.

For the resulting bounds on a graph G^h_{ren} see Corollary 6.6 for Ref. 3. The final conclusion is: any EQFT in d>2 dimensions with dimensionless interaction is both UV- and IR-renormalizable.

Now the counterterms δV in (55) or (76) are introduced as a sum (54c) over trees. One might ask whether this GN definition of δV agrees with the conventional BPHZ prescription, in which graphs are not decomposed as a sum over trees. To see that the answer is yes, let δV_n be all the terms of (54c) which are of order n in λ, and let L_n be the part of the localization operator L which produces terms of order n. We also write $\delta V_{\leq n} = \sum_{j=2}^{n} \delta V_j$, etc.

Corollary 10. For $n \geq 1$

$$\begin{aligned}\delta V_{n+1}(\Phi^e) = &- L^0_{n+1}[\log \mathcal{E}^{[0,U]}(e^{(V+\delta V_{\leq n})(\Phi^{[0,U]}+\Phi^e)})]_0 \\ &- L^+_{n+1}[\log \mathcal{E}^{[I,U]}(e^{(V+\delta V_{\leq n})(\Phi^{[I,U]}+\Phi^e)})]_0 .\end{aligned} \tag{80}$$

Proof. Set r= -1 in (76) and apply L^0. The C-graphs on the RHS drop out because of the $\chi(h_f \geq 0)$ in (74d), and the R-graphs drop out because

$$L^0 R = L^0(1-L) = L^0 - L^0 = 0 .$$

Therefore we obtain

$$L^0 V_{ren,-1} = L^0 V \ . \qquad (81a)$$

Then set r=e(i.e. I-1) in (76) and apply L^+. Now the C-graphs on the RHS of (76) drop out because of the $\chi(h_F \leq r)$ in (74d), and we obtain

$$L^+ V_{ren,e} = L^+ V \ . \qquad (81b)$$

Substituting (76) into (81) gives

$$L^0 \log \mathcal{E}^{[0,U]} e^{(V+\delta V)} = L^0 V \qquad (82a)$$

and

$$L^+ \log \mathcal{E}^{[I,U]} e^{(V+\delta V)} = L^+ V. \qquad (82b)$$

(80) follows from projecting (82) onto the order (n+1) of perturbation theory. ∎

Remarks. 1. The advantage of the GN tree definition (54c) of counterterms is that the cancellations against graphs needing renormalization is manifest, whereas the advantage of the representation (80) of the counterterms is that symmetries are manifest (see the discussion of gauge invariance below).

2. From (81) we see that dimensionless parameters are fixed by a normalization condition at scale -1 (analogous to having external momenta of order 1), whereas dimensionful parameters (such as mass) are fixed by a normalization condition at scale I-1 → -∞ (i.e. at zero external momenta). Any graph contributing to such a normalization condition would be IR divergent (difficulty i)) and consequently has zero as an exceptional momentum.

3. In particular, (81b) says that if Φ_i is a massless field then its renormalized mass is also zero. The contribution to the left side of (81b) from mass graphs in the IR region is finite. But this finite piece is essential both for fixing the renormalized mass at 0 and for convergence of the expansion.

LOCAL BOREL SUMMABILITY

Having such a well-oiled machine, I'd like to show you one thing it's good for, namely, bounds on the large order contributions of perturbation theory. Such bounds were first obtained for ϕ_4^4 by de Calan and Rivasseau[6]. Consider a dimensionless EQFT in d≥3 dimensions, possibly involving massless fields. Let $V_{e,p}^{(n)}$ be the contribution of nth order perturbation theory to the pth degree part (in Φ^e) of the effective potential V_e (or to a p-point Schwinger function S_p, or...). In the sense of formal power series,

$$V_{e,p} \sim \sum_{n=0}^{\infty} \lambda^n V_{e,p}^{(n)}.$$

There will be many graphs contributing to $V_{e,p}^{(n)}$ but combining our bounds on the contribution of an individual graph (as in (56)) with a bound of

Giovanni Felder (Appendix F of Ref.1) on the number of graphs, we have proved[3]:

Theorem 11. (Large order bounds) There are constants K (independent of n) and R>0 (independent of p and n) such that

$$|V_{e,p}^{(n)}| \leq K\, R^{-n} (n!)^{\frac{m}{2}-1}$$

where m = deg V.

Hence, if m ≤ 4, the Borel transform

$$\tilde{V}_{e,p} = \sum_{n=0}^{\infty} \frac{\lambda^n}{n!} V_{e,p}^{(n)}$$

of $V_{e,p}$ exists and is analytic in at least the disk $|\lambda| < R$. That is, the model is locally Borel summable.

GAUGE INVARIANCE

Consider the example of QED_4, given in (2). Applying the renormalization theorems developed above, we conclude that QED_4 is UV- and IR-renormalizable. But here's the catch: the local, Euclidean invariant terms of dimension 4 or less that arise as counterterms are

$$F^2,\ \bar{\psi}\psi,\ -i\bar{\psi}\not{\partial}\psi,\ \bar{\psi}\not{A}\psi \tag{83}$$

and

$$(\partial\cdot A)^2,\quad A^2,\quad A^4, \tag{84}$$

where $F^2 = \sum_{\mu,\nu} (\partial_\mu A_\nu - \partial_\nu A_\mu)^2$. However, the terms in (84) are gauge-variant and are forbidden in the renormalization of QED. Inasmuch as the GN procedure is inherently gauge variant we cannot rule out these forbidden counterterms.

To understand this dilemma let us examine the Ward identities. In terms of the (formal) expectation (4) define

$$W(\eta,\bar{\eta},B) = [\log \int e^{-\int dx(e\bar{\psi}B\psi + \bar{\psi}\eta + \bar{\eta}\psi)} e^{V(\Phi)} d\nu_f(\psi,\bar{\psi})\, d\nu_b(A)]_0$$

where B_μ is a vector source, η and $\bar{\eta}$ spinor sources, and we have written $d\nu_0(\Phi) = d\nu_f d\nu_b$. Since $S^{-1} = -i\not{\partial}+m$ satisfies

$$e^{-ie\chi} S^{-1} e^{ie\chi} = S^{-1} + e\not{\partial}\chi \tag{85a}$$

we have

$$d\nu_f(e^{ie\chi}\psi,\ e^{-ie\chi}\bar{\psi}) = e^{-e\int\bar{\psi}\not{\partial}\chi\psi} d\nu_f(\psi,\bar{\psi}) \tag{85b}$$

and so

$$W(\eta,\bar{\eta},B) = W(e^{-ie\chi}\eta,\ e^{ie\chi}\bar{\eta},\ B + \partial\chi) \tag{86}$$

These are the Ward identities. Note that the Ward identities depend on the relations (85) for $d\nu_f(\psi,\bar{\psi})$ and not on properties of $d\nu_b(A)$ (which is not gauge invariant since $\mathcal{L}_0$ has mass and gauge fixing terms).

Suppose we knew that the counterterms $\delta V(\psi,\bar{\psi},A)$ satisfied the Ward identities (86) (with $\eta,\bar{\eta},B$ replaced by $\psi,\bar{\psi},A$). Then we could conclude that the forbidden counterterms (84) do not occur and that only the counterterms (83) occur (with the last two in the combination $-i\bar{\psi}\partial\!\!\!/\psi + e\psi A\!\!\!/\bar{\psi}$). Unfortunately there seems to be no way of defining the scale propagators $S^{(h)}$ and regularized propagator

$$S^{(\leq N)} = \sum_{h=0}^{N} S^{(h)} \tag{87}$$

so as to satisfy (22b) and anything like (85a). This is the basic conflict between the GN procedure and the requirements of gauge invariance. Actually this problem arises with any renormalization scheme based on BPHZ ideas.

To resolve this conflict we introduce an auxiliary regularization involving S which does not disturb the Ward identities. Here, as in Ref. 3, I shall use loop regularization. Another possibility is dimensional regularization[7]. Instead of defining loop regularization for a general graph I'll simply illustrate it for the "vacuum polarization" graph

G = 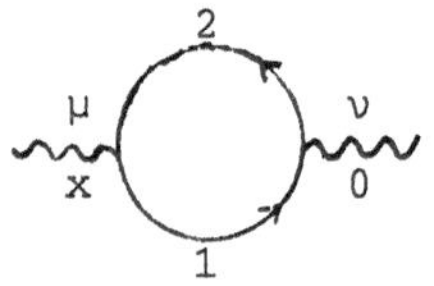

.

Here, as usual, the solid line is a fermion line S and the wavy leg with index μ is a photon field A_μ. As in (37) and (57) the kernel of G is

$$K^h = c \int d\alpha_1 d\alpha_2\, \zeta(\alpha,h)\, e^{-\beta\Delta^2/4}\ \mathrm{tr}\, \gamma^\mu (i\Delta\!\!\!/\beta_1/2+m)\gamma^\nu(i\Delta\!\!\!/\beta_2/2+m)$$

where $c = (4\pi)^{-4}$, $\beta = \beta_1+\beta_2$, $\Delta = x$, and

$$\zeta(\alpha,h) = \prod_{\ell=1}^{2} \chi_\ell(\alpha_\ell,h)\, \alpha_\ell^{-2} e^{-m^2\alpha_\ell} \equiv \chi(\alpha,h) \prod_\ell \alpha_\ell^{-2} e^{-m^2\alpha_\ell} .$$

Evaluating the trace, we have

$$K^h = c \int d\alpha\, \chi\, \alpha_1^{-3}\alpha_2^{-3} e^{-m^2(\alpha_1+\alpha_2)-\beta x^2/4} [2x^\mu x^\nu - \delta^{\mu\nu}x^2 - 4\alpha_1\alpha_2 m^2 \delta^{\mu\nu}]. \tag{88}$$

It is not hard to see that

$$\|K^h\|_1 \sim M^{2h}$$

so that without the cutoff function χ (88) would diverge because of the singularity at $\alpha_1 = \alpha_2 = 0$. On the other hand, χ spoils the Ward identity that we'd like K^h to satisfy:

$$\partial_{x^\mu} K^h = c \int d\alpha\, \chi\, \alpha_1^{-3}\alpha_2^{-3}\, e^{-m^2(\alpha_1+\alpha_2)-\beta x^2/4} [8 - \tfrac{1}{2}\beta x^2 + 2m^2(\alpha_1+\alpha_2)]x^\nu$$
$$\neq 0.$$

Loop regularization consists of taking a second difference of (88) with respect to m^2. Let $m(s)^2 = m^2 + s\Lambda^2$ where Λ is a large parameter. Then the loop regularization of a function $H(m^2)$ is

$$H(m^2+2\Lambda^2) - 2H(m^2+\Lambda^2) + H(m^2) = \Lambda^4 \int_0^1 ds_1 \int_0^1 ds_2\, H''(m^2(s_1+s_2)). \quad (89)$$

It's easy to check that $\partial^2_{m^2}K^h$ has two extra powers of α_1 and α_2 so that $\|\partial^2_{m^2}K^h\|_1 < \infty$, uniformly in h. Moreover, with the cutoff χ omitted, $\partial_{x^\mu}\partial^2_{m^2}K=0$; i.e. the desired Ward identity holds.

The standard way to implement loop regularization is to introduce fictitious spinor fields $\psi_j,\bar{\psi}_j$, j=1,2,3, where ψ_1 is a fermi field with mass $M_1=m(2)$ and ψ_2, ψ_3 are bose fields with mass $M_2=M_3=m(1)$. That is, the Lagrangian $\mathcal{L}$ of (2) is replaced by

$$\mathcal{L}_\Lambda = \mathcal{L} + \sum_{j=1}^{3} :\bar{\psi}_j\,(-i\partial\!\!\!/ + M_j + \lambda A\!\!\!/)\,\psi_j:. \quad (90)$$

The effect of these extra terms in $\mathcal{L}$ is to replace each graph in the unrenormalized tree expansion for V_e by a sum of similar graphs where each fermi line is labelled j=0,1,2, or 3 (j=0 being the original real fermi line) and each fermi loop is regularized as in (89).

Together with a UV-cutoff U>0 and an IR-cutoff I≤0 on the photon propagator, the loop regularization Λ renders the (unrenormalized) graphs contributing to V_e finite (see Lemma 3.1 of Ref. 3). Unfortunately, loop regularization cannot be used to define a sufficiently sensitive scale decomposition of S. The difficulty is that it does not allow a decomposition of each fermi propagator in a loop independently, as is necessary for the introduction of the correct counterterms. Accordingly, we still decompose S as in (87) (with N = ∞) and we are obliged to accept forbidden counterterms at each scale. However, when these forbidden counterterms are summed up over all scales they add up to zero by virtue of the Ward identities (which are not disturbed by the regularizations Λ,U,I).

Why do the summed counterterms verify the Ward identities? Let $\mathcal{E}^{(>r),\Lambda}$ denote the Gaussian expectation for the fields $A^{(r,U]}$, $\Psi = (\psi,\psi_1,\psi_2,\psi_3)$ and $\bar{\Psi}$; let $A^{(\leq U)} = A^e + A^{[I,U]}$; let

$$V = -\lambda \sum_{j=0}^{3} \int :\bar{\psi}_j A\!\!\!/^{(\leq U)}\psi_j:dx$$

be the interaction with the regularizations U,I,Λ in place; and let $\delta V_{\leq n}$ be the sum of all counterterms given by the GN procedure up to $O(\lambda^n)$. Define

$$W_{r,n}(A^{(\leq r)}) = [\log \mathcal{E}^{(>r),\Lambda}(e^{(V+\delta V_{\leq n})})]_0.$$

By Corollary 10, the sum of the purely bosonic counterterms of order

λ^{n+1} are obtained by taking appropriate local parts of $W_{r,n}$: the marginal counterterms of order λ^{n+1} are given by $-L^0_{n+1}W_{-1,n}$ and the non-marginal counterterms by $-L^+_{n+1}W_{I-1,n}$. As in (86), these counterterms are gauge-invariant if the counterterms $\delta V_{\le n}$ are. By this inductive reasoning we conclude that the renormalized effective potential $V^{I,U,\Lambda}_{ren}$ involves only gauge invariant counterterms.

There is a drawback to loop regularization that is perhaps not widely recognized, namely, its fragility with respect to renormalization. In defining the counterterms δV we use the correct counterterms $\bar\psi\psi$, $\bar\psi\partial\!\!\!/\psi$ and $\bar\psi A\!\!\!/\psi$ for the real spinor fields, but we cannot do so for the fictitious spinor fields. The reason is that loop regularization is a rather delicate subtraction that works only if the masses M_j satisfy

$$M_1^2 - M_2^2 - M_3^2 + m^2 = 0$$

and if the terms $\bar\psi(-i\partial\!\!\!/+\lambda A\!\!\!/)\psi$ and $\bar\psi_j(-i\partial\!\!\!/+\lambda A\!\!\!/)\psi_j$ in (90) occur with the same coefficient. To maintain these cancellations we are obliged to choose fictitious field counterterms that are matched to the real field counterterms and are incorrect insofar as renormalization cancellations are concerned. At first this use of incorrect counterterms seems disastrous. But when U and I are finite the graphs that appear to require fictitious field counterterms are actually finite, uniformly in Λ and N (see §5 of Ref. 3).

Thus, we establish that the renormalized graphs of QED_4 are finite provided we remove the regularizations in the right order. Our construction of the renormalized effective potential proceeds through the following sequence of effective potentials:

	$V^{I,U,\Lambda,N}_{ren,e}$	Run the tree expansion, including forbidden counterterms.
$N\to\infty$	$\downarrow$	
	$V^{I,U,\Lambda}_{ren,e}$	Ward Identities are recovered and show that the forbidden counterterms sum to 0.
$\Lambda\to\infty$	$\downarrow$	
	$V^{I,U}_{ren,e}$	The dependence on fictitious fields drops out as their masses $\to\infty$.
$U\to\infty$, $I\to-\infty$	$\downarrow$	
	$V_{ren,e}$	The standard power counting applies.

We emphasize that only gauge invariant counterterms are introduced in $V^{I,U,\Lambda}_{ren}$ but that the cancellations throughout the construction are performed in a gauge variant way.

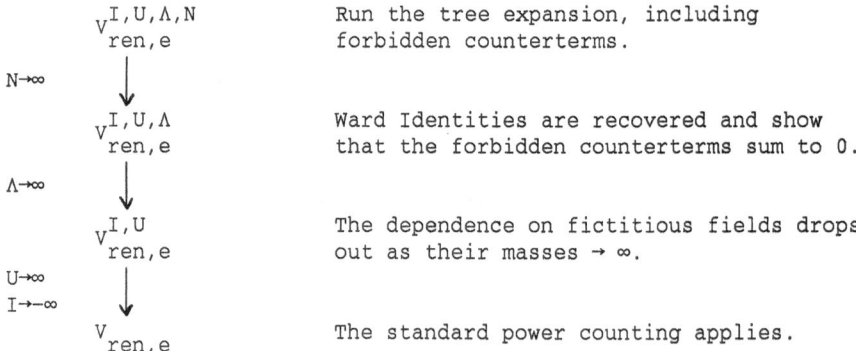

REFERENCES

1. G. Gallavotti, Rev. Mod. Phys. **57**, 471-562 (1985).

2. G. Gallavotti and F. Nicolò, Commun. Math. Phys. **100**, 545-590 (1985) and **101**, 247-282 (1986).

3. J. Feldman, T. Hurd, L. Rosen and J. Wright, QED: A Proof of Renormalizability, Springer Lecture Notes in Physics 312, Springer-Verlag, Heidelberg, 1988.

4. K. Hepp, Commun. Math. Phys. **2**, 301-326(1966).

5. W. Zimmermann, Commun. Math. Phys. **15**, 208-234 (1969).

6. C. de Calan and V. Rivasseau, Commun. Math. Phys. **82**, 69-100 (1981).

7. L. Rosen and J. Wright, Dimensional Regularization and Renormalization of QED, preprint.

CRITICAL PROPERTIES OF SOME

DISCRETE RANDOM SURFACE MODELS

Bergfinnur Durhuus

Mathematics Institute
University of Copenhagen
Universitetsparken 5
DK-2100 Copenhagen Ø

1. INTRODUCTION

Detailed insight into the statistical mechanics of random surfaces, and, in particular, the universality properties of such systems, is important for understanding a variety of phenomena in both condensed matter and high energy physics, of which we may note the following (see also refs. [1,2,3,4]):

1) Crystal growth and properties of crystalline surfaces.
2) Interfaces separating different phases of a system, appearing e.g. in wetting phenomena or as domain walls in ferromagnets.
3) Lipid bilayers, which are also interesting because of their similarity with certain biological membranes. These may be characterized as fluid membranes, meaning that their molecular constituents behave as a two-dimensional fluid.
4) Micro-emulsions, f.ex. mixtures of oil and water with a surfactant added. An essential feature of these is a very tiny surface tension, and the sizes of oil/water regions are of the order of 100 Å, i.e. much smaller than in ordinary emulsions where they are typically around 10^{-5} m.
5) Membranes with crystalline (or hexatic) order, as opposed to fluid membranes.
6) Random surface expansions in lattice gauge theory. Examples of such are strong conpling expansions and large -N expansions.
7) Effective theory for QCD flux tubes, or hadronic strings. Here the surfaces appear as world sheets of propagating strings. Using Feynman's approach to quantum mechanics one is lead to consider expressions of the form

$$\int \mathcal{D}S e^{i/hA(S)}$$

for transition amplitudes between string states. Here $A(S)$ is an action attributed to each surface S and $\mathcal{D}S$ is a measure on a suitable set of surfaces.

Constructive Quantum Field Theory II, Edited by G. Velo and
A. S. Wightman, Plenum Press, New York, 1990

One way to give a rigorous meaning to such an integral is by constructing continuum limits of corresponding well defined discrete expressions, as we shall discuss in the following.

8) (Super-) string theory, viewed as a fundamental theory of elementary particles.

9) Two-dimensional quantum gravity.

Although related to some of the problems listed here, the models to be considered in the following are to some extent chosen because of their accessibility to analytical investigations. In particular, the strongest results are obtained for surfaces without selfavoiding constraints. Naturally, such constraints are common for applications in condenced matter physics, but they may in some cases not be essential.

We shall discuss two classes of models. Section 2 deals with surfaces embedded in the hypercubic lattice $\mathbf{Z}^d$, while in section 3 we discuss piecewise linear surfaces embedded into $\mathbf{R}^d$. Section 4 contains some concluding remarks.

2. LATTICE SURFACES

2.1 Random Walk in $\mathbf{Z}^d$

In order to motivate some of the concepts introduced below for random surfaces let us briefly recall some of the properties of random walks in a hypercubic lattice $\mathbf{Z}^d$.

We define the two-point function $G_\beta(i,j)$ for $i,j \in \mathbf{Z}^d$ by

$$G_\beta(i,j) = \sum_{\omega:i\to j} e^{-\beta|\omega|}, \tag{1}$$

where the sum is over all connected random paths ω in $\mathbf{Z}^d$ starting at i and ending at j, $|\omega|$ is the length of ω and $\beta > ln2d$ is a coupling constant. Note that the number of paths of length n starting at i is $(2d)^n$, and it is an easy exercise to show that fixing the endpoint to be j only modifies this number by a multiplicative power of n. Thus, denoting by $N_{i,j}(n)$ the number of paths of length n from i to j, we have

$$\lim_{n\to\infty} \frac{\log N_{i,j}(n)}{n} = \log 2d \equiv \beta_1 \tag{2}$$

for arbitrary $i,j \in \mathbf{Z}^d$. This implies that the sum in (1) is convergent for $\beta > \beta_1$ and divergent for $\beta < \beta_1$.

It is easy to verify by inspection that

$$(-\Delta_L + m^2(\beta))G_\beta(\cdot,j) = \delta_j, \tag{3}$$

where Δ_L is the lattice Laplace operator defined by

$$(\Delta_L f)(i) = \sum_{\nu=1}^{d}(f(i+\underline{e}_\nu) + f(i-\underline{e}_\nu) - 2f(i)) \tag{4}$$

for $f : \mathbf{Z}^d \to \mathbf{R}$. Here, and in the following, $\underline{e}_1,\ldots,\underline{e}_d$ are unit vectors along the positive coordinate axes in $\mathbf{Z}^d$, and

$$m^2(\beta) = e^\beta - 1 \tag{5}$$

is the square of the dimensionless lattice mass.

Thus $G_\beta(i,j)$ is the kernel of the operator $(-\Delta_L + m^2(\beta))^{-1}$ on $L_2(\mathbf{Z}^d)$. By well known properties of this operator (see e.g. [5]) the mass $m(\beta)$ is determined by

$$m(\beta) = \lim_{n\to\infty} -\frac{\log G_\beta(0, n\underline{e}_\nu)}{n}. \tag{6}$$

The susceptibility $\chi(\beta)$ is defined as

$$\chi(\beta) = \sum_{j\in\mathbf{Z}^d} G_\beta(0,j), \tag{7}$$

and is easily calculable. One gets

$$\chi(\beta) = \sum_{n=0}^{\infty}(2d)^n e^{-\beta n} = (1 - e^{\beta_1-\beta})^{-1}. \tag{8}$$

From (5) and (8) we see that

$$m(\beta) \sim (\beta - \beta_1)^{\frac{1}{2}} \text{ as } \beta \searrow \beta_1, \tag{9}$$

$$\chi(\beta) \sim (\beta - \beta_1)^{-1} \text{ as } \beta \searrow \beta_1. \tag{10}$$

Here, and in the following, $A \sim B$ as $\beta \searrow \beta_1$ means that $c_1 B \le A \le c_2 B$ for β close to β_1, where c_1 and c_2 are finite positive constants.

The exponents $\nu = \frac{1}{2}$ and $\gamma = 1$ are the critical exponents of the mass and susceptibility, respectively.

A third critical exponent is the anomalous dimension η defined by

$$G_\beta(i,j) \sim |i-j|^{-(d-2+\eta)} \text{ for } 1 << |i-j| << m(\beta)^{-1} \tag{11}$$

as $\beta \searrow \beta_1$, where $|\cdot|$ denotes the euclidean distance. It is well known that $\eta = 0$ for the random walk.

Knowing the exponents ν and η is sufficient in order to construct a continuum limit of the random walk: We choose β as a function $\beta(a)$ of a lattice spacing a in physical units in such a way that

$$\frac{m(\beta(a))}{a} \to m_* > 0 \text{ as } a \to 0. \tag{12}$$

Then, choosing e.g. $a = 2^{-n} \equiv a_n$ and $x, y \in \cup_{n=1}^{\infty} 2^{-n}\mathbf{Z}^d$ we have that $a_n^{-1}x$, $a_n^{-1}y$ $\in \mathbf{Z}^d$ for n sufficiently large, and we may define the continuum limit G_* of the two-point function as

$$G_*(x,y) = \lim_{n\to\infty} a_n^{-(d-2)} G_{\beta(a_n)}(a_n^{-1}x, a_n^{-1}y). \tag{13}$$

When extended by continuity to $\{(x,y) \in \mathbf{R}^{2d} \mid x \neq y\}$ G_* is proportional to the kernel of $(-\Delta + m_*^2)^{-1}$ on $\mathbf{R}^d$, i.e. the euclidean free particle propagator.

Thus we have carried out the construction of a continuum limit for the lattice random walk. The following sections may be viewed as decribing results pertaining to a similar construction for the considerably less trivial case of lattice random surfaces.

2.2 Random Surfaces in $\mathbf{Z}^d$

An oriented lattice surface S in $\mathbf{Z}^d$ is given by:

1) A finite set $V(S)$ called the vertex set.
2) A set $P(S)$ of cyclically ordered subsets of $V(S)$, called plaquettes, each containing four elements, such that $\cup_{p \in P(S)} p = V(S)$. -An oriented link in S is a pair (ij) such that there exists a $p = (ijlm) \in P(S)$, where the notation $(ijlm)$ is used to indicate the cyclic ordering of the elements i, j, l, m in p. Note that e.g. $(ijlm) = (mijl)$. We require that each oriented link is contained in exactly one $p \in P(S)$. An interiour link in S is a set $\{i, j\}$ such that both (ij) and (ji) are oriented links in S. If only (ij), and not (ji), is an orientet link in S we call $\{i, j\}$ a boundary link and i and j are called boundary vertices. By $L(S)$ we denote the set of all (interiour and boundary) links in S.
3) A mapping $V(S) \to \mathbf{Z}^d$, which we also denote by S, such that plaquettes are mapped onto (the corners of) elementary plaquettes in $\mathbf{Z}^d$ in cyclic order.

Two surfaces S_1 and S_2 are considered identical if there is a bijective mapping $\varphi : V(S_1) \to V(S_2)$, which maps the vertices of each $p_1 \in P(S_1)$ onto the vertices of a $p_2 \in P(S_2)$ preserving cyclic order, and such that $S_1 = S_2 \circ \varphi$.

For a given lattice plaquette $\tilde{p}$ in $\mathbf{Z}^d$ we define the multiplicity of $\tilde{p}$ in S as the number of plaquettes in $P(S)$ mapped onto $\tilde{p}$ by S. Similarly we define the multiplicities of lattice links and points in $\mathbf{Z}^d$.

A surface is said to be selfavoiding if no vertex (and hence no link or plaquette) has multiplicity > 1.

A path in S is a sequence of links in S of the form

$$\{i_0, i_1\}, \{i_1, i_2\}, \{i_2, i_3\}, \ldots, \{i_{n-1}, i_n\}.$$

The vertices $i_0, \ldots, i_n$, of course, determine the path, and we say that i_0 and i_n are connected by this path. A surface S is said to be connected if any two vertices in S can be connected by a path in S. *In the following all surfaces are assumed to be connected.*

The boundary ∂S of a surface S is the one-dimensional complex consisting of boundary links and vertices. We shall denote the number of connected components in ∂S (which, of course, may be considered as closed paths in $\mathbf{Z}^d$) by $b(S)$.

The Euler characteristic $\chi(S)$ is defined by

$$\chi(S) = |S| - l(S) + v(S), \tag{14}$$

where $|S|$, $l(S)$ and $v(S)$ denote the number of plaquettes, links and vertices in S, respectively. The genus $h(S)$ (or number of handles) is then given by

$$\chi(S) = 2(1 - h(S)) - b(S). \tag{15}$$

For a given vertex $i \in V(S)$ we let σ_i denote the order of i, i.e. the number of links (or plaquettes) in S containing i. Noting that

$$\sum_{i \in V(S)} \sigma_i = 2l(S) \tag{16}$$

and

$$4|S| = 2l(S) - |\partial S|, \tag{17}$$

where $|\partial S|$ is the length of (i.e. number of links in) ∂S, we get that

$$\chi(S) = \frac{1}{4} \sum_{i \in V(S)} (4 - \sigma_i) - \frac{1}{4}|\partial S|, \tag{18}$$

which is the lattice analog to the Gauss-Bonnet formula, since the deficiency index

$$\delta_i = 4 - \sigma_i \tag{19}$$

is a measure for the intrinsic curvature of S at i.

We now proceed to define the analogs of the two-point function $G_\beta(i,j)$ in (1). The endpoints i, j are replaced by a set of oriented lattice loops $\gamma_1, \ldots, \gamma_n$ in $\mathbf{Z}^d$, and we let

$$G_\beta(\gamma_1, \ldots, \gamma_n) = \sum_{\partial S = \gamma_1 \cup \cdots \cup \gamma_n} e^{-\beta|S|}. \tag{20}$$

Without any restriction on S this sum, however, diverges for all $\beta \in \mathbf{R}$, since the number of surfaces with area A and fixed boundary grows faster than $(A!)^q$, $q > 0$. F.eks. if $n = 1$ and γ_1 is the boundary of a plaquette $\tilde{p}$ in $\mathbf{Z}^d$ the number of surfaces consisting of A copies of $\tilde{p}$ and with boundary γ_1 is $\geq ((\frac{A-1}{2})!)^2$, as the reader may easily verify [6].

Clearly, there is a number of natural ways to restrict the class of surfaces, depending on the physical nature of the system under consideration. We shall confine ourselves to only two of these: 1) The surfaces are required to be selfavoiding, and 2) the surfaces are required to have a fixed topology, i.e. a fixed number of handles.

The topological restriction in 2) is inspired partly by the topological expansion of scattering amplitudes in string theory and partly by large-N expansions in gauge theory. As we shall see below, the critical behaviour of this model can be analyzed in great detail.

That the above mentioned restrictions actually are suitable in relation to the proposed definition (20) is implied by the next two lemmas.

LEMMA 2.1. *Let $\gamma_1, \ldots, \gamma_n$ be a non-selfintersecting set of loops in $\mathbf{Z}^d$ and let $\mathcal{U}(\gamma_1 \ldots, \gamma_n)$ be the set of selfavoiding surfaces with boundary components $\gamma_1, \ldots, \gamma_n$. Then the number*

$$N_{\gamma_1,\ldots,\gamma_n}(A) = \sharp\{S \in \mathcal{U}(\gamma_1, \ldots, \gamma_n) \mid |S| = A\}, \tag{21}$$

of surfaces in $\mathcal{U}(\gamma_1, \ldots, \gamma_n)$ with area A fulfills

$$\lim_{A \to \infty} \frac{\log N_{\gamma_1,\ldots,\gamma_n}(A)}{A} = \delta_0, \tag{22}$$

where $\delta_0 \in \mathbf{R}_+$ and is independent of n and $\gamma_1, \ldots, \gamma_n$.

REMARK 2.2: Clearly, this lemma shows that

$$G_\beta^{\mathcal{U}}(\gamma_1, \ldots, \gamma_n) = \sum_{S \in \mathcal{U}(\gamma_1,\ldots,\gamma_n)} e^{-\beta|S|} \tag{23}$$

is convergent and analytic for $\beta > \delta_0$ while the sum diverges for $\beta < \delta_0$. The loop correlation functions $G_\beta^{\mathcal{U}}(\gamma_1, \ldots, \gamma_n)$ define the selfavoiding random surface (SARS) model.

Next we have

LEMMA 2.3. *Let $\gamma_1, \ldots, \gamma_n$ be arbitrary loops in $\mathbf{Z}^d$ and denote by $\mathcal{S}^h(\gamma_1, \ldots, \gamma_n)$ the set of surfaces in $\mathbf{Z}^d$ with boundary components $\gamma_1, \ldots, \gamma_n$ and with h handles. Then the number*

$$n^h_{\gamma_1,\ldots,\gamma_n}(A) = \sharp\{S \in \mathcal{S}^h(\gamma_1, \ldots, \gamma_n) \mid |S| = A\} \tag{24}$$

fulfills

$$\lim_{A\to\infty} \frac{\log n^h_{\gamma_1,\ldots,\gamma_n}(A)}{A} = \beta_0, \tag{25}$$

where $\beta_0 \in \mathbf{R}_+$ and is independent of n and $\gamma_1, \ldots, \gamma_n$ and h.

REMARK 2.4: This lemma shows that

$$G^h_\beta(\gamma_1, \ldots, \gamma_n) = \sum_{S\in\mathcal{S}^h(\gamma_1,\ldots,\gamma_n)} e^{-\beta|S|}, \tag{26}$$

converges and is analytic for $\beta > \beta_0$ while it is divergent for $\beta < \beta_0$. For $h = 0$, which is the case we shall mainly be concerned with in the following, the loop correlation functions $G^0_\beta(\gamma_1, \ldots, \gamma_n)$, for which we shall use the notation $G_\beta(\gamma_1, \ldots, \gamma_n)$, define the planar random surface (PRS) model.

Since the proofs of the two lemmas are quite similar, we shall give a proof of the second one only. For a proof of the former we refer to [7].

PROOF OF LEMMA 2.3.

Independence of $\gamma_1, \ldots, \gamma_n$: Observe that if $\gamma'_1, \ldots, \gamma'_n$ and $\gamma''_1, \ldots, \gamma''_n$ are two sets of loops, we can connect γ'_i and γ''_n, $i = 1, \ldots, n$, by cylinders S_i, i.e. $\partial S_i = \gamma'_i \cup \gamma''_i$ and $h(S_i) = 0$. By gluing these along $\gamma'_1 \ldots, \gamma'_n$ to any $S \in \mathcal{S}^h(\gamma'_1, \ldots, \gamma'_n)$ we get that

$$n^h_{\gamma''_1,\ldots,\gamma''_n}(A + |S_1| + \cdots + |S_n|) \geq n^h_{\gamma'_1,\ldots,\gamma'_n}(A), \tag{27}$$

which clearly implies that existence and value of the limit in (25) for a given n is independent of $\gamma_1, \ldots, \gamma_n$.
Independence of n: If γ_1 and $\gamma'_1, \ldots, \gamma'_n$ are arbitrary loops in $\mathbf{Z}^d$ we choose a fixed $S_0 \in \mathcal{S}^0(\gamma_1, \gamma'_1, \ldots \gamma'_n)$. Gluing S_0 along γ_1 to an arbitrary $S \in \mathcal{S}^h(\gamma_1)$ we get

$$n^h_{\gamma'_1,\ldots,\gamma'_n}(A + |S_0|) \geq n^h_{\gamma_1}(A). \tag{28}$$

On the other hand, given n loops $\gamma_1, \ldots, \gamma_n$ in $\mathbf{Z}^d$ we may choose fixed surfaces $S_i \in \mathcal{S}^0(\gamma_i)$, $i = 2, \ldots, n$. By gluing these to an arbitrary $S \in \mathcal{S}^h(\gamma_1, \ldots, \gamma_n)$ we get

$$n^h_{\gamma_1}(A + |S_2| + \cdots + |S_n|) \geq n^h_{\gamma_1,\ldots,\gamma_n}(A). \tag{29}$$

Clearly, inequalities (28) and (29) imply that existence and value of β_0 is independent of n.

We also note that if $\mathcal{S}^h$ denotes the set of closed surfaces with h handles and containing a fixed point in $\mathbf{Z}^d$, say 0, and if

$$n^h(A) = \sharp\mathcal{S}^h(A), \tag{30}$$

where

$$\mathcal{S}_h(A) = \{S \in \mathcal{S}^h \mid |S| = A\}, \tag{31}$$

then the limit β_0 in (25) exists if and only if

$$\lim_{A\to\infty} \frac{\log n^h(A)}{A} = \beta_0. \tag{32}$$

This is seen as follows. Choose γ_1 to be the boundary of a plaquette in $\mathbf{Z}^d$ containing 0. By adding this plaquette to any $S \in \mathcal{S}^h(\gamma_1)$ we obtain a surface $S' \in \mathcal{S}^h$ and, clearly, at most $|S'|$ different surfaces in $\mathcal{S}^h(\gamma_1)$ can give rise to the same surface in $\mathcal{S}^h$. Thus

$$n^h(A+1) \geq \frac{n^h_{\gamma_1}(A)}{A+1}. \tag{33}$$

On the other hand, given $S' \in \mathcal{S}^h$, it contains at least one plaquette containing 0. Removing one of these we obtain a surface $S \in \mathcal{S}^h(\gamma_i)$, where γ_i is the boundary of one of the $2d(d-1)$ plaquettes in $\mathbf{Z}^d$ containing 0, and, clearly, different elements in $\mathcal{S}^h(A+1)$ give rise to different elements in $\cup_{i=1}^{2d(d-1)} \mathcal{S}^h(\gamma_i)$. Thus we obtain

$$2d(d-1)n^h_{\gamma_1}(A) \geq n^h(A+1). \tag{34}$$

From (33) and (34) the claim follows.

Existence of the limit (32): We first consider the case $h = 0$ and apply a subadditivity argument. For $S_1 \in \mathcal{S}^0(A_1)$, $S_2 \in \mathcal{S}^0(A_2)$ let l_1 be a link in S_1 both of whose endpoints have a maximal x_1-coordinate and let l_2 be a link in S_2 both of whose endpoints have a minimal x_1-coordinate. Now rotate S_2 around the x_1-axis such that l_2 becomes parallel to l_1, and then translate S_2 in such a way that l_2 is moved to the link $l_1 + \underline{e}_1$ (i.e. l_1 translated by one lattice unit in the positive 1-direction). Now cut the two surfaces open along the links l_1 and $l_1 + \underline{e}_1$ and glue them together by adding the cylinder $S(l_1, l_1 + \underline{e}_1)$ consisting of two copies of the plaquette containing l_1 and $l_1 + e_1$, and whose boundary is $\gamma(l_1) \cup \gamma(l_1 + \underline{e}_1)$, where $\gamma(l)$ denotes the loop of length 2 consisting of two copies of the link l. We have now obtained a surface $S_3 \in \mathcal{S}^0(A_1 + A_2 + 2)$ and it is clear that if (S_1, S_2) and (S_1', S_2'), with $|S_1| = |S_1'| = A_1$ and $|S_2| = |S_2'| = A_2$, give rise to the same surface S_3, then $S_1 = S_1'$ and S_2' can be obtained from S_2 by first translating it and then rotating it around the 1-axis. The number of possible translations of one surface $S_2' \in \mathcal{S}^0(A_2)$ to another one $S_2 \in \mathcal{S}^0(A_2)$ is at most $v(S_2) = A_2 + 2$. Thus, denoting by c_d the number of lattice rotations around the x_1-axis, we conclude that

$$n^0(A_1 + A_2 + 2) \geq \frac{n^0(A_1)n^0(A_2)}{c_d(A_2+2)}. \tag{35}$$

Equivalently, if we set

$$f(A) = -\log(c_d^{-1} n^0(A-2)),\ A \geq 2, \tag{36}$$

we have

$$f(A_1 + A_2) \leq f(A_1) + f(A_2) + \log A_2. \tag{37}$$

This shows that f is subadditive up to a logarithmic term which, however, does not violate the standard proof of the existence of the limit of $f(A)/A$ as $A \to \infty$ (see e.g. [8]). This proves that the limit (32) exists for $h = 0$.

The argument actually proves the following more general inequality

$$n^{h_1+h_2}(A_1 + A_2 + 2) \geq \frac{n^{h_1}(A_1)n^{h_2}(A_2)}{c_d(A_2 + 2)}. \tag{38}$$

Setting $h_1 = 0$, taking logs, dividing by A_1 and letting $A_1 \to \infty$, we obtain

$$\liminf_{A\to\infty} \frac{\log n^h(A)}{A} \geq \lim_{A\to\infty} \frac{\log n^0(A)}{A}. \tag{39}$$

To prove that

$$\limsup_{A\to\infty} \frac{\log n^h(A)}{A} \leq \lim_{A\to\infty} \frac{\log n^0(A)}{A} \tag{40}$$

involves cutting up handles and is somewhat complicated from a combinatorial point of view. Since we shall not need the independence of β_0 on h in the following, we refrain from giving this argument her.

Finally, we have to prove that β_0 is positive and finite.

Lower bound on β_0: By considering random walks whose steps are surfaces of the form $S(l, l + \underline{e}_\nu)$, where $\underline{e}_\nu$ is orthogonal to l, it follows that

$$n^h_{\gamma(l_0)}(2A) \geq (2(d-1))^A$$

and thus

$$\beta_0 \geq \frac{1}{2}\log 2(d-1). \tag{41}$$

Upper bound on β_0: Let $S \in S^h(A)$ be given. We may construct S by successively gluing on plaquettes and gluing links together in the following way: Let S_1 be the surface consisting of one of the plaquettes in S containing 0. Choose a cyclic ordering of the vertices in its boundary and set $n_1 = 1$. Assume that we have constructed (S_i, n_i), where S_i is a subsurface of S with no handles and one boundary component, whose vertices $O^i_1, \ldots, O^i_{m_i}$ are ordered cyclically along ∂S_i, and such that $n_i < m_i (= |\partial S_i|)$ and each link $(O^i_j O^i_{j+1})$, with $j = 1, \ldots, n_i - 1$, has the property that (considered in S) it has to be glued to another link in ∂S_i, but not to its predecessor in ∂S_i (which is $(O^i_{j-1} O^i_j)$ if $j \geq 2$ and $(O^i_{m_i} O^i_1)$ if $j = 1$).

The pair (S_{i+1}, n_{i+1}) is then constructed as follows. Consider the link $l_1 = (O^i_{n_i} O^i_{n_i+1})$ in ∂S_i. There are now three possibilities:

1) If a new plaquette in S (i.e. one which is not in S_i already) has to be glued onto l_i we let S_{i+1} be the surface obtained by gluing this plaquette onto S_i along l_i and we set $O^i_1 = O^{i+1}_1$ and $n_{i+1} = n_i$ (see fig. 1)

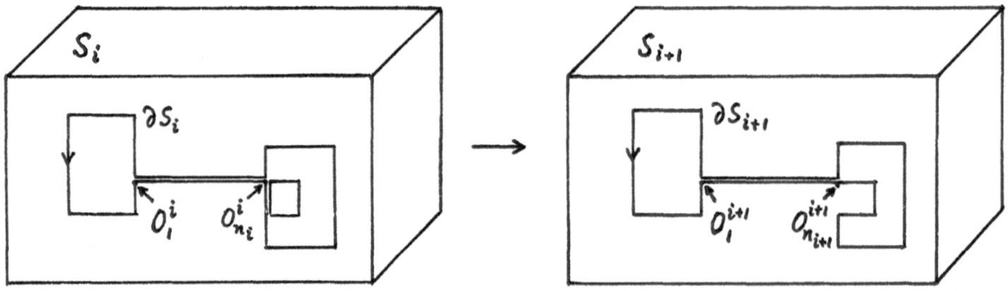

Figure 1

2) If the link l_i overlaps its predecessor and has to be glued to it (in S) we let S_{i+1} be the surface so obtained from S_i, and we let $O_1^{i+1} = O_1^i$ and $n_{i+1} = n_i - 1$ in case $n_i \geq 2$, whereas, if $n_i = 1$, we let $n_{i+1} = 1$ and $O_1^{i+1} = O_2^i$ (see fig. 2)

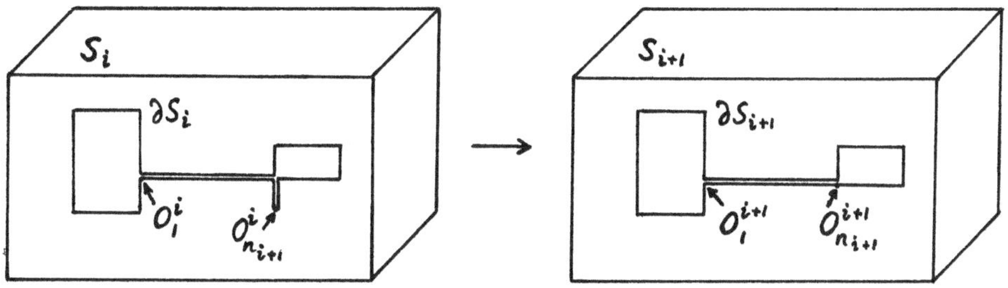

Figure 2

3) If the link l_i has to be glued (in S) to another link in ∂S_i but not to its predecessor, we let $S_{i+1} = S$, $O_1^{i+1} = O_1^i$ and $n_{i+1} = n_i + 1$ (see fig. 3).

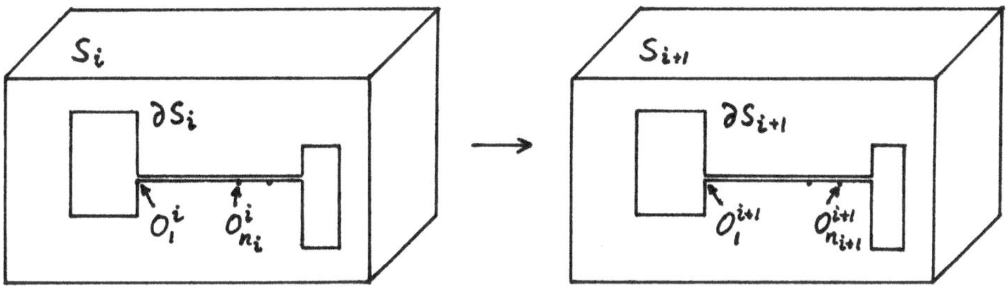

Figure 3

Clearly, after a number of steps $i_0 \leq 4A$ we encounter a situation where $n_{i_0} = m_{i_0}$. We have then obtained a surface S_{i_0} containing all the plaquettes in S, and S may be obtained from S_{i_0} by gluing pairwise the links in ∂S_{i_0} with the restriction that at most one link (namely $(O_1^i O_2^i)$) can be glued to its predecessor in ∂S_{i_0}. Since S_{i_0} has no handles while S has h handles the final gluings of the links in ∂S_{i_0} have to give rise to exactly h handles. It is easy to see by using formulas (14) and (15) and the restricition that at most one pair of neighbouring links in ∂S_{i_0} may be glued to each other, that from a given S_{i_0} at most $A^{c_1 h + c_2}$ different surfaces with h handles may be constructed in this way, where c_1 and c_2 are constants. F.eks. if $h = 0$ it is trivial to see that there is only one way to glue the links in ∂S_{i_0}.

Combining this observation with the fact that operation 1) can be performed in $2(d-1)$ ways depending on the orientation of the plaquette which is glued on, while 2) and 3), of course, each may be performed in only one way, such that at each step there are at most $2d$ possibilities, it follows that

$$n^h(A) \leq A^{c_1 h + c_2}(2d)^{4A}$$

showing that

$$\beta_0 \leq 4 \log 2d,$$

which is the desired bound.

This concludes the proof of Lemma 2.3. □

REMARK 2.5: The argument may be refined to yield the bound

$$\beta_0 \le \log 54(d-1). \tag{42}$$

It has been argued in [9] by using a large-N reduction technique that

$$\beta_0 \le \log 24(d-1).$$

REMARK 2.6: It is an interesting problem to find suitable topological restrictions for higher dimensional (discrete) manifolds, such that their number increases exponentially as a function of the volume.

REMARK 2.7: As mentioned above the proof of Lemma 2.1 is similar to the above proof. The combinatorial part is quite trivial, while the independence of δ_0 on n and $\gamma_1, \ldots, \gamma_n$ is slightly more involved (see [7]).

REMARK 2.8: In Section 2.3 we shall also need to consider the PRS_2-model whose n-loop function is defined by

$$G_{2,\beta}(\gamma_1, \ldots, \gamma_n) = \sum_{S \in \mathcal{S}_2^0(\gamma_1, \ldots, \gamma_n)} e^{-\beta|S|},$$

where $\mathcal{S}_2^0(\gamma_1, \ldots, \gamma_n)$ denotes the set of surfaces in $\mathcal{S}^0(\gamma_1, \ldots, \gamma_n)$, which contain no loops of length 2 (i.e. no loops of the form $\gamma(l)$) exept possibly in the boundary. All the results proven here for the PRS-model are valid also for the PRS_2-model, with trivial modifications of the proofs. The value of β above which $G_{2,\beta}$ is finite is denoted by β_2. By scaling a surface in $\mathcal{S}^0(\gamma_1, \ldots, \gamma_n)$ with area A by a factor 2 we obtain a surface in $\mathcal{S}_2^0(2\gamma_1, \ldots, 2\gamma_n)$ with area $4A$. From this it follows that

$$\frac{1}{4}\beta_0 \le \beta_2 \le \beta_0.$$

Invoking universality we expect that the critical behaviours, and in particular the critical exponents to be defined below, are identical in the PRS- and the PRS_2-model.

Having defined our models we proceed to define the analogs of the mass and susceptibility for the random walk.

Let γ_0 be an arbitrary loop in $\mathbf{Z}^d$ and let γ_n be a translate of γ_0 by n lattice units along a coordinate axis. We define

$$m(\beta) = -\lim_{n\to\infty} \frac{\log G_\beta(\gamma_0, \gamma_n)}{n}. \tag{43}$$

It is easy to see that $m(\beta) \le 2\beta$ in the PRS-model and $m(\beta) \le 4\beta$ in the $SARS$-model, provided the limits exist. We shall prove the existence of the limit (43) for the PRS-model below.

The susceptibility is defined as

$$\chi(\beta) = \sum_{l \in \mathbf{Z}^d} G_\beta(\gamma(l_0), \gamma(l)), \tag{44}$$

where the sum is over all links in $\mathbf{Z}^d$ and l_0 is some fixed link. We note that by gluing together the two copies of l in $\gamma(l)$ in a surface $S \in \mathcal{S}^h(\gamma(l_0), \gamma(l))$ (resp. $\mathcal{U}(\gamma(l_0), \gamma(l))$) one obtains a surface $S' \in \mathcal{S}^h(\gamma(l_0))$ (resp. $\mathcal{U}(\gamma(l_0))$), and it is easy to see that a given surface S' can be obtained in this way from at most $|S'|$ and at least $\frac{1}{2}|S'|$ different surfaces S. Thus, since

$$-\frac{d}{d\beta}G_\beta(\gamma(l_0)) = \sum_{S \in \mathcal{S}^h(\gamma(l_0))} |S|e^{-\beta|S|}, \tag{45}$$

we have

$$-\frac{1}{2}\frac{d}{d\beta}G_\beta(\gamma(l_0)) \le \chi(\beta) \le -\frac{d}{d\beta}G_\beta(\gamma(l_0)). \tag{46}$$

As concerns the critical behaviour to be discussed below we might thus as well have defined

$$\chi(\beta) = -\frac{d}{d\beta}G_\beta(\gamma(l_0)). \tag{47}$$

In the following we shall make use of the most convenient definition in any case at hand.

A third quantity, which has no analog in the case of a random walk, is the string- or surface tension $\tau(\beta)$. It is defined as follows. Let $\gamma_{R,T}$ be a rectangle with edges of length R and T in a coordinate plane. Then

$$\tau(\beta) = -\lim_{R,T\to\infty} \frac{\log G_\beta(\gamma_{R,T})}{RT}. \tag{48}$$

The existence of this limit in the PRS-model is proven below. It is easy to see that $\tau(\beta) \le \beta$, if it exists.

Now the critical exponents ν, γ and μ, provided they exist, are defined by

$$m(\beta) - m(\beta_0+) \sim (\beta - \beta_0)^\nu, \tag{49}$$

$$\chi(\beta) \sim (\beta - \beta_0)^{-\gamma}, \tag{50}$$

$$\tau(\beta) - \tau(\beta_0+) \sim (\beta - \beta_0)^\mu, \tag{51}$$

as $\beta \searrow \beta_0$. In this definition it is assumed that $m(\beta_0+)$ and $\tau(\beta_0+)$ are finite and that $\chi(\beta) \nearrow \infty$ as $\beta \searrow \beta_0$. If $\chi(\beta_0+)$ is finite we define γ by

$$\chi(\beta_0+) - \chi(\beta) \sim (\beta - \beta_0)^{-\gamma} \tag{52}$$

as $\beta \searrow \beta_0$. Thus $\gamma \le 0$ in this case, since clearly χ is a decreasing function of β. It is also clear that m and τ are non-decreasing functions of β.

If $m(\beta_0+) = 0$ we define the anormalous dimension η, provided it exists, by

$$G_\beta(\gamma_0, \gamma_n) \sim n^{-(d-2+\eta)} \text{ for } 1 << n << m(\beta)^{-1}, \tag{53}$$

as $\beta \searrow \beta_0$. In this case we may define the continuum limit of the two-loop function as follows. We set

$$G_\beta(i,j) = G_\beta(\gamma(l_i), \gamma(l_j)), \tag{54}$$

where i and j are the midpoints of the links l_i and l_j. Letting $a_n = 2^{-n}$ and $x, y \in \bigcup_{n=1}^{\infty} 2^{-n}\mathbf{Z}^d_{\frac{1}{2}}$, where $\mathbf{Z}^d_{\frac{1}{2}}$ is the set of all midpoints of links in $\mathbf{Z}^d$ (i.e. $\mathbf{Z}^d_{\frac{1}{2}} = \bigcup_{\nu=1}^{d}(\mathbf{Z}^d + \frac{1}{2}\underline{e}_\nu)$), we choose β as a function of the lattice spacing a such that

$$a^{-1}m(\beta(a)) \to m_* > 0 \text{ as } a \to 0,$$

and define

$$G_*(x,y) = \lim_{n\to\infty} a_n^{-(d-2+\eta)} G_{\beta(a_n)}(a_n^{-1}x, a_n^{-1}y), \tag{55}$$

provided the limit exists.

The continuum string- or surface tension is then

$$\tau_* = \lim_{a\to 0} a^{-2}\tau(\beta(a)). \tag{56}$$

Finiteness of both m_* and τ_* is an essential feacture of a genuine continuum string- or surface theory. Clearly, this can be obtained only if

$$\frac{\tau(\beta)}{m^2(\beta)} \to \text{ const. as } \beta \searrow \beta_0. \tag{57}$$

We have here only defined the continumm limit of the two-loop function in the case where the loops shrink to points, corresponding to continuum surfaces with two punctures. Continuum limits of n-loop functions corresponding to continuum surfaces with n punctures is straight foreward (see refs. [7, 10, 11]), whereas the case of surfaces with macroscopic boundary components requires additional renormalizations, and will not be discussed here.

2.3 Analysis of the *PRS*-model.

We shall from now on concentrate on the *PRS*-model. Along the way we shall comment on which of our results may be extended to the *SARS*-model. First we establish the existence of the limits (43) and (48) and some elementary properties of the functions $m(\beta)$ and $\tau(\beta)$.

PROPOSITION 2.9. *The limits $m(\beta)$ and $\tau(\beta)$ exist and are positive, concave, increasing functions of β for $\beta > \beta_0$. On the other hand, χ is a positive, decreasing convex function of β for $\beta > \beta_0$.*

PROOF: Assume that the corners of $\gamma_{R,T}$ have coordinates $(0,0,0,\ldots,0)$, $(T,0,0,\ldots,0)$, $(T,R,0,\ldots,0)$ and $(0,R,0,\ldots,0)$. From $S_1 \in \mathcal{S}^0(\gamma_{R,T_1})$ and $S_2 \in \mathcal{S}^0(\gamma_{R,T_2})$, where $T_1, T_2 \in \mathbf{N}$, we may construct a surface $S_3 \in \mathcal{S}^0(\gamma_{R,T_1+T_2+1})$ by first translating S_2 by T_1+1 units in the positive 1-direction and then gluing S_1 to this translated surface by adding a $R \times 1$-rectangle as indicated on fig. 4. The gluing piece is added in order to ensure that S_3 uniquely determines S_1 and S_2.

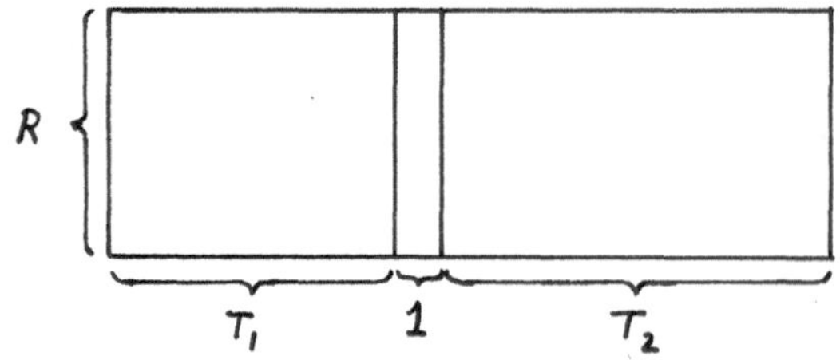

Figure 4

Using this observation it follows that

$$G_\beta(\gamma_{R,T_1+T_2+1}) \geq G_\beta(\gamma_{R,T_1})G_\beta(\gamma_{R,T_2})e^{-\beta R}, \tag{58}$$

which implies that the function

$$g_\beta(R,T) = -\log G_\beta(\gamma_{R-1,T-1})e^{-\beta(R+T-1)} \tag{59}$$

is subadditive as a function of T for fixed R, and by symmetry also as a function of R for fixed T. It thus follows that

$$-\lim_{R,T\to\infty}\frac{1}{RT}\log G_\beta(\gamma_{R,T}) = \lim_{R,T\to\infty}\frac{1}{RT}g_\beta(R,T) = \inf_{R,T}\frac{1}{RT}g_\beta(R,T),$$

proving the existence of $\tau(\beta)$. As mentioned previously, $\tau(\beta) \leq \beta$.

For a function $f : \mathcal{S}^0(\gamma_{R,T}) \to \mathbf{R}$ we let

$$< f >_{R,T} = \frac{1}{G_\beta(\gamma_{R,T})}\sum_{S\in\mathcal{S}^0(\gamma_{R,T})} f(S)e^{-\beta|S|},\ \beta > \beta_0. \tag{61}$$

If

$$\tau_{R,T}(\beta) = -\frac{1}{RT}\log G_\beta(\gamma_{R,T}), \tag{62}$$

we have

$$\frac{d\tau_{R,T}}{d\beta} = \frac{1}{RT} < |S| >_{R,T} \geq 1, \tag{63}$$

since $|S| \geq RT$ for any $S \in \mathcal{S}^0(\gamma_{R,T})$. From (63) it follows that τ is increasing with

$$\tau(\beta') - \tau(\beta'') \geq \beta' - \beta'' \ \text{ for } \beta' > \beta'' > \beta_0.$$

Furthermore,

$$\frac{d^2\tau_{R,T}}{d\beta} = -\frac{1}{RT} < (|S| - < |S| >_{R,T})^2 >_{R,T} \leq 0,$$

from which it follows that $\tau_{R,T}$ and hence τ is concave.

Next we show that $\tau(\beta) > 0$ for $\beta > \beta_0$. By (63) it is sufficient to show that $\tau(\beta) \geq 0$. Given $S_1, S_2 \in \mathcal{S}^0(\gamma_{R,T})$ we may construct a surface $S_3 \in \mathcal{S}^0(\gamma(l))$, where l is some link in $\gamma_{R,T}$, by first translating S_2 by one unit in some direction and then adding the cylinder with area $2(R+T)$ connecting the boundaries of S_1 and the the translated surface, leaving, however, the link l unglued to the corresponding link in the cylinder. From this we get that

$$G_\beta(\gamma(l)) \geq e^{-2\beta(R+T)}(G_\beta(\gamma_{R+T}))^2,$$

and hence $\tau(\beta) \geq 0$ for each $\beta > \beta_0$, since $G_\beta(\gamma(l))$ is independent of R, T.

The part of the proposition concerning m is proven in a similar way (for details see [7, 11], and the part concerning χ is trivial.

REMARK 2.10: For the $SARS$-model we have at present no proof of the existence of the limits (43) and (48). Assuming, however, their existence the proof of the other properties of m, τ and χ except for the positivity of τ carries over. Positivity of $\tau(\beta)$ for β sufficiently large is easy to prove, but it is not known whether it stays positive for β close to δ_0.

REMARK 2.11: The proof of Proposition 2.9 also applies to the PRS_2-model. We shall use the notation m_2, τ_2 and χ_2 for the mass, string tension and susceptibility in that model.

Note also that concavity of m and τ implies that $0 \leq \mu, \nu \leq 1$.

For the construction of a scaling limit the question whether $m(\beta_0+) = 0$ is crucial. The following proposition relates this question to the divergence of χ.

PROPOSITION 2.12. *If $\chi(\beta) \nearrow \infty$ as $\beta \searrow \beta_0$ then $m(\beta) \searrow 0$ as $\beta \searrow \beta_0$.*

PROOF: If $m(\beta_0+) > 0$ we have that

$$G_\beta(i,j) \leq c(\beta)e^{-M|i-j|} \tag{64}$$

for some constant $M > 0$ and some function $c(\beta)$. An elementary, but slightly technical argument (see [11]), shows that $c(\beta)$ may be chosen to be bounded as $\beta \searrow \beta_0$. From (64) and (44) it then follows that $\chi(\beta) \leq$ const. as desired.

It has not been proven that χ diverges at β_0. Numerical simulations in $d = 2$ and $d = 3$ (see [12]) indicate that it is the case with $\gamma \simeq .4 \pm .2$. Also, a large-d approximation (see [11, 12, 13]) yields $\gamma = \frac{1}{2}$. Thus it seems reasonable to *conjecture that $\chi(\beta)$ diverges at β_0 for all d*, and hence that β_0 is a critical point in the sense that $m(\beta)$ vanishes at β_0.

The next result implies that if this conjecture holds then the values of the critical exponents are those of mean field theory (see [11]) and the continuum string tension τ_* is infinite.

THEOREM 2.13. *If χ diverges at β_0 and χ_2 diverges at β_2 then $\gamma = \frac{1}{2}$, $\nu = \frac{1}{4}$, $\eta = 0$ and $\tau(\beta_0+) > 0$.*

PROOF: We introduce the notation

$$G(\beta) = G_\beta(\gamma(l)) \ , \ G_2(\beta) = G_{2,\beta}(\gamma(l)),$$

which are, of course, independent of the link l.

Given a surface $S' \in \mathcal{S}_2^0(\gamma_{R,T})$, $R, T \in \mathbf{N}$, and a link l in S', we can build a surface in $\mathcal{S}^0(\gamma_{R,T})$ by cutting S' open along l, if l is an interiour link, and gluing on a surface $S_l \in \mathcal{S}^0(\gamma(l))$ along $\gamma(l)$, or, if l is a boundary link, by just gluing S_l to S' along one of the copies of l in $\gamma(l)$. By performing such a construction for a subset of links in S' we obtain a surface $S \in \mathcal{S}^0(\gamma_{R,T})$, and it is clear that S determines S' and S_l , $l \in L(S')$, uniquely. Here S_l may be empty, meaning that no gluing is performed at l.

Thus we have

$$\begin{aligned} G_\beta(\gamma_{R,T}) &= \sum_{S' \in \mathcal{S}_2^0(\gamma_{R,T})} e^{-\beta|S'|}(1+G(\beta))^{l(S')} \\ &= (1+G(\beta))^{R+T} \sum_{S' \in \mathcal{S}_2^0(\gamma_{R,T})} e^{-\beta|S'|}(1+G(\beta))^{2|S'|} \\ &= (1+G(\beta))^{R+T} G_{2,\varphi(\beta)}(\gamma_{R,T}), \end{aligned} \tag{65}$$

where

$$\varphi(\beta) = \beta - 2\log(1+G(\beta)), \tag{66}$$

and we have used that $l(S') = 2|S'| + R + T$ for $S' \in \mathcal{S}^0(\gamma_{R,T})$.

We note two consequences of (65) and (66). First, since $G_\beta(\gamma_{R,T}) < \infty$ for $\beta > \beta_0$, it follows that $\varphi(\beta) > \beta_2$ and hence that $G(\beta) \leq$ const. for $\beta > \beta_0$ by (66), proving that *the one-loop correlation functions $G_\beta(\gamma)$ do not diverge at β_0*. Second, taking logs at both sides in eq. (65), dividing by RT and letting $R, T \to \infty$ we obtain

$$\tau(\beta) = \tau_2(\varphi(\beta)) \ , \ \beta > \beta_0. \tag{67}$$

For $G_\beta(\gamma(l))$ the analog of eq. (65) is

$$G(\beta) = G_2(\varphi(\beta)), \ \beta > \beta_0. \tag{68}$$

The derivation is the same as that for $G_\beta(\gamma_{R,T})$, except that one has to pay special attention to interiour links in $S \subset S^0(\gamma(l))$ which are copies of l. We leave the details for the reader. Differentiating (68) and using

$$\frac{d\varphi}{d\beta}(\beta) = 1 + \frac{2\chi(\beta)}{1+G(\beta)}, \tag{69}$$

we find

$$\chi(\beta) = \chi_2(\varphi(\beta)) \left(1 - \frac{2\chi_2(\varphi(\beta))}{1+G(\beta)}\right)^{-1}. \tag{70}$$

Since χ and χ_2 are positive we get from (70) that

$$\frac{2\chi_2(\varphi(\beta))}{1+G(\beta)} < 1 \text{ for } \beta > \beta_0. \tag{71}$$

Thus, since $G(\beta) \leq$ const. for $\beta > \beta_0$, it follows that $\chi_2(\varphi(\beta_0+)) < \infty$ and hence

$$\varphi(\beta_0+) > \beta_2, \tag{72}$$

since χ_2 has been assumed to diverge at β_2.

Now (67) and (72) imply that

$$\tau(\beta_0+) = \tau_2(\varphi(\beta_0+)) > 0, \tag{73}$$

since $\tau_2(\beta) > 0$ for $\beta > \beta_2$ by Prop. 2.9 applied to the PRS_2-model.

Next, differentiating eq. (70) w.r.t. β and using (69) we obtain

$$-\chi' = [-\chi_2' - \frac{2\chi_2^3}{(1+G(\beta))^2}] \left(\frac{\chi}{\chi_2}\right)^3, \tag{74}$$

where $\chi = \chi(\beta)$, $\chi_2 = \chi_2(\varphi(\beta))$, $\chi' = \chi'(\beta)$ and $\chi_2' = \chi_2'(\varphi(\beta))$. By (71) the expression in square bracket is bounded from below by

$$\begin{aligned} -\chi_2' - \frac{1}{2}\chi_2 &= \sum_{S \in S_2^0(\gamma(l))} (|S|^2 - \frac{1}{2}|S|)e^{-\varphi(\beta)|S|} \\ &\geq \frac{1}{2}\chi_2, \end{aligned}$$

and from above it is trivially bounded by $-\chi_2'$. Thus, using that $\chi_2(\varphi(\beta_0+)) < \infty$ and $\chi_2'(\varphi(\beta_0^+)) < \infty$ as a consequence of (72), we conclude from (74) that

$$-\chi'(\beta) \sim (\chi(\beta))^3 \text{ as } \beta \searrow \beta_0. \tag{75}$$

Using (50) this implies that $\gamma = \frac{1}{2}$ as desired.

The value of ν is deduced from a similar analysis of the two-loop correlation function and its Fourier transform. In fact, one can prove that the continuum limit of the two-loop function is just the kernel of $(-\Delta + m_*^2)^{-1}$, and, in particular, $\eta = 0$. Intuitively, the reason that the continuum limit equals that of the random walk is, that the continuum string tension given by (56) is infinite, since $\tau(\beta_0+) > 0$. This implies that the surfaces collaps into narrow polymerlike objects, which are also the ones that dominate in the large-d limit (see [11, 12, 13]). We refer to refs. [11, 14] for detailed arguments.

REMARK 2.14: 1) Note that the exponents γ, ν, η fulfill the scaling relation

$$\gamma = \nu(2 - \eta).$$

See ref. [11] for a heuristic derivation of this relation.

2) Assuming that

$$n^h_{\gamma(l)}(A) \sim A^{\varepsilon_h} e^{\beta_0 A} \tag{76}$$

as $A \to \infty$, it is easy to see that

$$\gamma = 2 + \varepsilon_h. \tag{77}$$

Thus we conclude that

$$\varepsilon_0 = -\frac{3}{2}, \tag{78}$$

under the assumptions of the theorem. We expect that ε_h has (approximately) a linear dependence on h.

The theorem implies that the PRS-model is neither suitable as a discrete approximation to a quantum mechanical string, since in the continuum limit the string collapses to a point, nor can it be used to describe statistical ensembles of surfaces with approximately vanishing surface tension (see section 1). These features make it desirable to investigate a broader class of models. Fortunately, it turns out that our methods can be applied to a rather wide spectrum of models which are our next subject of discussion.

2.3 GENERALIZED PRS-MODELS

In order to define the generalized action we need to introduce some more notation. An interiour link l in S is called an edge link if the two plaquettes sharing l are either orthogonal or overlap each other (i.e. the two plaquettes do not make up a flat piece of the surface). By $E(S)$ we denote the set of all edgelinks in S and by $e(S)$ their number. A corner in S is a vertex from which at least two edgelinks emerge, which are either orthogonal or overlap each other. By $C(S)$ we denote the set of corners in S. For $j \in C(S)$ we let $n_S(j)$ be the number of edgelinks emerging from j and set

$$N(S) = \sum_{j \in C(S)} n_S(j). \tag{79}$$

For $\underline{\lambda} = (\lambda_0, \lambda_1, \lambda_2) \in \mathbf{R}^3$ and $S \in \mathcal{S}^0(\gamma_1, \ldots, \gamma_n)$, where $\gamma_1, \ldots, \gamma_n$ are loops in $\mathbf{Z}^d$, we then define

$$A(S) = \lambda_2 |S| + \lambda_1 e(S) + \lambda_0 N(S) \tag{80}$$

and

$$G_{\underline{\lambda}}(\gamma_1, \ldots, \gamma_n) = \sum_{S \in \mathcal{S}^0(\gamma_1, \ldots \gamma_n)} e^{-A(S)}. \tag{81}$$

Note, that, contrary to the simple area action, the action (or energy) given by (80) depends on the way the surface S is embedded in $\mathbf{Z}^d$. In particular, the second term in (80) is a discrete analog of extrinsic curvature and represents a bending energy, and the last term is closely related to the intrinsic curvature (in fact, for $d = 3$ we could just as well use the numerical value of the defect angle (19) in stead of $N(S)$ for the following). As a consequence the surfaces are expected to become smoother on the average with increasing λ_1 or λ_0, and we may thus hope to suppress the singular surfaces dominating the simple PRS-model. Continuum analogs of the action (80) have been discussed in refs. [4, 15].

The generalized PRS-model defined by the loop correlation functions (81) is defined in a region $\mathcal{B} \subset \mathbf{R}^3$ given by the following theorem.

THEOREM 2.15. *There exists a non-empty convex open region $\mathcal{B} \subset \{\underline{\lambda} \in \mathbf{R}^3 \mid \lambda_2 > 0\}$ such that for all sets of loops $\gamma_1, \ldots, \gamma_n$ in $\mathbf{Z}^d$ the loop correlation functions $G_{\underline{\lambda}}(\gamma_1, \ldots \gamma_n)$ are well defined and finite for $\underline{\lambda} \in \mathcal{B}$, while they are divergent outside $\overline{\mathcal{B}}$.*

PROOF: The reader can easily verify that $G_{\underline{\lambda}}(\gamma_1, \ldots, \gamma_n)$ diverges if $\lambda_2 \leq 0$. By the same arguments as in the proof of Lemma 2.3 the convergence region is independent of n and $\gamma_1, \ldots, \gamma_n$. Furthermore, $\mathcal{B} \neq \emptyset$ since it contains the line $\{\underline{\lambda} \mid \lambda_2 > \beta_0 \,,\, \lambda_1 = \lambda_0 = 0\}$ according to Lemma 2.3. That $\mathcal{B}$ is convex is proven by a simple application of Hölder's inequality.

The mass $m(\underline{\lambda})$, the surface tension $\tau(\underline{\lambda})$ and the susceptibility $\chi(\underline{\lambda})$ for the generalised model are defined as previously. Positivity, concavity and monotonicity in λ_2 of m and τ may be proven by the same arguments as for Prop. 2.9. Let us also note that

$$\tau(\underline{\lambda}) \leq \lambda_2 \tag{82}$$

and

$$m(\underline{\lambda}) \leq 2(\lambda_2 + \lambda_1) \tag{83}$$

for $\underline{\lambda} \in \mathcal{B}$, since, by taking into-account only the minimal surfaces in $\mathcal{S}^0(\gamma_{R,T})$ and $\mathcal{S}^0(\gamma_0, \gamma_n)$, we have $G_{\underline{\lambda}}(\gamma_{R,T}) \geq e^{-\lambda_2 RT}$ and $G_{\underline{\lambda}}(\gamma_0, \gamma_n) \geq e^{-2(\lambda_2+\lambda_1)n}$, if $\gamma_0 = \gamma(l)$.

The critical indices ν, μ, γ and η are also defined as previously, except that $\beta - \beta_0$ is replaced by the distance from $\underline{\lambda}$ to a point $\underline{\lambda}_0 \in \partial\mathcal{B}$, and usually it is assumed that $\underline{\lambda}$ approaches $\partial\mathcal{B}$ transversally. Thus at any $\underline{\lambda}_0 \in \partial\mathcal{B}$, at which the mass $m(\underline{\lambda})$ vanishes, a continuum limit may be constructed.

REMARK 2.16: The PRS_2-model may similarly be extended to an open convex region $\mathcal{B}_2 \subset \{\underline{\lambda} \mid \lambda_2 > 0\}$ containing $\mathcal{B}$. Assuming that $\chi(\underline{\lambda})$ diverges at any $\underline{\lambda}_0 \in \partial\mathcal{B}$ and that $\chi_2(\underline{\lambda})$ diverges at any $\underline{\lambda}_2 \in \partial\mathcal{B}_2$ the arguments in the proof of Theorem 2.13 may be extended (see ref. [16]) to conclude that the critical indices are the same as those in the simple PRS-model when approaching the boundary $\partial\mathcal{B}$ transversally, and that the continuum limit constructed at any finite $\underline{\lambda}_0 \in \partial\mathcal{B}$ again is a free field. As we explain below there are, however, reasons to believe that the assumptions do not hold for sufficiently large λ_1.

From (82) we see that $\tau(\underline{\lambda})$ attains arbitrarily small values if $\partial\mathcal{B}$ approaches the plane $\{\underline{\lambda} \mid \lambda_2 = 0\}$ as λ_1 or λ_0 tends to ∞. That this actually is the case is ensured by the following theorem.

THEOREM 2.17. *Let*

$$n^0(A, e, N) = \sharp\{S \in \mathcal{S}^0(A) \mid e(S) \leq e, N(S) \leq N\}. \tag{84}$$

Then, for every $\varepsilon > 0$, there exists a convex region $\mathcal{B}(\varepsilon)$ in the (λ_0, λ_1)-plane containing the intervals

$$I_0 = \{(\lambda_0, \lambda_1) \mid \lambda_1 = 0\,,\, B_0(\varepsilon) < \lambda_0 < \infty\}$$
$$I_1 = \{(\lambda_0, \lambda_1) \mid \lambda_0 = 0\,,\, B_1(\varepsilon) < \lambda_1 < \infty\}$$

for some finite constants $B_0(\varepsilon), B_1(\varepsilon)$, such that for $(\lambda_0, \lambda_1) \in \mathcal{B}(\varepsilon)$

$$n^0(A, e, N) \leq e^{\varepsilon A + \lambda_1 e + \lambda_0 N}. \tag{85}$$

The theorem may be proven by a rather straight foreward extension of the proof of the upper bound on β_0 in Lemma 2.3, by exploiting the fact that the gluings along non-edgelinks are unique. The detailed arguments are rather lengthy and may be found in ref. [17]. Alternatively, one may apply a renormalization group argument, which is an extension of the one-step coarse-grainning argument in the proof of Theorem 2.13. In this context the action (80) turns out to be natural, in the sense that it approximately reproduces itself under renormalization (see ref. [17]).

The theorem ensures the existence of trajectories $\underline{\lambda}(t)$, $t \in \mathbf{R}_+$, in $\mathcal{B}$ tending to $(\infty, 0, 0)$ as $t \to \infty$. It furthermore follows from (82) and (83) that both m and τ tend to zero as $(\infty, 0, 0)$ is approached. It is an important problem to decide whether there exist such trajectories for which (57) if fulfilled as $(\infty, 0, 0)$ is approached, since in that case one may attempt to construct a continuum limit with finite mass and string tension. At present this is not known.

On the other hand, for trajectories approaching $(0, \infty, 0)$ it follows from (82) that τ tends to zero, whereas the upper bound (83) on m diverges. In fact, under certain reasonable assumptions it can be proven that $m(\underline{\lambda}) \geq c \cdot \lambda_1$ for some positive constant c and for λ_1 sufficiently large (see ref. [17]). In particular, it follows that $m(\underline{\lambda})$ does not vanish on $\partial\mathcal{B}$ for λ_1 large enough, and hence the susceptibility $\chi(\underline{\lambda})$ does not diverge here as a consequence of Prop. 2.12, which is also valid for the generalized PRS-model. We do not know the behaviour of $\tau(\underline{\lambda})$ in this part of $\partial\mathcal{B}$ (except at $(0, \infty, 0)$).

To sum up, we have found that the phase-diagram of the generalized PRS-model looks as indicated on fig. 5.

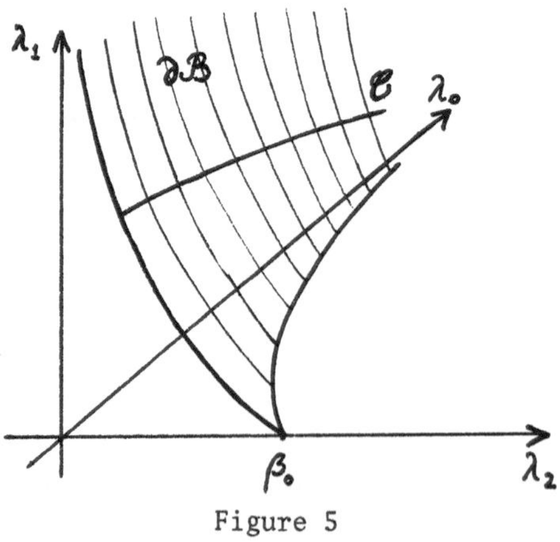

Figure 5

Above the curve $\mathcal{C}$ in $\partial\mathcal{B}$ the mass does not vanish. Below $\mathcal{C}$ the mass vanishes, whereas the surface tension is positive. The point $(\infty, 0, 0)$ is most likely the only one at which both m and τ vanish and hence the only one at which a continuum surface theory may be constructed, but for this purpose a more detailed understanding of the behaviour of m and τ, in the first place, is needed.

It is also an interesting problem to investigate the nature of the transition taking place at $\mathcal{C}$, and, in particular, to understand its relation to transitions observed in crystalline membranes (see [18]).

REMARK 2.18: A random walk analog of the generalized action (80) is given by

$$A(\omega) = \lambda_1|\omega| + \lambda_0 N(\omega),$$

where $N(\omega)$ is the number of corners in ω (i.e. the number of points in ω at which two orthogonal or overlapping links meet). The model so defined can be solved explicitly (see [19]) and one finds a convex open domain of definition $\mathcal{A} \subset \{(\lambda_0, \lambda_1) \mid \lambda_1 > 0\}$ whose boundary approaches the λ_0-axis as $\lambda_0 \to \infty$. Scaling limits constructed at points in $\mathbf{R}^2 \cap \partial\mathcal{A}$ coincide with the one for the simple random walk with $\lambda_0 = 0$. At $(\infty, 0)$, however, different continuum limits with $\eta = 1$ can be constructed. It turns out that those are not euclidean invariant, because the continuum limits of expectation values of local functions of the discrete set of stepvariables are finite and non-trivial in this case.

This may be taken as an indication that there might also be problems with euclidean invariance in the generalised PRS-model at $(0, \infty, 0)$ (and possibly also at $(\infty, 0, 0)$). In the next section we shall see how to discretize random surfaces in a manifestly euclidean invariant way.

3. TRIANGULATED SURFACES

We shall now discuss a class of euclidean invariant discrete random surface models, obtained by triangulating the surfaces. Thus the surfaces are embedded in d-dimensional euclidean space, the embedding being determined by a mapping S from the vertices of a triangulation into $\mathbf{R}^d$. As such the surface has an induced metric from $\mathbf{R}^d$. On the other hand, the triangulation itself determines an independent metric on the surface by assigning length one to each link in the triangulation. Thus a summation over triangulations is similar in spirit to the integration over an independent metric in the Polyakov approach to the string functional integral [20].

In the context of physical membranes (or bilayers) a summation over triangulations is appropriate for fluidlike membranes, since for those the linking numbers of molecules, represented by the orders of vertices, may vary. On the other hand, for crystalline membranes a fixed triangulation is dictated by the structure of the crystal. We shall primarily discuss the former alternative, while the latter will be discussed briefly in section 4.

Since the methods of proof of the results in this section are not too different in spirit from those in section 2 our presentation will to some extent be descriptive without detailed proofs.

We begin by defining the models. By $\mathcal{T}^h(m_1, \ldots, m_n)$ we denote the set of oriented, connected, two-dimensional abstract simplicial complexes with h handles and n boundary components containing $m_1, \dots, m_n$ vertices, respectively. For $T \in \mathcal{T}^h(m_1, \dots, m_n)$ we let T also denote the set of triangles (2-simplices), $L(T)$ the set of links (1-simplices) and $V(T)$ the set of vertices (0-simplices) in T, and we set $|T| = \sharp T$, $l(T) = \sharp L(T)$ and $v(T) = \sharp V(T)$. Furthermore, for a vertex $i \in V(T)$ we let σ_i denote the order of i.

Let now $\gamma_1, \dots, \gamma_n$ be oriented polygonal loops in $\mathbf{R}^d$ with $m_1, \dots, m_n$ corners, respectively. A surface with boundary components $\gamma_1, \dots, \gamma_n$ triangulated by $T \in \mathcal{T}^h(m_1, \dots, m_n)$ is a mapping $S : V(T) \to \mathbf{R}^d$ such that the vertices in the i'th boundary component of T containing m_i vertices, $i = 1, \dots, n$, are mapped onto the corners in γ_i preserving the cyclic ordering.

Given an action (or energy) functional $A(S)$ we set

$$G^T(\gamma_1 \dots, \gamma_n) = \int \prod_{i \in T \setminus \partial T} dS(i) e^{-A(S)}, \tag{86}$$

where ∂T is the boundary of T and $dS(i)$ is the uniform measure in $\mathbf{R}^d$ w.r.t. the variable $S(i)$. We then define the loop correlation function as

$$G(\gamma_1, \dots, \gamma_n) = \sum_{T \in \mathcal{T}^h(m_1, \dots, m_n)} \rho(T) G^T(\gamma_1, \dots, \gamma_n), \tag{87}$$

where $\partial(T)$ is some positive weight function on triangulations to be specified later.

We note that a simple version of the argument used for the upper bound on β_0 in the proof of Lemma 2.3 yields

$$\sharp\{T \in \mathcal{T}^h(m_1, \dots, m_n) \mid |T| = A\} \leq A^{c_1 h + c_2} 8^A \tag{88}$$

for suitable constants c_1 and c_2. Moreover, for $h = 0$ the exact asymptotic behaviour of this number is known (see ref. [21]):

$$\sharp\{T \in \mathcal{T}^0(m_1, \dots, m_n) \mid T| = A\} \sim A^{-5/2} \left(\frac{4^4}{3^3}\right)^{A/2}. \tag{89}$$

For simplicity we shall consider mainly the case $h = 0$ in the following and drop the superscript h.

We now have to make a choice of action and discuss the convergence properties of (87). In view of the previous section a natural candidate for the action $A(S)$ is the area $|S|$ of the surface S, i.e. the sum of the areas of the images $S(\Delta)$ under S of the triangles Δ in T. It turns out, however, that this is not a good choice, since the model is unstable in the sense that some points in the surface wander off to ∞ with large probability, as is seen from the following theorem, whose proof may be found in [22].

THEOREM 3.1. *If $A(S) = \beta|S|$, $\beta > 0$, we have for any triangulation $T \in \mathcal{T}^h(m_1, \dots, m_n)$ that either $G^T(\gamma_1, \dots, \gamma_n) = \infty$ or there exists a vertex $i_0 \in V(T)$ and an integer $N_{d,h}$ such that*

$$\int \prod_{i \in T \setminus \partial T} dS(i) |S(i_0)|^N e^{-A(S)} = \infty$$

for $N \geq N_{d,h}$.

REMARK 3.2: If T is a regular triangulation all of whose vertices are of order 6, there are numerical indications that $G^T(\gamma_1,\ldots,\gamma_n) < \infty$ (see ref. [23]). For examples of triangulations for which $G^T(\gamma_1,\ldots,\gamma_n) = \infty$ see ref. [22].

Thus we have to find another analog of the area term in (80). The quantity

$$\sum_{(ij)\in L(T\setminus\partial T)} |S(i) - S(j)|^p \quad ,p > 0, \tag{90}$$

does, in fact, serve this purpose, as is seen by noting that $|S|$ could be replaced by $l(S)$ in (80) as a consequence of (17). The choice of p is not expected to affect the critical properties of the model.

In order to write analogs of the two other terms in (80) we introduce the following notation: If Δ and Δ' are two triangles in T sharing the link $l \in L(T\setminus\partial T)$ we denote by $\theta_l(S)$ the angle between $S(\Delta)$ and $S(\Delta')$, i.e. the angle between the triangles in the subspace of $\mathbf{R}^d$ (of dimension ≤ 3) spanned by the edges of $S(\Delta)$ and $S(\Delta')$, defined such that it is π if $S(\Delta)$ and $S(\Delta')$ overlap. Moreover, if $i \in V(T\setminus\partial T)$ we define the defect angle

$$\delta_i(S) = 2\pi - \sum_{\Delta\ni i} \varphi_{\Delta,i}(S), \tag{91}$$

where the sum is over triangles Δ containing i and $\varphi_{\Delta,i}(S)$ is the angle in $S(\Delta)$ at the corner $S(i)$.

We then define

$$A(S) = \lambda_2 \sum_{(ij)\in L(T\setminus\partial T)} |S(i) - S(j)|^p + \lambda_1 \sum_{l\in L(T\setminus\partial T)} f(\theta_l(S)) + \lambda_0 \sum_{i\in V(T\setminus\partial T)} g(\delta_i(S)), \tag{92}$$

where $f : [0,\pi] \to \mathbf{R}$ and $g :]-\infty, 2\pi] \to \mathbf{R}$ are non-negative continuous functions which vanish only at 0. These requirements on f and g ensure that flat surfaces dominate in the limits $\lambda_1 \to \infty$ or $\lambda_0 \to \infty$.

We shall denote by $G^T_{\underline{\lambda}}(\gamma_1,\ldots,\gamma_n)$ and $G_{\underline{\lambda}}(\gamma_1,\ldots,\gamma_n)$ the loop correlation functions (86) and (87) if $A(S)$ is given by (92). Their convergence properties are given by the following theorem.

THEOREM 3.3. *If*

$$\rho(T) = \prod_{i\in T\setminus\partial T} \varphi(\sigma_i), \tag{93}$$

where

$$0 < const. \leq \varphi(\sigma_i) \leq const.(\sigma_i)^\alpha \quad ,\alpha \geq 0, \tag{94}$$

there exists a non-empty, open, convex domain $\mathcal{D} \subseteq \{\underline{\lambda}|\lambda_2 > 0\}$ *such that all* $G_{\underline{\lambda}}(\gamma_1,\ldots,\gamma_n)$ *are finite if* $\underline{\lambda} \in \mathcal{D}$ *and are divergent if* $\underline{\lambda} \notin \overline{\mathcal{D}}$.

PROOF: That $\mathcal{D}$ is independent of $\gamma_1,\ldots,\gamma_n$ and n is proven in a similar way as the corresponding statement for the lattice case (see [22]). It is also easy to see that $G_{\underline{\lambda}}(\gamma_1,\ldots,\gamma_n)$ diverges if $\lambda_2 \leq 0$.

In order to show that $\mathcal{D} \neq \emptyset$ it is enough to verify that for fixed $m_1,\ldots,m_n$

$$\rho(T) \leq c_1^{|T|} \quad ,T \in \mathcal{T}(m_1,\ldots,m_n), \tag{95}$$

and that for fixed $\gamma_1, \ldots, \gamma_n$ and $\lambda_1, \lambda_0 \geq 0$ and $\lambda_2 > 0$

$$G^T_{\underline{\lambda}}(\gamma_1, \ldots, \gamma_n) \leq (c_2/\lambda_2)^{c_3|T|} \tag{96}$$

for some positive constants c_1, c_2, c_3, since (87) is then convergent for λ_2 sufficiently large by (89).

Setting $N = v(T)$ we have

$$\left(\prod_{i \in V(T)} \sigma_i\right)^{1/N} \leq \frac{1}{N} \sum_{i \in V(T)} \sigma_i = 6 - \frac{2|\partial T| + 6\chi(T)}{N} \leq const., \tag{97}$$

where $\chi(T)$ is defined by the analog of eq.(14), $|\partial T|$ is the number of vertices in ∂T (which equals $m_1 + \cdots + m_n$) and we have used an analog of eq.(18) for the triangulation T. Using the upper bound in (94) and that

$$v(T) = \frac{1}{2}|T| - \frac{1}{2}|\partial T| + \chi(T), \tag{98}$$

which is easily verified, we obtain (95) from (97).

If T' is a spanning tree for T we clearly have for $\lambda_0, \lambda_1, \lambda_2 \geq 0$

$$G^T_{\underline{\lambda}}(\gamma_1, \ldots, \gamma_n) \leq G^{T'}_{(0,0,\lambda_2)}(\gamma_1, \ldots, \gamma_n), \tag{99}$$

where, of course, $G^{T'}_{(0,0,\lambda_2)}(\gamma_1, \ldots, \gamma_n)$ is obtained from $G^T_{(0,0,\lambda_2)}(\gamma_1, \ldots, \gamma_n)$ by leaving out the terms in $A(S)$ corresponding to links (ij) that are not in $L(T')$. $G^{T'}_{(0,0,\lambda_2)}(\gamma_1, \ldots, \gamma_n)$ is then estimated as follows. If there is a vertex $i \in V(T \setminus \partial T)$, which is of order 1 in T', we can perform the integration over $S(i)$. Each such integration yields a factor

$$k(\lambda_2) = \int_{\mathbf{R}^d} dx e^{-\lambda_2 |x|^p} = const.\lambda_2^{-d/p}. \tag{100}$$

If, on the other hand, all vertices in T' of order 1 belong to ∂T we may leave out the corresponding term in the action defining $G^{T'}_{(0,0,\lambda_2)}(\gamma_1, \ldots, \gamma_n)$ thereby obtaining an upper bound since $\lambda_2 > 0$.

Repeating this process we obtain

$$G^{T'}_{(0,0,\lambda_2)}(\gamma_1, \ldots, \gamma_n) \leq (k(\lambda_2))^{v(T \setminus \partial T)}. \tag{101}$$

Combining (98-101) we get (96).

Finally, convexity of $\mathcal{D}$ follows from a simple application of Hőlder's inequality.

REMARK 3.4: The theorem is valid for arbitrary h and for a larger class of weight functions. Those fulfilling (93) ad (94), are, however, the most natural candidates for ρ (see [22]), and are the ones we shall consider in the following.

The method of successive integrations used above to prove that $\mathcal{D} \neq \emptyset$ may quite easily be refined to prove the following analog of Theorem 2.17.

THEOREM 3.5. *For every $\epsilon > 0$ there exist finite constants $D_1(\epsilon), D_2(\epsilon)$ such that $\underline{\lambda} = (\lambda_0, \lambda_1, \lambda_2)$ belongs to $\mathcal{D}$ if $\lambda_2 \geq \epsilon$ and $\lambda_1 \geq D_1(\epsilon)$, $\lambda_0 \geq 0$ or $\lambda_1 \geq 0$, $\lambda_0 \geq D_0(\epsilon)$.*

The details of the proof, which is considerably simpler than the corresponding argument for lattice surfaces, can be found in [24].

We next discuss the mass, surface tension and susceptibility. If $m_1 = \cdots = m_n = 1$ and $T \in \mathcal{T}(1, \ldots, 1)$ the boundary components $\gamma_1, \ldots, \gamma_n$ of surfaces triangulated by T degenerate to n points $x_1, \ldots, x_n$. These surfaces are said to be pinched at $x_1, \ldots, x_n$ and we denote in this case $G_{\underline{\lambda}}(\gamma_1, \ldots, \gamma_n)$ by $G_{\underline{\lambda}}(x_1, \ldots, x_n)$.

The mass $m(\underline{\lambda})$ is then defined by

$$m(\underline{\lambda}) = - \lim_{|x-y|\to\infty} \frac{1}{|x-y|} \log G_{\underline{\lambda}}(x, y). \tag{102}$$

This definition is convenient for the following, but we might also have used literally the same definition as in the lattice case using non-degenerate loops.

To define the surface tension we let $\gamma_{L,\kappa}$, where $L \in \mathbf{N}, \kappa \in \mathbf{R}_+$, denote a loop with $4L$ edges, each of length κ, making up a square with edgelength $L\kappa$ in a plane. We then define

$$\tau(\underline{\lambda}) = - \lim_{L\to\infty} \frac{1}{(L\kappa)^2} \log G_{\underline{\lambda}}(\gamma_{L,\kappa}), \tag{103}$$

where κ is fixed. We expect $\tau(\underline{\lambda})$ to be independent of κ.

The susceptibility is defined as

$$\begin{aligned} \chi(\underline{\lambda}) &= -\frac{\partial}{\partial \lambda_2} G_{\underline{\lambda}}(0) \\ &= \frac{d}{\lambda_2 p} \sum_{T\in\mathcal{T}(1)} (v(T)-1) G_{\underline{\lambda}}^T(0) \rho(T) \end{aligned} \tag{104}$$

where the second equality follows from

$$G^T_{(\lambda_0,\lambda_1,\lambda_2)}(0) = (\lambda_2)^{-\frac{d}{p}(v(T)-1)} G^T_{(\lambda_0,\lambda_1,1)}(0), \tag{105}$$

which is obtained by changing variables $S(i) \to \lambda^{1/p} S(i)$ in (86) and using that the last two terms in the action (92) are scale invariant.

Using (104) it follows that

$$\chi(\underline{\lambda}) \sim \int_{\mathbf{R}^d} dy G_{\underline{\lambda}}(0, y) \tag{106}$$

in analogy with (44).

Having defined m, τ and χ the critical indices ν, μ, γ are defined as in the lattice case. Moreover, from the two-point function $G_{\underline{\lambda}}(x, y)$ the anomalous dimension η is defined as in (11) or (53) at points in $\partial\mathcal{D}$ where m vanishes and the continuum limit is defined as in (55) with $\underline{\lambda}$ replacing β(except that x and y may now be arbitrary points in $\mathbf{R}^d$).

We note the following analog of Proposition 2.9.

PROPOSITION 3.6. *The functions $m(\underline{\lambda})$ and $\tau(\underline{\lambda})$ are (provided the limits defining them exist) positive, concave functions on $\mathcal{D}$, which are increasing in λ_2 and non-decreasing in λ_1 and λ_0. χ is a positive, convex function on $\mathcal{D}$, which is decreasing in each variable.*

Only the proofs of positivity of m and τ are different from the lattice case. We refer to [24] for these.

The generalization of Theorem 2.13 to the triangulated models considered here presents some technical obstacles, but may be proven (at least for $\lambda_0 = \lambda_1 = 0$) by making certain plausible assumptions in addition to the assumption that $\chi(\underline{\lambda})$ diverges on $\partial\mathcal{D}$ and that the susceptibility $\chi_2(\underline{\lambda})$ in a certain coarse grained model (defined in a region $\mathcal{D}_2 \subset \mathbf{R}^3$), corresponding to the PRS_2-model in section 2, diverges on $\partial\mathcal{D}_2$ (see ref. [25]). On the other hand, it can be proven directly (without any assumptions on divergence of susceptibilities) that $\tau(\underline{\lambda})$ does not vanish on $\partial\mathcal{D}$ for $\lambda_0 = \lambda_1 = 0$, as the following theorem shows.

THEOREM 3.7. *If $p = 2$,*

$$\tau(\underline{\lambda}) \leq const.\lambda_2 \text{ for all } \underline{\lambda} \in \mathcal{D}, \tag{107}$$

and

$$\tau(\underline{\lambda}) \geq const.\lambda_2 \text{ if } \lambda_2 > \overline{\lambda}_2, \lambda_1 \geq 0, \lambda_0 \geq 0, \tag{108}$$

where $\overline{\lambda}_2 = \inf\{\lambda_2 | (0,0,\lambda_2) \in \mathcal{D}\}$.

If $p = 1$,

$$m(\underline{\lambda}) \leq const.\lambda_2 \text{ for all } \underline{\lambda} \in \mathcal{D}. \tag{109}$$

PROOF: The two upper bounds (107) and (109) are obtained by taking into account in (87) only the contribution from surfaces triangulated by regular triangulations and which are small perturbations of the minimal surfaces contributing to $G_{\underline{\lambda}}(\gamma_{L,\kappa})$ and $G_{\underline{\lambda}}(x,y)$, respectively. This means that the vertices $S(i)$ are allowed to vary only in small balls around the vertices of the minimal surface. By making an optimal choice of the size of the balls (which depends on p) one easily obtains (107) and (109) (see [24] for details).

Since $\tau(\underline{\lambda})$ is non-decreasing in λ_0 and λ_1 it is sufficient to prove (108) for $\lambda_1 = \lambda_0 = 0$. If $p = 2$ the action is gaussian in the $S(i)$ and hence for any $T \in \mathcal{T}(4L)$ we have

$$A(S) = A(S_{\min}) + A(S - S_{\min}), \tag{110}$$

where $S_{\min}$ is a surface with boundary $\gamma_{L,\kappa}$ triangulated by T at which $A(S)$ takes its minimal value. Clearly, $(S - S_{\min})(i) = 0, \quad i \in \partial T$, for any surface S with boundary $\gamma_{L,\kappa}$. Thus $S - S_{\min}$ is a surface triangulated by T, whose boundary $\gamma_L \equiv \gamma_{L,0}$ has length zero and contains $4L$ vertices all of which are equal.

It is easy to verify that $A(S) \geq 2(L\kappa)^2$ for all S with boundary $\gamma_{L,\kappa}$. In particular, $A(S_{\min}) \geq 2(L\kappa)^2$ and we get from (110) that

$$G_{(0,0,\lambda_2)}(\gamma_{L,\kappa}) \leq e^{-2\lambda_2(L\kappa)^2} G_{(0,0,\lambda_2)}(\gamma_L). \tag{111)}$$

A slightly technical argument [26] shows that $G_{(0,0,\lambda_2)}(\gamma_L)$ is bounded by $(const.)^L$, the constant depending on λ_2. This is, in fact, also a consequence of the expected independence of $\tau(\underline{\lambda})$ on κ.

Thus we have

$$G_{(0,0,\lambda_2)}(\gamma_{L,\kappa}) \leq e^{-2\lambda_2(L\kappa)^2+const.L}$$

from which (108) follows.

As remarked earlier, it is reasonable to expect that the critical behaviour of the models discussed here is independent of p. Assuming this, it follows from (107) and (109) that both $m(\underline{\lambda})$ and $\tau(\underline{\lambda})$ vanish for $\lambda_2 \to 0$ and $\underline{\lambda} \in \mathcal{D}$. This is contrary to our expectations for the lattice models described in section 2, in which case we predicted the transition line $\mathcal{C}$ in fig. 5. It is tempting to conjecture that there is no such transition line in corresponding phase diagram for the triangulated models, and that m vanishes on all of $\partial\mathcal{D}$. The continum limits at points of $\partial\mathcal{D} \cap \mathbf{R}^3$ are then presumably gaussian as in the lattice case, and it should be possible to extend the proof of the lower bound (108) in Theorem 3.7 to all of $\mathcal{D}$.

It should be mentioned here that this picture is consistent with the results of perturbative or mean field calculations for the corresponding continuum models with $\lambda_0 = 0$ (see refs. [15]). In this case one finds that the β-function for the bare rigidity modulus $\alpha = \lambda_1^{-1}$ to lowest order in α or d^{-1} is given by

$$\beta(\alpha) = \Lambda\frac{d\alpha}{d\Lambda} = -\frac{d}{4\pi}\alpha^2,$$

where Λ is a momentum cutoff. Unless higher order corrections give rise to additional zeroes for the β-function this shows that in order to obtain a finite positive effective rigidity modulus, α has to approach zero, i.e. $\lambda_1 \to \infty$ as $\Lambda \to \infty$. This means that the continum model is obtained by approaching $(0, \infty, 0)$.

It is still possible that a transition line is present also in the phase diagram for the triangulated model, but that the nature of the transition at this line is different from the corresponding one in the lattice case, due to the continuous versus discrete nature of the normals to the surfaces in the two models, as was also alluded to at the end of section 2.

Thus a sufficiently detailed picture of the critical behviour in these models has to await closer investigation. In particular, it is important to clarify the nature of the continum limits that possibly can be constructed at $(0, \infty, \lambda_0)$ and $(0, \lambda_1, \infty)$. In this connection it may be mentioned that the upper bounds (107) and (109) may be generalised to other values of p, but these bounds are not sufficiently accurate to decide about existence of trajectories in $\mathcal{D}$ tending to $\partial\mathcal{D}$ along which e.g. (57) is fulfilled (see ref. [24]).

REMARK 3.8: A discussion of random walk models analogous to the models treated in this section can be found in [19].

We end this section by briefly discussing the case $\lambda_1 = \lambda_0 = 0$ and $p = 2$. The action $A(S)$ is in this case quadratic in $S(i)$ and the model is defined for $\lambda_2 > \overline{\lambda}_2$, where $\overline{\lambda}_2$ is defined in Theorem 3.7. This model was first considered in [22, 27, 28].

For $T \in \mathcal{T}(1)$ with $\partial T = \{i_o\}$ we have

$$\int \prod_{i\in\partial T\setminus\{i_o\}} \frac{dS(i)}{\pi^{d/2}} e^{-\sum\limits_{(ij)\in L(T)} |S(i)-S(j)|^2} = (det(-\Delta_T))^{-d/2}, \qquad (112)$$

where the symmetric $(v(T) - 1) \times (v(T) - 1)-$ matrix Δ_T is defined by

$$(x, -\Delta_T x) \equiv \sum_{i,j\in V(T)\setminus\{i_0\}} x_i(-\Delta_T)_{ij}x_j = \sum_{(ij)\in L(T)} (x_i - x_j)^2 \qquad (113)$$

for all $x = (x_i)_{i \in V(T)} \in \mathbf{R}^{v(T)}$ with $x_{i_0} = 0$. Thus

$$(\Delta_T)_{ij} = \begin{cases} 1 & \text{if } (ij) \text{ is a link in } T \setminus \{i_0\} \\ -\sigma_i & \text{if } i = j \\ 0 & \text{otherwise.} \end{cases} \tag{114}$$

Setting $e^{-\beta} = (\pi/\lambda_2)^{d/2}$, $G_\beta(x_1, \ldots, x_n) = G_{(0,0,\lambda_2)}(x_1, \ldots, x_n)$ and $\chi(\beta) = \chi(0, 0, \lambda_2)$ we get from (86), (87) and (112) that

$$G_\beta(0) = \sum_{T \in \mathcal{T}(1)} e^{-\beta(v(T)-1)} \rho(T) \left(det(-\Delta_T)\right)^{-d/2} \tag{115}$$

and

$$\chi(\beta) \sim \sum_{T \in \mathcal{T}(1)} (v(T) - 1) e^{-\beta(v(T)-1)} \rho(T) \left(det(-\Delta_T)\right)^{-d/2}. \tag{116}$$

It is easy to see that for any $d \in \mathbf{R}$ there exists a $\beta_0' \in \mathbf{R}$ such that the sums (115) and (116) are convergent for $\beta > \beta_0'$ and divergent for $\beta < \beta_0'$. Thus $G_\beta(0)$ and $\chi(\beta)$ can naturally be defined for arbitrary values of $d \in \mathbf{R}$. More generally, the Fourier transforms of the n-point functions $G_\beta(x_1, \ldots, x_n)$ can be continued to non-physical values of d.

It is well known that

$$det(-\Delta_T) = \sharp\{\text{ spanning trees for } T\} \tag{117}$$

(see e.g. [29]). Using this representation for $det(-\Delta_T)$ and (116) it can be shown (see [30, 31]) that

$$-\frac{d\chi}{d\beta} \geq const.\chi^3(\beta) \text{ for } \beta > \beta_0'. \tag{118}$$

Recall from the proof of Theorem 2.13 that this inequality is also valid (without any assumptions about divergent susceptibilities) in the lattice case. From (118) it follows that

$$\gamma \leq \frac{1}{2}. \tag{119}$$

Numerical calculations (see e.g. [27, 30-35]) indicate that $\gamma > 0$ if $d \gtrsim 1$ and $\varphi(\sigma) = \sigma^\alpha$ with $\alpha \geq const.d$. As discussed above we expect that mean field theory is exact in this range of the parameters α and d. On the other hand, for $d \lesssim 1$ there is numerical evidence that $\gamma < 0$. In particular, for $d = \alpha = 0$ we have according to (89) and (116)

$$\chi(\beta) \sim \sum_{A=2}^{\infty} A^{-3/2} \left(\frac{4^4}{3^3}\right)^{A/2} e^{-\beta A/2}, \tag{120}$$

from which we see that $\beta_0' = \log \frac{4^4}{3^3}$ and

$$\chi(\beta) - \chi(\beta_0'+) \sim (\beta - \beta_0')^{1/2}$$

as $\beta \searrow \beta_0'$. Hence $\gamma = -\frac{1}{2}$ for $d = 0$ (see also [36]).

As observed in [28], also for $d = -2$ and $\alpha = 0$, in which case

$$\chi(\beta) \sim \sum_{T \in \mathcal{T}(1)} e^{-\beta(v(T)-1)}(v(T)-1)det(-\Delta_T),$$

one can make use of (117) and evaluate β_0' and γ explicitly. One finds $\gamma = -1$ in this case.

For the case $d = 1, \alpha = 0$ it has been argued in [37] that $\gamma = 0$. Thus $d = 1$ seems to be a critical dimension below which $\chi(\beta_0'+)$ is finite.

Recently, critical indices in some continuum models of two-dimensional gravity coupled to matter fields have been computed exactly in [38]. For a scalar field X defined on a manifold M the action of the model is

$$A(g, X) = \frac{1}{2}(X, -\Delta_g X)_g, \tag{121}$$

where $(\cdot,\cdot)_g$ denotes the inner product on the space of square integrable functions on M determined by the metric g on M and Δ_g is the corresponding covariant lapacian. The formal similarity between (113) and (121) is obvious. In particular, the summation over triangulations in (115) and (116) corresponds to integration over the metric g in the corresponding continuum functional integral.

For d scalar fields on a manifold without handles the formula found in [38] (see also [39]) for γ is

$$\gamma = \frac{d - 1 - \sqrt{(1-d)(25-d)}}{12} \tag{122}$$

for $d \leq 1$. This formula fits with the exactly computable indices for $d = 0, -2, 1$, and it implies that $\gamma < 0$ for $d < 1$. The validity of (122) breaks down for $d > 1$. Note that the righthand side is complex for $1 < d < 25$. This indicates once more that a transition occurs at $d = 1$ and that the properties of the model in physical dimensions is radically different from the properties for $d < 1$.

Another interesting but more qualitative argument in favour of a transition from a "crumpled" to a "branched polymer" phase at $d = 1$ has been given in [40] (see also [41]).

Whether it is possible to obtain a non-trivial continuum theory for $d > 1$ without introducing curvature terms with fine-tuned couplings as discussed above, remains to be seen.

4. DISCUSSION

It was our hope, originally, that, in the spirit of ordinary quantum field theory, a first quantized continuum string could be obtained in a reparametrization invariant way from a simple discrete approximation, f.ex. one whose action was a straightforeward discretization of the Nambu-Goto (i.e. the area) action. As we have seen the problem seems to be considerably more involved, at least in the framework we have been working, and more work has to be done in order to materialize the potential candidates for a continuum string or surface theory.

A different approach, in which the parameter space is discretized and a light-cone gange is chosen has been used in [42, 43]. In the continuum limit one obtains the standard bosonic string (see e.g. [44]). It was clear a priori that the models considered in these notes could not yield the standard bosonic string as a continuum limit, since they satisfy the principle of reflection positivity [45] (the triangulated models, however, only approximately), thereby excluding the occurence of a tachyon. One might speculate, however, whether the bosonic string could be obtained from e.g. the PRS-model at a critical point in the complex β-plane. But the existence of such a critical point, at which one would like τ to vanish through positive values and $m^2(\beta)$ to vanish through negative values, does not seem to be easy to establish.

The models we have discussed in these notes, of course, constitute a rather special class of models. A number of other models have been discussed in the litterature, which deserve being mentioned here:

(1) The well known SOS-model has been widely discussed in the physics litterature. For an account of rigorous results we refer to [46] and references therein.
(2) Models of random tubes have been treated from a rigorous point of view in [47].
(3) Wheras we have discussed only a single random surface, results on a model of several random surfaces can be found in [48].
(4) Models of crystalline surfaces, which are identical to our triangulated models except that the triangulations are restricted to be regular (and $\lambda_0 = 0$), have recently received attention. Numerical evidence (at least in low dimensions) [18] has been found for the occurence of a phase transition at a value $\lambda_{1c} > 0$ of the extrinsic curvature coupling. In particular, a jump in the Hansdorff dimension, which for our models may be defined as $d_H = \nu^{-1}$, from ∞ for $\lambda_1 < \lambda_{1c}$ to a finite value for $\lambda > \lambda_{1c}$ is observed. This is in contrast to our expectations in the randomly triangulated case in physical dimensions, as discussed in section 3. From a rigorous point of view not much is understood about the nature or mechanism of this transition (see, however, [18]).
(5) Related to the random surface models are field theories on a random lattice, viewed either as discretizations of matter fields in flat space or as matter fields coupled to gravitational fields (see e.g. refs. [49 -53]). Various aspects of such models have been discussed. In particular, in [53] an Ising model on a random graph is solved exactly by mapping it onto an exactly solvable two-matrix model. A transition is found to occur and the susceptibility exponent at the transition may be computed and equals the value found for a Majorana fermion coupled to a gravitational field in [38]. For related work see [54].
(6) A model closely related to the PRS-model, but where surfaces containing nearest neighbour overlapping plaquettes are forbidden, was investigated numerically in [55].

ACKNOWLEDGEMENTS

I want to thank J. Ambjørn, J. Fröhlich, T. Jónsson and P. Orland for the pleasant collaboration I have had with them on the subjects discussed in these notes.

REFERENCES

1. H.J. Leamy, G.H. Gilmer, K.A. Jackson, in: "Surface Physics of Crystalline Materials", J.M. Blakely (ed.), New York, Academic Press, 1976.

2. J. Fröhlich, C. Pfister, T. Spencer, in: "Springer Lecture Notes in Physics" 173, Berlin-Heidelberg-New York, Springer-Verlag, 1982.
3. P.G. De Gennes, C. Taupin "J. Phys. Chem.," 86 1982, p.2294.
4. W. Helfrich "Z. Naturforsch.," 28c, 1973, p.693.
5. J. Glimm, A. Jaffe, "Quantum Physics," New York-Heidelberg-Berlin, Springer Verlag, 1981.
6. D. Weingarten, "Phys.Lett." 90**B**, 1980, p.280.
7. B. Durhuus, J. Fröhlich, T. Jónsson, "Nucl. Phys." **B**225 [FS9], 1983, p.185.
8. D. Ruelle, "Statistical Mechanics, Rigorous Results," New York, Benjamin 1969.
9. T. Eguchi, H. Kawai, "Phys. Lett." 114**B**, 1982, p.247.
10. B. Durhuus, J. Fröhlich, T. Jónsson, "Phys. Lett." 137**B**, 1984, p.93.
11. B. Durhuus, J. Fröhlich, T. Jónsson "Nucl. Phys." **B**240 [FS12], 1984, p.453.
12. H. Kawai, Y. Okamoto, "Phys.Lett." 130**B**, 1983, p.415.
13. J. Drouffe, G. Parisi, N. Sourlas, "Nucl. Phys." **B**161, 1980, p397.
14. T. Jónsson, "Comm.Math.Phys." 106, 1986, p.679.
15. W. Helfrich, "J. Physique 46, 1985, p.1263; L. Peliti, S. Leibler, " Phys.Rev.-Lett." 54, 1985, p.1690; D. Förster, "Phys. Lett." 114**A**, 1986, p.115; A.M. Polyakov, "Nucl. Phys." **B**268, 1986, p. 406; H. Kleinert, "Phys. Lett." 174**B**, 1986, p.335.
16. B. Durhuus, T. Jónsson, "Phys. Lett." 180**B**, 1986, p.385.
17. J. Ambjørn, B. Durhuus, J. Fröhlich, T. Jónsson, in: preparation.
18. Y. Kantor, D.R. Nelson, "Phys.Rev.Lett." 58, 1987, p.2774; D.R. Nelson, L. Peliti, "J. Physique" 48, 1987, p.1085; F. David, E. Guitter, "Europhys. Lett." 5, 1988, p.709; M. Baig, D. Espriu, J.F. Wheater, Univ. of Oxford preprint, 1988; J. Ambjørn, B. Durhuus, T. Jónsson, to appear in Nucl. Phys. B.
19. J. Ambjørn, B. Durhuus, T. Jónsson, "Journ. Phys." A 21, 1988, p.981.
20. A.M. Polyakov, "Phys. Lett." 103**B**, 1981, p.207.
21. W.T. Tutte, "Can. J. Math." 14, 1962, p.21.
22. J. Ambjørn, B. Durhuus, J. Fröhlich, "Nucl. Phys." **B**257, 1985, p.433.
23. A. Billoire, D. Gross, E. Marinari, "Phys. Lett." 139**B**, 1984, p.75.
24. J. Ambjørn, B. Durhuus, J. Fröhlich, T. Jónsson, Nucl. Phys." **B**290 [FS20], 1987, p.480.
25. J. Fröhlich, in: "Recent Developments in Quantum Field Theory", eds. J. Ambjørn, B. Durhuus, J.L. Petersen, Amsterdam-Oxford-New York-Tokyo, North-Holland, 1985.
26. J. Ambjørn, B. Durhuus, "Phys. Lett." 188**B**, 1987, p.253.
27. F. David, "Nucl. Phys." **B**257 [FS14], 1985, p.543.
28. V. Kazakov, I. Kostov, A.A. Migdal, "Phys. Lett." 157**B**, 1985, p.295.
29. N. Nakanishi, "Graph Theory and Feynman Integrals", Gordon and Breach 1971.
30. J. Ambjørn, B. Durhuus, J. Fröhlich, P. Orland, "Nucl. Phys." **B**270 [FS16], 1986, p.457.
31. J. Ambjørn, B. Durhuus, J. Fröhlich, "Nucl. Phys." **B**275 [FS17], 1986, p.161.
32. D.V. Boulatov, V. Kazakov, I. Kostov, A.A. Migdal, "Nucl. Phys." **B**275, [FS17] p.641.
33. F. David, J. Jurkiewicz, A. Krzywicki, B. Peterson. "Nucl. Phys." **B**290, [FS20] 1987, p.218.
34. A. Billoire, F. David, "Nucl. Phys." **B**275, 1986, p.617.

35. J. Ambjørn, Ph. De Forcrand, F. Koukiou, D. Petritis, "Phys. Lett." 197**B**, 1987, p.548.
36. F. David, "Nucl. Phys." **B**257, 1985, p.45.
37. V. Kazakov, A.A. Migdal, Niels Bohr Institut preprint NBI-HE-88-28, 1988.
38. A.M. Polyakov, "Mod.Phys. Lett." **A**2, 1987, p.893; V. Knizhnik, A.M. Polyakov, A.B. Zamolodchikov, "Mod. Phys. Lett." **A**3, 1988, p.819.
39. F. David, Saclay preprint SPhT/88-132, 1988.
40. M.E. Cates communication through F. David.
41. F. David, E. Guitter, "Nucl. Phys." **B**295 [FS21], 1988, p.332.
42. R. Giles, C. Thorn, " Phys. Rev." **D**16, 1977, p.366.
43. P. Orland, "Nucl. Phys" **B**278, 1986, p.790.
44. S. Mandelstam, "Phys. Rep." 13, 1974, p.259.
45. B. Durhuus, J. Fröhlich, T. Jónsson, "Nucl. Phys." **B**257 [FS14], 1985, p.779.
46. J. Fröhlich, T. Spencer, "Comm. Math. Phys." 81, 1981, p.527.
47. D.B. Abraham, J.T. Chayes, L. Chayes, "Phys. Rev." **D**30, 1984, p.841, "Comm. Math. Phys." 96, 1984, p.439.
48. R. Schrader, "Comm. Math. Phys." 102, 1985, p.31.
49. N. Christ, R. Friedberg, T.D. Lee, " Nucl. Phys." **B**202, 1982, p.89, B210 [FS6] 1982, pp.310, 317.
50. M. Bander, C. Itzykson, in:"Springer Lecture Notes in Physics" 226, Berlin, Springer Verlag, 1985.
51. J.M. Drouffe, H. Kluberg-Stern, "Nucl. Phys." **B**260, 1985, p.253.
52. M.A. Bershadski, A.A. Migdal, "Phys. Lett." 174**B**, 1986, p.393.
53. V. Kazakov, "Phys. Lett." 119**A**, 1986, p.140.
54. B. Duplantier, I. Kostov, Saclay preprint SPhT/88-119, 1988.
55. B. Baumann, B. Berg, "Phys. Lett." **B**164, 1985, p.131; B. Baumann, "Nucl. Phys." **B**285 [FS19], 1987, p.391; B. Baumann, B. Berg, G. Münster, DESY preprint, 1988.

QUANTUM PHYSICS AND GRAVITATION

Rudolf Haag

II. Inst. f. Theoret. Phys.
Univ. Hamburg

1. INTRODUCTION

The subject matter of my lectures is not closely related to the main topic of this school: constructive quantum field theory and, if anyone expects from the title that I shall talk about superstrings I have to disappoint him right away. I share with many the expectation that after 60 years of continuous development we stand on the verge of a revolutionary change of the basic concepts of physical theory, that – in the terminology of Kuhn – a new paradigm will be created, a change as radical as the transition from classical to quantum physics. I believe that an essential component in this transition will involve a synthesis between ideas of general relativity and quantum physics. Whether superstrings will be the vehicle to achieve this seems to me at this stage less obvious, though I am quite ready to believe that some of the mathematical structures encountered in the study of superstrings may provide important clues. Nonetheless we should not underrate the importance of a critical understanding of the principles and recipes of today's (or yesterday's) paradigms. Where do they clash? What is the strength of their support? How natural are they? In short, I want to emphasize the role of natural philosophy and, since this is a summer school, I think it is legitimate to spend some time reviewing well known things to form a picture.

The best existing theory of gravity is Einstein's theory of general relativity. It is the last step in the persistent pursuit of the principle of locality in classical physics, a paradigm of immense fruitfulness, beginning with Faraday, implemented by Maxwell in his theory of electromagnetism, sharpened by Einstein in the special theory of relativity and ultimately in its most stringent form in the general theory of relativity. Loosely speaking the locality principle states that all physical effects propagate from point to neighboring point in space, governed by laws which involve only the physical situation in the immediate proximity. In its sharpened form, provided by special relativity, it further assumes that the velocity of propagation of effects is limited. As a consequence the comparison of times at different places in space is not intrinsically well defined. It needs a convention for the synchronization of clocks and there is a certain amount of arbitrariness in this convention. This prevents us from considering

Constructive Quantum Field Theory II, Edited by G. Velo and
A. S. Wightman, Plenum Press, New York, 1990

space and time as two distinct, objectively meaningful, concepts. Instead we must consider the 4-dimensional manifold of possible point-like *events* in space *and* time. The distinction of past, present and future of an event remains objectively meaningful but is not given by the magnitude of a time parameter but by a causal structure of the space-time manifold, assigning to each event a forward and backward cone. In special relativity the causal structure (and the related metric) is a priori imposed on the whole manifold. It is this last relic of global laws which is removed in general relativity where. following again the locality principle, this structure is determined by a metric field whose development is governed by partial differential equations relating its change to the local energy-momentum density.

Turning to quantum theory: the conceptual and mathematical structure, the philosophical (epistemological) aspects have been most thoroughly discussed within quantum *mechanics*, (the extrapolation to quantum field theory is not entirely straightforward in all respects). In quantum mechanics one singles out a "physical system" characterized by its material content, i.e. a collection of electrons, protons or other particles and one considers on the other side an observer with his instrument who may decide to make measurements on the system resulting in unambiguous phenomena (events). One essential lesson of quantum mechanics is the fact that a "measuring result" can not be interpreted as revealing a property of the system which existed (though unknown to us) prior to the act of measurement. Take the example of a position measurement on an electron. The position coordinates at a given time are not attributes of the electron. "Position" is a property of the *event* i.e. of the interaction process between the electron and the appropriately chosen measuring instrument (e.g. a screen), not of the electron alone. It is tempting to note that - though in quite different ways - special relativity and quantum mechanics suggest the primary importance of the concept of an event i.e. a phenomenon placed in space *and* time rather than the concept of a mass point in space which shifts as time goes on. A particle appears as (the simplest) causal tie between events.

The need to subdivide the world into an observed system and an observer has been especially emphasized by Bohr and Heisenberg ("Heisenberg cut"). Bohr points out that since "we must be able to tell our friends what we have done and what we have learnt" it is necessary to describe everything on the observer side, the instruments, the phenomena, in common language which may, for brevity and without harm, be supplemented by the use of classical physics because of its unambiguous correspondence to common experience and the deterministic character of its laws. Heisenberg showed in examples that to some extent the cut between system and observer may be shifted. Part of the observer's equipment may be included in the definition of the "system". But he stressed that always a cut must be made somewhere. A quantum theory of the universe is a contradiction in itself. This is one source of uneasiness most drastically felt in quantum field theory where one would like to take the universe as the system under consideration. Does the insistence on the observer mean that the ultimate cut is between the physical world and the world of the mind to which physical laws are not applied? Alternatively speaking, does an event never become factual until the mind intercedes? It seems clear that a flash on a scintillation screen can be considered in physics with a high level of confidence as a distinguished event which happened or did not happen irrespective of the presence of an observer. But can one isolate events with absolute precision. can on subdivide complex phenomena into sequences of elementary events? This is a question of the degree of idealization and it must be recognized that the limitations of such idealizations are not adequately understood. In quantum mechanics the idealization is achieved by introducing the above mentioned cut, limiting the class of events considered to the interaction of the "microscopic system" with a macroscopic measuring device. Other idealizations could be considered and may be useful depending on the regime of physics under study.

Another general aspect of quantum mechanics is that its law allow only statistical predictions for the occurrence of events. Given some knowledge about the past history of the system, subsumed in the notion of "state" and realized operationally for instance by describing the "source" (i.e. the instruments preparing the system) we have ordinary probability theory for the outcome of any subsequent measurement. Specifically, a measuring instrument (an "observable") defines an event space $\sum$ i.e. a Boolean algebra. A state defines a probability measure on it. The characteristic feature of the quantum mechanical formalism concerns the fact that for one system we have many event spaces (one for each observable) and many states (one for each source), that we can vary freely the combination of source and detector and that the total set of states available for a system, the state space $\mathcal{S}$, and the total set of available event spaces has a very specific mathematical structure: All the Boolean algebras of different possible event spaces are embedded in a single noncommutative W^* - algebra R, and the state space $\mathcal{S}$ consists of all (normal) normalized positive linear forms on R. For practical purposes in quantum mechanics R is the algebra of all bounded, linear operators acting on a Hilbert space, $\mathcal{S}$ is the set of all positive operators with unit trace, an event space is the Boolean algebra generated by a family of commuting orthogonal projectors containing the unit element. The probability of observing an event represented by the projector P in a state $\rho \in \mathcal{S}$ is given by $Tr\rho P$.

The question as to whether this specific mathematical structure can be understood from simpler principles has occupied quite a number of physicists and mathematicians. It is not within the scope of these lectures to give due credit to all authors who contributed to this subject. I shall quote only a few who emphasized different aspects and may serve as a source for other references [1] ... [6] and I shall add a few simple comments. The probabilistic setting implies that the state space is a convex body i.e. linear combinations of states with positive coefficients adding up to 1 give again possible states. This is the mixing process of ensembles. The linearity is physically significant because it is respected in the probabilities for the results of any subsequent measurement. By lifting the restriction that the coefficients must be positive and add up to 1 one embeds the state space in a real linear space with distinguished positive cone. Events are elements of the positive cone of the dual space. A convex cone is characterized by its facial structure (a "face" is a subcone stable under purification). The first remarkable feature of the quantum mechanical formalism is the correspondence between faces in state space and events (elements of the dual space). Physically this reflects a symmetry between preparing instruments and analyzing instruments, related in some sense to time reflection invariance [1]. Due

[1] If restricted to "pure states" and "finest events" it becomes the symmetry between "bras" and "kets" in Dirac's language.

to the symmetry states can be related to events and the probability function defines a metric in the linear space in which state space is embedded. In order to arrive at the complex linear structure and metric of a Hilbert space one must be able to "draw the square root" and this needs further structural properties. There are several approaches which lead to this result and it is hard to judge which one is the most natural. One requirement is the "facial homogeneity" of the cone, introduced by A. Connes and discussed in [6]; another one is the existence of "filtering projections" [5]; a third one is the consideration of composite events defining a noncommutative multiplication structure in event space resulting in a Baer – semigroup [3]...

The purpose of this lengthy discussion of quantum mechanics was to convey a certain amount of skepticism with regard to the persistence of the specific mathematical structure (Hilbert space, algebras ...) and specific terminology (observables ...) in future theories. We should be flexible and remember "it ain't necessarily so".

The next task will be the incorporation of the principle of locality into quantum theory. This is fairly natural on the level of special relativity because there the causal structure of space-time is a prior given and therefore events, whose causal relations we want to describe, have to be classified according to their support in space-time. If we take over from quantum mechanics the idea that events are mathematically represented by projectors of a W^* - algebra then we arrive at the following picture: to each region [2] $\mathcal{O}$ of Minkowski space we must have a corresponding W^* - algebra $R(\mathcal{O})$. The set of its projectors is physically interpreted as the set of events which may be produced in $\mathcal{O}$. In some sense we could say that we regard now regions of Minkowski space as "systems" [3]. Instead of a single algebra we have a *net of algebras*, an association of one algebra $R(\mathcal{O})$ to each region $\mathcal{O}$.

Due to its interpretation $R(\mathcal{O})$ should be a subalgebra of $R(\hat{\mathcal{O}})$ whenever $\mathcal{O} \subset \hat{\mathcal{O}}$. So for any pair of regions we can define $R(\mathcal{O}_1) \vee R(\mathcal{O}_2)$ as the W^*-algebra generated by the two algebras because there is always a region containing both $\mathcal{O}_1$ and $\mathcal{O}_2$ and therefore the two algebras are contained as subalgebras in a larger one.

Since one is more accustomed to dealing with a single algebra than with a net it is helpful to define a total object, the "quasilocal algebra" $\mathcal{A}$ as the smallest C^*-

[2] To avoid complications we shall understand by a "region" always an open, contractible subset of Minkowski space with compact closure.

[3] A W^*-algebra is an involutive algebra (an algebra equipped with a $*$ - operation) which in its role as a linear space is the dual of a Banach space, the latter arising physically from the set of allowed states. It is thus equipped with two significant topologies, a norm topology making it a C^* - algebra and a weak topology arising from its predual. A W^*-algebra is isomorphic to a v. Neumann algebra, an algebra of bounded operators acting on a Hilbert space, stable under the adjoint operation and equipped with the "ultra-weak" operator topology. To the best of present day knowledge each local algebra $R(\mathcal{O})$ should be isomorphic to the hyperfinite factor of type III_1 (unique up to equivalence), see [7]. The physical significance of this is that, if we consider $\mathcal{O}$ as a "system" it is necessarily an open system, coupled to its environment.

algebra containing all the $R(\mathcal{O})$ [4]. One must remember, however, that $\mathcal{A}$ is equipped with a distinguished family of subalgebras, realizing the correspondence $\mathcal{O} \rightarrow R(\mathcal{O})$ and that all physical information rests on this subdivision of $\mathcal{A}$.

The physical interpretation together with the principles of special relativity demand the following properties of the net:

(i) *Causality*

a) $R(\mathcal{O}_1)$ commutes with $R(\mathcal{O}_2)$ if $\mathcal{O}_1$ and $\mathcal{O}_2$ lie space-like.

b) $R(\mathcal{O}_1) \supset R(\mathcal{O}_2)$ if $\mathcal{O}_2$ is contained in the causal influence zone of $\mathcal{O}_1$.

Postulate a) results from the demand that space-like separated events can have no causal influence on each other, they are compatible. Note that this does not exclude correlations between them in certain states because they may be tied to common events in the past (a prime example is the Einstein-Podolski-Rosen effect). Postulate b) results from the demand that statistical predictions from the past to the future should be possible within the causal influence zone.

(ii) *Poincaré Symmetry*

The Poincaré group $\mathcal{P}$ is represented by a group of automorphisms of the net. To an element $g \in \mathcal{P}$ corresponds an automorphism α_g so that

$$\alpha_g(R(\mathcal{O})) = R(g\mathcal{O})$$

where $g\mathcal{O}$ denotes the image region of $\mathcal{O}$ under g.

We may consider representations of the net (of $\mathcal{A}$) by operators acting on a Hilbert space [5]. Attached to each representation there is a family of states, namely those which can be described by density operators (positive operators with trace 1) on the representation space. This family is called the set of normal states of the representation or the "folium" of states associated with the representation. It depends only on the (quasi)-equivalence class of the representation. Conversely, a state (normalized positive linear form) determines a representation by the Gelfand-Naimark-Segal (GNS)-construction. A (global) state ω provides an expectation functional for each local algebra and we shall call the restriction of ω to $R(\mathcal{O})$ the "partial state in $\mathcal{O}$".

In the setting described there will be many inequivalent representations of the net but the restrictions of any one of them to a single local algebra will be the same universal object, independent of the region, independent of the theory, namely the hyperfinite III_1-factor. This emphasizes once more that all physically interesting information is contained in the net structure, the relation between different local algebras of specified space-time regions. It is this which provides the identification of

---

[4] It is essential that the norm topology is used in this inductive limit because only then can we still interpret the elements of $\mathcal{A}$ as being essentially localized in a finite region.

[5] The rest of this paragraph is intended as a minimal sketch of some concepts, terminology and relations used. For details see [8] (especially its appendix) and [9].

algebraic elements with physical procedures and, in fact, the identification is complete in the sense that any two algebraic elements can be distinguished by their geometric description if the net does not have "inner symmetries" i.e. automorphisms which transform each $R(\mathcal{O})$ into itself.

Further properties of the net are needed to define a physically reasonable theory. One usually assumes:

(iii) There exists a distinguished state ω_0, the vacuum, which is invariant under all Poincaré transformations.

In the corresponding representation π_0, arising from ω_0 by the GNS-construction, the Poincaré group is implementable by unitary operators $U(g)$:

$$U(g)\pi_0(A)U(g)^{-1} = \pi_0(\alpha_g A); \quad A \in \mathcal{A}$$

The generators of the translation subgroup are interpreted as the energy-momentum operators P_μ and one can express the essential stability requirement for the theory as

(iv) *Positivity of Energy*
In the representation π_0 the vacuum is the state of lowest energy.

Comments

1) The frame sketched appears to be very natural and, as long as one accepts the basic tenets of special relativity and quantum theory without modification, the net structure with properties (i) and (ii) seems inescapable. The other two requirements, (iii) and (iv), might be parts of a not fully understood deeper structure. In particular it would be desirable to understand the local aspect of stability.

2) Every net with properties (i) through (iv) is a specific physical theory, complete with interpretation. To what extend a particular such theory agrees with experience is another matter. It is comforting, however, that the mentioned properties, together with a few other requirements which can be simply expressed [6)], suffice to guarantee quite a number of general features observed in high energy physics: the interpretation of phenomena in terms of collision processes of particles satisfying Bose- or Fermi-statistics (more precisely para-Bose or para-Fermi of a definite order), the existence of antiparticles, the role of charge quantum numbers, the connection of spin and statistics [7)].

3) It is still not entirely clear whether any nontrivial theory really is allowed in this frame. To answer this question by "yes" has been the central aim of constructive quantum field theory. I understand that we shall learn from Dr. Balaban at this school that the situation looks very hopeful.

---

6) The most important seems to be the "nuclearity" resulting if both the spacial extension and the energy are limit [10].

7) For the analysis of particles and collision processes in this frame see [11,12,13], for the analysis of statistics and charge quantum numbers see [14,15,16,17].

4) The impact of experiments on the development of quantum field theory in the past three decades has led from simple field theoretical models such as the Yukawa theory of strong interactions and the Fermi theory of weak interactions (in which the basic elements are fields associated to space-time points) to local gauge theories which may characterize by the feature that as basic elements (from which local algebras of observables are constructed) we need besides point-like objects also links, plaquettes, etc. It still remains a challenge to incorporate this lesson as a clearly stated principle into the properties required of the net. Together with a few other requirements such as nuclearity, asymptotic freedom this could render the structure tight enough so that a useful classification of possible theories can be attempted.

Moving now towards an inclusion of gravitational effects there is a first stage, in which we will describe gravitation by a classical (background) metric field and address ourselves only to the task of extending quantum field theory from the underlying Minkowski space to a given non-flat manifold. Obviously the net structure with properties (i) carries over, global Poincaré symmetry is lost and a replacement of properties (iii), (iv) has to be found. The most fascinating application of such an intermediate theory is the Hawking radiation associated with the gravitational collapse of a star. We shall address ourselves to this problem in the next section.

While this stage can be considered as reasonably well understood by now this is certainly not the case for quantum gravity. We shall discuss in section 3 the aspect of general covariance and the problem of constructing a theory from strictly local information.

2. HAWKING TEMPERATURE AND HAWKING RADIATIONS

Classical general relativity leads to the conclusion that very massive stars ultimately end by gravitational collapse, leading at some stage to the formation of a black hole from whose inside no signal can reach an outside observer. Furthermore, for the outside world the black hole has properties of a thermodynamic system in equilibrium ("no hair theorems", entropy) [18], [19]. In an admirable paper [20] Hawking argued that the association of a temperature

$$T = \frac{c^3\hbar}{8\hbar MG} = \frac{a}{2\hbar}\frac{\hbar}{c} \tag{1}$$

with the black hole surface could be understood by considering a quantum field theory in the curved space-time given by the gravitational field of the collapsing star. M is the mass of the star, G the gravitation constant and a the acceleration at the surface of the black hole. this can be exemplified in the simplest model, where the collapse is spherically symmetric and the quantum field theory is taken as that of a free scalar field ϕ obeying the linear covariant field equation

$$\Box_g\phi = 0; \quad \Box_g = |g|^{-\frac{1}{2}}\partial_\mu g^{\mu\nu}|g|^{\frac{1}{2}}\partial_\nu. \tag{2}$$

Here $g_{\mu\nu}$ is the given metric field computed from some classical model of stellar collapse e.g. the Oppenheimer-Snyder model, $|g|$ is the absolute value of the determinant of the metric tensor. The details of the metric field inside the star matter will turn out to be irrelevant for the effect studied. Neither will the quantum field theoretical model affect the value of the temperature though it does determine the radiation emitted.

Outside of the star and of the black hole the metric field is well known if we put the energy-momentum tensor equal to zero there (i.e. neglect the effect of the emitted radiation on the metric). Spherical symmetry together with the boundary conditions at space-like infinity (asymptotic flatness and given total mass M) allow only one solution of the Einstein equations, the Schwarzschild metric. It possesses a time-like Killing vector field. Using the corresponding time coordinate t (Schwarzschild time) and standard polar coordinates r, ϑ, φ the line element is

$$ds^2 = -\left(1 - \frac{r_0}{r}\right) dt^2 + \left(1 - \frac{r_0}{r}\right)^{-1} dr^2 + r^2(d\vartheta^2 + \sin^{-2}\vartheta d\phi^2) \tag{3}$$

Here

$$r_0 = \frac{2MG}{e^2} \tag{4}$$

is the Schwarzschild radius of the stellar mass. The singularity at $r = r_0$ results from the fact that the Schwarzschild time is not a good coordinate for the description of the collapse process; it approaches infinity when the black hole is formed i.e. when the stellar radius reaches the Schwarzschild radius. We shall make use of some other coordinate in the outside region:

$$\begin{aligned} r^* &= r + r_0 ln\left(\frac{r}{r_0} - 1\right) \\ v &= t + r^* \\ u &= t - r^* \end{aligned} \tag{5}$$

and a time coordinate

$$\tau = v - r = t + r^* - r \tag{6}$$

which remains finite on the horizon and can be extended to the inside. The origin of the τ-axis we choose so that the formation of the black hole occurs at $\tau = 0$. Fig. 1 illustrates the collapse in a $\tau - r$-diagram.

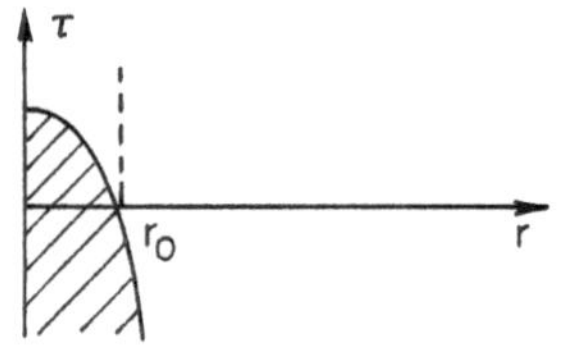

Figure 1

At the transition point $\tau = 0, r = r_0$ v remains finite whereas $t \to +\infty, r^* \to -\infty, u \to +\infty$.

The state ω of the quantum field is described by the expectation values of products of fields (Wightman distributions)

$$W^{(n)}(x_1, \dots x_n) = \langle \Phi(x_1) \dots \Phi(x_n) \rangle_\omega.$$

In the free field setting we may be content with the discussion of the 2-point-function $W^{(2)}$. We shall be interested in the partial state of a finitely extended region around a point $(T, R, \vartheta_0, \phi_0)$ where $R \ll r_0$ and T is very large; it is the region where the detector is to be placed which shall register the radiation [8]. It is, of course, not surprising that the changing gravitational field during the collapse phase is the source of an outgoing radiation. The interesting feature of Hawking radiation is that (as long as one neglects the back reaction) this radiation remains persistent, it becomes stationary as $T \to \infty$ (keeping R fixed), although the change of the metric for $\tau > 0$ occurs only in a region from which no causal influence can reach the detector. In this idealization the source of the radiation is a neighborhood of the transition point $\tau = 0, r = r_0$ shrinking more and more as T is taken larger. One of our main objectives must therefore be to verify that in this model there remains an asymptotically constant detector rate as $T \to \infty$. Of course this is physically impossible. The energy radiated away must be compensated by a loss of stellar mass ("black hole evaporation") and this means that the Schwarzschild radius must shrink and therefore the horizon, which was given by $r = r_0$ for $\tau > 0$ in the idealized model, becomes slightly inclined. This remedies also the causality paradox; the origin of the radiation received will then move along the horizon to increasing τ as T gets larger. But this correction of the simple model demands a self consistent treatment of the radiation and its back reaction on the metric, a task which has not yet been achieved. We shall discuss the simple model, adding at the end a few remarks on the status of the back reaction problem.

To compute the partial state at late times we would need, in principle, no new ideas which are not familiar from special relativistic quantum field theory. At very early times the star has an essentially constant radius, large compared to r_0, the Schwarzschild time can be extended to the inside of the star and the metric is everywhere practically stationary. In this region we have time translation invariance of the field theory and can define the field energy and compute the "vacuum" state of the field as the state of lowest energy. Thus $W^{(2)}$ for very early times is known and, since $W^{(2)}(x_1, x_2)$ obeys the field equation in both arguments the computation of its development for later time demands just the solution of an initial value problem. In his first paper on the subject Hawking treated the problem from this point of view using a geometrical optics approximation. The conclusions drawn have the ring of truth; they are more convincing than the computation itself. It is very hard to verify to what extent the approximations are trustworthy since the light rays pass through regions of extremely fast changing index of refraction and one is interested in a tiny effect after very long time. Moreover, the effect should have some universality, a fundamental significance, because it suggests the mechanism by which quantum theory may succeed in forbidding the formation of singularities which are a generic feature of classical general relativity.

To understand the significance of the temperature (1) in a more direct way subsequent papers have concentrated on the discussion of a permanent, static black hole. I do not want to discuss these in detail. A survey of the relevant literature is

---

[8] For large R it is irrelevant whether we take the Schwarzschild time or τ as the time coordinate.

given in [21]. The essential point seems to me that all allowed states have a common property in the small which may be viewed as the local version of the requirement (iv) of the special relativistic situation. This fixes the leading singularity in the short distance behavior of the Wightman distributions to be the same for all allowed states. In the case of a free field we have

$$W^{(2)}(x_1, x_2) = \frac{1}{\sigma_\varepsilon} + \widehat{W}^{(2)} \tag{7}$$

where σ_ε is the square geodesic distance between x_1 and x_2 with an imaginary part $i\varepsilon(\varepsilon \to 0)$ added to the time difference to regulate the behavior for lightlike distances. The dependence on the state is only in $\widehat{W}^{(2)}$ which is less singular. In the case of a static black hole the outside region is stable under time translation (with respect to the Schwarzschild time) and one can compute the thermal equilibrium states (Gibbs states, KMS-states) for this region for arbitrary temperatures. It turns out that of all these states only the one with the temperature (1) has the correct leading singularity on the horizon. For the others the relation between the real and the imaginary part of the most singular term is wrong [22]. Physically this may be interpreted as saying that of all thermal states outside only the one with the Hawking temperature can be extended to the inside of the black hole.

Why should the state be thermal (or rather semi-thermal with an outgoing radiation emanating from the black hole)? To answer this we have to return to the formation process sketched in fig. 1. I shall follow the line of argument of [23]. We can simulate the detector by an observable Q^*Q where $Q = \phi(h)$ is a smeared out field with a test function having its support in the observation region. Using the field equation we can express Q in terms of the field and its time derivative on the space-like surface $\tau = 0$. This is most easily done by remarking that we can write Q as

$$Q = \int_{\Sigma} \phi(x) \frac{\partial}{\partial x^\mu} f(x) d\sigma^\mu \tag{8}$$

where $\sum$ is an arbitrary space-like surface with surface element $d\sigma^\mu$ and f is a smooth c-number solution of the wave equation. The right hand side is independent of the choice of $\sum$. If we choose $\sum = \sum_1$, the surface $t = T$, then f and $\frac{\partial f}{\partial t}$ have spatial support in some neighborhood of R, ϑ_0, ϕ_0. We may then choose $\sum = \sum_2$ as the surface $\tau = 0$. To determine f there we have the problem of solving the wave equation for the numerical function f between the surfaces $\sum_1$ and $\sum_2$ with the given "final conditions" at $t = T$. In the whole relevant region the support of f will be outside of the star matter and outside of the black hole due to the causal structure prevailing. Therefore we need for this problem only to study the behavior of solutions of the wave equation in the Schwarzschild metric for very large time differences. This is qualitatively known (see e.g. [24]). As T gets larger and larger the function f on the surface $\tau = 0$ decomposes into

$$f = f_1 + f_2 \tag{9}$$

where f_1 is a wave packet concentrating more and more on the horizon and f_2 is a wave packet moving outwards towards $r \to \infty$. In evaluation the asymptotic counting rate as $T \to \infty$ we need to know $w^{(2)}(x_1, x_2)$ at $\tau = 0$ for r-values which are either "infinitely close" to $r = r_0$ or extremely large. The former contribute in the limit only because of the singular part of $w^{(2)}$ for $x_1 \to x_2$ at the transition point. This is universal for all allowed states and gives the Hawking radiation. The contribution where one of the points is at the horizon, the other at spatial infinity vanishes. The contribution where both points are at spatial infinity for $\tau = 0$ is unaffected by the collapse. It gives the expectation value of Q^*Q in the situation of the permanently static star (vanishing if Q^*Q is a detector of radiation).

Let us look at the behavior of the function f. Using the coordinates t, r^*, ϑ, ϕ and separating off the angular motion by expanding

$$f(t, r^*, \vartheta, \phi) = r^{-1} \sum_{l,m} f_{lm}(t, r^*) Y_l^m(\vartheta, \phi) \tag{10}$$

the wave equation becomes

$$\left(\frac{\partial^2}{\partial t^2} - \frac{\partial^2}{\partial r^{*2}} + V_l \right) f_{lm} = 0 \tag{11}$$

with

$$V_l = \left(1 - \frac{r_0}{r}\right) \left(\frac{l(l_1)}{r^2} + \frac{r_0}{r_3} \right). \tag{12}$$

Fig. 2 shows V_l as a function of r^*. It decreases exponentially for $r^* \to -\infty$.

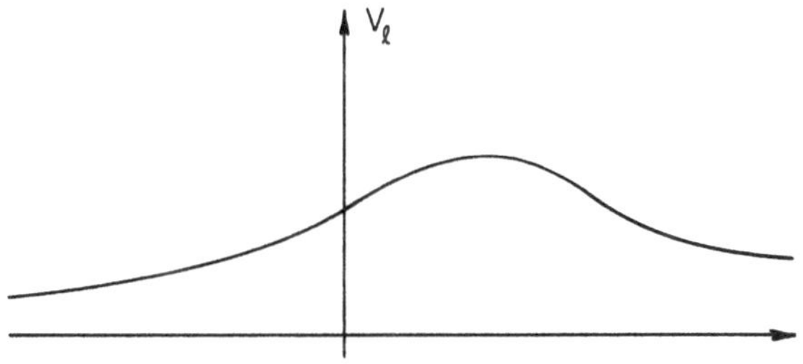

Figure 2

The values of f_{lm} and $\frac{\partial f_{lm}}{\partial t}$ at $t = T$ characterize the nature of the detector. If the detector is inward directed $\frac{\partial f}{\partial t} = -\frac{\partial f}{\partial r^*}$. We are interested in the solution of (11) subject to the conditions

$$f_{lm}(T_1 r^*) = F_{lm}(r^*); \frac{\partial f_{lm}}{\partial t}(T_1 r^*) = -\frac{\partial F_{lm}}{\partial r^*} \tag{13}$$

letting $T \to \infty$, keeping $F_{lm}(r^*)$ fixed. We shall omit the indices l, m in the following. The solution for large $T - t$ has the form (9) with

$$f_1(t, r^*) = (2\pi)^{-1} \int_{-\infty}^{\infty} a(\omega) l^{i\omega(r^* - t + T)} d\omega \tag{14}$$

and

$$a(\omega) = D(\omega^2)\int l^{-i\omega r^*} F(r^*)dr^* \equiv D(\omega^2)\widetilde{F}(\omega), \tag{15}$$

where D is the transmission amplitude for the 1-dimensional Schrödinger Problem with the potential barrier of fig. 2 and "energy" ω^2. For $\tau = 0$ and $T \to \infty$ f_1 is of the form

$$f_1 = \psi\left(\frac{r - r_0}{\lambda}\right), \quad \lambda \equiv l^{-\frac{T}{2r_0}} \to 0. \tag{16}$$

So we have a scaling of the radial coordinate to $r = r_0$. Inserting (14), (15) in (8) (with $\sum$ chosen as the surface $\tau = 0$) and evaluating Q^*Q in a state with the 2-point function (7) one finds that the (universal) $\frac{1}{\sigma}$-singularity leads to an asymptotically constant detection rate

$$\lim_{T\to\infty}\langle Q_T^* Q_T\rangle = \sum_{lm}\int |D_l(\omega^2)|^2\ |\widetilde{F}_1(\omega)|^2(1 - l^{-\beta\omega})^{-1}\omega d\omega \tag{17}$$

with

$$\beta = 4\pi r_o. \tag{18}$$

This is the asymptotic counting rate produced by an outgoing radiation of temperature $(4\pi r_0)^{-1}$ but modified by the barrier penetration effect.

The less singular part of $w^{(2)}$ does not contribute to the counting rate in the limit $T \to \infty$. Therefore we do not need any detailed knowledge about the state to derive the expression (17) for the outgoing radiation. The only information needed is the leading term in the short distance behavior at $\tau = 0, r = r_0$. The arguments why this should be given by $\frac{1}{\sigma_\epsilon}$ will be discussed in the next section.

The whole discussion so far was based on the quantum theory of a free field in the curved space-time provided by the metric of the stellar collapse. For a full understanding there remain two highly nontrivial problems. First, we must ask what changes if instead of the free field we take a realistic theory, say the standard model of elementary particle physics. Secondly there is the back reaction problem, the modification of the metric due to the energy radiated away which becomes crucial in the late stages.

We believe that the value of the Hawking temperature and the Bekenstein entropy are universal, that they do not depend on the specific dynamics of the quantum field. The strongest argument for this is probably the analogy of the Schwarzschild black hole with the Rindler wedge in Minkowski space. For the latter, Bisognano and Wichmann [25] have shown that the ordinary vacuum state of quantum field theory in Minkowski space looks in its restriction to the Rindler wedge like a thermal state, where Lorentz boosts replace the time translations. This result and the value of the temperature are derived in [25] from general principles without specifying the dynamics of the theory. It is amusing to note that this work was not motivated by black hole physics. Its relation to the Hawking effect and, in particular, for Unrah's discussion of the counting rate of a detector moving with constant acceleration through the vacuum [26] was noticed much later. One must recognize, however, that the Hawking temperature does not have a direct observational significance. This temperature relates the state on the surface of the black hole to a vector field of time translation which vanishes on this surface and becomes physically relevant far away. The observable effect is the outgoing radiation and this depends significantly on the dynamics of the quantum field. In the free field case the modification of thermal radiation is simply given by the transmission coefficients $|D|^2$ through the potential barrier (12). If we follow the same reasoning as above for the case of, say, Quantum Chromodynamics then we would conclude that the asymptotic counting rate at large times can be related to some scaling limit on the horizon. In this scaling limit the theory will degenerate to a theory of free quarks and gluons. The determination of the correspondence between observables describing detectors of physical particles far outside and observables in this limit theory is a nontrivial task which needs a fragmentation model.

THE BACK REACTION PROBLEM

The energy radiated away must cause a decrease of the mass, a shrinking of the Schwarzschild radius with increasing time, a modification of the metric in the outside region. To obtain a first order approximation for this effect one would like to solve the Einstein equations

$$R_{\mu\nu} - \frac{1}{2} R g_{\mu\nu} = -8\pi G \langle T_{\mu\nu} \rangle \tag{19}$$

where $\langle T_{\mu\nu} \rangle$ is the expectation value of the energy momentum tensor in the state of the quantum field described by the 2-point function $W^{(2)}$ considered above. Given $w^{(2)}(x, x')$, what is the expectation value of $T_{\mu\nu}$? There exists an extensive literature devoted to this problem, see e.g. [27]-[34]. But in my opinion no convincing answer has been given.

Starting from the classical expression

$$T_{\mu\nu} = \frac{\partial\phi}{\partial x^\mu}\frac{\partial\phi}{\partial x^\nu} - \frac{1}{2} g_{\mu\nu} \left(\frac{\partial\phi}{\partial x^\rho}\frac{\partial\phi}{\partial x^\sigma} g^{\rho\sigma} \right) \tag{20}$$

and restricting the short distance behavior of the states more stringently than in (7) to the so-called "Hadamard form"

$$\frac{1}{2}\langle \phi(x_1)\phi(x_2) + \phi(x_2)\phi(x_1) \rangle = P\frac{A}{\sigma} + B ln|\sigma| + C \tag{21}$$

where A, B and C are smooth functions ofx_1 and x_2 one can analyze the divergent parts of $\langle T_{\mu\nu} \rangle$ taking the arguments of the field derivatives on the right hand side of (20) to be slightly different ("point splitting method"). The divergent parts come from A and B and these two functions are determined by the field equations as definite expressions in the local metric and its derivatives. Thus the naive expression for $\langle T_{\mu\nu} \rangle$ has a divergent part which is universal for all Hadamard states. Different states differ only in the smooth function C and the difference between $\langle T_{\mu\nu} \rangle$ in two such states is finite. The divergent part of $\langle T_{\mu\nu} \rangle$ has a leading 4th order term (in $|x_1 - x_2|^{-1}$) which may be regarded as vanishing upon directional averaging; the next term is of the form $\alpha R_{\mu\nu} + \beta g_{\mu\nu}$ where α and β are quadratically diverging. If one modifies the field equations e.g. by introducing a diverging bare mass then β may be arranged to become $-\frac{1}{2}\alpha R$ so that this part may be considered as a "renormalization of the gravitational constant". This leaves linearly and logarithmically divergent parts. They might be cancelled by further modifications of the field equation but this remains an open area for speculation.

Another strategy is to search for a (locally) distinguished state, analogous to the vacuum in the Minkowski space theory (specifying the function C up to second order in $|x_1 - x_2|$ for this state). Again, to do this in a natural and consistent manner, one cannot ignore the possibly necessary modifications of the field equations and I feel that no convincing answer has been given which relates $\langle T_{\mu\nu}\rangle$ to the C-function.

3. GENERALLY COVARIANT QUANTUM FIELD THEORY AND SCALING LIMITS

One of the salient features of classical general relativity is the generally covariant formulation of the laws emphasizing the independence from the choice of a coordinate system on the manifold. To carry this over to local quantum physics one would like to postulate that the group of local diffeomorphism of the manifold is realized by a group of automorphisms of the net of algebras $\{R(\mathcal{O})\}$ (as in (ii) of section 1, but with g denoting now any local diffeomorphism). Speaking of diffeomorphisms instead of coordinate transformations suggests an interpretation of the group as symmetries in the active sense i.e. mappings leading from one possible physical situation to a different one which is also in accordance with the laws. This would mean, however, that we loose all predictive power for future events from the knowledge of the past because a "physical situation" determines macroscopically a metric field and we may consider diffeomorphisms which act trivially in the past of some space-like surface in this metric and nontrivially in the future. In the classical setting this problem is resolved by noting that the covariant field equations (e.g. the Einstein-Maxwell-system) do not yet define a Cauchy problem. One has to supplement them by four coordinate conditions and these break the general covariance. In quantum physics the most natural approach seems to me to try to build up the theory from its germs. This divides the problem into two stages. First the characterization of a germ i.e. the restriction of the net to an arbitrarily small neighborhood of a point, secondly the rule by which the local information is patched together. In both stages we meet many open questions. In [35] we started from an algebra on which the diffeomorphism group is naturally acting but which does not yet contain dynamical information. Specifically we took the tensor algebra of test functions of a scalar field. A state is characterized by the hierarchy of Wightman distributions and, given a state we can reconstruct a Hilbert space representation and a net $R(\mathcal{O})$ of von Neumann algebras.

The tensor algebra has many 2-sided ideals and this will manifest itself by the fact that taking an arbitrary state the representation constructed will have a central decomposition, i.e. the von Neumann algebra $R(\mathcal{O})$ will have nontrivial centers. A "primary state" is one for which the algebras $R(\mathcal{O})$ have no center (besides multiples of the identity). The representation generated by such a state is one in which some maximal ideal is the original tensor algebra is equated to zero. The task of specifying the laws is tantamount to the characterization of the allowed class of primary states. One aspect of this is the short distance behavior described by a scaling limit at each point. I can sketch here only the general idea and results and refer for precise definitions etc. to [35]. For each point P one considers 1-parameters subgroups of diffeomorphisms ("dilations") which contract neighborhoods of P to the point P as the parameter $\lambda \to 0$. Subjecting a given state to a certain limiting procedure with respect to such contractions one obtains in an intrinsic way a reduction of the theory on the manifold to a field theory in the tangent space at P with a distinguished state. The resulting distinguished state $\widehat{\omega}_P$ on the "tangent space algebra" (the tensor algebra of functions on tangent space) is the same for all original states in one primary folium and it is automatically invariant under translations in tangent space. The set of all allowed limit states $\{\widehat{\omega}_P\}$ at P in a generally covariant theory results from a single one by general linear transformations in the tangent space. It is invariant under some subgroup $\mathcal{G}$ on GLR (4). The original primary state (or rather its folium) determines a section in a bundle with fibre $\{\widehat{\omega}_P\}$ (isomorphic to GLR $(4)/\mathcal{G}$). What is the stability group $\mathcal{G}$? One interesting possibility is that it is the Lorentz group. In that case each allowed primary folium of states determines a classical metric field, the tangent space theory is a quantum field theory in a Minkowski space with a distinguished vacuum state. This looks similar to the effective field treatment of gravity used in the last section, but essential parts for a full theory are missing. On the one hand the tangent space theory contains only an asymptotic limit of the dynamical law, it is not yet a germ of the theory. On the other hand there remains the problem of patching germs together.

Another interesting avenue may start from the assumption that at distances below the Planck length the concept of metric gets lost and, if a scaling limit exists at all, it should be quite structureless, corresponding to a large stability subgroup like $SLR(4)$. The formation of an effective metric must then be explained as a cooperative effect, as a property of local equilibrium states in regions large compared to the Planck length. I think that this approach would be quite promising but it has not been explored at all.

Let us look, finally, at the other stage of the problem, the construction of the theory at large from local information. This is a very crucial problem since in the Minkowski space theory the commutativity of observables at space-like distances (no matter how large) is an essential part of the input and this is not contained in the local information. The natural approach to this problem must proceed in analogy to the theory of manifolds. The local information is given by a system of charts. A single chart $\Gamma = (\widehat{\mathcal{O}}, \mathcal{A}, \pi\mathcal{H})$ consists of an open region $\widehat{\mathcal{O}} \subset \mathbf{R}^4$, a net of algebra $\mathcal{A}(\mathcal{O})$ for $\mathcal{O} \subset \widehat{\mathcal{O}}$ and a representation π of this net by operator algebra acting in a Hilbert space $\mathcal{H}$. The net should separate the points of $\widehat{\mathcal{O}}$ i.e. for sufficiently small neighborhoods of two distinct points the algebras should only have multiples of the identity in common. The question is now how to combine two "adjoining charts" $\Gamma_i, i = 1,2$ to a theory in a larger domain. The word "adjoining" means that there are subregions $\mathcal{O}_1^{(2)} \subset \widehat{\mathcal{O}}_1$ and $\mathcal{O}_2^{(1)} \subset \widehat{\mathcal{O}}_2$ within which the nets are to be identified. So there should exist a unitary map U_{21} from $\mathcal{H}_1$ to $\mathcal{H}_2$ such that

$$U_{21}\pi_1(\mathcal{A}(\mathcal{O}))U_{21}^{-1} = \pi + 2(\mathcal{A}(\phi_{21}\mathcal{O})), \quad \text{for} \quad \mathcal{O} \subset \mathcal{O}_1^{(2)} \tag{22}$$

where ϕ_{21} is an ordinary transition function i.e. a diffeomorphism from $\mathcal{O}_1^{(2)}$ to $\mathcal{O}_2^{(1)}$. The transition operator U_{21} determines the transition function due to the separation property of the net. A system of charts together with transition operators satisfying the appropriate compatibility conditions will define a "theory" i.e. a manifold with a net of operator algebras acting in a single Hilbert space. But the transition operators cannot, in general, be regarded as parts of the local information. They are determined by the transition charts only up to a unitary in the commutant of $\pi_1(\mathcal{A}(\mathcal{O}_1^{(2)}))$. The choice of this operator will influence the commutation relations between observables in $\widehat{\mathcal{O}}_1, \widehat{\mathcal{O}}_2$ outside of the identified parts and, if one has a chain of such charts, the commutation relations at large distances. There are two situations in which U_{21} is locally determined.

One is, if instead of the representation π_i we give a specific (partial) state ω_i in each chart (which, of course, must coincide on he common parts) and which has a property of Reeh-Schlieder type. The construction of the theory at large from this kind of local information has been discussed in [35].

Another, at first sight adventurous, possibility is that the commutant of $\pi_1(\mathcal{A}(\mathcal{O}_1^{(2)}))$ is trivial i.e. that all local algebras are weakly dense in the algebra of all bounded operators [36]. In this case they can, of course, not be W^*-algebras but must be closed under a much stronger topology. From the physical point of view this possibility is rather natural since commutativity is related to causal disjointness and, if the metric fluctuates, we cannot predict which regions are strictly causally disjoint. An analogue of this situation is provided by the algebras of space-time regions in nonrelativistic theory where due the unlimited propagation velocity we also have no strict causal independence.

To sum up: there are many open questions and only a few answers have been attained. Perhaps someone in the audience will take up the challenge to think about some aspects in this fascinating area.

REFERENCES

1. P. Jordan, J. v.Neumann and E.P. Wigner, On an algebraic generalization of the quantum mechanical formalism, *Ann. of Math. 35*: 29-64 (1934); G. Birkhoff and J. v.Neumann, The logic of quantum mechanics, *Ann. of Math. 37*: 823-843 (1936); D. Finkelstein, J.M. Jauch, S. Schiminovich, and D. Speiser, Foundations of quaternion quantum mechanics, *J. Math. Phys. 3*: 207-220 (1962).
2. G. Ludwig, Versuch einer axiomat. Grundlegung der Quantenmechanik und allgemeinerer physikal Theorien, *Z. Phys. 181*: 233-260 (1964); Attempt of an axiomat. foundation ... II, *Commun. Math. Phys. 4*: 331-348 (1967); C. Piron, Axiomatique quantique, *Helv. Phys. ACta 37*: 439-468 (1964).
3. J.C.T. Pool, Baer-semigroups and the logic of quantum mechanics, *Commun. Math. Phys. 9*: 118-141 (1968).
4. B. Mielnik, Generalized quantum mechanics, *Commun. Math. Phys. 37*: 221-256 (1974).
5. H. Araki, On a characterization of the state space of quantum mechanics, *Commun. Math. Phys. 75*: 1-24 (1980); E.M. Alfsen and F.W. Shultz, State spaces of Jordan algebras, *Acta Math. 140*: 155-190 (1978).

6. J. Bellissard and B. Iochum, Homogeneous self-dual cones versus Jordan algebras, *Ann. Inst. Fourier Grenoble 28*: 27-67 (1978); B. Iochum, Cónes autopolaires at algebras de Jordan, *Lecture Notes in Mathematics*, Springer Berlin Heidelberg, NY (1984).
7. K. Fredenhagen, On the modular structure of local algebras of observables, *Commun. Math. Phys. 97*: 79-89 (1985).
8. R. Haag and D. Kastler. An algebraic approach to quantum field theory, *J. Math. Phys. 5*: 848 (1964).
9. R. Haag, R.V. Kadison and D. Kastler, Nets of C^*-algebras and classification of states, *Commun. Math. Phys. 16*: 81-104 (1970).
10. D. Buchholz and E.H. Wichmann, Causal independence and the energy level density of states in local quantum field theory, *Commun. Math. Phys. 106*: 321-344 (1986).
11. R. Haag and J.A. Swieca, When does a quantum field theory describe particles?, *Commun. Math. Phys. 1*: 308-320 (1965).
12. H. Araki and R. Haag, Collision cross sections in terms of local observables, *Commun. Math. Phys. 4*: 77-91 (1967).
13. D. Buchholz, Particles, infraparticles and the problem of asymptotic completeness, in *Proc. of VIIIth Intern. Congress of Mathematical Physics*, Marseille 1986. World Scientific, Singapore (1987).
14. S. Doplicher, R. Haag and J.E. Roberts, Fields, observables and gauge transformations II, *Commun. Math. Phys. 15*: 173-200 (1969); Local observables and particle statistics I and II, *Commun. Math. Phys. 23*: 199-230 (1971); *35*: 49-85 (1974).
15. D. Buchholz and K. Fredenhagen, Locality and the structure of particle states, *Commun. Math. Phys. 84*: 1-54 (1982).
16. K. Fredenhagen, On the existence of antiparticles, *Commun. Math. Phys. 79*: 141-151 (1981).
17. D. Buchholz and H. Epstein, Spin and statistics of quantum topological charges, *Fysica 17*: 329-343 (1985).
18. S.W. Hawking and G.F.R. Ellis, The large scale structure of space-time, Cambridge University Press 1980.
19. J.D. Bekenstein, Black holes and entropy, *Phys. Rev. D7*: 2333-2346 (1973).
20. S.W. Hawking, Particle creation by black holes, *Commun. Math. Phys. 43*: 199-220 (1975).
21. B.S. Kay and R.M. Wald, Some recent developments related to the Hawking effect, in *Proc. XVth Intern. Conf. on Differential Geom. Methods in Theoret. Phys.*, Clausthal 1986, ed. by H.D. Doebner and J.D. Henning, World Scientific, Singapore 1987.
22. R. Haag, H. Narnhofer and U. Stein, On quantum field theory in gravitational background, *Commun. Math. Phys. 94*: 219-238 (1984).
23. K. Fredenhagen and R. Haag, On the derivation of Hawking radiation associated with the formation of a black hole, to be published in *Commun. Math. Phys.*
24. J. Dimock and B.S. Kay, Classical and quantum scattering theory for linear scalar fields on the Schwarzschild metric, *Ann. Phys. (NY) 175*: 366-426 (1987).
25. J.J. Bisognano and E.H. Wichmann, On the duality condition for a Hermitean scalar field, *J. Math. Phys. 17*: 303 (1976).

26. W.G. Unruh, Notes on black hole evaporation, *Phys. Rev. D14*: 870-891 (1976).
27. S. Adler, J. Liebermann and Y.J. Ng, Regularization of the stress-energy tensor for vector and scalar particles propagating in a general background metric, *Ann. Phys. (NY) 106*: 279-321 (1977).
28. R.M. Wald, The back reaction effect in particle creation in curved space-time, *Commun. Math. Phys. 54*: 1-19 (1977); *Phys. Rev. D17*: 1477 (1978).
29. P.C.W. Davies and S.A. Fulling, Quantum vacuum energy in two-dimensional space-times, *Proc. Roy. Soc. (London) A354*: 59-77 (1977).
30. P. Candelas, Vacuum polarization in Schwarzschild space-time, *Phys. Rev. D21*: 2185-2202 (1980).
31. S.A. Fulling, M. Sweeney and R.M. Wald, Singularity structure of the two-point function in quantum field theory in curved space-time, *Commun. Math. Phys. 63*: 257-264 (1981).
32. M.R. Brown and A.C. Ottewill, The energy-momentum operator in curved space-time, *Proc. Roy. Soc. (London) 389A*: 379-403 (1983).
33. N.D. Birrell and P.C.W Davies, Quantum fields in curved space, Cambridge University Press 1982.
34. R.M. Wald, General relativity, Chicago University Press 1984.
35. K. Fredenhagen and R. Haag, Generally covariant quantum field theory and scaling limits, *Commun. Math. Phys. 108*: 91-115 (1987).
26. R. Haag, in *Quantum fields and geometry*, ed. by M. Cahen and M. Flato, Kluwer Acad. Publishers, Dordrecht, Boston, London 1988.

GEOMETRY OF SUPERSYMMETRY[1]

Arthur Jaffe and Andrzej Lesniewski

Harvard University
Cambridge, MA 02138 U.S.A.

CHAPTER I. SUPERSYMMETRIC QUANTUM FIELD THEORY

I.1. Introduction

These lectures are divided into three parts. The latter two chapters are mathematical, in the sense that all definitions are precise and results are formulated as theorems. The present chapter plays a different role. It provides a mixture of motivation and formal calculation. Some of the materials is or can be made mathematical; the presentation, however, is distinctly that from physics. A mathematician may wish to read this chapter for an overview; he will be more familiar with the style of Chapters II and III.

We mainly concentrate on two-dimensional problems. Thus our discussion of the field theory is especially oriented toward problems of field theories on a compact manifold (as a torus or a Riemann surface). The *infinite volume limit*, namely the study of field theory on non-compact spaces will not be treated. Many constructive field theory examples of infinite-volume field theories have been constructed on two-dimensional flat space. We refer the reader to the literature [GJ]. Infinite volume of supersymmetric fields have been studied recently [We], and much work remains to be done in that area.

A standard quantum field theory is described in one of several ways. The two major approaches are the Hilbert space formulation and the functional integral formulation. We discuss both formulations here. In the case of flat space-time the two formulations are equivalent — this is a theorem of Osterwalder and Schrader. In fact, the mathematics of the constructive field program to investigate the Hilbert space formulation relies, in an essential way, on estimates which are proved using the functional integral formulation.

[1] Supported in part by the National Science Foundation under Grant DMS/PHY 88-16214.

Supersymmetry has a simple expression in the Hilbert space formulation. The Hilbert space of a supersymmetric theory has a natural $\mathbb{Z}_2$ -grading given by the operator $\Gamma = (-1)^{N_f}$, where N_f is the Fermi number operator. There is a self-adjoint, local, odd operator Q whose square is the Hamiltonian H. Here *local* means that Q can be expressed as the integral of a density, $Q = \int Q(x)\,dx$, and *odd* means that

$$\{Q, \Gamma\} = 0 \ .$$

More precisely, supersymmetry entails the existence of a pair of operators Q_1, Q_2 such that

$$\{Q_\alpha, \bar{Q}_\beta\} = 2\gamma^\mu_{\alpha\beta} P_\mu \ . \tag{I.1.1}$$

Here $\gamma^\mu, \mu = 0, 1$ are the two-dimensional Dirac matrices in Majorana representation,

$$\gamma^0 = \begin{pmatrix} 0 & -i \\ i & 0 \end{pmatrix}, \qquad \gamma^1 = \begin{pmatrix} 0 & i \\ i & 0 \end{pmatrix},$$

and $\bar{Q} = Q\gamma^0$. In (I.1.1), $P_0 = H$ and $P_1 = P$, the infinitesimal generator of translations. As a consequence, it follows that the spectral condition,

$$0 \le H \pm P \ , \tag{I.1.2}$$

holds. Thus supersymmetry provides a powerful tool also in understanding positivity estimates. It turns out that (I.1.2) alone is not sufficient to control the construction of such models. Somewhat stronger estimates, which for example in Hamiltonian form have the structure that H is bounded from below by a uniformly elliptic operator, are needed. For example, we use the "Gårding inequality,"

$$\varepsilon N_\tau \le H + I \ , \tag{I.1.3}$$

where N_τ is the quantization of $(-d^2/dx^2 + 1)^{\tau/2}$, $\tau < 1$, and where $\varepsilon > 0$ is sufficiently small (depending on τ). Such estimates are established by functional integral methods.

In this chapter we forego estimates altogether; rather we concentrate on the formal structure of the supersymmetric field theory problems. In particular these formal constructions provide extremely useful insights as motivations for mathematical constructions.

I.2. The Loop Space Picture

Here we discuss two-dimensional quantum field theory from the point of view of analysis on loop space. The bosonic field can be analyzed in terms of a space of maps from space S or from space-time Σ to a target M. In particular the constant time maps yield the Schrödinger picture, in which case for $S = S^1$, the maps form the loop space of M. Fermionic fields arise as differential 1-forms on the loop space. This point of view was proposed by Witten [W1,2].

Thus we consider the space of smooth maps φ between two smooth manifolds:

$$\text{Map}\,(\Sigma, M) = \{\varphi : \Sigma \to M,\ \varphi \text{ smooth}\} \quad . \tag{I.2.1a}$$

There are two generic situations: M may be flat, but have a potential. Alternatively M may have a natural curvature arising from a metric. Examples of the flat theories are the φ^{2n} models which arise from $M = \mathbb{R}$ and their generalization to target spaces $\mathbb{R}^N$. These are the classic examples in constructive quantum field theory. For gauge theories, M is the Lie algebra of a compact Lie group G. The non-linear σ-models on a Riemann surface arise from the choice that Σ is a Riemann surface and M is a Riemannian manifold. The space Map (Σ, M) is itself an (infinite dimensional) Frechet manifold if Σ is compact [H].

In many cases in the physics literature, the space-time manifold Σ has a product structure $\Sigma = T \times S$, where T denotes the time direction and S the space manifold. In this case

$$\text{Map}\,(\Sigma, M) = \ \text{Path Space of Map}\,(S, M) \quad . \tag{I.2.1b}$$

There is an intimate connection between measure theory on Map (Σ, M) and harmonic analysis on Map (S, M) given by the Feynman-Kac formula. Even if Σ is not a product, the measure theory on Map (Σ, M) can be used to define a quantum field theory. The *loop space* corresponds to the special case

$$\Omega M = \text{Map}\,(S^1, M) \quad .$$

The standard σ-models arise from this choice, along with $\Sigma = \mathbb{R} \times S^1$. Let $E \to \Omega M$ be a vector bundle over ΩM and let $\bigwedge(E)$ denote its exterior algebra. The study of elliptic complexes on $\bigwedge(E)$ gives rise to the supersymmetric σ-models.

Let us illustrate this construction with a simple example. Let M be a (finite-dimensional) Riemannian manifold with metric tensor g and let $E = T^*\Omega M$ be the (complexified) cotangent bundle of its loop space. Assume the existence of a suitable measure on ΩM so that we can define

$$\mathcal{H} = L_2\Big(\Gamma(\textstyle\bigwedge(E))\Big) \quad ,$$

namely the space of L_2 sections of $\bigwedge(E)$ with respect to the L_2 norm on ΩM. Let

$$u^j(\sigma), \qquad \sigma \in S^1,\ j = 1, 2, \ldots, \dim M$$

be a basis for $T^*_\varphi \Omega M$, $\varphi \in \Omega M$. These basis elements are infinite dimensional analogs of the basis 1-forms dx^k in finite dimensional differential geometry. The $u^j(\sigma)$'s act on $\mathcal{H}$ by exterior multiplication (of 1-forms) on $\mathcal{H}$. As a consequence of the anticommutativity of the exterior product, they satisfy

$$\{u^j(\sigma),\ u^k(\sigma')\} = 0\ . \tag{I.2.2a}$$

Furthermore, the adjoints $u_j(\sigma)^*$ can be calculated. They are given by interior multiplication on $\mathcal{H}$, namely as annihilation operators in the language of physics. Thus

$$\{u^j(\sigma),\ u^k(\sigma')^*\} = g^{jk}(\varphi(\sigma))\delta(\sigma-\sigma')\ . \tag{I.2.2b}$$

We set

$$\psi_1^j(\sigma) = \frac{1}{\sqrt{2}}(u^j(\sigma) + u^j(\sigma)^*)\ , \tag{I.2.3a}$$

$$\psi_2^j(\sigma) = \frac{i}{\sqrt{2}}(u^j(\sigma) - u^j(\sigma)^*)\ , \tag{I.2.3b}$$

and note that as a consequence of (I.2.2),

$$\{\psi_\mu^j(\sigma), \psi_\nu^k(\sigma')\} = \delta_{\mu\nu}g^{jk}(\varphi(\sigma))\delta(\sigma-\sigma')\ . \tag{I.2.4}$$

In the physics literature the self-adjoint fields ψ are called the *Majorana Fermi fields.* They arise naturally in field theories with N = 1 supersymmetry and we will work with them in Section I.3 of these notes. In certain problems it is more natural to use the *Dirac Fermi fields.* Let us briefly comment on their geometric origin. In order to define the Dirac Fermi fields we double the number of fermionic degrees of freedom. Specifically, we take two copies of $\bigwedge(E)$ and set

$$\mathcal{H} = L_2\left(\Gamma(\bigwedge(E))\right) \oplus L_2\left(\Gamma(\bigwedge(E))\right)\ . \tag{I.2.5}$$

Let

$$u^j(\sigma)\ ,\quad v^j(\sigma)\ ,\qquad j = 1,2,\ldots,\ \dim M\ .$$

be the corresponding basis sections. Then we have

$$\{u^j(\sigma), v^j(\sigma')\} = \{u^j(\sigma), v^j(\sigma')^*\} = 0\ . \tag{I.2.6}$$

We define the Dirac Fermi fields by

$$\psi_1^j(\sigma) = \frac{1}{\sqrt{2}}\left(u^j(\sigma) + v^j(\sigma)^*\right)\ , \tag{I.2.7a}$$

$$\psi_2^j(\sigma) = \frac{i}{\sqrt{2}}\left(u^j(\sigma) - v^j(\sigma)^*\right) \ . \tag{I.2.7b}$$

Thus

$$\{\psi_\mu^j(\sigma),\ \psi_\nu^k(\sigma')\} = 0 \ , \tag{I.2.8a}$$

and

$$\{\psi_\mu^j(\sigma),\ \psi_\nu^k(\sigma')^*\} = \delta_{\mu\nu} g^{jk}(\varphi(\sigma))\delta(\sigma - \sigma') \ . \tag{I.2.8b}$$

We would like now to define the exterior derivative

$$d:\ \bigwedge^k(E) \to \bigwedge^{k+1}(E) \ . \tag{I.2.9}$$

In finite dimensions we have

$$d = dx^j \frac{\partial}{\partial x^j} \ ,$$

which in infinite dimensions suggests

$$d = \int_{S^1} u^j(\sigma)\frac{\delta}{\delta\varphi^j(\sigma)}\, d\sigma \ . \tag{I.2.10}$$

(We use the summation convention for repeated indices throughout.) However, because of the circle action

$$S^1 \ni \tau \to \varphi_\tau^j(\sigma) = \varphi^j(\sigma + \tau) \tag{I.2.11}$$

on ΩM, formula (I.2.10) does not define an operator. To get a meaningful object we consider the generator of the circle action, namely its Killing vector,

$$K^j(\sigma) = \frac{d}{d\tau}\varphi^j(\sigma+\tau)|_{\tau_0} = \frac{d}{d\sigma}\varphi^j(\sigma) \ . \tag{I.2.12}$$

The operator

$$\begin{aligned} d_K &= \int_{S^1}\left(u^j(\sigma)\frac{\delta}{\delta\varphi^j(\sigma)} + u^j(\sigma)K_j(\sigma)\right) d\sigma \\ &= \int_{S^1}\left(u^j(\sigma)\frac{\delta}{\delta\varphi^j(\sigma)} + u^j(\sigma)\frac{d}{d\sigma}\varphi_j(\sigma)\right) d\sigma \end{aligned} \tag{I.2.13}$$

is well defined (at least in the case of M flat). In finite dimensions, the operator d_K gives rise to the equivariant de Rham cohomology. Note that

$$\left[\frac{\delta}{\delta\varphi^j(\sigma)}\ ,\ K^k(\sigma')\right] = 0 \ . \tag{I.2.14}$$

However,

$$d_K^2 \neq 0 \ . \tag{I.2.15}$$

In other words, d_K is not a coboundary operator. In fact, d_K^2 is the Lie derivative in direction K, as was pointed out in [W2]. Let

$$Q_1 = \frac{1}{\sqrt{2}}(d_K + d_K^*) \ , \tag{I.2.16a}$$

$$Q_2 = \frac{i}{\sqrt{2}}(d_K - d_K^*) \ . \tag{I.2.16b}$$

The self-adjoint operators Q_μ are called *supersymmetry generators*. We define the self-adjoint operator $Q = \frac{1}{\sqrt{2}}(Q_1 + Q_2)$. This is a Dirac-like operator in infinitely many dimensions. Its square $H = Q^2$ is the Laplace operator which is also called the Hamiltonian of the supersymmetric σ-model. In the case that $M = \mathbb{R}^n$, the operator H is the Hamiltonian of a supersymmetric free field.

We can also consider a modified version of (I.2.13–16). For V a smooth function, define

$$\mathcal{V} = \int_{S^1} V(\varphi(\sigma)) \, d\sigma \ , \tag{I.2.17}$$

and

$$d_{K,V} = e^{-\mathcal{V}} d_K e^{\mathcal{V}} \ . \tag{I.2.18}$$

As above, we construct

$$Q_{V,1} = \frac{1}{\sqrt{2}}\left(d_{K,V} + d_{K,V}^*\right) \ , \tag{I.2.19a}$$

$$Q_{V,2} = \frac{i}{\sqrt{2}}\left(d_{K,V} - d_{K,V}^*\right) \ , \tag{I.2.19b}$$

and $Q_V = \frac{1}{\sqrt{2}}(Q_{V,1} + Q_{V,2})$. The corresponding Laplace operator is $H_V = Q_V^2$. We have studied in detail the example $M = \mathbb{R}^n$ and V polynomial in φ, which is also called the *chiral Wess-Zumino model*. We will come back to this model in the next section presenting it from a different point of view.

Studying various physical properties of supersymmetric field theory models boils down to studying topological properties of ΩM. An example is the question of supersymmetry breaking. We say that supersymmetry is unbroken, if there is a vector $\omega_0 \in \mathcal{H}$ such that

$$Q\omega_0 = 0 \ , \tag{I.2.20}$$

or equivalently

$$H\omega_0 = 0 \ . \tag{I.2.21}$$

Similarly, in the Wess-Zumino models, unbroken supersymmetry means $Q_V\omega_0 = 0$. A partial answer to this question is provided by studying the index of Q. We write

$$\mathcal{H} = \mathcal{H}_+ \oplus \mathcal{H}_- \ , \tag{I.2.22}$$

where $\mathcal{H}_+$ is the completion of $\Gamma\left(\bigoplus_{k\geq 0} \bigwedge^{2k}(E)\right)$ and where $\mathcal{H}_-$ is the completion of $\Gamma\left(\bigoplus_{k\geq 0} \bigwedge^{2k+1}(E)\right)$. Note that $Q : \mathcal{H}_\pm \to \mathcal{H}_\mp$ and therefore

$$Q = \begin{pmatrix} 0 & Q_- \\ Q_+ & 0 \end{pmatrix} \ , \tag{I.2.23}$$

where $Q_+ = Q_-^*$. We define the index of Q to be the difference between the number of linearly independent "bosonic" ground states and the number of linearly independent "fermionic" states,

$$i(Q_+) = \dim\ker Q_+ - \dim\ker Q_- \ . \tag{I.2.24}$$

If, by some methods, we can establish that $i(Q_+) \neq 0$, this implies that the supersymmetry is not broken. On the other hand, if $i(Q_+) = 0$, we cannot draw any conclusions about the breaking of supersymmetry. It may happen that there are as many bosonic as fermionic ground states. We come back to the index problem for the Wess-Zumino model in Chapter II.

I.3 Calculus of Superfields

The most natural way of formulating supersymmetric theories is to use the formalism of *superfields*. This leads to studying the properties of infinite dimensional supermanifolds and places supersymmetric field theory into the realm of "supergeometry" [M]. Super geometry is a new fascinating subject of great beauty and elegance and a big potential for interesting results.

We consider the case where the target manifold is flat, $M = \mathrm{R}^n$. Let γ^0 and γ^1 be the Euclidean Dirac matrices in Majorana representation

$$\gamma^0 = \begin{pmatrix} 0 & -i \\ i & 0 \end{pmatrix} \ , \quad \gamma^1 = \begin{pmatrix} 0 & 1 \\ 1 & 0 \end{pmatrix} \ . \tag{I.3.1}$$

They obey the algebra

$$\{\gamma^\mu, \gamma^\nu\} = 2\delta^{\mu\nu} \ . \tag{I.3.2}$$

We consider a $(2|2)$-dimensional torus $T^{2|2}$, see [M]. By σ_μ, where $\mu = 0, 1$, we denote even coordinates on $T^{2|2}$. Similarly, by θ_α, where $\alpha = 1, 2$, we denote odd coordinates. A (scalar) superfield on $T^{2|2}$ has the form

$$\Phi^j(\sigma, \theta) = \phi^j(\sigma) + \bar{\theta}\psi^j(\sigma) + \frac{1}{2}\bar{\theta}\theta F^j(\sigma) \ . \tag{I.3.3}$$

Here ϕ^j is a real Euclidean Bose field, ψ^j is a Euclidean Majorana field, and F^j is the auxiliary field, a scalar Bose field. Also, we use the notation

$$\bar{\chi}\eta := (\chi, \gamma^0 \eta) = i(\chi_2 \eta_1 - \chi_1 \eta_2) \ . \tag{I.3.4}$$

We wish to construct an action functional on the space of superfields. We define the following covariant derivatives

$$D_\alpha := \frac{\partial}{\partial \bar{\theta}_\alpha} \quad (\gamma^\mu \theta \partial_\mu)_\alpha \ . \tag{I.3.5}$$

They obey the following algebra

$$\{D_1, D_2\} = 0 \ , \tag{I.3.6}$$

$$D_1^2 = i\bar{\partial} \ , \tag{I.3.7}$$

$$D_2^2 = i\partial \ , \tag{I.3.8}$$

where

$$\partial := \frac{\partial}{\partial \sigma_1} - i\frac{\partial}{\partial \sigma_0} \ , \tag{I.3.9}$$

$$\bar{\partial} := \frac{\partial}{\partial \sigma_1} + i\frac{\partial}{\partial \sigma_0} \ , \tag{I.3.10}$$

We now define the free action functional $A_0(\Phi)$ by

$$A_0(\Phi) := \frac{i}{4} \int_{T^{2|2}} \bar{D}\Phi^j(\sigma, \theta) D\Phi_j(\sigma, \theta) \, d^2\sigma d^2\theta \ , \tag{I.3.11}$$

where the Berezin integral is defined as usual by

$$\int d^2\theta = \int \theta_\alpha \, d^2\theta = 0 \ , \qquad \int \theta_1 \theta_2 \, d^2\theta = 1 \ . \tag{I.3.12}$$

Writing Φ^j in terms of components (I.3.3) and using (I.3.12) to integrate the θ's we can write $A_0(\Phi)$ as a function of ϕ, ψ and F on the base manifold, namely as

$$A_0(\Phi) = \frac{1}{2}\int_{T^2} \left(\partial^\mu \phi^j \partial_\mu \phi_j + i\bar{\psi}^j \gamma^\mu \partial_\mu \psi_j + F^j F_j\right) d^2\sigma \ . \tag{I.3.13}$$

We can use $A_0(\Phi)$ to define a Gaussian functional on the space of superfields. (In the physics literature this is called a Gaussian measure.) In order to identify the covariance of this functional we wish to write A_0 in (I.3.11) as a Dirichlet form on superfields. In other words, we integrate by parts to obtain an expression in terms of a Laplacian,

$$A_0(\Phi) = \frac{1}{2}\int_{T^{2|2}} \Phi^j(\sigma,\theta)(-\Delta_s)\Phi_j(\sigma,\theta)\, d^2\sigma d^2\theta \ . \tag{I.3.14}$$

Here Δ_s is the super Laplacian, which we now identify. In fact,

$$\begin{aligned}\frac{i}{4}\int \bar{D}\Phi^j D\Phi_j\, d^2\sigma d^2\theta &= \frac{1}{2}\int D_1\Phi^j D_2\Phi_j\, d^2\sigma d^2\theta \\ &= -\frac{1}{2}\int \Phi^j D_1 D_2 \Phi_j\, d^2\sigma d^2\theta \ ,\end{aligned}$$

where the second identity is a consequence of the integration by parts formula on superspace. Therefore

$$\Delta_s = D_1 D_2 \ , \tag{I.3.15}$$

or explicitly

$$\Delta_s = \frac{\partial^2}{\partial\theta_2\partial\theta_1} + i\theta_1\frac{\partial}{\partial\theta_2}\partial - i\theta_2\frac{\partial}{\partial\theta_1}\bar{\partial} + \theta_1\theta_2\Delta \ , \tag{I.3.16}$$

where Δ is the ordinary Laplace operator

$$\Delta = \partial\bar{\partial} = \frac{\partial^2}{\partial\sigma_0^2} + \frac{\partial^2}{\partial\sigma_1^2} \ . \tag{I.3.17}$$

We claim that on $\ker(\Delta)^\perp$,

$$\begin{aligned}\Delta_s^{-1} &= \Delta^{-1}\Delta_s \\ &= \frac{\partial^2}{\partial\theta_2\partial\theta_1}\Delta^{-1} + i\theta_1\frac{\partial}{\partial\theta_2}\bar{\partial}^{-1} - i\theta_2\frac{\partial}{\partial\theta_1}\partial^{-1} + \theta_1\theta_2 \ .\end{aligned} \tag{I.3.18}$$

In fact, as a consequence of (I.3.7,8)

$$\begin{aligned}D_1^{-1} &= (i\bar{\partial})^{-1}D_1 \ , \\ D_2^{-1} &= (i\partial)^{-1}D_2\end{aligned} \tag{I.3.19}$$

on $\ker(\Delta)^{\perp}$. This yields

$$\Delta_s^{-1} = D_2^{-1} D_1^{-2} = -\Delta^{-1} D_2 D_1 = \Delta^{-1} D_1 D_2 \ ,$$

which proves (I.3.18).

In physicists language, we wish to define a Gaussian functional on the space of superfields with the form

$$d\mu_{\mathcal{C}}(\Phi) \sim e^{-A_0(\Phi)} \mathcal{D}\Phi \ .$$

In order to make this mathematical, we define

$$\mathcal{C} = \begin{cases} \Delta_s^{-1} & \text{on } \ker(\Delta)^{\perp}, \\ 0 & \text{on } \ker(\Delta), \end{cases} \tag{I.3.20}$$

and use $\mathcal{C}$ to define a Gaussian functional on the space of superfields. We set

$$d\mu_{\mathcal{C}}(\Phi) := d\mu_C(\phi) \otimes d\mu_S(\psi) \otimes d\mu_0(F) \ . \tag{I.3.21}$$

This is a perfectly well-defined mathematical expression, where $d\mu_C(\phi)$ is a Gaussian measure with covariance

$$C = \begin{cases} \Delta^{-1} & \text{on } \ker(\Delta)^{\perp}, \\ 0 & \text{on } \ker(\Delta), \end{cases} \tag{I.3.22}$$

where $d\mu_S(\psi)$ is a Gaussian fermionic functional with covariance

$$S = \begin{cases} (i\gamma^0\gamma^\mu\partial_\mu)^{-1} & \text{on } \ker(\Delta)^{\perp}, \\ 0 & \text{on } \ker(\Delta), \end{cases} \tag{I.3.23}$$

and where $d\mu_0(F)$ is a white noise measure. White noise is the Gaussian measure with zero mean and with the identity covariance. It is easy to verify that the usual rules of the calculus of functional integrals [GJ] hold also for $d\mu_{\mathcal{C}}(\Phi)$. The conditional expectation with respect to F, and with ϕ, ψ fixed is the free field functional

$$d\mu_0(\phi, \psi) = d\mu_C(\phi) d\mu_S(\psi) \ .$$

A standard way to construct non-Gaussian measures is to perturb a Gaussian measure by a local potential. Similarly, in the context of analysis on superfields this procedure leads to the Wess-Zumino models which we discussed in Section I.2 from the loop-space point of view.

We now make this connection in the framework of Euclidean field theory. Let V be a polynomial (called the superpotential) in n variables. We set

$$A_V(\Phi) := \int_{T^{2|2}} :V(\Phi(\sigma,\theta)): d^2\sigma d^2\theta \ , \tag{I.3.24}$$

where the colons around V mean Wick ordering with respect to $d\mu_C$. It is easy to express $A_V(\Phi)$ in terms of the component fields. In fact,

$$\begin{aligned} V(\Phi) &= V(\phi) + \partial_j V(\phi)(\bar{\theta}\psi^j + \frac{1}{2}\bar{\theta}\theta F^j) + \frac{1}{2}\partial^2_{jk} V(\phi)(\bar{\theta}\psi^j + \frac{1}{2}\bar{\theta}\theta F^j)(\bar{\theta}\psi^k + \frac{1}{2}\bar{\theta}\theta F^k) \\ &= (-i\partial_j V(\phi) F^j + \frac{i}{2}\partial^2_{jk} V(\phi)\bar{\psi}^j\psi^k)\theta_1\theta_2 + \dots \ , \end{aligned} \tag{I.3.25}$$

where $\dots$ means terms of lower order in θ_α. Consequently, the lower order terms vanish after integration over $T^{2|2}$ and we are reduced to

$$A_V(\Phi) = \frac{i}{2}\int_{T^2} (-2{:}\partial_j V(\phi){:}F^j + {:}\partial^2_{jk} V(\phi){:}\ {:}\bar{\psi}^j\psi^k{:})d^2\sigma \ . \tag{I.3.26}$$

We set

$$d\mu_V(\Phi) = Z_V^{-1} \exp\{-A_V(\Phi)\}\, d\mu_C(\Phi) \ , \tag{I.3.27}$$

where

$$Z_V = \int \exp\{-A_V(\Phi)\}\ d\mu_C(\Phi) \ . \tag{I.3.28}$$

The main mathematical problem is to establish the existence of

$$\langle \mathcal{F} \rangle_V := \int \mathcal{F}(\Phi)\ d\mu_V(\Phi) \tag{I.3.29}$$

for a suitable class of functions $\mathcal{F}$, *e.g.*, polynomials. As an example,

$$\langle \Phi(x,\theta)\Phi(y,\eta)\rangle_0 = \Delta_s^{-1}(x,\theta;y,\eta) \ .$$

Another important class of expectations is of functions $\mathcal{F}(\phi,\psi)$ which are independent of the auxiliary field F. The existence of these expectations has been established in certain cases:

(α) $n = 1$, any polynomial $V(x)$ [JLW3].

(β) $n = 2$, any harmonic polynomial $V(x,y)$ [JLW1,2], [JL1] (see also [JL2] for a review).

The analysis of the general V is a subtle question because of the occurrence of "flat directions" in $\partial V(x_1,\dots,x_n)$, *i.e.*, the directions in the $(x_1,\dots,x_n)$ space along which

$$\frac{|\partial V(x)|}{|x|^m} \longrightarrow 0, \quad \text{as} \quad |x| \to \infty \ , \tag{I.3.30}$$

where $m =$ algebraic degree of ∂V.

Let us now evaluate this conditional expectation for $\mathcal{F}$ as above. We note that the F integral in (I.3.29) is Gaussian and it can be done explicitly. This leads to studying the measure

$$Z_V^{-1} \exp\left\{-\frac{1}{2}\int\left(-i : \partial^2_{jk}V(\phi) :: \bar{\psi}^j\psi^k : + | : \partial V(\phi) : |^2\right) d^2\sigma\right\} d\mu_C(\phi) d\mu_S(\psi) \ , \tag{I.3.31}$$

for integrals of functions of ϕ and ψ. The bosonic moments have the form

$$\langle\phi(\sigma_1)\ldots\phi(\sigma_k)\rangle_V = \int \phi(\sigma_1)\ldots\phi(\sigma_k) d\mu_{WZ}^{Bose}(\phi) \ ,$$

where

$$d\mu_{WZ}^{Bose}(\phi) = Z_V^{-1} \mathrm{Pf}(i\gamma^0 : \partial^2 V(\phi) :, \ i\gamma^0\gamma^\mu\partial_\mu) \exp\left\{-\int |\partial V(\phi)|^2 d^2\sigma\right\} \ d\mu_C(\phi) \ . \tag{I.3.32}$$

Here $\mathrm{Pf}(A,B)$ is the relative Pfaffian of two skew symmetric operator A and B [JLW4]. The main property of $\mathrm{Pf}(A,B)$ is that

$$\mathrm{Pf}(A,B)^2 = \det(I - AB) \ , \tag{I.3.33}$$

where $\det$ means the Fredholm determinant. As it stands, formula (I.3.32) is merely a formal expression, as it is plagued by non-integrable local singularities. These singularities, however, cancel each other and the right hand side of (I.3.32) can be given a meaning as

$$d\mu_{WZ}^{Bose}(\phi) = Z_V^{-1} \mathrm{Pf}_3(i\gamma^0 : \partial^2 V(\phi) :, \ i\gamma^0\gamma^\mu\partial_\mu) \exp\left\{-\mathfrak{A}_V(\phi)\right\} \ d\mu_C(\phi) \ , \tag{I.3.34}$$

where $\mathrm{Pf}_3(A,B)$ is the regularized relative Pfaffian [JLW4], and where $\mathfrak{A}_V(\phi)$ is a well-defined random variable. For precise formulations and proofs of these facts see [JLW1-4], [JL1]. The unique feature of the case (β) listed above is its finiteness: the Wick ordering dots are superfluous! This leads to classical field equations [JLW5].

A far richer structure is provided by supersymmetric σ-models. Let M be a Riemannian manifold with metric tensor g_{jk}. The action of the $N = 1$ supersymmetric σ-model with target manifold M is defined by

$$A_\sigma(\Phi) = \frac{i}{4}\int_{T^{2|2}} g_{jk}(\Phi)\bar{D}\Phi^j D\Phi^k \ d^2\sigma d^2\theta \ . \tag{I.3.35}$$

A tedious computation shows that in terms of the component fields

$$A_\sigma(\Phi) = \frac{1}{2}\int \Big\{ g_{jk}(\phi)\left(\partial^\mu \phi^j \partial_\mu \phi^k + i\bar{\psi}^j \gamma^\mu D_\mu \psi^k\right) + \frac{1}{6} R_{jklm}(\phi)\bar{\psi}^j \psi^l \bar{\psi}^k \psi^m \Big\}\, d^2\sigma \;, \tag{I.3.36}$$

where

$$D_\mu \psi^a := \partial_\mu \psi^a + \Gamma^a_{bc}(\phi)\partial_\mu \phi^b \psi^c \;. \tag{I.3.37}$$

Here Γ is the connection on M and R_{jklm} is the Riemann curvature tensor. The corresponding measure on the space of superfields is formally given by

$$d\mu_\sigma(\Phi) = Z^{-1} \exp\{-A_\sigma(\Phi)\}\, \mathcal{D}\Phi \;. \tag{I.3.38}$$

No rigorous results about the existence of (I.3.38) have so far been proved. Defining the measure (I.3.38) is a very important but also difficult problem.

Let us make a proposal for one possible way to approach this problem: assume that M is a hypersurface in $\mathbb{R}^n$ given by the equation $f(x) = 0$. One could then try to obtain (I.3.38) as a suitable limit of Wess-Zumino type measures with superpotentials critical on M. To be more specific, consider $M = S^2$. The Wess-Zumino action written in curvilinear coordinates has the form

$$A_{WZ}(\Phi) = \frac{i}{4}\int g_{jk}(\Phi)\bar{D}\Phi^j D\Phi^k \, d^2\sigma d^2\theta + \int V(\Phi)\, d^2\sigma d^2\theta \;. \tag{I.3.39}$$

Choosing polar coordinates (ρ, Θ, φ) and

$$V(\Phi) = \lambda(\Phi^2 - 1)^2 \;, \tag{I.3.40}$$

we find

$$A_{WZ}(\rho,\Theta,\varphi) = \frac{i}{4}\int \bar{D}\rho D\rho\, d^2\sigma d^2\theta + \lambda\int (\rho^2-1)^2\, d^2\sigma d^2\theta + \frac{i}{4}\int \rho^2(\bar{D}\Theta D\Theta + \sin^2\Theta \bar{D}\varphi D\varphi)\, d^2\sigma d^2\theta \;. \tag{I.3.41}$$

Set

$$Z_\lambda = \int \exp\left\{-\frac{i}{4}\int \bar{D}\rho D\rho\, d^2\sigma\, d^2\theta - \lambda\int (\rho^2-1)^2\, d^2\sigma d^2\theta\right\} \mathcal{D}\rho \;. \tag{I.3.42}$$

Formally

$$Z_\lambda^{-1} \exp\left\{-\lambda \int (\rho^2 - 1)^2 \, d^2\sigma d^2\theta\right\} \longrightarrow \delta(\rho^2 - 1) \ , \tag{I.3.43}$$

as $\lambda \longrightarrow \infty$, so we expect that

$$d\mu_{WZ} \longrightarrow d\mu_\sigma \ . \tag{I.3.44}$$

This argument is too naive to be sufficient. We hope, however, that when combined with a renormalization group type analysis, it should yield the existence of (I.3.38).

CHAPTER II. INDEX THEORY

As explained in Section I.2 studying the index of the supersymmetry generator Q is of great interest for both mathematics and physics. In this section we recall some basic facts from index theory and formulate the index theorems for the Wess-Zumino models.

Let $\mathcal{H}_1$ and $\mathcal{H}_2$ be Hilbert spaces and let $A : \mathcal{H}_1 \to \mathcal{H}_2$ be a densely defined linear operator. A is said to be Fredholm, if

(i) $\dim\ker(A) < \infty, \quad \dim\ker(A^*) < \infty,$

(ii) $\mathrm{Im}(A)$ is closed. The index of a Fredholm operator is

$$i(A) = \dim\ker(A) - \dim\ker(A^*) \ . \tag{II.1}$$

A large class of Fredholm operators arise from the infinite-dimensional examples discussed in Section I. We can formulate them abstractly as follows. Consider the quadruple $(\mathcal{H}, \Gamma, Q, H)$ where (a) $\mathcal{H}$ is a Hilbert space. (b) Γ is a $\mathbb{Z}_2$ grading of $\mathcal{H}$, *i.e.*, $\Gamma^2 = \Gamma = \Gamma^*$. We write

$$\mathcal{H} = \mathcal{H}_+ \oplus \mathcal{H}_- \ , \tag{II.2}$$

where $\Gamma = \pm I$ on $\mathcal{H}_\pm$. (c) Q is a self-adjoint operator on $\mathcal{H}$ such that

$$\{\Gamma, Q\} = 0 \ . \tag{II.3}$$

It follows that

$$Q = \begin{pmatrix} 0 & Q_- \\ Q_+ & 0 \end{pmatrix} \tag{II.4}$$

with respect to (II.2). (d) $H = Q^2$. Clearly

$$[\Gamma, H] = 0 \ . \tag{II.5}$$

We assume that

$$\mathrm{Tr}(e^{-tH}) < \infty \ , \tag{II.6}$$

for all $t > 0$.

Theorem II.1. *Under the assumptions (a)–(d):*

(i) $Q_\pm$ *are Fredholm operators.*

(ii) The following representation of the index holds,

$$i(Q_+) = \mathrm{Tr}(\Gamma e^{-tH}), \quad \textit{for all} \quad t > 0 \ . \tag{II.7}$$

(iii) If X *is a topological space and the mapping*

$$X \ni x \to Q(x) \tag{II.8}$$

is continuous in the metric of Hilbert-Schmidt heat kernels

$$d(Q, Q') := \| \, e^{-H} - e^{-H'} \, \|_{\mathrm{HS}} \ , \tag{II.9}$$

then $i(Q(x)_+)$ *is locally constant.*

For the proof of this theorem see e.g. [JL2].

Our next theorem deals with the Wess-Zumino models.

Theorem II.2. *The Wess-Zumino models* (α) *and* (β) *satisfy assumptions (a)–(b). Furthermore we can compute the index and find that*

$$i(Q_+) = \begin{cases} \pm(m-1) \bmod 2, & \textit{in case } (\alpha), \\ m-1, & \textit{in case } (\beta), \end{cases} \tag{II.10}$$

where $m =$ *algebraic degree of* V. *There is a path integral representation of* $i(Q_+)$*:*

$$\begin{aligned} i(Q_+) &= \int \mathrm{Pf}_3\left(i\gamma^0 : \partial^2 V(\phi) :, \ i\gamma^0\gamma^\mu\partial_\mu\right) \exp\left\{-\mathfrak{A}_V(\phi)\right\} \, d\mu_C(\phi) \\ &= \int d\mu_{WZ}^{Bose}(\phi) \ . \end{aligned} \tag{II.11}$$

Here we assume periodic boundary conditions for the Fermionic covariance S. For further explanation and for proofs see [JLW1-4], [JL1,2]. The most important ingredient in the computation of $i(Q_+)$ is the homotopy invariance of the index, part (iii) of Theorem II.1.

It allows us to reduce the computation to a finite dimensional problem.

The central problem of index theory on loop space is to get a similar control over various supersymmetric σ-models. Many index theorems connected with various elliptic complexes over the loop space ΩM have been conjectured by Witten [W1-4]. Their proofs require a better understanding of the ultraviolet problems of the supersymmetric σ-models.

CHAPTER III. NONCOMMUTATIVE DIFFERENTIAL GEOMETRY

III.1 Introduction

The basic source of noncommutative differential geometry is the observation that all properties of a (compact) differentiable manifold can be encoded into purely algebraic structures. The manifold itself can be identified with the spectrum of the algebra $C(M)$ of bounded, continuous functions on M. Studying vector bundles over M amounts to studying the K-theory group $K_0(C(M))$. Therefore the structure of M as a point set becomes irrelevant and questions about M can be translated into questions about the commutative C*-algebra $C(M)$.

Noncommutative differential geometry is a far reaching generalization of this observation. The commutative algebra $C(M)$ is replaced by a C*-algebra $\mathfrak{A}$. This algebra correspond to a manifold. Thus noncommutative differential geometry is a natural tool in studying the properties of geometric structures which are more general than ordinary manifolds.

Recently Connes gave a natural setting for noncommutative differential geometry in terms of cyclic cohomology. The original formulation of Connes [C1] was intrinsically finite dimensional and was based on the assumption that certain operators were trace class. Later this assumption was relaxed ([C2], [JLO1,2], [K2], [JLW6]) to deal with infinite dimensional structures. Examples of such structures arise from supersymmetric quantum field theory discussed in the previous sections. In this section we describe a general setup based on the notion of a quantum algebra and a super-KMS functional. We believe that this setup provides a natural framework for studying global geometrical and topological questions arising in quantum field theory.

III.2. Quantum Algebras

We define a *quantum algebra* as a quadruple $(\mathfrak{A}, \Gamma, \alpha_t, d)$ satisfying the following four conditions (i) – (ii). (i) $\mathfrak{A} = \mathfrak{A}_+ \oplus \mathfrak{A}_-$, is a initial, $\mathbb{Z}_2$-graded C*-algebra. Let $a = a_+ + a_-$ denote the decomposition of $a \in \mathfrak{A}$ into its homogeneous components under the grading. (ii) Γ is an involutive, $*$-automorphism of $\mathfrak{A}$ given by $a \to a^\Gamma = a_+ - a_-$. (iii) α_t is a norm-continuous, one-parameter group of $*$-automorphisms of $\mathfrak{A}$. We assume that $[\Gamma, \alpha_t] = 0$. Let D denote the generator of α_t,

$$D = -i \frac{d}{dt} \alpha_t \bigg|_{t=0} . \tag{III.2.1}$$

(iv) d is a densely defined superderivation of $\mathfrak{A}$, namely

$$(da)^\Gamma = -da^\Gamma, \qquad d(ab) = da \cdot b + a^\Gamma db, \qquad a, b \in \text{Domain}(d) \ . \tag{III.2.2}$$

We assume that d is a square root of D, *i.e.*,

$$d^2 = D \ . \tag{III.2.3}$$

An example of a quantum algebra arises in the following way. We assume that $\mathfrak{A}$ is a $\mathbb{Z}_2$-graded C^*-algebra of operators on a $\mathbb{Z}_2$-graded Hilbert space $\mathcal{H} = \mathcal{H}_+ \oplus \mathcal{H}_-$, and that Γ is a unitary self-adjoint operator such that $\Gamma = \pm I$ on $\mathcal{H}_\pm$. Then $a^\Gamma = \Gamma a \Gamma$. Let Q be a self-adjoint operator on $\mathcal{H}$ such that $\{Q, \Gamma\} = 0$. We assume that the graded commutator $da := Qa - a^\Gamma Q$ is defined on a dense /subspace of $\mathfrak{A}$. Clearly, d is a superderivation. We set $H := Q^2$ and require that

$$\text{tr}\,(e^{-tH}) < \infty, \qquad \text{for all } t > 0 \ . \tag{III.2.4}$$

The group of automorphisms α_t is then given by

$$\alpha_t = \exp\{it\text{Ad}(H)\} \ . \tag{III.2.5}$$

Such a quantum algebra is said to be Θ-*summable*.

Examples of Θ-summable quantum algebras arise from studying Dirac-type operators on finite dimensional compact manifolds. In the infinite dimensional situation, Θ-summable quantum algebras arise from supersymmetric quantum field theories on compact space. If we study Dirac operators on a noncompact manifold, or supersymmetric field theory on a noncompact space (such as the infinite volume limit), the resulting quantum algebra is not in general Θ-summable.

III.3. Super-KMS Functionals

We call an $a \in \mathfrak{A}$ entire, if $t \to \alpha_t(a)$ extends to an entire $\mathfrak{A}$-valued analytic function. Clearly, the set $\mathfrak{A}_\alpha$ of entire elements is a subalgebra of $\mathfrak{A}$. Continuity of α_t in t ensures that $\mathfrak{A}_\alpha$ is norm-dense in $\mathfrak{A}$.

A super-KMS (sKMS) functional on the quantum algebra $(\mathfrak{A}, \Gamma, \alpha_t, d)$ is a continuous linear functional on $\mathfrak{A}$ such that for all $a, b \in \mathfrak{A}_\alpha$,

$$\omega(da) = 0, \quad \text{and} \quad \omega(ab) = \omega(b^\Gamma \alpha_i(a)) \ . \tag{III.3.1}$$

It is easy to construct an sKMS state in the case of a Θ-summable quantum algebra.

We set ω equal to the supertrace functional, namely

$$\omega(a) = \mathrm{Tr}(\Gamma a e^{-H}) \ . \tag{III.3.2}$$

It is straightforward to verify that conditions (III.3.1) are satisfied.

In general, if the quantum algebra is not Θ-summable, there is not canonical way of defining an sKMS functional. In certain situations, such as for analysis on non-compact manifold, one can construct sKMS functionals on such a quantum algebra by means of a limiting procedure, as limits of functionals of the form (III.3.2).

III.4 Entire Cyclic Cohomology of a Quantum Algebra

Let $C^n(\mathfrak{A})$ denote the space of densely defined, $(n+1)$-linear functional on $\mathfrak{A}$ which are continuous with respect to the norm $\| \cdot \|_*$ given by

$$\| a \|_* := \| a \| + \| da \| \ . \tag{III.4.1}$$

Let $\| f_n \|_*$ denote the norm of $f_n \in C^n(\mathfrak{A})$ with respect to $\| \cdot \|_*$. Define $C(\mathfrak{A})$ as the space of sequences $f = (f_0, f_1, f_2, \ldots)$, where $f_n \in C^n(\mathfrak{A})$, which satisfy the following Connes' growth condition:

$$n^{1/2} \| f_n \|_*^{1/n} \to 0, \quad \text{as} \quad n \to \infty \ . \tag{III.4.2}$$

We call the elements of $C(\mathfrak{A})$ *entire cochains*. The space of entire cochains can be decomposed into components which are even or odd under the action of Γ,

$$C(\mathfrak{A}) = C_+(\mathfrak{A}) \oplus C_-(\mathfrak{A}) \ . \tag{III.4.3}$$

We introduce two coboundary operators b and B [C2] such that

$$b : C^n(\mathfrak{A}) \to C^{n+1}(\mathfrak{A}), \quad B : C^n(\mathfrak{A}) \to C^{n-1}(\mathfrak{A}) \ . \tag{III.4.4}$$

We set

$$\begin{aligned}(bf)_{n+1}(a_0, \ldots, a_{n+1}) = \sum_{j=0}^{n} (-1)^j f_n(a_0, \ldots, a_j a_{j+1}, \ldots, a_{n+1}) \\ + (-1)^{n+1} f_n(a_{n+1}^{\gamma} a_0, a_1, \ldots, a_n) \ ,\end{aligned} \tag{III.4.5}$$

and

$$(Bf)_{n-1}(a_0,\dots,a_{n-1}) = \sum_{j=0}^{n-1}(-1)^{(n-1)j}\Big(f_n(\mathbf{1},a^\gamma_{n-j},\dots,a^\gamma_{n-1},a_0,\dots,a_{n-j-1})$$
$$+(-1)^{n-1}f_n(a^\gamma_{n-j},\dots a^\gamma_{n-1},a_0,\dots,a_{n-j-1},\mathbf{1})\Big) \quad \text{(III.4.6)}$$

Here

$$a^\gamma = \begin{cases} a^\Gamma, & \text{if } \ f_n \in \mathcal{C}^n_+(\mathfrak{A}) \ , \\ a, & \text{if } \ f_n \in \mathcal{C}^n_-(\mathfrak{A}) \ . \end{cases} \quad \text{(III.4.7)}$$

It can be then verified [Cl, EFJL] that b and B are in fact coboundary operators, *i.e.*,

$$b^2 = B^2 = 0 \ , \quad \text{(III.4.8)}$$

and that

$$bB + Bb = 0 \ . \quad \text{(III.4.9)}$$

As a consequence, the operator

$$\partial := b + B \quad \text{(III.4.10)}$$

is a coboundary operator. We write now

$$\mathcal{C}(\mathfrak{A}) = \mathcal{C}^e(\mathfrak{A}) \oplus \mathcal{C}^o(\mathfrak{A}) \ , \quad \text{(III.4.11)}$$

where $\mathcal{C}^e(\mathfrak{A})$ is the space of sequences $f = (f_0, f_2, f_4, \dots)$, and where $\mathcal{C}^o(\mathfrak{A})$ is the space of sequences $f = (f_1, f_3, f_5, \dots)$.Clearly

$$\partial : \mathcal{C}^e(\mathfrak{A}) \to \mathcal{C}^o(\mathfrak{A}), \quad \partial : \mathcal{C}^o(\mathfrak{A}) \to \mathcal{C}^e(\mathfrak{A}) \ . \quad \text{(III.4.12)}$$

We call the cohomology of the complex

$$\dots \to \mathcal{C}^e(\mathfrak{A}) \xrightarrow{\partial} \mathcal{C}^o(\mathfrak{A}) \xrightarrow{\partial} \mathcal{C}^e(\mathfrak{A}) \longrightarrow \dots \quad \text{(III.4.13)}$$

the *entire cyclic cohomology* and denote the corresponding cohomology group by $\mathcal{H}^e(\mathfrak{A})$ and $\mathcal{H}^o(\mathfrak{A})$.

III.5. The Chern Character

Let ω denote an sKMS functional. For $a_0, \dots, a_n \in \mathfrak{A}_\alpha$ we define

$$\tau_n(a_0, a_1, \dots, a_n) = i^{\varepsilon_n} \int_{\sigma_n} \omega\Big(a_0 \alpha_{is_1}(da_1^\Gamma)\alpha_{is_2}(da_2)\dots\alpha_{is_n}(da_n^{\Gamma^n})\Big)\, ds_1 \dots ds_n \ , \quad \text{(III.5.1)}$$

where $\varepsilon_n = n(\text{mod } 2)$ and where σ_n denotes the simplex

$$\{s \in \mathbb{R}^n : 0 \leq s_1 \leq s_2 \leq \ldots \leq s_n \leq 1\} \ . \tag{III.5.2}$$

We claim that τ is an entire cyclic cocycle.

Theorem III.1 *(i) The sequence $\tau = (\tau_0, \tau_1, \ldots)$ is an element of $C(\mathfrak{A})$. Moreover, there is $c < \infty$ such that*

$$\| \tau_n \|_* \leq \frac{1}{n!} |\omega|(1) e^{cn} \ . \tag{III.5.3}$$

(ii) The cochain τ is a cocycle,

$$\partial\tau = 0 \ . \tag{III.5.4}$$

For the (nontrivial) proof of this theorem we refer the reader to the original literature ([JLO1], [K2], [JLO2], [JLW6]).

Why is τ called a "Chern character?" In the usual, finite-dimensional differential geometry of a manifold M, the Chern character provides an isomorphism between two standard topological objects, namely $K_{\mathbb{C}}(M)$ and $H^{\text{even}}(M)$. Here $K_{\mathbb{C}}(M)$ is the Grothendieck group associated with the equivalence classes of vector bundles over M, and $H^{\text{even}}(M) = \oplus_i H^{2i}(M)$ is the even part of the cohomology ring. For details, see for example [K1].

In Connes' framework of non-commutative geometry, $K_{\mathbb{C}}(M)$ is replaced by the group $K_0(\mathfrak{A}_+)$, the Grothendieck group associated with the equivalence classes of idempotents in $\text{Mat}(\mathfrak{A}_+)$. Furthermore, $H^{\text{even}}(M)$ is replaced by the even part of the cyclic cohomology $H^{\text{even}}(\mathfrak{A}_+)$. Then the cocycle τ provides a pairing $< \cdot \ , \ \cdot >$ between $\mathcal{H}^{\text{even}}(\mathfrak{A}_+)$ and $K_0(\mathfrak{A}_+)$. In particular, for $e \in K_0(\mathfrak{A}_+)$,

$$\psi(e) = \sum_k \tau_{2k}(e, e, \ldots, e)(-1)^k \frac{(2k)!}{k!} = < \tau, e >$$

depends only on the cohomology class of τ. This pairing extends to the Θ-summable case [C2], as well as to the case of super-KMS functionals. In general, $\psi(e)$ can be interpreted as the index of a superderivation d^e, given by the restriction of d to $e\mathfrak{A}e$ [JL3].

Note that in the case of a Θ-summable quantum algebra (III.5.1) takes the following form

$$\tau_n(a_0, a_1, \ldots, a_n) = i^{\varepsilon_n} \int_{\sigma_n} \text{Tr}\Big(\Gamma a_0 e^{-s_1 H} da_1^{\Gamma} e^{-(s_2 - s_1)H} \ldots e^{-(s_n - s_{n-1})H} da_n^{\Gamma^n} e^{-(1-s_n)H}\Big) ds_1 \ldots ds_n \ . \tag{III.5.5}$$

III.6. Stability of the Chern Character

In this section we address the following question. Does the cohomology class of τ change if we perturb slightly the superderivation d? We describe one class of deformations of d under which the cohomology class of τ is invariant.

For $q \in D(d) \cap \mathfrak{A}_-$ we set

$$\delta_q(a) = qa - a^\Gamma q \ , \quad a \in \mathfrak{A} \ , \tag{III.6.1}$$

and define the perturbed derivation by

$$d_q := d + \delta_q \ . \tag{III.6.2}$$

Then d_q is a bounded perturbation of d. Let

$$D_q := d_q^2 = d^2 + \mathrm{Ad}(\Omega) \ , \tag{III.6.3}$$

where

$$\Omega = dq + q^2 \ . \tag{III.6.4}$$

We can thus define a new quantum algebra $(\mathfrak{A}, \Gamma, \alpha_t^q, d_q)$, where

$$\alpha_t^q = \exp\{itD_q\} = \exp\{it(D + \mathrm{Ad}(\Omega))\} \ . \tag{III.6.5}$$

Let ω be an sKMS functional on $(\mathfrak{A}, \Gamma, \alpha_t, d)$. We define the following functional

$$\omega^q(a) = \omega(a\gamma_i^q(1)) \ , \tag{III.6.6}$$

where

$$\gamma_t^q(a) := \exp\{it(D + \Omega)\}(a) \ . \tag{III.6.7}$$

Theorem III.2. [JLW6] *(Stability of the sKMS property). The functional* ω^q *is an sKMS functional on* $(\mathfrak{A}, \Gamma, \alpha_t^q, d_q)$.

Let τ^q be the cocyle constructed of the deformed sKMS functional ω^q. Assume now that $[0,1] \ni \lambda \to q(\lambda) \in \mathfrak{A}_- \cap D(d)$ is continuously differentiable.

Theorem III.3. [JLW6] *(Homotopy invariance of the Chern character). There is a cochain* $G^\lambda \in C(\mathfrak{A})$ *such that*

$$\frac{d}{d\lambda}\tau^{q(\lambda)} = \partial G^{\lambda} \quad . \tag{III.6.8}$$

In particular, the cohomology class of τ^{λ} *is independent of* λ.

Acknowledgement

We wish to thank M. Bues, P. Feng, and S. Klimek for helpful comments on these notes.

References

[C1] Connes, A.: Noncommutative Differential Geometry, *Publ. Math. IHES* **62** (1985), 257–360.

[C2] Connes, A.: Entire Cyclic Cohomology of Banach Algebras and Characters of θ-Summable Fredholm Modules, *K-Theory*, **1** (1988), 519–548.

[EFJL] Ernst, K., Feng, P., Jaffe, A., and Lesniewski, A.: Quantum K-Theory II. Homotopy Invariance of the Chern Character, *Jour. Funct. Anal.*, to appear.

[GS] Getzler, E. and Szenes, A.: On the Chern character of theta-summable Fredholm modules, *J. Funct. Anal.*, to appear.

[GJ] Glimm, J. and Jaffe, A.: *Quantum Physics*, Second Edition, Springer: New York 1987.

[H] Hamilton, R.: The Inverse Function Theorem of Nash and Moser, *Bull. Amer. Math. Soc.* **7** (1982), 65–222.

[JL1] Jaffe, A. and Lesniewski, A.: A priori estimates for N=2 Wess-Zumino models on a cylinder, *Commun. Math. Phys.* **114** (1988), 553–575.

[JL2] Jaffe, A. and Lesniewski, A.: Supersymmetric Field Theory and Infinite Dimensional Analysis, in *Nonperturbative Quantum Field Theory*, G. 't Hooft *et al.*, Editors, Plenum, New York, 1988.

[JL3] Jaffe, A. and Lesniewski, A.: An Index Theorem for Superderivations, *Commun. Math. Phys.*, to appear.

[JLL] Jaffe, A., Lesniewski, A., and Lewenstein, M.: Ground State Structure in Supersymmetric Quantum Mechanics, *Ann. Phys.* **178** (1987), 313–329.

[JLO1] Jaffe, A., Lesniewski, A., and Osterwalder, K.: Quantum K-Theory I: The Chern Character, *Commun. Math. Phys.* **118** (1988), 1–14.

[JLO2] Jaffe, A., Lesniewski, A., and Osterwalder, K.: On Super-KMS Functionals and Entire Cyclic Cohomology, *K-Theory*, to appear.

[JLW1] Jaffe, A., Lesniewski, A., and Weitsman, J., Index of a Family of Dirac Operators on Loop Space, *Commun. Math. Phys.* **112**, 75–88 (1987).

[JLW2] Jaffe, A., Lesniewski, A., and Weitsman, J.: The Two-Dimensional, N=2 Wess-Zumino Model On a Cylinder. *Commun. Math. Phys.* **114** (1988), 147–165.

[JLW3] Jaffe, A., Lesniewski, A., and Weitsman, J.: Pfaffians on Hilbert Space, *J. Funct. Anal.*, **83** (1989), 348–363.

[JLW4] Jaffe, A., Lesniewski, A., and Weitsman, J.: The Loop Space $S^1 \to \mathbb{R}$ and Supersymmetric Quantum Fields, *Annals of Physics,* **183** (1988), 337–351.

[JLW5] Jaffe, A., Lesniewski, A., and Wieczerkowski, C.: *A Priori* Quantum Field Equations, *Ann. Phys.*, to appear.

[JLW6] Jaffe, A., Lesniewski, A., and Wisniowski, M.: Deformation of Super-KMS Functionals, *Commun. Math. Phys.*, **121** (1989), 527–540.

[K1] Karoubi, M.: *K-Theory,* Springer: New York 1978.

[K2] Kastler, D.: Cyclic Cocycles from Graded KMS Functionals, *Commun. Math. Phys.*, **121** (1989), 345–350.

[M] Manin, Yu.: Complex Geometry and Gauge Fields, Springer: 1988.

[We] Weitsman, J.: A Supersymmetric Field Theory in Infinite Volume, PhD Thesis, Harvard University 1988.

[W1] Witten, E.: Constraints on Supersymmetry Breaking, *Nucl. Phys.* **B202** (1982), 253–316.

[W2] Witten, E.: Supersymmetry and Morse Theory, *J. Diff. Geom.* **17** (1982), 661–692.

[W3] Witten. E.: Global Anomalies in String Theory, in *Anomalies, Geometry and Topology*, 61–99, W. Bardeen and A. White, Eds., World Scientific, Singapore (1985).

[W4] Witten, E.: Elliptic Genera and Quantum Field Theory, *Commun. Math. Phys.* **109** (1987), 525–536.

CONFORMAL FIELD THEORY IN STRING THEORY

Jeffrey A. Harvey

Physics Department
Princeton University
Princeton, NJ 08544

The role of conformal field theory in string theory is briefly reviewed. Topics discussed include the formulation of first quantized string theory, simple string compactifications, and equivalences between conformal field theories.

INTRODUCTION

It is now four years since the rebirth of string theory. I have been asked in these lectures to review the current status of string theory. To do that in full would be impossible due to limitations in both time and competence. What I would like to do instead is to review some of those aspects of string theory which are related to conformal field theory and the search for classical solutions to string theory which correspond to compactification of the extra spatial dimensions. There are several reasons for this choice. First, conformal field theory is a rich and beautiful subject in its own right. The lectures here by Gawedski and Fröhlich cover different aspects of conformal field theory. The intense interest in conformal field theory is due in part to its connection with string theory, hence some explanation of this connection seems appropriate. Second, those aspects of string theory related to the compactification of internal dimensions have undergone intense development and there is now a large body of material to choose from.

The material that I will cover in these lectures is as follows. I will first describe how two-dimensional conformal transformations arise in the classical theory of string propogation in Minkowski space. I will then discuss first quantization of bosonic string theory and the role that conformal invariance plays in identifying physical states. Interactions and the correspondence between states and operators will then be introduced. As an example of the formalism I will treat in detail the case of string compactification on a circle. Orbifolds and fermionic formulations of string compactification will then be discussed with emphasis on the equivalence between different formulations. A few topics related to the classification of conformal field theories will then be presented. I will end with a brief discussion of the extension of some of these ideas to superstring compactifications.

Constructive Quantum Field Theory II, Edited by G. Velo and A. S. Wightman, Plenum Press, New York, 1990

CLASSICAL STRING THEORY

We will begin by discussing the classical propagation of a string in Minkowski space with $D-1$ space dimensions. The simplest choice for the action describing this propagation was discussed by Nambu and Goto. In analogy to the propagation of a point particle, they argued that the action should be given by the invariant area of the world sheet that is swept out by the propagating string. Denote the world sheet coordinates by (τ, σ) where τ is time-like and runs from $-\infty$ to $+\infty$ and σ is space-like and by convention runs from 0 to π. We will be considering only closed strings so that the string coordinates $X^\mu(\tau,\sigma)$ obey the boundary condition $X^\mu(\tau,\sigma+\pi) = X^\mu(\tau,\sigma)$. We will also use the notation $\dot{X} = \partial X/\partial\tau$ and $X' = \partial X/\partial\sigma$. We can now write down the Nambu-Goto action for string propagation. The Minkowski metric of spacetime induces a metric on the world sheet such that

$$ds^2 = (\dot{X}^2 d\tau^2 + 2\dot{X}X' d\tau d\sigma + X'^2 d\sigma^2). \tag{1}$$

The action is then given by the integral of the square root of the determinant of the induced metric

$$S[X] = -\frac{T}{2}\int d\tau d\sigma \sqrt{\dot{X}^2 X'^2 - (\dot{X}X')^2}. \tag{2}$$

Here T is a parameter with units of energy per length which is introduced to give the action the correct dimensions. It will be related to the energy per unit length of a macroscopic string.

It is clear from the form of (2) that the physics is invariant under reparametrization of the world sheet coordinates. (2) is a perfectly fine starting point for analyzing classical solutions, but the square roots make the quantum treatment unpleasant. The standard procedure is to choose an action which is classically equivalent to (2) but which removes the square roots by introducing new gauge degrees of freedom. This procedure should be familiar from the problem of writing down the action for a massless point particle. In that case one introduces an affine parameter which can be thought of as a sort of "einbein" on the world line of the particle and the resulting action is invariant under reparametrizations of the affine parameter. We thus introduce an independent metric $h_{\alpha\beta}$ on the world sheet and write the action as

$$S[h, X] = -\frac{T}{2}\int d^2\sigma \sqrt{h} h^{\alpha\beta}\partial_\alpha X^\mu \partial_\beta X_\mu, \tag{3}$$

with $(\sigma^0, \sigma^1) = (\tau, \sigma)$. If we eliminate $h^{\alpha\beta}$ by solving its classical equation of motion we get back the action (2) so that (3) and (2) are classically equivalent. We don't really want the physics to depend on a choice of h so we should check that the components of h are really gauge degrees of freedom. The local symmetries of (3) consist of reparametrization of the world sheet coordinates, $\delta\sigma^\alpha = \xi^\alpha(\sigma^\beta)$, and Weyl rescalings of the metric, $\delta h^{\alpha\beta} = \Lambda(\sigma^\alpha)h^{\alpha\beta}$. Since there are three independent gauge functions, and h has three independent components, it is clear that at least locally we can choose h as we please by a suitable choice of gauge. This counting shows that introducing the metric h does not introduce any new local degrees of freedom, at least classically.

In first-quantized string theory, interactions are introduced by summing over world-sheets of different topology, corresponding to the breaking and rejoining of strings. When the world-sheet has genus $g > 0$ there exist global diffeomorphisms or gauge transformations which are not connected to the identity. Invariance of the theory under these global diffeomorphisms imposes important constraints on the theory, particularly for chiral theories such as the heterotic string. At genus $g = 1$ these transformations take the form of modular transformations:

$$\tau \longrightarrow \frac{a\tau + b}{c\tau + d} \qquad a, b, c, d \quad \text{integers}, \qquad ad - bc = 1, \tag{4}$$

where τ is a complex parameter which labels conformal equivalence classes of the torus. Modular invariance plays a crucial role in many deeper aspects of conformal field theory, which unfortunately we will not have time to pursue.

The action (3) also possesses global symmetries corresponding to transformations of X^μ under the Poincaré group, $\delta X^\mu = a^\mu_\nu X^\nu + b^\mu$. This global symmetry will be used to identify the spacetime quantum numbers of various states of the string.

Varying (3) with respect to $h^{\alpha\beta}$ gives an equation requiring that the world-sheet stress tensor vanish:

$$T_{\alpha\beta} = \partial_\alpha X^\mu \partial_\beta X_\mu - \frac{1}{2} h_{\alpha\beta} h^{\gamma\delta} \partial_\gamma X^\mu \partial_\delta X_\mu = 0. \tag{5}$$

Note that $T_{\alpha\beta}$ is automatically traceless as a consequence of Weyl symmetry. We can use the gauge symmetries of reparametrization and Weyl invariance to choose a simple form for $h^{\alpha\beta}$ in which case we must impose (5) as a constraint on the possible classical solutions. Since we are considering the time evolution of a single closed string, and for now neglecting interactions represented by splitting and rejoining of the string, the string world-sheet has the topology of a cylinder. The simplest possiblility is then to choose $h_{\alpha\beta} = \eta_{\alpha\beta}$ where $\eta_{\alpha\beta}$ is the flat two-dimensional Minkowski metric. The action (3) is then just

$$S = -\frac{T}{2} \int d^2\sigma \eta^{\alpha\beta} \partial_\alpha X^\mu \partial_\beta X_\mu, \tag{6}$$

and the X equation of motion becomes simply the two-dimensional free wave equation

$$\left(\partial^2/\partial\sigma^2 - \partial^2/\partial\tau^2\right) X^\mu = 0. \tag{7}$$

What we have accomplished by introducing h and then gauge fixing is to put the non-linearities of the problem in the constraints rather than in the equations of motion. Of course this is not much progress unless it is possible to solve the constraints. This turns out to be possible in both the classical and quantum theories. To understand what is going on it is useful to introduce two-dimensional light-cone coordinates $\sigma^\pm = \tau \pm \sigma$. In these coordinates the tracelessness of T is the statement $T_{+-} = T_{-+} = 0$ and the constraint equations are $T_{++} = T_{--} = 0$.

In this language the equation of energy-momentum conservation is $\partial_- T_{++} = \partial_+ T_{--} = 0$. As a result we can construct an infinite number of conserved currents by choosing arbitrary functions of σ^+ and σ^- and using

$$\begin{aligned} \partial_-(f(\sigma^+) T_{++}) &= 0, \\ \partial_+(g(\sigma^-) T_{--}) &= 0. \end{aligned} \tag{8}$$

If we imagine expanding $f(\sigma^+)$ and $g(\sigma^-)$ in power series then we see that the conserved charges are given by moments of T_{++} and T_{--}. Since we have an infinite number of conserved charges we would expect to find an infinite symmetry group with the symmetry transformations being generated by the Poisson brackets of the charges. It is not hard to see what this symmetry group is. Having fixed the gauge in which $h_{\alpha\beta} = \eta_{\alpha\beta}$ there is a residual symmetry consisting of those reparametrizations whose effect on the metric can be undone by a Weyl rescaling, namely reparametrizations $\xi^\alpha(\sigma)$ satisfying

$$\partial^\alpha \xi^\beta + \partial^\beta \xi^\alpha = \Lambda(\sigma)\eta^{\alpha\beta} \tag{9}$$

for some function $\Lambda(\sigma)$. Equation (9) implies that ξ^+ is a function only of σ^+ and ξ^- is a function only of σ^-.

The generators of conformal transformations can be constructed in terms of the moments of the stress tensor as

$$\begin{aligned} L_n &= \frac{T}{2}\int_0^\pi e^{-2in\sigma} T_{--} d\sigma, \\ \bar{L}_n &= \frac{T}{2}\int_0^\pi e^{-2in\sigma} T_{++} d\sigma. \end{aligned} \tag{10}$$

The Poisson brackets of these generators satisfy the classical (i.e. without central extension) Virasoro algebra

$$\begin{aligned} [L_n, L_m]_{P.B.} &= i(n-m)L_{n+m}, \\ [\bar{L}_n, \bar{L}_m]_{P.B.} &= i(n-m)\bar{L}_{n+m}. \end{aligned} \tag{11}$$

So far we have been using two-dimensional coordinates for a cylindrical string world sheet with signature $(-,+)$. These are natural coordinates for describing the propagation of a single free closed string. In many aspects of string theory however the formalism is simplified by first continuing to a Euclidean space with coordinates $(t,\sigma) = (-i\tau, \sigma)$ and then by mapping the resulting Euclidean cylinder onto the complex plane via

$$z = e^{2(t+i\sigma)}. \tag{12}$$

Under this mapping the origin $z = 0$ corresponds to the infinite past while $|z| = \infty$ corresponds to the infinite future. Time translation on the cylinder is thus replaced by radial dilations on the complex z plane. The non-vanishing components of the stress tensor become $T(z) \equiv T_{++}(z)$ and $\bar{T}(\bar{z}) \equiv T_{--}(\bar{z})$. The generators L_n, $\bar{L}_n$ of conformal transformations $z \to z + \epsilon z^{n+1}$, $\bar{z} \to \bar{z} + \bar{\epsilon}\bar{z}^{n+1}$ can then be expressed as contour integrals

$$\begin{aligned} L_n &= \oint \frac{dz}{2\pi i} z^{n+1} T(z), \\ \bar{L}_n &= \oint \frac{d\bar{z}}{2\pi i} \bar{z}^{n+1} \bar{T}(\bar{z}). \end{aligned} \tag{13}$$

Note in particular that $L_0 + \bar{L}_0$ generates scale transformations and thus plays the role of the Hamiltonian.

In the gauge $h_{\alpha\beta} = \eta_{\alpha\beta}$ the equation of motion for X^μ,

$$\partial_z \partial_{\bar{z}} X^\mu = 0, \tag{14}$$

has the solution

$$X^\mu(z,\bar{z}) = \frac{1}{2}(X_L^\mu(z) + X_R^\mu(\bar{z})) \tag{15}$$

where X_L and X_R can be expanded in normal modes as

$$\begin{aligned} X_L^\mu =& q^\mu + p^\mu \ln z + i \sum_{n\neq 0} \frac{\tilde{\alpha}_n^\mu}{n} z^{-n}, \\ X_R^\mu =& q^\mu + p^\mu \ln \bar{z} + i \sum_{n\neq 0} \frac{\alpha_n^\mu}{n} \bar{z}^{-n}. \end{aligned} \tag{16}$$

The Virasoro generators can then be constructed as

$$\begin{aligned} L_n =& \frac{1}{2} \sum_{-\infty}^{+\infty} \alpha_{n-m} \cdot \alpha_m, \\ L_0 =& \frac{1}{2}\alpha_0^2 + \sum_1^\infty \alpha_{-n} \cdot \alpha_n, \end{aligned} \tag{17}$$

where we have written $\alpha_0^\mu \equiv \frac{1}{2}p^\mu$. The $\bar{L}_n$ are constructed in the same way with $\alpha \to \tilde{\alpha}$. The Virasoro generators (17) can be seen to satisfy the Poisson brackets (11). As we will see in the next section, the quantum mechanical version of the Virasoro algebra plays a central role in the structure of first-quantized string theory. Finally, note that the constraint (5) requires that $L_n = \bar{L}_n = 0$. For $n = 0$, remembering that $\alpha_0^\mu \alpha_{0\mu} = p^\mu p_\mu / 4 = -M^2/4$ with M the mass, this gives the classical mass formula

$$M^2 = 8N = 8\bar{N}, \tag{18}$$

with

$$N = \sum_{n=1}^\infty \alpha_{-n} \cdot \alpha_n, \qquad \bar{N} = \sum_{n=1}^\infty \tilde{\alpha}_{-n} \cdot \tilde{\alpha}_n, \tag{19}$$

the number operators for right-moving and left-moving excitations respectively.

FIRST QUANTIZATION

Having discussed the classical theory of strings we are in a position to discuss first quantized string theory. We will first discuss the quantum mechanics of a single free bosonic string. In the next section we will discuss the correspondence between states and operators in the first quantized theory and the role that this plays in string compactification. There are a variety of approaches to quantizing string theory. Each has its advantages and drawbacks. Since we will not have time to delve into this subject very deeply, let me just summarize one of the possible approaches. Since we want to quantize the theory the first thing we do is to replace Poisson Brackets with commutators. In particular this implies that the oscillators appearing in the string normal mode expansion satisfy the commutation relations

$$[\alpha_n^\mu, \alpha_m^\mu] = n\eta^{\mu\nu}\delta_{n+m,0} \tag{20}$$

and similarly for the $\tilde{\alpha}_n$. We can thus identify the α_n and $\tilde{\alpha}_n$ with negative n as creation operators and those with positive n as annihilation operators. The vacuum is simply the ground state for all the oscillators, i.e. it is annihilated by all the annihilation operators.

There are two important issues of consistency that we must then face at the quantum level. The first has to do with implementing the classical constraints $L_n = \bar{L}_n = 0$ in the quantum theory. The other problem we must deal with is the apparent existence of states with negative norm such as $\alpha^0_{-1}|0\rangle$ which involve the timelike components of the oscillators.

The quantum mechanical analog of the classical constraint of the vanishing of the moments of the stress-energy tensor are that the positive frequency parts of these moments, regarded as operators in the first quantized theory, should annihilate physical states. That is

$$L_n|\phi\rangle = \bar{L}_n|\phi\rangle = 0 \qquad n > 0, \tag{21}$$

for $|\phi\rangle$ a physical state. Because of normal ordering ambiguities the conditions on L_0 and $\bar{L}_0$ become $(L_0 - a)|\phi\rangle = (\bar{L}_0 - a)|\phi\rangle = 0$ with a an undetermined constant, and the classical mass formula (18) is modified to read

$$M^2 = 8(N - a) = 8(\bar{N} - a). \tag{22}$$

We might hope that these conditions are not satisfied by the negative norm states which would thus be removed from the spectrum. This turns out to be true, but the proof is rather involved, and works only if certain conditions on a and the number of spacetime dimensions are satisfied.

To see how the dimension of spacetime enters first note that we must normal order the Virasoro generators L_n and $\bar{L}_n$ in order to obtain well defined operators. This normal ordering introduces an anomalous contribution in the commutation relations of the L_n which modifies their commutation relations to

$$[L_n, L_m] = (n - m)L_{n+m} + \frac{c(m)}{12}\delta_{n+m,0}. \tag{23}$$

The central term $c(m)$ must have the from $c(m) = cm^3 + bm$ in order to be consistent with the Jacobi identity [1]. The linear term b can be changed by constant shifts in L_0. The constant c is the central charge of the Virasoro algebra. For the case considered here, when the world-sheet fields are D free bosons, one has $c = D$ where D is the number of spacetime dimensions and it is conventional to take $b = -D$. With these commutation relations the famous no ghost theorem then says that the constraints (21) are sufficient to remove all negative norm states if $D = 26$ and $a = 1$ or if $D < 25$ and $a < 1$. Only the case of $D = 26$ and $a = 1$ leads to a consistent theory when interactions are introduced however.

With these facts in hand it is straightforward to work out the spectrum. States will be classified according to their mass and "spin", that is their transformation properties under $SO(D-2)$ for massless states and under $SO(D-1)$ for massive states, corresponding to the presence of the global Poincaré symmetry. Using (21) and (22) we see that the ground state of the bosonic string consists of a scalar tachyon with $M^2 = -8$. The first excited states with $N = \bar{N} = 1$ are more interesting. They are massless states transforming as a symmetric traceless tensor of $SO(24)$ (corresponding to the graviton in 26 spacetime dimensions), a antisymmetric tensor of $SO(24)$ (often called the Kalb-Ramond field), and a scalar of $SO(24)$ called the dilaton. The existence of these massless states is striking and is not limited to the bosonic string. All known string theories contain massless particles coupling as gravitons in the critical dimension. The presence of the antisymmetric tensor and dilaton states is also generic in string theory. Upon compactification to four dimensions, the antisymmetric tensor field is dual to a pseudo-scalar field, and in superstring theory this pseudo-scalar field has the same couplings as the axion field, first introduced to solve the strong CP problem. The dilaton field plays an even more special role, and is the source of much mystery and difficulty in trying to match string theory onto low-energy physics. It can be shown that in string theory there is really not a free coupling constant, but rather the value of the coupling constant can be shifted by shifting the vacuum expectation value of the dilaton field. Thus the question of whether the theory is strongly or weakly coupled is itself a dynamical question. This situation is peculiar to string theory and will probably have to be understood more deeply if string theory is to be a useful model of the real world.

The crucial ingredients in the approach outlined here are the constraints (21) and the commutation relations (23)with $c = D = 26$. In particular, it is not necessary that the Virasoro generators be constructed out of the free fields X^μ. If we can construct a conformal field theory with conformal anomaly $c = d$ with d an integer less than 26, then we can construct a string theory which is "$26 - d$-dimensional" in the sense that it only has Poincaré symmetry in $26 - d$ dimensions and is constructed by tensoring this theory with $26 - d$ free bosons X^μ which represent the string coordinates. This then leads us to the subject of string compactification.

STRING COMPACTIFICATION

There are various ways that one can go about trying to construct a conformal field theory to describe string compactification. In the σ-model approach one considers the propagation of string in the background geometry appropriate to the compactified space. If the space has a metric $G^{ab}(Y)$ with Y the string coordinates on the d compactified dimensions, then the action for string propagation on this space would be

$$S = \int d^2\sigma G^{ab}(Y)\partial_\alpha Y_a \partial^\alpha Y_b \qquad (24)$$

Consistency of string propagation requires conformal invariance of the σ-model. This requires that the β functional for the metric G^{ab} vanish, which in turn can be seen to give equations of motion which must be satisfied by the background fields. To lowest order in perturbation theory this just yields Einstein's equation for the metric, but there are string corrections to this which appear at higher orders in σ model perturbation theory.

In these lectures I will pursue a perhaps less general but more explicit approach.

The idea is to construct conformal field theories explicitly which have enough structure to be interesting. Most of these constructions are based on modifying actions for free fields either by changing their boundary conditions or by inserting background charges.

Before pursuing these explicit constructions we can analyze in some generality the correspondence between various operators in the CFT and the states of the resulting string theory.

Let me first remind you of how this works for the closed bosonic string in 26-dimensional Minkowski space. From the previous section the mass squared of a string state is given by

$$\frac{(\mathrm{Mass})^2}{4} = N + \tilde{N} - 2 \tag{25}$$

where N and $\tilde{N}$ are number operators for right and left-moving oscillators. Thus the ground state is a tachyon with squared mass equal to -8 (We are working in string units with $\alpha' = 1/2\pi T = 1/2$). The next mass level consists of a graviton, an antisymmetric tensor field, and a massless scalar called the dilaton. Explicitly these states are given by $\alpha^{\mu}_{-1}\tilde{\alpha}^{\nu}_{-1}|0\rangle$ for appropriate choices of the spacetime indices $(\mu\nu)$. Massive modes are constructed by exciting the vacuum with more creation operators.

For each state of the string there must be a corresponding local operator. In the mapping between the cylinder and the plane the initial state of the string at time $-\infty$ gets mapped to the origin of the plane. There must therefore exist a local operator $O(z,\bar{z})$ which creates this state when acting at $z = 0$. In a similar way we can imagine a string scattering process with tubes stretching off to plus or minus infinity which carry information about the initial and final states of the various strings involved in the scattering process. By an appropriate mapping we can again map the initial and final states to finite points on some Riemann surface. We can thus reduce the calculation of string scattering amplitudes at order g in perturbation theory to the calculation of correlation functions of the form

$$\langle O(z_1,\bar{z}_1)\ldots O(z_n,\bar{z}_n)\rangle \tag{26}$$

on a Riemann surface of genus g, where the $O(z,\bar{z})$ are local operators in a conformal field theory. At string tree level one is concerned with the case $g = 0$.

If an operator O is to create a state which has *spacetime* momentum k then under translation of the spatial coordinates $X^{\mu} \to X^{\mu} + a^{\mu}$ it should transform as $O \to e^{ik\cdot a}O$. Operators which create the lowest lying states of the bosonic string are $O_{tach} = e^{ik\cdot X}$ which creates the tachyon state and $O_{grav} = \xi_{\mu\nu}\partial_z X^{\mu}\partial_{\bar{z}}X^{\nu}e^{ik\cdot X}$ with $\xi_{\mu\nu} = \xi_{\nu\mu}$ and $k^{\mu}\xi_{\mu\nu} = 0$ which creates the graviton state. In string theory we construct vertex operators by integrating these local operators over the world sheet of the string

$$V = \int d^2z\, O(z,\bar{z}). \tag{27}$$

In calculations of string scattering this means that we integrate the probability of emitting the string state at a particular point over all points on the string world sheet.

The vertex operator V must respect the scale symmetry of the theory which is generated by $L_0 + \bar{L}_0$. So if we rescale $z \to \alpha z$, V will be invariant only if $O \to \alpha^{-2} O$, which says that O must be a dimension two operator. At first sight this seems impossible for the tachyon operator $e^{ik \cdot X}$ since from the two-dimensional point of view X is just a dimensionless free scalar field. We are used to the idea that operators can pick up anomalous dimensions as a result of interactions, but we don't expect operators to have departures from naive scaling in a free theory. Nonetheless, the following simple calculation shows that this happens here.

If an operator O has dimension d in a scale invariant theory, then its two point function must have the form

$$\langle O^\dagger(z,\bar{z}) O(w,\bar{w}) \rangle \propto |z-w|^{-2d}. \tag{28}$$

So we compute

$$\langle e^{ik \cdot X(z)} e^{-ik \cdot X(w)} \rangle = e^{k^2 G(z,w)/2} \tag{29}$$

where $G(z,w) = ln|z-w|$ is the two-dimensional Green function. We thus see that the tachyon operator has dimension $k^2/4$ and the corresponding vertex operator is scale invariant only if $k^2 = 8$. We thus determine the tachyon mass to be $(\text{mass})^2 = -k^2 = -8$ in string units. A similar calculation shows that the graviton is massless as it should be. The fact that exponentials of free fields can have non-trivial scaling dimension in two dimensions is due to the bad infrared behavior of the theory. In order to make the theory well defined we should really put an infrared cutoff μ into the Green function and then show that μ cancels out of physical correlation functions. See [2] for details.

When we compactify the theory we can split the operators up into a piece that depends on the remaining spacetime coordinates and a piece that refers to the internal conformal field theory:

$$O(z,\bar{z}) = (\text{Polynomial in } \partial_z X, \partial_{\bar{z}} X, \partial_z^2 X, (\partial_{\bar{z}} X)^2 \cdots) e^{ik \cdot X} O_{int}(z,\bar{z}). \tag{30}$$

We now want to explore the correspondence between the world-sheet properties of the operators O_{int} and the space-time properties of the states they create.

Consider for example the simplest operator which creates a spacetime scalar $O(z,\bar{z}) = e^{ik \cdot X} O_{int}(z,\bar{z})$. If O_{int} has dimension d, then O will have dimension $k^2/4 + d$ which should equal 2 by scale invariance. We thus see that O_{int} with dimension 2 correspond to massless scalars in the uncompactified dimensions. Gravitons in the uncompactified dimensions correspond to using the previous graviton vertex operator and taking O_{int} to be the identity operator. Before compactification the bosonic string does not contain any states that correspond to massless spacetime vector fields, i.e. gauge fields. Once we compactify the theory it is possible for such states to arise. In order to see how this happens it is useful to refine our notion of scale invariance.

Recall that $L_0 + \bar{L}_0$ generates scale transformations. Since both L_0 and $\bar{L}_0$ are good symmetry generators we can look at the behavior of operators under separate scalings of z and $\bar{z}$ generated by L_0 and $\bar{L}_0$. In general we will say that an operator has dimension $(h, \bar{h})$ and total dimension $d = h + \bar{h}$ if it transforms with a factor of

$\alpha^{-h}\bar{\alpha}^{-\bar{h}}$ under the transformation $z \to \alpha z$, $\bar{z} \to \bar{\alpha}\bar{z}$. Then massless spacetime scalars correspond to dimension $(0,1)$ operators in the internal theory. We can construct operators which create massless spacetime vectors as

$$O = \zeta_\mu \partial_{\bar{z}} X^\mu e^{ik\cdot X} O_{(1,0)}(z), \tag{31}$$

or

$$O = \zeta_\mu \partial_z X^\mu e^{ik\cdot X} O_{(0,1)}(\bar{z}). \tag{32}$$

We will see that some of these gauge bosons are similar to those that arise in Kaluza-Klein theory while others are intrinsically "stringy".

We can continue this procedure to enumerate the possible states in the theory and their spacetime properties. For the moment though let us concentrate on the massless scalar fields since these parametrize the infinitesimal perturbations of the vacuum of the theory. The connection between the spacetime properties and the world-sheet properties is as follows. First let us consider an operator O_{int} of dimension $(d/2, d/2)$ which will create a spacetime scalar with $m^2 = 4(d-2)$. If we call the corresponding spacetime field ϕ then ϕ will have a classical potential

$$V(\phi) = \frac{m^2}{2}\phi^2 + \frac{\mu}{3}\phi^3 + \frac{\lambda}{4}\phi^4 + \cdots \tag{33}$$

with the higher point couplings determined by higher point correlation functions in the internal conformal field theory.

To determine the minimum of $V(\phi)$ in the spacetime picture we shift $\phi \to \phi + \phi_c$ and demand that

$$V'(\phi_c) = 0 = m^2\phi_c + \mu\phi_c^2 + \cdots \tag{34}$$

¿From the world-sheet point of view we perturb the Lagrangian for the internal conformal field theory:

$$\mathcal{L} \to \mathcal{L} + g\int d^2z O_{int}(z,\bar{z}), \tag{35}$$

with O_{int} the operator that creates the scalar state and g corresponds to ϕ_c in the case that ϕ_c is translationally invariant. The equation of motion for ϕ_c translates into the statement that the perturbed theory should still be conformally invariant. This means that the beta function for g should vanish:

$$\beta(g) = 0 = (d-2)g + bg^2 + \cdots \tag{36}$$

where b is determined by the three point correlation function $\langle O_{int}O_{int}O_{int}\rangle$. We thus see that the the beta function vanishes to lowest order if the scalar is massless. Vanishing at higher orders requires that higher order terms in the potential also vanish. If the potential vanishes altogether then there is a flat direction and any constant value of ϕ_c is allowed. In conformal field theory an operator O for which the corresponding beta function vanishes identically is called a critical marginal operator.

$c = 1$ CONFORMAL FIELD THEORY

I would now like to illustrate this general discussion with some examples. One of the simplest examples is given by $c = 1$ conformal field theories. These are all based on the theory of a $1+1$ dimensional free boson. In spite of the simplicity of the theory we will find that the space of $c = 1$ theories contains $SU(2)$ Kac-Moody algebra, orbifolds, and free fermions.

We will first consider compactification of the bosonic string on a circle of radius R. Let X denote the string coordinate on this circle, $X \equiv X + 2\pi R$. The mode expansion of X is then given by

$$X(\sigma,\tau) = x + 2LR\sigma + \frac{M}{R}\tau + \frac{i}{2}\sum_{n\neq 0}\frac{\alpha_n}{n}e^{-2in(\tau-\sigma)} + \frac{i}{2}\sum_{n\neq 0}\frac{\tilde{\alpha}_n}{n}e^{-2in(\tau+\sigma)}, \tag{37}$$

where L counts the number of times the string wraps around the circle, $X(\sigma+\pi) = X(\sigma) + 2\pi RL$, and $p = M/R$, M integer, is the momenta, quantized so that e^{ipX} is invariant under $X \to X + 2\pi R$.

It is convenient to split X into right-movers and left-movers as

$$X(\sigma,\tau) = \frac{1}{2}(X_L(\tau+\sigma) + X_R(\tau-\sigma)) \tag{38}$$

with

$$X_L = x + p_L(\tau+\sigma) + \text{oscillators} \tag{39}$$

$$X_R = x + P_R(\tau-\sigma) + \text{oscillators} \tag{40}$$

where $(p_L, p_R) = (M/2R + LR, M/2R - LR)$.

The Virasoro generators are constructed as moments of the left- and right-moving stress tensors

$$T(z) = -\frac{1}{2}\partial_z X\partial_z X \tag{41}$$

$$\bar{T}(\bar{z}) = -\frac{1}{2}\partial_{\bar{z}} X\partial_{\bar{z}} X. \tag{42}$$

This gives a conformal field theory with $c = 1$. L_0 and $\bar{L}_0$ are given explicitly as

$$\bar{L}_0 = \frac{1}{2}p_L^2 + \sum_{n>o}\tilde{\alpha}_{-n}\tilde{\alpha}_n \tag{43}$$

$$L_0 = \frac{1}{2}p_R^2 + \sum_{n>0}\alpha_{-n}\alpha_n. \tag{44}$$

We want to focus on the $(1,1)$ and $(1,0)$ operators in this theory. For any value of R there is one $(1,1)$ operator, one $(1,0)$ operator and one $(0,1)$ operator. These are $\partial_z X\partial_{\bar{z}} X$, $\partial_z X$, and $\partial_{\bar{z}} X$ respectively. In Kaluza-Klein language these correspond to two $U(1)$ gauge bosons coming from the components $G_{\mu 25}$ and $B_{\mu,25}$ of the metric and the antisymmetric tensor field (taking the compactified dimension to be the one with index 25 in the bosonic string) and a radial dilaton mode of the metric, $G_{25,25}$. If we perturb the conformal field theory by the one available dimension $(1,1)$ operator

$$\mathcal{L} \to \mathcal{L} + g\int d^2 z\partial_z X\partial_{\bar{z}} X, \tag{45}$$

it is clear that $\beta(g) \equiv 0$ to all orders since this is still the Lagrangian for a free field theory. If we define a new field $X' = \sqrt{1+g}X$ then the Lagrangian has the canonical normalization, but the periodicity is now $X' \equiv X' + 2\pi R\sqrt{1+g}$ so that perturbing by this operator changes the radius of the circle. We thus seem to have a one parameter family of $c = 1$ conformal field theories with the parameter corresponding to the compactification radius. However there are some stringy effects which modify this picture.

First note that the spectrum and all interactions are invariant under the duality transformation $R \to 1/2R$ accompanied by an interchange of momenta and winding numbers, $L \leftrightarrow M$. This equivalence of conformal field theories tells us that the real line $R \in [0, \infty]$ is really a double cover of the parameter space of inequivalent theories given by $R \in [1/\sqrt{2}, \infty]$.

Now let's look for (1,0), (0,1), or (1,1) operators which appear only at special values of R. These must be of the form $O = e^{i(p_L X_L + p_R X_R)}$ with dimension $(p_L^2/2, p_R^2/2)$. Demanding that this operator have dimension (1,0) gives the conditions $R = 1/\sqrt{2}$ and $L = M = \pm 1$. In fact at this self-dual point there are also two new dimension (0,1) operators as well as three dimension (1,0) operators. From the previous discussion these correspond to six massless spacetime vector fields or gauge bosons which turn out to be gauge fields of $SU(2)_L \otimes SU(2)_R$.

In general given some collection of dimension (1,0) fields, or currents $j^a(z)$, we can construct conserved charges

$$Q^a = \oint \frac{dz}{2\pi i} j^a(z). \tag{46}$$

If we write the mode expansion of the currents as

$$j^a(z) = \sum_n j_n^a z^{-n-1}, \tag{47}$$

then one can show that the normal modes obey the commutation relations

$$[j_n^a, j_m^b] = i f^{abc} j_{n+m}^c + nk\delta^{ab}\delta_{n+m,0}, \tag{48}$$

where the f^{abc} are structure constants of a Lie Algebra and k is a constant known as the level of the two-dimensional current algebra or affine Lie algebra. In the simple case considered above we have $f^{abc} = \epsilon^{abc}$ and $k = 1$. The basic lesson of this example is that current algebra on the string world sheet can be promoted to spacetime gauge fields.

If we look for (1,1) operators at special values of R we find that R must be a multiple of $1/\sqrt{2}$, but that the beta function for these operators is non-vanishing due to three-point interactions except at $R = 1/\sqrt{2}$ or $R = \sqrt{2}$ [3]. At the self-dual point $R = 1/\sqrt{2}$ there are in fact 9 (1,1) operators constructed as bilinears in the $SU(2)$ currents as $j^a(z)j^b(\bar{z})$, $a, b = 1 \ldots 3$. Since they are equivalent under $SU(2)_L \otimes SU(2)_R$ transformations they only give one inequivalent marginal perturbation which is the one which corresponds to changing R.

At $R = \sqrt{2}$ the situation is more interesting. There are 4 marginal operators in addition to the one which changes R. These have $L = \pm 1$, $M = 0$ and $L = 0, M = \pm 4$ and are all equivalent under the $U(1)_L \otimes U(1)_R$ symmetry generated by $\partial_z X$ and $\partial_{\bar{z}} X$. There are thus two inequivalent marginal directions in which the theory can be perturbed. One corresponds to changing the radius of the circle, but the other indicates the existence of a new line of conformal field theories which are inequivalent to the theory of a free boson.

This new line of theories corresponds to string propagation on what is known as an orbifold [4]. An orbifold is similar to a manifold except that it is allowed to have singularities where it looks locally like Euclidean space after identification of points by a discrete group of symmetry transformations. In the case at hand the orbifold is S^1/Z_2 where the Z_2 acts on the coordinate of the circle by $X \to -X$. The resulting orbifold is simply a line segment running from 0 to πR. In this special case the orbifold is just a manifold with boundary. More generally we can construct orbifolds on which string propagation is conformally invariant by starting with a d-dimensional torus $T^d = R^d/\Lambda$ where R^d is d-dimensional Euclidean space and Λ is the lattice which defines the torus. The simplest types of orbifolds can be constructed by generalizing the construction of the torus to include a discrete group of rotations, or a point group in the nomenclature of crystallography. We thus take $O^d = T^d/P$ where P is the point group. As an example take a two torus with basis vectors $\hat{e}_1 = (0,1), \hat{e}_2 = (1/2, \sqrt{3}/2)$ and take the point group to be Z_2 acting on the complex coordinate of the two-torus as $Z \to -Z$. The point group leaves 4 points fixed up to transformations by the lattice generated by $\hat{e}_1$ and $\hat{e}_2$. Folding up the edges in the way dictated by the Z_2 identifications shows that this orbifold is isomorphic to the surface of a tetrahedron with the vertices of the tetrahedron being the four fixed points.

The basic idea behind formulating string propagation on an orbifold is simple, although implementing it can be somewhat subtle. The basic strategy is to

1) Project out the states on T^d which are not invariant under P.

2) Add strings that are closed on the orbifold but not on T^d and again project out those not invariant under P.

The first step is just what we would do in formulating the quantum mechanics of a point particle on the quotient space T^d/P. That is, we would take the Hilbert space to be the subspace of the original Hilbert space which was invariant under the action of P. In closed string theory this is not sufficient because there are new closed string states on the orbifold, namely strings which close only up to an identification by an element of the point group P. Such string states are called twisted states. It is also possible to see that the existence of these states is required by modular invariance.

In the $c = 1$ example we can make an orbifold S^1/Z_2 with the Z_2 acting by $X \to -X$. States in the untwisted sector which are invariant under the Z_2 are then states built on momentum states $(|p_L, p_R\rangle + |-p_L, -p_R\rangle)$ by acting with an even number of oscillators or states built on the momentum states $(|p_L, p_R\rangle - |-p_L, -p_R\rangle)$ by acting with an odd number of oscillators. To construct states in the twisted sector we first write down the mode expansion of the string coordinate X with the boundary condition that the string closes only up to the Z_2 identification, $X(\sigma + \pi, \tau) = -X(\sigma, \tau)$,

$$X = x_0 + \frac{i}{2} \sum_{n \in Z+1/2} \left(\frac{\alpha_n}{n} e^{-2in(\tau-\sigma)} + \frac{\tilde{\alpha}_n}{n} e^{-2in(\tau+\sigma)} \right) \tag{49}$$

where $x_0 = 0, \pi R$ are the two fixed points. Note that these twisted strings are localized at the fixed points and have no momentum. We then have a Fock space created by acting on the vacuum with the half-integer moded oscillators. The physical states are those invariant under the Z_2 action, namely those states constructed by acting with an even number of oscillators. In this way one may construct an alternate one-parameter family of conformal field theories with $c = 1$. An examination of the spectrum of the orbifold theory at $R = 1/\sqrt{2}$ shows that it coincides with that of the circle theory at $R = \sqrt{2}$ which supports the identification of this line of theories with the marginal perturbation of the circle theory at $R = \sqrt{2}$.

I will not have time to delve into the details of orbifold conformal field theory. The basic construction is described in [4] and interactions are discussed in [5] and [6]. The original construction was inspired by the fact that orbifolds arise as singular limits of certain Calabi-Yau spaces, but the construction is more general and in fact can be applied to any conformal field theory with a discrete symmetry group G whether or not there is an underlying geometrical interpretation for the construction (see e.g. [7] and references therein.).

EQUIVALENCES BETWEEN CFT'S AND DECONSTRUCTION

One of the interesting but confusing features of conformal field theory is the existence of different ways of describing the same conformal field theory. For $c = 1$ one example of this is the equivalence between the conformal field theory given by a circle of radius $R = \sqrt{2}$ and that given by a Z_2 orbifold of the circle at radius $R = 1/\sqrt{2}$. We will also see that there are equivalences between $c = 1$ theories and theories of free fermions. These different formulations make it clear that in string theory it is not the geometry of the internal space which is of primary importance but rather the structure of the conformal field theory associated to the space which can be the same for different geometries. I will give some explicit examples of equivalent formulations at $c = 1$ and then discuss a general procedure for searching for a particular class of equivalences which was worked out in collaboration with Lance Dixon [8].

In the last section I described the construction of the line of circle and orbifold $c = 1$ conformal field theories. The existence of a new marginal perturbation in the circle theory at radius $R = \sqrt{2}$ suggests that the two lines may join at this point. This can be shown to be true by exploiting the $SU(2)_L \otimes SU(2)_R$ symmetry of the theory at $R = 1/\sqrt{2}$. We consider making an orbifold out of two possible symmetries. One is the Z_2 symmetry A: $X \to -X$ and the other is the Z_2 symmetry B: $X \to X + \pi R$ where R is the radius. Twisting by A gives the orbifold theory at radius R while twisting by B gives the circle theory back but at half the radius. To see this note that the projection onto invariant states restricts the theory to even momenta M while the twisted sector adds states with half-integral winding numbers L. This spectrum is just what you would get in the theory at radius $R/2$. Now in general these two twists are inequivalent, but at the point $R = 1/\sqrt{2}$ they are equivalent under the $SU(2)_L \otimes SU(2)_R$ symmetry. This then gives the identification of the circle theory at radius $R = 1/2\sqrt{2}$, or $R = \sqrt{2}$ by duality, with the orbifold theory at $R = 1/\sqrt{2}$, thus justifying the earlier identification of the marginal operator with the line of orbifold theories.

It is also possible to construct a $c = 1$ conformal field theory in terms of free fermion fields. A Majorana fermion in two dimensions has $c = 1/2$. The action is given by

$$S = \int d^2\sigma(\psi\partial_-\psi + \bar{\psi}\partial_+\bar{\psi}) \tag{50}$$

where $\psi(\tau+\sigma)$ is the left-moving part of the fermion field and $\bar{\psi}(\tau-\sigma)$ is the right-moving part.

By taking two copies of this theory with $\psi^1, \psi^2, \bar{\psi}^1, \bar{\psi}^2$, it would seem possible to construct a $c = 1$ theory with no geometrical interpretation in terms of string compactification. We might be tempted to call the string theory with one of the coordinates replaced by this $c = 1$ CFT an intrinsically lower-dimensional string theory. However this would not be accurate. Although the geometry is not obvious, this theory is in fact equivalent to string propagation on a circle of a particular radius.

To see the equivalence let us first look at the $(1,0)$, $(0,1)$, and $(1,1)$ operators in this theory. Since ψ has dimension $(1/2, 0)$ and $\bar{\psi}$ has dimension $(0, 1/2)$, we have $\psi^1\psi^2$ with dimension $(1,0)$, $\bar{\psi}^1\bar{\psi}^2$ with dimension $(0,1)$ and $\psi^1\psi^2\bar{\psi}^1\bar{\psi}^2$ with dimension $(1,1)$. This should be reminiscent of the circle theory at an arbitrary radius where there are two currents corresponding to $U(1)_L \otimes U(1)_R$ and one marginal operator which changes the radius. In the fermion theory, perturbing by the $(1,1)$ operator would give rise to a Thirring coupling $g \int \psi^1\psi^2\bar{\psi}^1\bar{\psi}^2$. It is known that such a coupling preserves conformal invariance. Thus we start to suspect that this line of theories is actually equivalent to the line of circle theories. One way to check this is to show that the spectrum and the interactions match exactly [9]. It is not too hard to do this, but I would like to explore a somewhat different route which has other applications. This involves a bit heavier use of some of the technology of CFT than the previous discussion.

First, recall that the Virasoro algebra

$$[L_n, L_m] = (n-m)L_{n+m} + \frac{c}{12}(n^3 - n)\delta_{n+m,0} \tag{51}$$

with

$$L_n = \int \frac{dz}{2\pi i} z^{n+1} T(z) \tag{52}$$

is equivalent to the operator product expansion (OPE)

$$T(z)T(w) \sim \frac{c/2}{(z-w)^4} + 2T(w)\frac{1}{(z-w)^2} + \partial_w T(w)\frac{1}{(z-w)} + \text{analytic.} \tag{53}$$

Primary fields $\phi(w, \bar{w})$ of dimension $(h, \bar{h})$ satisfy the OPE

$$T(z)\phi(w,\bar{w}) \sim \frac{h\phi(w,\bar{w})}{(z-w)^2} + \partial_w\phi\frac{1}{(z-w)}, \tag{54}$$

with a corresponding equation with $T(z)$ replaced by $\bar{T}(\bar{z})$ and h replaced by $\bar{h}$.

Now if we take two conformal field theories with conformal anomalies c_1 and c_2 we can construct another conformal field theory with $c = c_1 + c_2$ and with a stress tensor which is just the sum of the constituent stress tensors, $T(z) = T_1(z) + T_2(z)$ (in what follows I will focus only on the holomorphic part of the theory, the anti-holomorphic part can be treated analogously). $T(z)$ itself is a dimension $(2,0)$ field but is not primary because of the anomaly term. However we can easily construct a primary dimension $(2,0)$ field namely $\Phi(z) = -c_2T_1(z) + c_1T_2(z)$. Thus whenever we form a conformal field theory as the sum of two theories in this sense, the new theory will contain a primary field of dimension $(2,0)$.

Before applying this idea to our example with $c = 1$ and $c_1 = c_2 = 1/2$ let us turn this around and ask if whenever we have a conformal field theory with primary dimension $(2,0)$ fields we can write its stress tensor as the sum of stress tensors with smaller conformal anomalies. Let us first consider a simple example. Suppose we have a CFT which has a single primary $(2,0)$ field Φ. We can then write down the most general OPE's involving the stress tensor and Φ:

$$\begin{aligned}
T(z)T(w) \sim&\frac{c/2}{(z-w)^4} + 2T(w)\frac{1}{(z-w)^2} + \partial_w T(w)\frac{1}{(z-w)}\\
T(z)\Phi(w) \sim&2\Phi(w)\frac{1}{(z-w)^2} + \partial_w\Phi(w)\frac{1}{(z-w)}\\
\Phi(z)\Phi(w) \sim&\frac{1}{(z-w)^4} + (b\Phi(w) + \frac{4}{c}T(w))\frac{1}{(z-w)^2} + \frac{1}{2}\partial_w(b\Phi(w) + \frac{4}{c}T(w))\frac{1}{(z-w)}.
\end{aligned} \tag{55}$$

In this expression the normalization of Φ has been chosen so that the leading singularity in the Φ Φ OPE is one, the coefficient of Φ in the subleading singularity is then an arbitrary coefficient, b. The coefficient of T in the subleading singularity of this OPE is determined from the first two OPE's using associativity of the operator product expansion.

We can now try to decompose the stress tensor for this theory into its constituent parts. Let us first try to construct a dimension $(2,0)$ operator which has the OPE of a stress tensor. We take

$$T'(z) = \alpha_0 T(z) + \alpha_1\Phi(z). \tag{56}$$

The OPE of T' with itself then follows from (55). We easily find that T' will have the proper OPE with conformal anomaly

$$c' = \alpha_0^2 c + 2\alpha_1^2 \tag{57}$$

provided that

$$\begin{aligned}
2\alpha_0 =&2\alpha_0^2 + \frac{4}{c}\alpha_1^2\\
2\alpha_1 =&b\alpha_1^2 + 4\alpha_0\alpha_1
\end{aligned} \tag{58}$$

Solving (58) we find two possible solutions for c'

$$c'_\pm = \frac{c}{2}\left(1 \pm \frac{1}{\sqrt{1+\frac{32}{b^2c}}}\right). \tag{59}$$

If we denote the corresponding stress tensors by $T'_\pm$ then it is easy to see that $T = T'_+ + T'_-$ and that T'_+ and T'_- have no singular terms in their OPE with each other. This shows that the original Virasoro generators can be written as the sum of two commuting sets of Virasoro generators with conformal anomalies $c'_\pm$. A special case of this "deconstruction" involving an arbitrary number of $(2,0)$ fields but no dimension $(1,0)$ or $(3,0)$ fields was first analyzed by Zamolodchikov [10]. The special case when the $(2,0)$ fields are constructed as bilinears in currents has been analyzed in detail in [11].

As an example of this "deconstruction" let us go back to our $c = 1$ example. Choose $R = 1$ which is the unique radius (up to duality) at which a primary dimension $(2,0)$ field appears. The field is $\frac{1}{\sqrt{2}}(e^{2iX} + e^{-2iX})$. The coefficient $b = 0$, so this theory deconstructs into two copies of $c = 1/2$. This is just what we were trying to find. It is not hard to use the new stress tensor to find an explicit mapping between operators in the $c = 1$ theory and operators in the $c = 1/2$ theories. The equivalence between the circle theory at radius $R = 1$ and the product of two free fermions is well known.

As a second example take a toroidal compactification with the spatial lattice given by $\frac{1}{\sqrt{2}}\Lambda_{weight}(G)$ and the momentum lattice by $\sqrt{2}\Lambda_{root}(G)$ where G is a simple laced Lie algebra (A,D,E). Then the roots of G correspond to dimension $(2,0)$ exponentials. Let $\phi_i = e^{ip_iX} + e^{-ip_iX}$ where $p_i = \sqrt{2}\alpha_i$ and α_i is a root. Then set $\phi = \frac{1}{\sqrt{N_r}}\sum_i \phi_i$ where N_r is the total number of roots. Then one finds $b = 2N_p/N_r$ where N_p is the number of ways of adding exactly two roots to get a given root. Also, $c_0 = r$ where r is the rank of G. For example, if $G = SU(N)$, then $N_r = N(N-1)$,$N_p = N-2$ and $r = N-1$ and one finds that this deconstructs to $c = \frac{N(N-1)}{N+2}$ and $c = \frac{2(N-1)}{N+2}$. The last value of c is that of the Nth parafermion CFT.

If there is more than one primary dimension $(2,0)$ field the procedure also works with minor changes. There will now also be dimension 1 and dimension 3 operators appearing in OPE of the dimension $(2,0)$ fields, but they cancel in the OPE of the new putative stress tensor which only depends on the coefficient of ϕ_k in the OPE of ϕ_i with ϕ_j. Call this coefficient b_{ijk}. Then we can take $T = \alpha_0 T_0 + \sum_i \alpha_i\phi_i$ and work out its OPE, and demand that it agree with (53) for some choice of c. This gives

$$\begin{aligned} \alpha_0 &= \alpha_0^2 + \frac{2}{c_0}\sum_i \alpha_i^2 \\ \alpha_i &= 2\alpha_0\alpha_i + \frac{1}{2}\sum_i \alpha_i\alpha_j b_{ijk} \\ c &= \alpha_0^2 c_0 + 2\sum_i \alpha_i^2 = \alpha_0 c_0. \end{aligned} \tag{60}$$

Unfortunately, it is hard to solve these equations analytically. Alternatively, one can deconstruct a theory into two pieces. First pick a $(2,0)$ field and deconstruct with respect to this field, then reorganize all the fields by calculating their conformal weights with respect to the component Virasoro algebras. When there is a new dimension $(2,0)$ operator with respect to a constituent Virasoro algebra, the procedure can be iterated to give one more step in the deconstruction. In this way, many of the solutions to (60) can be found. This has been done in detail for the $c = 2$ theory based on $SU(3)$ and one finds that this $c = 2$ theory deconstructs to $1/2 + 7/10 + 4/5$. (But it is not the product of the diagonal modular invariants).

This method is useful not only for finding equivalences between different conformal field theories, but also for searching for new conformal field theories. The GKO or coset construction is basically a special case of the construction given here when the dimension $(2,0)$ operators are constructed out of bilinears in currents [12].

The equivalences between apparently different conformal field theories also emphasizes the fact that it is the CFT and not the geometry of the underlying space that is of primary importance. One of the outstanding problems in string theory is to find a generalization of geometry which can unify these different points of view. Although it is not clear precisely what set of ideas this will involve, it is clear that conformal field theory will be one of the essential tools.

DISCUSSION

The previous sections have been concerned with the simplest compactifications of the simplest string theory, the bosonic string. Much of the current interest in string theory however is due to the fact that there exist compactifications of the heterotic string which at low-energies reduce to supersymmetric grand unified theories. The study of superstrings and their compactifications is much more complicated and much richer than for bosonic strings. Unfortunately, I do not have the time to discuss this subject.

However, many of the topics discussed here have generalizations to superstring theory and superstring compactifications. In superstring theory one considers two-dimensional superfields with components (X^{μ}, ψ^{μ}) coupled to two-dimensional supergravity. After gauge fixing one is left with a residual super-conformal symmetry group which plays a role analogous to that of the conformal group in the bosonic string. In particular, there are negative norm states associated with the time-like excitations of both X and ψ which are removed by the supersymmetric generalization of the constraints (21) for $D = 10$. Many of the features of the simple compactifications discussed here have extensions to compactifications of the superstring described by super-conformal field theories. For example, at $c = 1$ we found a one-dimensional moduli space of conformal field theories labelled by the radius, and a discrete symmetry, $R \to 1/2R$ which acted on this moduli space. Superstring compactifications have a much richer geometry, for example in the sigma-model approach they correspond to Ricci-flat Kähler manifolds. The structure of the moduli spaces of these manifolds and their discrete symmetries is a question of current mathematical interest which also has implications for low-energy applications of string theory [13]. The "deconstruction" described above also has interesting generalizations. As an example, in [14] Gepner constructed tensor products of certain minimal conformal field theories which correspond to compactification of four or six dimensions of the superstring. He was able to argue convincingly that these theories in fact correspond to certain points in the moduli space of certain Ricci-flat Kähler manifolds. They thus provide exact solutions for string propagation on spaces for which even the metric is not known!

Finally, let me mention that although conformal field theory has been an invaluable tool in understanding string theory, it is only a description of the classical solutions to string theory. Even if the semi-classical approximation is valid in string theory, quantum corrections and/or non-perturbative corrections are bound to be important. The problem of extending the formalism of conformal field theory to deal with such effects is a fascinating and non-trivial one which is bound to attract a great deal of attention in the following years.

REFERENCES

[1] see e.g. P. Goddard and D. Olive. Int. J. Mod. Phys. **A1** (1986) 303.

[2] M.Green, J.Schwarz, and E.Witten, Superstring Theory: 1, Cambridge University Press (1987).

[3] R. Dijkgraaf, E. Verlinde, and H. Verlinde, Comm. Math. Phys. **115** (1988) 649.

[4] L. Dixon, J. Harvey, C. Vafa, E. Witten, Nucl. Phys. **B261** (1985) 678; Nucl. Phys. **B274** (1986) 285.

[5] L. Dixon, D. Friedan,E. Martinec, and S. Shenker, Nucl. Phys. **B282** (1987) 13.

[6] S. Hamidi and C. Vafa, Nucl. Phys. **B279** (1987) 465.

[7] P.Ginsparg, "Applied Conformal Field Theory", 1988 Les Houches Lectures, HUTP-88/A054.

[8] L. Dixon and J.Harvey, "Deconstructive Conformal Field Theory", unpublished.

[9] J. Bagger, D. Nemeschansky, N. Seiberg, and S. Yankielowicz, Nucl. Phys. **B289** (1987) 53.

[10] A. Zamolodchikov, Theor. Math. Phys. **65** (1985) 1205.

[11] M. B. Halpern and E. Kiritsis, Mod. Phys. Lett. **A4** (1989) 1373; Erratum ibid. A4 (1989) 1797.

[12] P.Goddard, A. Kent, and D. Olive, Comm. Math. Phys. **103** (1986) 105.

[13] see e.g. N. Seiberg, IASSNS/HEP-87/71 and references therein

[14] D. Gepner, Phys. Lett. **B199** (1987) 380; Nucl. Phys. **B296** (1988) 757.

MULTIPARTICLE STRUCTURE AND WILSON SHORT-DISTANCE EXPANSION IN FIELD THEORIES

D. Iagolnitzer

Service de Physique Théorique de Saclay
Laboratoire de l'Institute de Recherche Fondamentale
du Commissariat a l'Energie Atomique
91191 Gif-sur-Yvette Cedex

Recent results on scattering (for theories with mass gap) and on large momentum properties are reviewed.

Results on scattering include [1,2] the definition, via cluster expansion "of order p" [1], $p \geq 1$, of general irreducible kernels and the proof of euclidean structural equations well suited [3] to provide, by analytic continuation, information on multi-particle structure and asymptotic completeness in higher and higher energy regions in Minkowski space (work in progress).

For the simplest case of 2-particle structure, results of [4] obtained via a general analysis of 2-particle convolution (issued from [5,6]) complete those previously known for the super-renormalizable $P(\phi)$ models in dimension 2 (possible occurrence of 2-particle bound states via threshold effect, ...) and extend them to other theories in dimension 2,3 such as $(\lambda\phi^4 + \lambda'\phi^3)_3$ and the (non super-renormalizable) massive Gross-Neveu (GN) model in dimension 2.

The (nontrivial) adaptation of methods to the non super-renormalizable case is made [2] through the definition (via phase space analysis) of (non intrinsic) irreducible kernels satisfying equations with fixed ultraviolet cut-off, in accordance with earlier axiomatic ideas [5]. An alternative method for the analyses of 2-particle structure in the GN model will be indicated in the last paragraph.

Proofs of large momentum properties (in euclidean space) and of Wilson-Zimmermann short-distance expansion are given in [7] for the GN model. (Methods are applicable more generally.) They are based on a suitable modification of the renormalization procedure and on the introduction of another type of "irreducible" kernels (with respect to the momentum slices that occur in phase-space analysis) and of corresponding structure equations.

Alternative proofs of Wilson expansion at first order [8] and of results on 2-particle structure [8,9], using the "true" 2-particle irreducible Bethe-Salpeter kernel B or renormalized BS kernel G can be given for the GN model (whose renormalization parts are 2 and 4-point functions), in accordance with ideas of [10,11] and [11,12] respectively. The analysis includes [8] the proof of simple large momentum properties of these kernels and of a la Symanzik "subtracted BS equations", or [9] of a related expansion of the 4-point function in terms of renormalized Feynman-type integrals with kernels G at each vertex.

REFERENCES

1. D. Iagolnitzer, J. Magnen, *Comm. Math. Phys. 110*: 51 (1987).
2. D. Iagolnitzer, J. Magnen, *Comm. Math. Phys. 111*: 81 (1987).
3. D. Iagolnitzer, *Fizika 17*: 361 (1985).
4. J. Bros, D. Iagolnitzer, 2-particle asymptotic completeness and bound states in weakly field theories, Saclay preprint SPhT/88-57 April, 1988, to be published in *Comm. Math. Phys.*
5. J. Bros, in this volume and references therein.
6. D. Iagolnitzer, *Comm. Math. Phys. 88*: 235 (1983).
7. D. Iagolnitzer, J. Magnen, Large momentum properties and Wilson short-distance expansion in nonperturbative field theories, Saclay preprint SPhT/88-030, April, 1988, to be published in *Comm. Math. Phys.*
8. D. Iagolnitzer, J. Magnen, Bethe-Salpeter kernel and short-distance expansion in the massive Gross-Neveu model, Saclay preprint SPhT/88-051, April, 1988, to be published in *Comm. Math. Phys.*
9. D. Iagolnitzer, Renormalized Bethe-Salpeter kernel and 2-particle structure, Saclay preprint, August, 1988.
10. K. Symanzik, *Comm. Math. Phys. 23*: 49 (1971); *34*: 7 (1973).
11. J. Bros, B. Ducomet, *Ann. Inst. H. Poincaré 45*: 173 (1986).
12. D. Iagolnitzer, *Comm. Math. Phys. 99*: 441 (1985).

VARIOUS ASPECTS OF THE BETHE-SALPETER STRUCTURE IN QUANTUM FIELD THEORY

J. Bros

SPRT-Cen Saclay
France

The Bethe-Salpeter structure of Green's functions has a deep and general connection with scattering theory, that has been studied from an axiomatic view point and illustrated in the models. In renormalizable theories, this structure is also involved in large-momentum analysis through a "renormalized Bethe-Salpeter formalism" not developed here (see [1 to 4]).

We review the following aspects of two-particle scattering, implied by the B.S. structure $F = B + F_0 B$ of the four-point function F of a massive field theory (where $F_0 B$ denoted a suitable regularized Feynman-type convolution = $\mathbf{F}$ = $\mathbf{G}$ = in the s-channel):

-i) Equivalence between irreducibility of B (i.e. $\Delta_s B = 0$ for $s < (3m)^2$) and asymptotic completeness for $s < (3m)^2$, and the corresponding ramified meromorphic structure of F around $s = 4m^2$ (see [5 to 8], and [9] for the general program).

-i)' Explicit construction of relevant B.S. kernels B and detailed anaysis of the poles of F (exhibiting bound or antibound states near $s = 4m^2$ by a kinematrical "threshold effect") for all small coupling models of constructive W.F.T. in dimensions 2 and 3. (See [10] and references therein.)

-ii) Close parallel between the (test-function and wave-function) Hilbert space descriptions of states and presentations of scattering theory respectively in the Q.F.T. and Wave-Mechanical frameworks (see [11]).

-iii) Angular momentum analyticity in axiomatic QFT: a Froissart, Gribov-type transform $\widetilde{F}$ of F, admitting a diagonalized B.S. structure with respect to an angular momentum variable λ, exhibits theoretically a joint mesomorphic structure in (s, λ) ([12]).

REFERENCES

1. K. Symanzik, Comm. Math. Phys. *23*, 49 (1971) and *34*, 7 (1973).
2. J. Bros, B. Ducomet, Ann. Inst. H. Poincaré *45*, 173 (1986).
3. D. Iagolnitzer, Comm. Math. Phys. *199*, 491 (1985).
4. D. Iagolnitzer, J. Magnen, Bethe-Salpeter kernel and short-distance expansion in the massive Gross-Neveu model, Saclad preprint, SPHT/88-051.
5. K. Symanzik, J. Math. Phys. *1*, 249 (1960).
6. J. Bros in *Analytic Methods in Math-Physics*, Gilbert, Newton eds, p. 85, New York, Gordon and Breach 1970.
7. J. Bros, M. Lassalle, Comm. Math. Phys. *54*, 33 (1977).
8. J. Bros, D. Iagolnitzer, Comm. Math. Phys. *85*, 197 (1982).
9. J. Bros, Pysis *124A*, 145 (1984), North Holland, Amsterdam.
10. J. Bros, D. Iagolnitzer, 2-particle asymptotic completeness and bound states in weakly coupled quantum field theories, Saclay preprint, SPHT/88-57.
11. J. Bros in *Prospect of Algebraic Analysis* (1988) (dedicated to Professor M. Sato for his 60th Birthday), to appear; Saclay preprint, SPHT/87-154.
12. J. Bros and G. Viano, in preparation.

SUPERSYMMETRY BREAKING IN WESS-ZUMINO MODELS

John Z. Imbrie *

Departments of Mathematics and Physics
Harvard University
Cambridge, MA 02138 USA

A combination of index theory [1,2] and multiphase analysis [3,4] should lead to an understanding of the vacuum structure of the weakly coupled two-dimensional Wess-Zumino model in infinite volume. The supersymmetric interaction is $|V'(\phi)|^2 + V''(\phi)\overline{\psi}\psi$, where for the $N = 1$ model ϕ, ψ are real, while for the $N = 2$ model ϕ, ψ are ocmplex. We take V to be a polynomial of degree n¡ and scale it as $V(\Phi) \to \lambda^{-2}V(\lambda\Phi)$ with λ small. If V' has $n-1$ distinct zeros, then $|V'|^2$ has $n-1$ minima at zero. At small λ a multiphase expansion should be possible, leading as in [4] to a construction of the stable vacua. The index theorems proven in [5] should help determine the number of stable vacua. For example, in the $N = 2$ model the periodic partition function is equal to $n-1$, so the free energy must be zero and so supersymmetry is unbroken. A careful analysis of the expansion should lead to the further conclusion that all $n-1$ possible vacua are stable thermodynamically [3]. These conclusions must be modified in the $N = 1$ case because the Pffaffian in the measure of alternates in sign from one minimum to the next.

REFERENCES

1. Jaffe, A., Lesniewski, A., these proceedings.
2. Witten, E.: Nucl. Phys. *B202*, 253 (1982).
3. Borgs, C., Imbrie, J.Z.: Multiphase Analysis in Quantum Field Theory, in preparation.
4. Imbrie, J.Z.: Commun. Math. Phys. *82*, 261 (1981) and *82*, 305 (1981).
5. Jaffe, A., Lesniewski, A., Weitzman, J.: Commun. Math. Phys. *112*, 75 (1987). The Loop Space $S' \to \mathbf{R}$ and Supersymmetric Quantum Fields, to appear.

* Alfred P. Sloan Research Fellow, Research Partially supported by the National Science Foundation under Grant PHY-87-06420.

FIRST ORDER PHASE TRANSITIONS IN LARGE N LATTICE HIGGS MODELS AND PIROGOV SINAI THEORY

A Short Summary

Christian Borgs

Theoretische Physik
EHT-Hönggerberg
CH-8093 Zürich, Switzerland

It is generally expected that the phase diagram of the $SU(N)$ lattice Higgs model with action

$$S(\Phi,U) = \frac{1}{2}\sum_{\langle xy\rangle} |\Phi_x - U_{xy}\phi_y|^2 + \frac{1}{g^2}\sum_p ReTrU_{\partial p} + \sum_x \left(\frac{\lambda}{N-1}\Phi_x^4 - \frac{\mu^2}{2}\Phi_x^2\right)$$

has the structure indicated in Fig. 1. While the order of the phase transition is still an object of controversy for small g^2, it is generally conjectured that it is first order for large g^2. For large values of N, the most convincing arguments for this conjecture have been given by Wolff [1], who calculated the effective potential for Φ^2 for $N = \infty$ and $g^2 = \infty$, and found a first order phase transition line with critical endpoint, see Fig. 2. In the region $\mu_-(\lambda) < \mu < \mu_+(\lambda)$ the effective potential has two local minima, with equal height for $\mu = \mu_0(\lambda)$.

In our recent work [2], Waxler and I have shown that the effective potential has the same structure for finite N. We then use a combination of cluster expansion methods and Pirogov Sinai theory to actually prove the existence of a first order phase transition [1] for large, but finite, N and g^2. The region where we could show the existence of a first order phase transition line is the shaded region in Fig. 1. Since the extension to finite but large g^2 is a straightforward combination of the method described in this talk with the strong coupling cluster expansion, I restrict myself to infinite g^2 in this talk.

For $g^2 = \infty$, the gauge fields in the partition function corresponding to $S(\Phi, U)$ can be integrated out exactly, leading to an effective action of the form

$$S_{eff}(\Phi) = n\sum_x V_n(r_x) + n\sum_{\langle xy\rangle} W_n(r_x, r_y)$$

[1] Characterized e.g. by a jump in the expectation value of Φ^2.

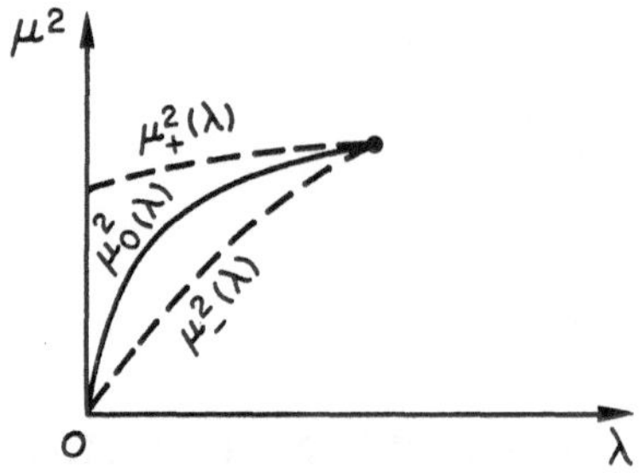

Figure 1. The phase diagram of the $SU(N)$ lattice.

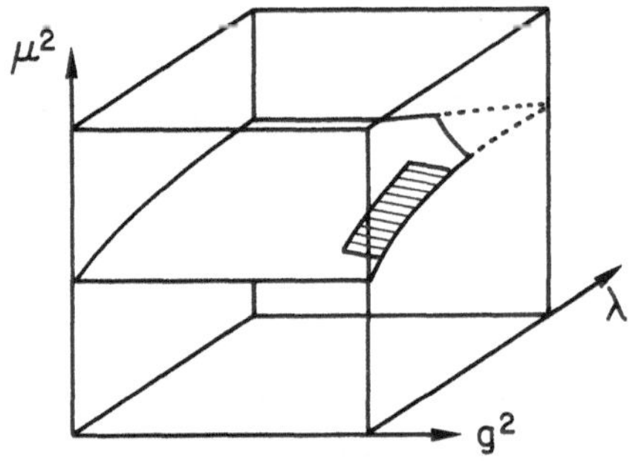

Figure 2. The phase diagram for $g^2 = N = \infty$.

with $n = N - 1$ and $r_x = |\Phi_x|^2/n$. V_n is the effective potential calculated in [1] and W_n is a kinetic energy type interaction behaving like $(r_x - r_y)^2$ for small values of $|r_x - r_y|$ [2].

Since the second term in the effective action suppresses fluctuations between the two minima of V_n, while the first one suppresses all fields r_x which do not lie in a small neighborhood of one of the minima, it seem natural to combine a Peierls contour expansion (controlling fluctuations between the two minima) with a mean field cluster expansion [3] (controlling fluctuations in a neighborhood of a given minimum) to prove the existence of a first order phase transition for large n.

Unfortunately the technique developed in [3] for the pure bosonic Φ^4 model relies heavily on the $\Phi \to -\Phi$ symmetry of this model, whereas the two minima of $V_n(r)$ are not related by a symmetry. Combining the techniques of [3] with the techniques of

Pirogov Sinai theory in the form [4], it is possibly, however, to overcome this problem. I cannot discuss this further in this short summary [2] and refer the reader to ref. [1] and [5] for further details.

REFERENCES

1. U. Wolff, The $SU(N)$ lattice Higgs model at strong gauge coupling, *Nucl. Phys.* *B280* [FS18]: 680 (1987).
2. C. Borgs, R. Waxler, preprint in preparation.
3. J. Glimm, A. Jaffe, T. Spencer, A convergent expansion about mean field theory, *Ann. Phys. 101*: 610 (1976).
4. M. Zahradnik, An alternative version of Pirogov Sinai theory, *Comm. Math. Phys. 93*: 559 (1984).
5. C. Borgs, R. Waxler, First order phase transitions in unbounded spin systems, preprint in preparation.

[2] A short introduction into Pirogov Sinai theory was given in the oral version of this talk.

THE COLEMAN-WEINBERG MECHANISM IN THE ABELIAN LATTICE HIGGS MODEL

Florian Nill

Freie Universitat Berlin
Institut für Theoretische Physik
D-1000 Berlin

ABSTRACT

The order of the Higgs phase transition in the Abelian lattice Higgs model is investigated by means of perturbation theory. Based on gauge invariant criteria such as generalized Clausius-Clapeyron and Ehrenfest equations it is shown that in the validity domain of the loop expansion, i.e., $e \to 0$ and Ne^2 fixed, there is no evidence for a first order transition up to two loops. Here e is the gauge coupling and λ the scalar self coupling constant. If, however, we perform a perturbation analysis as $e \to 0$ and Ne^4 is kept fixed, the Coleman-Weinberg-Linde result on a first order phase transition can be rederived gauge invariantly. This leads to an intimate relation between Linde's lower bound on the Higgs mass in a continuum theory and the possible existence of a tricritical line separating the first order from the second order phase transition in the lattice model. In tems of renormalized parameters the tricritical line is given explicitly up to second order. I close with a tentative discussion of different scenarios as to where the tricritical line may be found in the space of bare coupling constants.

REFERENCES

1. F. Nill, Untersuchungen zur phasenstruktur abelscher Higgs modelle in der Gittereichtheorie, Ph.D. Thesis, Universität Munchen, 1987, Max-Planck-Institut Preprint MPI-PHE/PTH 31/8.
2. F. Nill, The Coleman-Weinberg mechanism in the Abelian lattice Higgs model: I) Generalized Clausius-Clapeyron and Ehrenfest equations. Paper in preparation.
3. F. Nill and H. Ortmanus, The Coleman-Weinberg mechanism in the Abelian lattice Higgs model: II) First order phase transition, tricritical line and the lower bound on the Higgs mass. Paper in preparation.

QUANTUM SOLITONS

P.A. Marchetti

We briefly report some results, obtained in a joint work with J. Fröhlich, in the general program of a rigorous (beyond semi-classical approximation) euclidean functional integral quantization of topological excitations (solitons) in Quantum Field Theory.

(1) Osterwalder-Schrader reconstruction of soliton field operators [1]. Basic objects are the *euclidean Green functions of solitons*; they are given by expectation values of *disorder fields.*

Suppose that the partition function, Z, of some Euclidean field theory can be expressed as a sum over configurations of closed line defects v (e.g. Peierls contours in ϕ_2^4, vortex loops in $(\text{Higgs})_3$), carrying a charge q with values in a discrete abelian group $\mathcal{Z}$, i.e.

$$Z = \sum_{v:\partial v=0} Z(v) \tag{1}$$

where $Z(v)$ is the Boltzmann weight of the configuration v. Then the correlation functions of the disorder fields are given by

$$\langle D(x_1, q_1, \ldots x_n, q_n)\rangle = \frac{1}{Z} \sum_{v:\partial v=\{x_1,q_1,\ldots x_n,q_n\}} Z(v) \tag{2}$$

Let us assume that the correlation functions with non zero total defect charge vanish and clustering holds, then the Hilbert space of states, $\mathcal{H}$, of the quantum field theory reconstructed from the Euclidean field theory decomposes into orthogonal sectors $\mathcal{H}_q$ labelled by the total defect charge $z \in \mathcal{Z}$; for $q \neq 0$ they are soliton sectors. These sectors, together with the soliton field operators $S_q(x)$, mapping the vacuum sector $\mathcal{H}_0$ to $\mathcal{H}_q$, can be obtained from the joint order-disorder correlation functions by analytic continuation in the time variables (O.S. reconstruction).

Remark 1) With suitable modification (in case in the lattice approximation) such construction applies e.g. to kinks in $(\phi_2^4, (\cos\phi)_2)$, solitons with fractional fermion number [2] (no index theorem is needed in the fully quantized theory [9]), $\mathbf{Z}_n$-parafermions in $P(\phi)_2$ theories, vortices (e.g. in (Higgs)$_3$) and "anyons" (particles with "any" real spin in $d = 3$, see [3]), monopoles and dyons (in $U(1)$ lattice gauge theories) in $d = 4$.

Remark 2) Using the excitation expansions method of [4] one can prove on the lattice (in a suitable range of coupling constants) that the soliton field operator of unit charge couples the vacuum to a one-particle state e.g. in ϕ_2^4, in the broken symmetry phase and in (Higgs)$_3$ models, in the superconducting phase [1,5]. Using renormalization group techniques combined with Peierls estimates one can show that monopoles and dyons in $U(1)$ gauge theories are infraparticles in the QED phase [6].

2) Relation between quantum solitons and fibre bundles [7,8]. Disorder correlation functions in the continuum have a natural interpretation in the language of differential geometry: they are originated integrating over sections of non trivial bundles. E.g. in ϕ_2^4 we consider a vector bundle $E_{\underline{x}}$ (*soliton bundle*) with structure group $\mathbf{Z}_2$ (corresponding to $\phi \to -\phi$), fiber $\mathbf{R}$, base space $\mathbf{R}^2\backslash\{\underline{x}\}, \underline{x} = \{x_1, \ldots, x_n\}$, and non trivial holonomy for the loops that encircle each singel point x_i. Let $\Delta_{\underline{x}}$ denote the corresponding covariant laplacian, then the Euclidean Green functions of kinks are formally given by

$$S_n(x_1, \ldots, x_n) = \frac{\int \mathcal{D}_E\tilde{\phi} \exp\left[-\int_{\mathbf{R}^2\backslash\{\underline{x}\}} \frac{1}{2}\tilde{\phi}(-\Delta_{\underline{x}} + 1)\tilde{\phi} + :P(\tilde{\phi}):\right]}{\int \mathcal{D}\phi \exp\left[-\int_{\mathbf{R}^2} \frac{1}{2}\phi(-\Delta + 1)\phi + :P(\phi):\right]} \tag{3}$$

where $\mathcal{D}_E\tilde{\phi}$ is a formal measure on the space of sections $\tilde{\phi}$ of $E_x, \mathcal{D}\phi$ a formal measure on the space of real functions on $\mathbf{R}^2$ and $:P(\phi):$ is a ϕ^4 Wick ordered polynomial.

Remark 1) A rigorous version of (3) in particular involves the substitution of $\mathcal{D}_E\tilde{\phi}\exp[\ldots]$ with a Gaussian measure with covariance $(-\Delta_{\underline{x}} + 1)^{-1}$ over the space of *section distributions* of $E_{\underline{x}}$.

Remark 2) The correlation function S_n are ultradistributions [7], e.g. $S_2(x,y) \sim \exp[log^2|x - y|]$ as $|x - y| \downarrow 0$, so that the quantum field theory of kinks in ϕ_2^4 is a Wightman theory for Jaffe fields [9].

Remark 3) The above geometric construction can be adapted to discuss (formally) theories of vortices and anyons in $d = 3$. In particular for anyons one can exhibit a spin-statistics connection and relate the statistics to unitary representations of the Braid group [10] (for application of anyons to fractional quantum Hall effect and high Tc superconductivity, see [11]).

REFERENCES

1. J. Fröhlich, P.A. Marchetti, *Commun. Math. Phys. 112*: 343 (1987).
2. R. Jackiw, C. Rebbi, *Phys. Rev. D13*: 3398 (1976).
3. F. Wilczek, *Phys. Rev. Lett. 48*: 1144 (1982).
4. J. Bricmont, J. Fröhlich, *Nucl. Phys. B251*: 517 (1985); *Comm. Math. Phys. 98*: 553 (1985).
5. P.A. Marchetti, preprint DFPD 16/87, to appear in *Comm. Math. Phys.*

6. J. Fröhlich, P.A. Marchetti, *Europhys. Lett.* *2*: 933 (1986).
7. P.A. Marchetti, *Europhys. Lett.* *4*: 663 (1987).
8. J. Fröhlich, P.A. Marchetti, *Comm. Math. Phys.* *116*: 127 (1988).
9. F. Constantinescu, W. Thalheimer, *Comm. Math. Phys.* *38*: 299 (1974).
10. J. Fröhlich, P.A. Marchetti, Quantum field theory of anyons, ETH preprint and Padova preprint 1988.
11. See e.g. *The Quantum Hall Effect*, ed. R.E. Prange, S.M. Girvin, Springer Verlag, 1987; A.M. Polyakov, Landau Institute preprint 1988; I.E. Dzialoshinsky, A.M. Polyakov and P.B. Wiegman, to appear in *Phys. Rev. Lett.*; R.B. Laughlin, preprint, 1988.

INDEX